UMBACH

Selected Problems in Discrete Mathematics

Г. П. Гаврилов, А. А. Сапоженко

Сборник задач по дискретной математике

Издательство «Наука» Москва

G.P. Gavrilov,
A.A. Sapozhenko

Selected Problems in Discrete Mathematics

Mir Publishers Moscow

Translated from Russian by
Ram S. Wadhwa and
Natalia V. Wadhwa

First published 1989
Revised from the 1977 Russian edition

На английском языке

Printed in the Union of Soviet Socialist Republics

ISBN 5-03-000522-6

Contents

Preface

This collection of problems is intended as an accompaniment to a course on discrete mathematics at the universities. Senior students and graduates specializing in mathematical cybernetics may also find the book useful. Lecturers can use the material for exercises during seminars.

The material in this book is based on a course of lectures on discrete mathematics delivered by the authors over a number of years at the Faculty of Mechanics and Mathematics, and later at the Faculty of Computational Mathematics and Cybernetics at Moscow State University.

The reader can use *Introduction to Discrete Mathematics* by S. Yablonsky as the main text when solving the problems in this collection.

The book consists of eight chapters. The first two chapters are devoted to Boolean algebra which forms the basis of discrete mathematics. About a quarter of the total teaching time during lectures and practicals at the Computational Mathematics and Cybernetics Faculty at Moscow University is devoted to Boolean algebra. The material in this part introduces the student to the concepts of discrete functions, superposition, and functionally complete sets. It also acquaints the student with various methods for specifying a discrete function (tables, polynomial representation, normal forms, geometrical representation using an n-dimensional unit cube, etc.). Methods for testing the completeness and closure of sets of functions are also considered.

The third chapter is devoted to k-valued logics. The problems presented are intended to acquaint the reader with the canonical expansions of k-valued functions, equivalent transformations of formulas, closed classes of the k-valued functions, and methods for testing the completeness and closure of functions. Several problems in the chapter illustrate the difference between k-valued logics ($k > 2$) and Boolean algebra.

The fourth chapter contains problems on the theory of directed and undirected graphs, and the network and circuit theory.The chapter describes the basic concepts, methods and terms of graph theory, which are widely used to describe and investigate the structural properties of objects in various branches of science and technology. The problems are intended to consolidate the basic concepts of graph theory, to illustrate the application of network and graph theory to the construction of circuits representing Boolean functions, to count the number of objects with a given geometrical structure, etc. The authors hope that the lecturer will also find problems in this chapter to help him demonstrate the mathematical rigor during the proof of geometrically "obvious" statements.

The fifth chapter describes the basic concepts of coding theory. The problems concern the properties of error correcting codes, alphabetical codes, and minimum redundancy codes.

The sixth chapter contains problems demonstrating different ways of describing discrete transformers (automatons). Problems aimed at revealing deterministic and boundedly deterministic automatons are also given. Other problems concern the different ways of representing automatons (diagrams, canonical equations, and schemes (circuits)), the investigation of the functional completeness and closure of sets of automaton mappings, and also the properties of operations involving such mappings.

The seventh chapter deals with the elements of algorithm theory and is intended to provide an idea about effective computability and complexity of computations. It is also about certain ways for specifying algorithms, such as Turing's machines and recursive functions.

The eighth chapter describes the elements of combinatorial analysis. While studying discrete mathematics, one frequently comes across questions concerning the existence, counting, and estimation of various combinatorial objects. Hence, combinatorial problems are included in the book.

For the sake of convenience, the authors have started each section with a theoretical background.

Hints and answers are provided for most (but not all)

problems. Solutions are given in a concise form in the form of notes, and trivial conclusions are omitted. In some cases, only the outlines of solutions are presented.

The exercises in the book have various origins. Most of the material is traditional and specialists on discrete mathematics are all too familiar with such problems. However, it is practically impossible to trace the origin of the problems of this kind. Most of the problems were conceived by the authors during seminars and practical classes, during examinations, and also while preparing this book. Some of the problems resulted from studying publications in journals, and a few have been borrowed from other sources. Several problems were passed on to us by staff at the Faculty and by other colleagues. The authors express their sincere gratitude to them all.

The authors are deeply indebted to S.V. Yablonsky for his persistent interest during the preparation of this book. His comments and suggestions played a significant role in determining the structure and scope of this book.

We are also grateful to our reviewers V.V. Glagolev and A.A. Markov for their critical comments and suggestions for improving the collection.

G.P. Gavrilov
A.A. Sapozhenko

Chapter One

Boolean Functions: Methods of Defining and Basic Properties

1.1. Boolean Vectors and a Unit n-Dimensional Cube[1]

A vector $(\alpha_1, \alpha_2, \ldots, \alpha_n)$ whose coordinates assume values from the set $\{0, 1\}$ is called a *binary*, or *Boolean*, *vector* (*tuple*). We shall denote such a vector by $\tilde{\alpha}^n$ or $\tilde{\alpha}$. The number n is called the *length of the vector*. The set of all Boolean vectors of length n is called a *unit n-dimensional cube* and is denoted by B^n. The vectors $\tilde{\alpha}^n$ are called the *vertices of the cube* B^n. The *weight* or *norm* $\| \tilde{\alpha}^n \|$ *of the vector* $\tilde{\alpha}^n$ is the number of coordinates of this vector that are equal to unity, i.e. $\| \tilde{\alpha}^n \| = \sum_{i=1}^{n} \alpha_i$. The set of all vertices of the cube B^n having a weight k is called *k-th stratum of the cube* B^n and is denoted by B_k^n. To each Boolean vector $\tilde{\alpha}^n$, there corresponds a number $\nu(\tilde{\alpha}^n) = \sum_{i=1}^{n} \alpha_i 2^{n-i}$, called the *number of the vector* $\tilde{\alpha}^n$. The tuple $\tilde{\alpha}^n$ is obviously a binary expansion of the number $\nu(\tilde{\alpha}^n)$. The *(Hamming) distance between the vertices* $\tilde{\alpha}$ *and* $\tilde{\beta}$ *of the cube* B^n is the number $\rho(\tilde{\alpha}, \tilde{\beta}) = \sum_{i=1}^{n} |\alpha_i - \beta_i|$, equal to the number of coordinates in which they differ. The Hamming distance is a metric, and the cube B^n is a metric space. The tuples $\tilde{\alpha}$ and $\tilde{\beta}$ from B^n are called *adjacent* if $\rho(\tilde{\alpha}, \tilde{\beta}) = 1$, and *opposite* if $\rho(\tilde{\alpha}, \tilde{\beta}) = n$. An unor-

---

[1] This section is auxiliary. We shall be using only problems 1.1.1.-1.1.6., 1.1.11., 1.1.14., 1.1.15., 1.1.31., 1.1.34., 1.1.35., and 1.1.44.

dered pair of adjacent vertices is called an *edge of the cube*. The set $B_k^n(\tilde{\alpha}) = \{\tilde{\beta}: \rho(\tilde{\alpha}, \tilde{\beta}) = k\}$ is called a *sphere*, while the set $S_k^n(\tilde{\alpha}) = \{\tilde{\beta}: \rho(\tilde{\alpha}, \tilde{\beta}) \leqslant k\}$ is a *ball of radius k with a centre at* $\tilde{\alpha}$. The tuple $\tilde{\alpha}^n$ is said *to precede* the tuple $\tilde{\beta}^n$ (notation: $\tilde{\alpha}^n \leqslant \tilde{\beta}^n$) if $\alpha_i \leqslant \beta_i$ for all $i = \overline{1, n}$. If in this case $\tilde{\alpha}^n \neq \tilde{\beta}^n$, the tuple $\tilde{\alpha}^n$ is said *to precede* $\tilde{\beta}^n$ *strictly* (notation: $\tilde{\alpha}^n < \tilde{\beta}^n$). If at least one of the relations $\tilde{\alpha}^n \leqslant \tilde{\beta}^n$ or $\tilde{\beta}^n \leqslant \tilde{\alpha}^n$ is satisfied, $\tilde{\alpha}^n$ and $\tilde{\beta}^n$ are called *comparable*. Otherwise, $\tilde{\alpha}^n$ and $\tilde{\beta}^n$ are said to be *incomparable*. The tuple $\tilde{\alpha}^n$ *directly precedes* $\tilde{\beta}^n$ if $\tilde{\alpha}^n < \tilde{\beta}^n$ and $\rho(\tilde{\alpha}^n, \tilde{\beta}^n) = 1$. The precedence relation between the tuples is the relation of partial order in B^n.

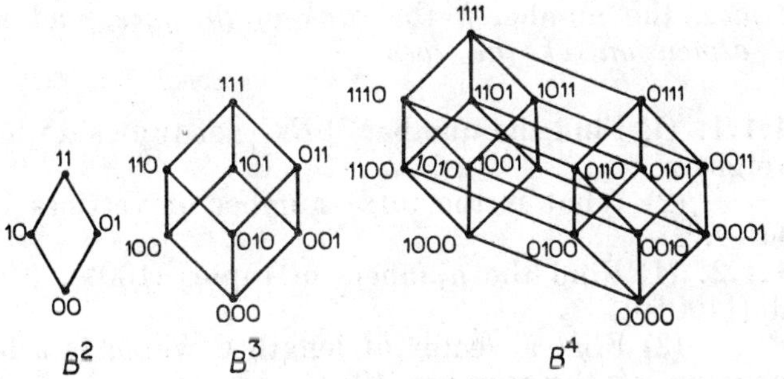

Fig. 1

Figure 1 shows the diagrams of partially ordered sets B^2, B^3 and B^4. The sequence of vertices of the cube $\{\tilde{\alpha}_0, \tilde{\alpha}_1, \ldots, \tilde{\alpha}_k\}$ is called a *chain connecting* $\tilde{\alpha}_0$ *and* $\tilde{\alpha}_k$ (notation: $[\tilde{\alpha}_0, \tilde{\alpha}_k]$) if $\rho(\tilde{\alpha}_{i-1}, \tilde{\alpha}_i) = 1$ $(i = \overline{1, k})$. The number k is called the *length of the chain* $[\tilde{\alpha}_0, \tilde{\alpha}_k]$. The chain $\{\tilde{\alpha}_0, \tilde{\alpha}_1, \ldots, \tilde{\alpha}_k\}$ is called an *ascending chain* if $\tilde{\alpha}_{i-1} < \tilde{\alpha}_i$ $(i = \overline{1, k})$. A chain z of the type $\{\tilde{\alpha}_0, \tilde{\alpha}_1, \ldots, \tilde{\alpha}_k\}$ is called a *cycle of length* k if $\tilde{\alpha}_0 = \tilde{\alpha}_k$. Let $\tilde{\alpha} = (\alpha_1, \alpha_2, \ldots, \alpha_n)$ and $\tilde{\beta} = (\beta_1, \beta_2, \ldots, \beta_n)$ be vectors from B^n. We denote by $\tilde{\alpha} \oplus \tilde{\beta}$ the vector $(\alpha_1 \oplus \beta_1, \alpha_2 \oplus \beta_2, \ldots,$

$\alpha_n \oplus \beta_n)$ obtained by exclusive sum of vectors $\tilde{\alpha}$ and $\tilde{\beta}$. By $\tilde{\alpha} \cup \tilde{\beta}$ we denote a vector whose i-th coordinate is equal to zero if and only if $\alpha_i = \beta_i = 0$, and by $\tilde{\alpha} \cap \tilde{\beta}$ a vector whose i-th coordinate is equal to 1 if and only if $\alpha_i = \beta_i = 1$. By $\bar{\tilde{\alpha}}$ we denote a vector (*opposite to* $\tilde{\alpha}$) whose i-th coordinate assumes the value 0 if $\alpha_i = 1$, and the value 1 if $\alpha_i = 0$. If $\sigma \in \{0, 1\}$, we put $\sigma\tilde{\alpha} = (\sigma\alpha_1, \sigma\alpha_2, \ldots, \sigma\alpha_n)$. The symbols $\tilde{0}$ and $\tilde{1}$ are used to denote vectors $(0, 0, \ldots, 0)$ and $(1, 1, \ldots, 1)$ respectively.

The set $B^{n, i_1, \ldots, i_k}_{\sigma_1 \ldots \sigma_k}$ of all tuples $(\alpha_1, \ldots, \alpha_n)$ from B^n for which $\alpha_{i_j} = \sigma_j$ $(j = \overline{1, k})$ is called a *face of the cube* B^n. The set $I = \{i_1, \ldots, i_k\}$ is called the *direction of the face*, the number k the *rank of the face*, and $n - k$ the *dimension of the face*.

1.1.1. (1) Find the number $|B^n_k|$ of tuples $\tilde{\alpha}^n$ having a weight k.

(2) What is the total number of vertices in the cube B^n?

1.1.2. (1) Find the numbers of tuples (1001), (01101), and (110010).

(2) Find a vector of length 6 which is a binary expansion of the number 19.

1.1.3. Find the number of tuples $\tilde{\alpha} \in B^n_k$ satisfying the condition $2^{n-1} \leqslant \nu(\tilde{\alpha}) < 2^n$.

1.1.4. Show that the following relations hold for any $\tilde{\alpha}, \tilde{\beta}, \tilde{\gamma}$ in B^n:

(1) $\rho(\tilde{\alpha}, \tilde{\gamma}) \leqslant \rho(\tilde{\alpha}, \tilde{\beta}) + \rho(\tilde{\beta}, \tilde{\gamma})$;

(2) $\rho(\tilde{\alpha}, \tilde{\gamma}) = \rho(\tilde{\alpha} \oplus \tilde{\beta}, \tilde{\gamma} \oplus \tilde{\beta})$;

(3) $\rho(\tilde{\alpha}, \tilde{\beta}) = \|\tilde{\alpha}\| + \|\tilde{\beta}\| - 2\|\tilde{\alpha} \cap \tilde{\beta}\|$;

(4) $\rho(\tilde{\alpha}, \tilde{\beta}) = \|\tilde{\alpha} \oplus \tilde{\beta}\|$.

1.1.5. (1) Find the number of unordered pairs of adjacent vertices of B^n.

(2) Find the number of unordered pairs of tuples $(\tilde{\alpha}^n, \tilde{\beta}^n)$, such that $\rho(\tilde{\alpha}^n, \tilde{\beta}^n) = k$.

1.1.6. Let $\tilde{\alpha}$ and $\tilde{\beta}$ be the vertices of a cube B^n, $\rho(\tilde{\alpha}, \tilde{\beta}) = m$. Find the number of vertices $\tilde{\gamma}$ satisfying the condition

(1) $\rho(\tilde{\alpha}, \tilde{\gamma}) + \rho(\tilde{\gamma}, \tilde{\beta}) = \rho(\tilde{\alpha}, \tilde{\beta})$;

(2) $\rho(\tilde{\alpha}, \tilde{\gamma}) = k, \quad \rho(\tilde{\beta}, \tilde{\gamma}) = r$;

(3) $\rho(\tilde{\alpha}, \tilde{\gamma}) \leqslant k, \quad \rho(\tilde{\beta}, \tilde{\gamma}) = r$;

(4) $\rho(\tilde{\alpha}, \tilde{\gamma}) \leqslant k, \quad \rho(\tilde{\beta}, \tilde{\gamma}) \geqslant r$.

1.1.7. Prove that the following systems of relations are incompatible for $\tilde{\alpha}$, $\tilde{\beta}$ and $\tilde{\gamma}$ in B^n, $n \geqslant 2$:

(1) $\rho(\tilde{\alpha}, \tilde{\beta}) > 2n/3, \quad \rho(\tilde{\beta}, \tilde{\gamma}) > 2n/3, \quad \rho(\tilde{\gamma}, \tilde{\alpha}) > 2n/3$;

(2) $\nu(\tilde{\alpha}) < \nu(\tilde{\beta} \oplus \tilde{\gamma}), \quad \nu(\tilde{\beta}) < \nu(\tilde{\gamma} \oplus \tilde{\alpha}), \quad \nu(\tilde{\gamma}) < \nu(\tilde{\alpha} \oplus \tilde{\beta})$;

(3) $\| \tilde{\alpha} \| > \| \tilde{\beta} \oplus \tilde{\gamma} \|, \ \| \tilde{\beta} \| > \| \tilde{\gamma} \oplus \tilde{\alpha} \|, \ \| \tilde{\gamma} \| > \| \tilde{\alpha} \oplus \tilde{\beta} \|, \ \| \tilde{\alpha} \cap (\tilde{\beta} \cap \tilde{\gamma}) \| = 0$;

(4) $\| \tilde{\alpha} \oplus \tilde{\beta} \oplus \tilde{\gamma} \| = 0, \ \| \bar{\tilde{\alpha}} \oplus \bar{\tilde{\beta}} \oplus \bar{\tilde{\gamma}} \| = n - 1$.

1.1.8. Let $\tilde{\alpha}$, $\tilde{\beta}$ and $\tilde{\gamma}$ be the vertices of a cube B^n. Show that:

(1) $\tilde{\alpha} \leqslant \tilde{\beta}$ is equivalent to $\tilde{\alpha} \cap \bar{\tilde{\beta}} = 0$;

(2) $\bar{\tilde{\alpha}} \cup \tilde{\beta} = \tilde{1}$ is equivalent to $\bar{\tilde{\beta}} \leqslant \bar{\tilde{\alpha}}$;

(3) $\tilde{\alpha} \cap (\tilde{\alpha} \cup \tilde{\beta}) = \tilde{\alpha}$;

(4) $\tilde{\alpha} \cup (\tilde{\alpha} \cap \tilde{\beta}) = \tilde{\alpha}$;

(5) $(\tilde{\alpha} \cup \tilde{\beta}) \cap (\tilde{\beta} \cup \tilde{\gamma}) = (\tilde{\alpha} \cap \tilde{\gamma}) \cup \tilde{\beta}$;

(6) $((\tilde{\alpha} \cap \tilde{\gamma}) \cup \tilde{\beta}) \cap \tilde{\gamma} = (\tilde{\alpha} \cap \tilde{\gamma}) \cup (\tilde{\beta} \cap \tilde{\gamma})$;

(7) $\tilde{\alpha} \leqslant \tilde{\gamma}$ leads to the relation $\tilde{\alpha} \cup (\tilde{\beta} \cap \tilde{\gamma}) = (\tilde{\alpha} \cup \tilde{\beta}) \cap \tilde{\gamma}$;

(8) $\tilde{\alpha} \leqslant \tilde{\gamma}$ is equivalent to $\tilde{\alpha} \cup (\tilde{\beta} \cap \tilde{\gamma}) \leqslant (\tilde{\alpha} \cup \tilde{\beta}) \cap \tilde{\gamma}$;

(9) $(\tilde{\alpha} \cap \tilde{\beta}) \cup (\tilde{\beta} \cap \tilde{\gamma}) \cup (\tilde{\gamma} \cap \tilde{\alpha}) = (\tilde{\alpha} \cup \tilde{\beta}) \cap (\tilde{\beta} \cup \tilde{\gamma}) \cap (\tilde{\gamma} \cup \tilde{\alpha})$.

1.1.9. How many vectors $(\alpha_1, \alpha_2, \ldots, \alpha_{12})$ in B_6^{12} satisfy the relation $\sum_{1\leqslant i\leqslant m} \alpha_i \leqslant m/2$ for all $m = \overline{1, 12}$?

1.1.10. Find the number of vectors $\tilde{\alpha}$ in B_k^n, $1 \leqslant k \leqslant n/2$, $1 \leqslant r \leqslant (n-k)/(k-1)$, which have at least r zero coordinates between two unit coordinates.

1.1.11. (1) Show that B^n contains a set consisting of $\binom{n}{[n/2]}$ pairwise incomparable vectors.

(2) Show that any subset containing not less than $n+2$ vectors includes a pair of incomparable vectors.

1.1.12. Let $0 \leqslant l < k \leqslant n$ and let $A(\tilde{\alpha})$ be a set of all vectors in B^n comparable with $\tilde{\alpha}$. Find the power of the set C:

(1) $C = A(\tilde{\alpha}) \cap B_k^n$, $\quad \tilde{\alpha} \in B_l^n$;

(2) $C = A(\tilde{\alpha}) \cap B_l^n$, $\quad \tilde{\alpha} \in B_k^n$;

(3) $C = A(\tilde{\alpha})$, $\quad \tilde{\alpha} \in B_k^n$.

1.1.13. Let $A \subseteq B_l^n$, and B be a set of all tuples in B_k^n that are comparable with at least one set in A. Prove that $\frac{|A|}{\binom{n}{l}} \leqslant \frac{|B|}{\binom{n}{k}}$.

1.1.14. (1) Show that B^n contains $n!$ pairwise different ascending chains of length n.

(2) Show that the number of pairwise different ascending chains of length n containing a fixed vertex $\tilde{\alpha}$ in B_k^n is equal to $k!(n-k)!$.

1.1.15*. (1) Show that the power of any subset of pairwise incomparable tuples of cube B^n does not exceed $\binom{n}{[n/2]}$.

(2) Show that if the subset $A \subseteq B^n$ consists of pairwise incomparable sets and $\|\tilde{\alpha}\| \leqslant k$ for any $\tilde{\alpha} \in A$, then $|A| \leqslant \binom{n}{k}$ for $k \leqslant n/2$.

1.1.16. Let $p_1, p_2, \ldots, p_n$ be pairwise different prime numbers, and N be the set of all numbers that can be presented in the form $p_1^{\alpha_1} p_2^{\alpha_2} \ldots p_n^{\alpha_n}$, where $\alpha_i \in \{0, 1\}$, $i = \overline{1, n}$. Let $A \subset N$ be a subset such that none of the numbers $a \in A$ is a divisor of any of the numbers $b \in A$ other than a. Prove that $|A| \leqslant \binom{n}{[n/2]}$.

1.1.17. Let $\tilde{\alpha}$ and $\tilde{\beta}$ be the vertices of a cube B^n, such that $\tilde{\alpha} < \tilde{\beta}$, $\rho(\tilde{\alpha}, \tilde{\beta}) = k$. Prove that the number of tuples $\tilde{\gamma}$ satisfying the relation $\tilde{\alpha} \leqslant \tilde{\gamma} \leqslant \tilde{\beta}$ is equal to 2^k.

1.1.18*. Prove that the cube B^n can be represented as a union of pairwise nonintersecting ascending chains having the following properties:

(1) the number of chains having a length $(n-2k)$ is equal to $\binom{n}{k} - \binom{n}{k-1}$, $k = \overline{0, [n/2]}$, the minimal tuple of each chain having a weight k and the maximal tuple a weight $n-k$;

(2) if $\tilde{\alpha}_i$, $\tilde{\alpha}_{i+1}$, and $\tilde{\alpha}_{i+2}$ are three consecutive vertices of an ascending chain having a length $n - 2k$, the vertex $\tilde{\beta}$ satisfying the relation $\tilde{\alpha}_i < \tilde{\beta} < \tilde{\alpha}_{i+2}$, $\tilde{\beta} \neq \tilde{\alpha}_{i+1}$ belongs to a certain chain of length $n - 2k - 2$.

1.1.19. Let $\tilde{\alpha}$ and $\tilde{\beta}$ be the vertices of the cube B^n such that $\rho(\tilde{\alpha}, \tilde{\beta}) = k$, and let $C(\tilde{\alpha}, \tilde{\beta})$ be a set of all vertices $\tilde{\gamma}$ for each of which there exists a chain $[\tilde{\alpha}, \tilde{\beta}]$ of length which is not more than $k + 2r$ containing this vertex $\tilde{\gamma}$. Find the power of the set $C(\tilde{\alpha}, \tilde{\beta})$.

1.1.20. The set $A \subseteq B^n$ is called a *complete set in* B^n if any vector $\tilde{\beta} \in B^n$ can be uniquely reconstructed under the condition that the distance $\rho(\tilde{\alpha}, \tilde{\beta})$ is known for each $\tilde{\alpha} \in A$. A complete set in B^n is called a *base set* if for each vector $\tilde{\alpha}$ in A, the set $A \setminus \{\tilde{\alpha}\}$ is not a complete set.

(1) Prove that any ascending chain $\tilde{\alpha}_0, \tilde{\alpha}_1, \ldots, \tilde{\alpha}_{n-1}$ in B^n forms a base set.

(2) Show that the sets B_1^n and B_{n-1}^n are complete in B^n for $n > 2$. Indicate a number $n > 2$ for which B_1^n is not a base set.

(3) For what n and k is the set B_k^n not complete in B^n?

(4) Prove that any base set A in B^n satisfies the condition $\frac{n}{\log_2 (n+1)} \leqslant |A| \leqslant n$.

(5) Prove that none of the faces having a dimension $n-2$ is a complete set in B^n.

(6) Prove that the number Ψ_n of base sets in B^n satisfies the inequalities $2(n!) \leqslant \Psi_n \leqslant \binom{2^n}{n}$.

1.1.21. Let φ be a one-to-one mapping of B^n onto itself. The *mapping* φ is said *to preserve the distance between vertices* if $\rho(\tilde{\alpha}, \tilde{\beta}) = \rho(\varphi(\tilde{\alpha}), \varphi(\tilde{\beta}))$ for all $\tilde{\alpha}$, $\tilde{\beta}$ in B^n. Prove that the mapping φ preserves the distance if and only if it can be obtained with the help of:

(1) a certain commutation of coordinates simultaneously in all vectors in B^n;

(2) a substitution of 0 by 1 and 1 by 0 in certain coordinates of all vectors.

1.1.22. The mapping φ of the set B^n onto itself is called *monotonic* if the condition $\nu(\tilde{\alpha}) \leqslant \nu(\tilde{\beta})$ leads to the inequality $\nu(\varphi(\tilde{\alpha})) \leqslant \nu(\varphi(\tilde{\beta}))$. Find the number of monotonic mappings of the cube B^n.

1.1.23. Let $\tilde{\alpha} \in B_k^n$. The set $M_k^n(\tilde{\alpha}) = \{\tilde{\gamma}: \nu(\tilde{\gamma}) \leqslant \nu(\tilde{\alpha}), \tilde{\gamma} \in B_k^n\}$ is called the *initial segment of the stratum* B_k^n. Let $A \subseteq B^n$; we denote by $Z_l^n(A)$ the set of all tuples $\tilde{\beta} \in B_l^n$ for each of which there exists an $\tilde{\alpha} \in A$ such that $\tilde{\beta} \leqslant \alpha$.

(1) Let $\tilde{\alpha} \in B_k^n$, and $i_1, i_2, \ldots, i_k$ $(1 \leqslant i_1 < i_2 < \ldots < i_k \leqslant n)$ be the numbers of those coordinates of the vector $\tilde{\alpha}$ which are equal to unity. Show that

$$|M_k^n(\tilde{\alpha})| = 1 + \binom{n-i_1}{k} + \binom{n-i_2}{k-1} + \ldots + \binom{n-i_k}{1}.$$

(2) Show that if $l \leqslant k$ and A is the initial segment of the stratum B_k^n, the set $Z_l^n(A)$ is the initial segment of the stratum B_l^n.

(3) Let $\tilde{\alpha} \in B_k^n$ and $i_1, i_2, \ldots, i_k$ $(1 \leqslant i_1 < i_2 < \ldots < i_k \leqslant n)$ be the numbers of those coordinates of the vector $\tilde{\alpha}$ which are equal to unity. Let $A = M_k^n(\tilde{\alpha})$ and $l \leqslant k$. Show that

$$|Z_l^n(A)| = 1 + \binom{n-i_1}{l} + \binom{n-i_2}{l-1} + \ldots + \binom{n-i_l}{1}.$$

(4)* Let $1 \leqslant m \leqslant \binom{n}{k}$. Prove that the minimum of the quantity $|Z_{k-1}^n(A)|$ for all $A \subseteq B_k^n$ satisfying the condition $|A| = m$ is attained at the initial segment of the stratum B_k^n.

(5) Let $1 \leqslant m \leqslant \binom{n}{k}$ and $l \leqslant k$. Prove that the minimum of the quantity $|Z_l^n(A)|$ for all $A \subseteq B_k^n$ satisfying the condition $|A| = m$ is attained at the initial segment of the stratum B_k^n.

(6) Let $a_0, a_1, a_2, \ldots, a_n$ be given numbers for which there exists a set A of pairwise incomparable tuples such that $|A \cap B_k^n| = a_k$ $(k = \overline{0, n})$. Then the minimum of the quantity $|Z_l^n(A)|$ for all such sets A is attained when the set $A \cap B_i^n$ is the initial segment of the stratum B_i^n for all $i > l$.

1.1.24*. Let $A \subseteq B^n$ be a set of all tuples such that there are no tuples $\tilde{\alpha}$, $\tilde{\beta}$, $\tilde{\gamma}$ in A for which $\tilde{\alpha} \cap \tilde{\beta} = \tilde{0}$ and $\tilde{\alpha} \cup \tilde{\beta} = \tilde{\gamma}$. Let $a_k = |A \cap B_k^n|$. Show that

$$a_{k+m} \Big/ \binom{n}{k+m} + a_k \Big/ \binom{n}{k} + a_m \Big/ \binom{n}{m} \leqslant 2$$

for all natural numbers k and m that do not exceed n.

1.1.25. The set $\Gamma(A) = \{\tilde{\alpha}: \tilde{\alpha} \in B^n \setminus A, \rho(\tilde{\alpha}, A) = 1\}$ is called the *boundary of the subset* $A \subset B^n$. Let $1 \leqslant i_1 < i_2 < \ldots < i_k \leqslant n$. We denote by $A^{i_1, i_2, \ldots, i_k}_{\sigma_1 \sigma_2 \ldots \sigma_k}$ the set of all tuples in A for which the coordinate with number i_j is equal to σ_j $(j = \overline{1, k})$.

The *centre of set* A is defined as a tuple whose i-th coordinate is equal to zero for $|A_0^i| \geqslant 1/2\, |A|$, and equal to unity otherwise. We denote by $\tilde{\alpha}^{i_1, i_2, \ldots, i_k}$ the tuple obtained from $\tilde{\alpha}$ by reversing the values of coordinates with numbers $i_1, i_2, \ldots, i_k$.

(1) Show that for any $A \subseteq B^n$ there exists an $A' \subseteq B^n$ such that $|A'| = |A|$, $|\Gamma(A')| = |\Gamma(A)|$ and the centre of A' is the vertex $\tilde{0} = (0, 0, \ldots, 0)$.

(2) We shall say that the set $A \subseteq B^n$ has the property I if for all $i = \overline{1, n}$ it follows from $\tilde{\alpha} \in A_1^i$ that $\tilde{\alpha}^i \in A$. Show that for all $A \subseteq B^n$ there exists a set A' with cen-

tre at $\tilde{0}$, such that $|A'| = |A|$, A' possesses the property I, and $|\Gamma(A')| \leqslant |\Gamma(A)|$.

(3) We shall say that the set $A \subseteq B^n$ possesses the property II if for all i, j $(1 \leqslant i < j \leqslant n)$ it follows from $\tilde{\alpha} \in A_{10}^{i,j}$ that $\tilde{\alpha}^{i,j} \in A$. Show that for any A having its centre at $\tilde{0}$ and possessing the property I, there exists an A' with centre at $\tilde{0}$ possessing the properties I, II, and such that $|A| = |A'|$, and $|\Gamma(A')| \leqslant |\Gamma(A)|$.

(4)* Show that the minimum of $|\Gamma(A)|$ for all $A \subseteq B^n$ satisfying the inequality $\sum_{i=0}^{k-1} \binom{n}{i} < |A| \leqslant \sum_{i=0}^{k} \binom{n}{i}$ is attained on the set $A = S_{k-1}^n(\tilde{0}) \cup M_k^n(\tilde{\alpha})$, where $M_k^n(\tilde{\alpha})$ is a certain finite segment of the stratum B_k^n (see Problem 1.1.23).

1.1.26. Let $\varphi(n)$ $(\varphi'(n))$ be the maximum power of the set $A \subseteq B^n$ such that $\|\tilde{\alpha} \cap \tilde{\beta}\| = 1$ (resp. $\|\tilde{\alpha} \cap \tilde{\beta}\| \geqslant 1$) for any two different vectors in A. Show that

(1) $\varphi(n) = n$; (2) $\varphi'(n) = 2^{n-1}$.

1.1.27. Let $F(n, k)$ be a family of subsets A of the set B^n such that $\|\tilde{\alpha} \cap \tilde{\beta}\| \geqslant k$ for any $\tilde{\alpha}, \tilde{\beta}$ in A. Let $\varphi(n, k) = \max_{A \in F(n,k)} |A|$. Prove the following statements:

(1) $\varphi(n, k) \geqslant \sum_{i=]\frac{n+k+1}{2}[}^{n} \binom{n}{i}$.

(2) Let $A \subset F(n, k)$; then $|A \cap B_l^n| + |A \cap B_{n-l+k-1}^n| \leqslant \binom{n}{l}$.

(3) For $l \geqslant \frac{n+k+1}{2}$ the equality $|A \cap B_l^n| + |A \cap B_{n-l+k-1}^n| = \binom{n}{l}$ holds only when $A \cap B_{n-l+k-1}^n = \varnothing$.

(4) $\varphi(n, k) = \sum_{i=\frac{n+k}{2}}^{n} \binom{n}{i}$ for even $n+k$,

$$\varphi(n, k) = \binom{n-1}{\frac{n+k-1}{2}} + \sum_{i=\frac{n+k+1}{2}}^{n} \binom{n}{i} \text{ for odd } n+k.$$

1.1.28. Let $F_r(n)$ be a family of subsets $A \subseteq B^n$ such that $\rho(\tilde{\alpha}, \tilde{\beta}) \leqslant 2r$ for all $\tilde{\alpha}$ and $\tilde{\beta}$ in A. Let $\varphi_r(n) = \max_{A \in F_r(n)} |A|$.

(1)* Show that the maximum of $|A|$ is attained for all $A \in F_r(n)$ on sets of the type $S_r^n(\tilde{\alpha})$ and is equal to $\sum_{k=0}^{r} \binom{n}{k}$.

(2) For odd n and $r = (n-1)/2$, give an example of a set $A \in F_r(n)$ which is not a ball of radius r, and for which $|A| = \varphi_r(n)$.

1.1.29*. The subset $A \subseteq B^n$ is said to possess the property 1 if any two vectors in A have a common unit coordinate, property 2 if for any $\tilde{\alpha} \in A$ the opposite vector $\tilde{\tilde{\alpha}}$ does not lie in A, and property 3 if it follows from $\tilde{\alpha} \in A$ and $\tilde{\alpha} \leqslant \tilde{\beta}$ that $\tilde{\beta} \in A$. Let $\psi(n)$ be the number of subsets $A \subseteq B^n$ possessing properties I and II simultaneously, and $\varphi(n)$ be the number of subsets $B \subseteq B^n$ possessing property 3. Show that:

(1) $\psi(n) \geqslant 2^{\binom{n-1}{[(n-1)/2]}}$; (2) $\psi(n) \leqslant \varphi^2(n-1)$.

1.1.30. Let $A \subset B^n$, and $R(A)$ be the number of such edges $(\tilde{\alpha}, \tilde{\beta})$ that $\tilde{\alpha} \in A$, $\tilde{\beta} \in B^n \setminus A$. Show that $|R(A)| \geqslant \min\{|A|, 2^n - |A|\}$.

1.1.31. Prove the following statements:

(1) The number of different faces in a fixed direction $\{i_1, i_2, \ldots, i_k\}$ is equal to 2^k.

(2) Two different faces in the same direction do not intersect.

(3) The union of all faces of the cube B^n having a given direction is the entire cube.

(4) The number of all faces of rank k in the cube B^n is equal to $\binom{n}{k} 2^k$.

(5) The total number of faces of the cube B^n is equal to 3^n.

(6) The number of faces of dimension k containing a given vertex $\tilde{\alpha}$ is equal to $\binom{n}{k}$.

(7) The number of faces of dimension k containing a given face of dimension l is equal to $\binom{n-l}{k-l}$.

(8) The number of k-dimensional faces intersecting a given l-dimensional face of the cube B^n is equal to $\sum_{j=0}^{\min(k,\, l)} \binom{l}{j} 2^{l-j} \binom{n-l}{k-j}$.

1.1.32. We define the *interval of the cube* B^n as a set of the form $\{\tilde{\gamma}: \tilde{\alpha} \leqslant \tilde{\gamma} \leqslant \tilde{\beta}\}$, where $\tilde{\alpha}$ and $\tilde{\beta}$ are certain vertices such that $\tilde{\alpha} \leqslant \tilde{\beta}$. The number $\rho(\tilde{\alpha}, \tilde{\beta})$ is called the *dimension of the interval*. Show that a face of dimension k is an interval of dimension k.

1.1.33. Let g_1, g_2, g_3 be the faces of the cube B^n. Show that the relations $g_1 \cap g_2 = \varnothing$, $g_2 \cap g_3 \neq \varnothing$, $g_3 \cap g_1 \neq \varnothing$ lead to the inequality $(g_1 \cap g_2) \cap g_3 \neq \varnothing$.

1.1.34. Let $n_1, n_2, \ldots, n_s$ be non-negative integers such that $\sum_{i=1}^{s} 2^{n_i} \leqslant 2^n$. In this case, B^n contains pairwise non-intersecting faces $g_1, g_2, \ldots, g_s$ whose dimensions are respectively equal to $n_1, n_2, \ldots, n_s$.

1.1.35. The faces g_1 and g_2 of the cube B^n are called *incomparable* if neither of the inclusions $g_1 \subseteq g_2$ and $g_2 \subseteq g_1$ is satisfied.

(1) Show that there exists a set of faces of the cube B^n consisting of $\binom{n}{[n/3]}\binom{n-[n/3]}{[n/3]}$ pairwise incomparable faces.

(2) Show that the power of any set of pairwise incomparable faces of the cube B^n does not exceed $\binom{n}{[n/3]} \times 2^{n-[n/3]}$.

1.1.36. Let $L(n, k)$ be the minimum number of vertices in B^n such that each k-dimensional face contains at least one of these vertices. Prove that

(1) $L(n, 1) = 2^{n-1}$;

(2) $L(n, n-1) = 2$;

(3) $L(n, 2) \leqslant [2^n/3]$;

(4*) $m \leqslant L(n, n-2) \leqslant m + 2$, where m is the smallest integer for which $\binom{m}{[m/2]} \geqslant n$;

(5) $L(n, k) \leqslant \sum_{i=0}^{[n/k]} \binom{n}{ki}$;

(6) $L(n, k) \geqslant 2^{(n-r)} L(r, k), \quad k \leqslant r \leqslant n$.

1.1.37. Show that tuples in B^n can be expanded into a sequence $\tilde{\alpha}_1, \tilde{\alpha}_2, \ldots, \tilde{\alpha}_{2n}$ which is a cycle.

1.1.38. Show that B^n does not contain cycles of odd length.

1.1.39. The binary vector $(\alpha_0, \alpha_1, \ldots, \alpha_{N-1})$ is called *n-universal* if for each vector $(\beta_0, \beta_1, \ldots, \beta_{n-1})$ in B^n there exists a number k such that $\beta_i = \alpha_{k \oplus i}$ ($i = \overline{0, n-1}$), where $k \oplus i = k + i \pmod N$. Will the vector $\tilde{\alpha}$ be n-universal if

(1) $\tilde{\alpha} = (0011), \quad n = 2$;

(2) $\tilde{\alpha} = (01011), \quad n = 2$;

(3) $\tilde{\alpha} = (00110), \quad n = 2$;

(4) $\tilde{\alpha} = (00011101), n = 3$?

1.1.40. Prove that for each natural number n there exists an n-universal vector of length 2^n, and no n-universal vector having a length smaller than 2^n.

1.1.41. The cycle Z of the cube B^n is called a *2d-cycle* if $|Z \cap S_d^n(\tilde{\alpha})| = 2d + 1$ for all $\tilde{\alpha} \in Z$, i.e. if for any vertex $\tilde{\alpha}$ in the cycle Z the set of vertices of the cycle situated at a Hamming distance not exceeding d coincides with the set of vertices at a distance not exceeding d "along the cycle".

(1) Let $l(n)$ be the maximum length of a 2-cycle in B^n. Find the value of $l(n)$ for $n = \overline{2, 5}$.

(2) Let Z be a 2-cycle in B^n. Show that any face of dimension 4 does not contain more than 8 vertices of the 2-cycle Z.

(3)* Show that the maximum length of a 2-cycle in B^n does not exceed 2^{n-1}, $n > 3$.

1.1.42. The pair $(\tilde{\alpha}_i, \tilde{\alpha}_{i+1})$ formed by two consecutive vertices of the cycle Z in cube B^n is called an *edge of the cycle*. Show that B^5 contains a cycle of length 32, any four consecutive edges of which have pairwise different directions.

1.1.43. Let $a_1, \ldots, a_n$ be such numbers that $a_i > 2$ $(i = \overline{1, n})$. Show that the number of sums

$$\sum_{i=1}^{n} (-1)^{\sigma_i} a_i, \quad \sigma_i \in \{0, 1\}, \; i = \overline{1, n},$$

satisfying the condition $\left| \sum_{i=1}^{n} (-1)^{\sigma_i} a_i \right| \leqslant 1$ does not exceed $\binom{n}{[n/2]}$.

1.1.44. Let $A \subseteq B^n$, $|A| > 2^{n-1}$. Show that at least n edges of the cube are completely contained in A.

1.2. Methods of Defining Boolean Functions. Elementary Functions. Formulas. Superposition Operation

We shall assign the symbol $\tilde{x}^n$ (or $\tilde{x}$) to the tuple of variables $(x_1, x_2, \ldots, x_n)$, and denote the set of the variables by X^n. The function $f(\tilde{x}^n)$, which is defined on the set B^n and which assumes values from the set $\{0, 1\}$, is termed a *Boolean function*. We shall denote the set of all Boolean functions of variables $x_1, x_2, \ldots, x_n$ by $P_2(X^n)$.

The Boolean function $f(\tilde{x}^n)$ can be presented in tabular form $T(f)$ (see Table 1). Here, the tuples $\tilde{\sigma}$ are arranged

Table 1

x_1	x_2	. . .	x_{n-1}	x_n	$f(x_1, x_2, \ldots, x_{n-1}, x_n)$
0	0	. . .	0	0	$f(0, 0, \ldots, 0, 0)$
0	0	. . .	0	1	$f(0, 0, \ldots, 0, 1)$
0	0	. . .	1	0	$f(0, 0, \ldots, 1, 0)$
. . .	. . .	. . .	. . .	. . .	. . . . . .
1	1	. . .	1	1	$f(1, 1, \ldots, 1, 1)$

in ascending order of their numbers. Assuming such an arrangement to be the standard procedure in the following, we shall define the function $f(\tilde{x}^n)$ through the vector $\tilde{\alpha}_f^{2^n} = (\alpha_0, \alpha_1, \ldots, \alpha_{2^n-1})$ in which the coordinate α_i is the value of the function $f(\tilde{x}^n)$ on the tuple $\tilde{\sigma}$ with number i ($i = 0, 1, \ldots, 2^n-1$).

The symbol N_f will be used to denote the set $\{\tilde{\sigma}: (\tilde{\sigma} \in B^n) \,\&\, (f(\tilde{\sigma}) = 1)\}$.

The Boolean function $f(\tilde{x}^n)$ can be also presented through the rectangular table $\Pi_{k,n-k}(f)$ (see Table 2) in which the value $f(\sigma_1, \sigma_2, \ldots, \sigma_n)$ of the function f lies at the intersection of the "row" $(\sigma_1, \sigma_2, \ldots, \sigma_k)$ and the "column" $(\sigma_{k+1}, \sigma_{k+2}, \ldots, \sigma_n)$, $1 \leqslant k < n$.

Table 2

	$0\ 0\ \ldots\ \sigma_{k+1}\ \ldots\ 1$	x_{k+1}
	$0\ 0\ \ldots\ \sigma_{k+2}\ \ldots\ 1$	x_{k+2}
	$\ldots\ldots\ldots\ldots$	$\ldots$
	$0\ 1\ \ldots\ \sigma_n\ \ldots\ 1$	x_n
$x_1\ x_2\ \ldots\ x_k$		
$0\ 0\ \ldots\ 0$		
$0\ 0\ \ldots\ 1$		
$\ldots\ldots$		
$\sigma_1\ \sigma_2\ \ldots\ \sigma_k$	$f(\tilde{\sigma})$	
$\ldots\ldots$		
$1\ 1\ \ldots\ 1$		

Boolean functions described in Tables 3 and 4 are assumed to be *elementary* functions.

Let us now consider the notation and names of these functions.

Table 3

x	0	1	f_1	f_2
0	0	1	0	1
1	0	1	1	0

Table 4

x_1	x_2	f_3	f_4	f_5	f_6	f_7	f_8	f_9
0	0	0	0	0	1	1	1	1
0	1	0	1	1	0	1	1	0
1	0	0	1	1	0	0	1	0
1	1	1	1	0	1	1	0	0

(1) The functions 0 and 1 are called *zero* (or 0-*function*) and *unity* (or 1-*function*) respectively.

(2) The function f_1 is called an *identity function* and is denoted by x.

(3) The function f_2 is called the *negation of* x, denoted by $\bar{x}$ or $\neg x$, and often read as "*not* x".

(4) The function f_3 is called the *conjunction of* x_1 *and* x_2, denoted by $x_1 \& x_2$, or $x_1 \cdot x_2$, or $x_1 x_2$, or $\min(x_1, x_2)$, and often read as "x_1 *and* x_2".

(5) The function f_4 is called the *disjunction of* x_1 *and* x_2, denoted by $x_1 \vee x_2$, or $x_1 + x_2$, or $\max(x_1, x_2)$, and often reads as "x_1 *or* x_2".

(6) The function f_5 is called the *exclusive sum of* x_1 *and* x_2, denoted by $x_1 \oplus x_2$, and often read as "x_1 *plus* x_2".

(7) The function f_6 is called the *equivalence of* x_1 *and* x_2, denoted by $x_1 \sim x_2$, or $x_1 \leftrightarrow x_2$, or $x_1 \equiv x_2$, and often read as "x_1 *is equivalent to* x_2".

(8) The function f_7 is called the *implication of* x_1 *and* x_2, denoted by $x_1 \to x_2$, or $x_1 \supset x_2$, and often read as "x_1 *implies* x_2".

(9) The function f_8 is called *Sheffer's stroke for* x_1 *and* x_2, denoted by $x_1 \mid x_2$, and often read as "*not both* x_1 *and* x_2".

(10) The function f_9 is called the *Peirce's arrow of* x_1 *and* x_2 denoted by $x_1 \downarrow x_2$, and often read as "*neither* x_1 *nor* x_2".

Sometimes, the functions 0 and 1 are treated as functions depending on the empty set of variables.

The symbols $\neg$, &, $\vee$, $\oplus$, $\sim$, etc. are used in the notations of elementary functions and are called *sentential connectives.*

Let us fix a certain (finite or countably infinite) *alphabet of variables* X. Let $\Phi = \{f_1^{(n_1)}, f_2^{(n_2)}, \ldots\}$ be a set of *functional symbols*, where the superscripts indicate the *number of sites* where the symbols can be placed. Sometimes the superscripts are omitted if the arity of the functional symbols is assumed to be known.

Definition 1.1. A *formula generated by the set* Φ is such (and only such) expression as

(1) f_k and $f_j(x_{i_1}, x_{i_2}, \ldots, x_{i_n})$, where f_k and f_j are functional symbols of zero and n arguments respectively, and $x_{i_1}, x_{i_2}, \ldots, x_{i_n}$ are variables in the set X;

(2) $f_m(\mathfrak{A}_1, \mathfrak{A}_2, \ldots, \mathfrak{A}_s)$, where f_m is a functional symbol of s arguments and $\mathfrak{A}_i$ is either a formula generated by Φ or a variable in X, $i = \overline{1, s}$.

In order to accentuate the fact that formula $\mathfrak{A}$ contains *only variables in X* (or *only functional symbols in* Φ), we shall write $\mathfrak{A}(X)$ (resp. $\mathfrak{A}[\Phi]$).

Sometimes, formulas of the type $f(x, y)$ are written in the form (xfy) or xfy, and formula $f(x)$ in the form (fx) or fx. The symbol f is called a *connective*.

Usually, connectives are denoted by symbols from the set $\mathfrak{S} = \{\neg, \&, \vee, \oplus, \sim, \rightarrow, \mid, \downarrow\}$.

Definition 1.2. A *formula generated by the* $\mathfrak{S}$ is such (and only such) expression as

(1) x, i.e. any variable from the set X;

(2) $(\neg\mathfrak{A})$, $(\mathfrak{A} \& \mathfrak{B})$, $(\mathfrak{A} \vee \mathfrak{B})$, $(\mathfrak{A} \oplus \mathfrak{B})$, $(\mathfrak{A} \sim \mathfrak{B})$, $(\mathfrak{A} \rightarrow \mathfrak{B})$, $(\mathfrak{A} \mid \mathfrak{B})$, $(\mathfrak{A} \downarrow \mathfrak{B})$, where $\mathfrak{A}$ and $\mathfrak{B}$ are formulas generated by $\mathfrak{S}$.

The following convention is usually adopted for abbreviating the notation of formulas generated by the set $\mathfrak{S}$ of connectives:

(a) the outer brackets in the formulas are omitted;

(b) formula $(\neg\,\mathfrak{A})$ is written in the form $\overline{\mathfrak{A}}$;

(c) formula $(\mathfrak{A} \& \mathfrak{B})$ is written in the form $(\mathfrak{A} \cdot \mathfrak{B})$ or $(\mathfrak{A}\mathfrak{B})$;

(d) it is assumed that the connective $\neg$ is *stronger* than any connective of two variables in $\mathfrak{S}$;

(e) the connective & is assumed to be *stronger* than any of the connectives $\vee$, $\oplus$, $\sim$, $\rightarrow$, $\mid$, $\downarrow$.

With the help of this convention, we can write, for example, the formula $((\neg x) \rightarrow ((x \& y) \vee z))$ in the form $\bar{x} \rightarrow (xy \vee z)$.

"*Mixed*" form of notation is also used, for example, $x \oplus f(y, z)$ or $x_1 f(x_2, 0, x_3) \vee \bar{x}_1 f(1, \bar{x}_2, x_3)$.

Suppose that each functional symbol $f_i^{(n_i)}$ in the set Φ has a corresponding function F_i: $B^{(n_i)} \to B$. The concept of the *function* $\varphi_{\mathfrak{A}}$, *represented by formula* $\mathfrak{A}$ *generated by the set* Φ, is defined by induction:

(1) if $\mathfrak{A} = f_i^{(n_i)}(x_{j_1}, x_{j_2}, \ldots, x_{j_{n_i}})$, then for each tuple $(\alpha_1, \alpha_2, \ldots, \alpha_{n_i})$ of values of the variables x_{j_1}, $x_{j_2}, \ldots, x_{j_{n_i}}$, the value of the function $\varphi_{\mathfrak{A}}$ is equal to $F_i(\alpha_1, \alpha_2, \ldots, \alpha_{n_i})$;

(2) if $\mathfrak{A} = f(\mathfrak{A}_1, \mathfrak{A}_2, \ldots, \mathfrak{A}_m)$, where $f \in \Phi$, $\mathfrak{A}_k = \mathfrak{A}_k(y_{k1}, y_{k2}, \ldots, y_{ks_k})$ is a formula generated by Φ or a variable in X, and on each tuple $(\alpha_{k1}, \alpha_{k2}, \ldots, \alpha_{ks_k})$ of the values of variables $y_{k1}, y_{k2}, \ldots, y_{ks_k}$ the function $\varphi_{\mathfrak{A}_k}$ is equal to β_k $(k = \overline{1, m})$, then

$$\varphi_{\mathfrak{A}}(\alpha_{11}, \alpha_{12}, \ldots, \alpha_{1s_1}, \ldots, \alpha_{k1}, \alpha_{k2}, \ldots, \alpha_{ks_k}, \ldots, \alpha_{m1}, \alpha_{m2}, \ldots, \alpha_{ms_m}) = F(\beta_1, \ldots, \beta_k, \ldots, \alpha_m)$$

(here F is the function corresponding to the functional symbol f).

If $\mathfrak{A} = f_i^{(n_i)}(x_{j_1}, x_{j_2}, \ldots, x_{j_{n_i}})$, $\varphi_{\mathfrak{A}}$ is usually denoted by $F_i(x_{j_1}, x_{j_2}, \ldots, x_{j_{n_i}})$. If, however,

$$\mathfrak{A} = f(\mathfrak{A}_1, \mathfrak{A}_2, \ldots, \mathfrak{A}_m),$$

then $\varphi_{\mathfrak{A}}$ is denoted by

$$F(\varphi_{\mathfrak{A}_1}(y_{11}, y_{12}, \ldots, y_{1s_1}), \ldots, \varphi_{\mathfrak{A}_m}(y_{m1}, y_{m2}, \ldots, y_{ms_m})).$$

The concept of the *function* $\varphi_{\mathfrak{A}}$ *represented by formula* $\mathfrak{A}$ *generated by the set of connectives* $\mathfrak{S}$ is introduced as follows:

(1) the formula $\mathfrak{A} = x$, where $x \in X$, is juxtaposed to the identity function $\varphi_{\mathfrak{A}}(x) = x$;

(2) if $\mathfrak{A} = (\neg\mathfrak{B})$ (or $\mathfrak{A} = (\mathfrak{B} \bigcirc \mathfrak{C})$, where $\bigcirc \in \{\&, \vee, \oplus, \sim, \to, |, \downarrow\}$), then $\varphi_{\mathfrak{A}} = \overline{\varphi_{\mathfrak{B}}}$ (resp. $\varphi_{\mathfrak{A}} = \varphi_{\mathfrak{B}} \bigcirc \varphi_{\mathfrak{C}}$, where the symbol $\bigcirc$ should now be considered as the notation for the corresponding elementary Boolean function, see Tables 3, 4).

Let $\{x_1, x_2, \ldots, x_n\}$ be a set of variables which are encountered in at least one of the formulas $\mathfrak{A}$ or $\mathfrak{B}$. Formulas $\mathfrak{A}$ and $\mathfrak{B}$ are called *equivalent* (notation: $\mathfrak{A} =$

$\mathfrak{B}$ or $\mathfrak{A} \equiv \mathfrak{B}$) if on each tuple $(\alpha_1, \alpha_2, \ldots, \alpha_n)$ of the values of variables $x_1, x_2, \ldots, x_n$ the values of functions $\varphi_{\mathfrak{A}}$ and $\varphi_{\mathfrak{B}}$ represented by formulas $\mathfrak{A}$ and $\mathfrak{B}$ respectively coincide.

Let Φ be a set of functional symbols (or sentential connectives), and P be the set of functions corresponding to them. The *superposition generated by the set* P can be defined as any function F that can be obtained through a formula of the set Φ.

1.2.1. What is the number of functions in $P_2(X^n)$ that assume the same values on opposite tuples?

1.2.2. Find the number of functions in $P_2(X^n)$ which assume opposite values on any pair of adjacent tuples.

1.2.3. Find the number of functions in $P_2(X^n)$ that assume the value 1 on less than k tuples in B^n.

1.2.4. Using the functions $f(x_1, x_2)$ and $g(x_3, x_4)$ specified in the vector form, write the function h in the vector form:

1) $\tilde{\alpha}_f = (1011)$, $\quad \tilde{\alpha}_g = (1001)$,
$$h(x_2, x_3, x_4) = f(g(x_3, x_4), x_2);$$

2) $\tilde{\alpha}_f = (1011)$, $\quad \tilde{\alpha}_g = (1001)$,
$$h(\tilde{x}^4) = f(x_1, x_2) \vee g(x_3, x_4);$$

3) $\tilde{\alpha}_f = (1000)$, $\quad \tilde{\alpha}_g = (0111)$,
$$h(\tilde{x}^4) = f(x_1, x_2) \mathbin{\&} g(x_3, x_4).$$

1.2.5. Let ν_1 be a number having its binary expansion in the form of the tuple (x_1, x_2), and let ν_2 be a number with its binary expansion in the form (x_3, x_4). Let $f_i(\tilde{x}^4)$ be the i-th order of the binary representation of the number $|\nu_1 - \nu_2|$, $i = 1, 2$. Construct a rectangular table $\Pi_{2,2}$ of the functions $f_1(\tilde{x}^4)$ and $f_2(\tilde{x}^4)$.

1.2.6. (1) The function $f(\tilde{x}^3)$ is defined as follows: it is equal to unity either for $x_1 = 1$, or when the variables x_2 and x_3 assume different values while the value of the variable x_1 is less than that of the variable x_3. Otherwise, the function is equal to zero. Compile the tables $T(f)$ and $\Pi_{1,2}(f)$ of the function $f(\tilde{x}^3)$ and write down the tuples of the set N_f.

(2) The function $f(\widetilde{x}^4)$ is defined as follows: it is equal to zero only on such tuples $\widetilde{\alpha} = (\alpha_1, \alpha_2, \alpha_3, \alpha_4)$ that satisfy the inequality $\alpha_1 + \alpha_2 > \alpha_3 + 2\alpha_4$. Compile the table $T(f)$ and write down the tuples of the set N_f of this function.

1.2.7. The function $f(\widetilde{x}^n)$ is called *symmetric* if $f(x_1, x_2, \ldots, x_n) = f(x_{i_1}, x_{i_2}, \ldots, x_{i_n})$ upon any substitution $\begin{pmatrix} 1 & 2 & \ldots & n \\ i_1 & i_2 & \ldots & i_n \end{pmatrix}$.

(1) Show that if $f(\widetilde{x}^n)$ is a symmetric function, the equality $\|\widetilde{\alpha}^n\| = \|\widetilde{\beta}^n\|$ leads to the equality $f(\widetilde{\alpha}^n) = f(\widetilde{\beta}^n)$.

(2) Find the number of symmetric functions in $P_2(X^n)$.

1.2.8. Let $f(\widetilde{x}^n)$ be an arbitrary Boolean function $(n \geqslant 1)$. We associate with it a symmetric function $S(y_1, y_2, \ldots, y_m)$, where $m = 2^n - 1$: $S(\widetilde{\alpha}^m) = f(\widetilde{\beta}^n)$, if $\|\widetilde{\alpha}^m\| = \nu(\widetilde{\beta}^n)$. Prove that

$$S(\underbrace{x_1, \ldots, x_1}_{2^{n-1}\text{ times}}, \underbrace{x_2, \ldots, x_2}_{2^{n-2}\text{ times}}, \ldots, \underbrace{x_i, \ldots, x_i}_{2^{n-i}\text{ times}}, \ldots, \underbrace{x_{n-1}, x_{n-1}}_{2\text{ times}}, x_n) = f(x_1, x_2, \ldots, x_i, \ldots, x_{n-1}, x_n).$$

1.2.9. Which of the following expressions are formulas generated by the set of sentential connectives $\{\neg, \&, \vee, \rightarrow\}$ if:

(1) $x \rightarrow y$;
(2) $(x \,\&)\ \neg y$;
(3) $(x \leftarrow y)$
(4) $(y \rightarrow (x))$;
(5) $(x \rightarrow (y \,\&\, (\neg\, x)))$;
(6) $(x \,\&\, y)\ \neg z$;
(7) $(\neg\, x \rightarrow z)$?

1.2.10. In how many ways can brackets be arranged in the expression A so that a formula generated by the set $\{\neg, \&\ \vee, \rightarrow\}$ of sentential connectives is obtained each time if:

(1) $A = \neg x \rightarrow y \,\&\, x$;
(2) $A = x \,\&\, y \,\&\, \neg\neg\, z \vee x$;
(3) $A = x \rightarrow \neg y \rightarrow z \,\&\, \neg x$?

1.2.11. The *complexity of a formula* generated by the set of sentential connectives $\mathfrak{S}$ is defined as the number of connectives in it. By induction with respect to the complexity of a formula, show that in a formula

(1) of nonzero complexity, there exists at least one pair of brackets;

(2) the number of left brackets is equal to the number of right brackets;

(3) two connectives cannot be put next to each other; and

(4) two variable symbols cannot be put next to each other.

1.2.12. The *connectivity index in a formula* is defined as the difference between the number of left brackets preceding the connective under consideration in the formula and the number of right brackets preceding this connective. Prove that any formula of nonzero complexity generated by the set $\{\neg, \&, \vee, \rightarrow\}$

(1) contains a single connective of index 1;

(2) can be uniquely presented in one of the following forms: $(\neg\mathfrak{A})$, $(\mathfrak{A} \& \mathfrak{B})$, $(\mathfrak{A} \vee \mathfrak{B})$, $(\mathfrak{A} \rightarrow \mathfrak{B})$, where $\mathfrak{A}$ and $\mathfrak{B}$ are formulas generated by $\{\neg, \&, \vee, \rightarrow\}$.

1.2.13. Let us consider the formulas generated by the set of connectives $\{\neg, \&, \vee, \rightarrow\}$ in a form without brackets. Instead of $(\neg x)$, $(x \& y)$, $(x \vee y)$ and $(x \rightarrow y)$, we shall write $\neg x$, $\& xy$, $\vee xy$, $\rightarrow xy$.

For example, the formula $((\neg x) \rightarrow (y \vee (\neg z)))$ will assume the form $\rightarrow \neg x \vee y \neg z$. Prove that if each of the connectives $\&$, $\vee$, $\rightarrow$ is evaluated as the number $+1$, each variable symbol as the number -1, and the connective $\neg$ as zero, the expression for A in a form without brackets will be a formula if and only if the sum of estimates of all occurrences of the symbols in A is equal to -1 and if this sum is non-negative in each proper initial segment of the expression A.

1.2.14. Find out if the expression A is a formula generated by the set Φ if

(1) $A = f^{(2)}(g^{(2)}(x, y), f^{(2)}(x))$, $\quad \Phi = \{f^{(2)}, g^{(2)}, \varphi^{(1)}\}$;

(2) $A =)x\,(\varphi^{(1)}$, $\quad \Phi = \{g^{(1)}, \varphi^{(1)}\}$;

(3) $A = \varphi^{(1)}(f^{(2)}(1, x))$, $\Phi = \{f^{(2)}, \varphi^{(1)}\}$;

(4) $A = g^{(2)}(\varphi^{(1)}, f^{(3)}(x, y, \varphi^{(1)}))$, $\Phi = \{f^{(3)}, g^{(2)}, \varphi^{(1)}\}$;

(5) $A = (\varphi^{(1)}(f^{(2)}(x, \varphi^{(1)}(x))))$, $\quad \Phi = \{f^{(2)}, \varphi^{(1)}\}$.

1.2.15. Find the vector $\tilde{\alpha}_\varphi$ of the function φ represented by the formula $\mathfrak{A}$ generated by the set $\Phi = \{f_1^{(1)}, f_2^{(2)}, g^{(2)}\}$ if the functional symbols $f_1^{(1)}$, $f_2^{(2)}$, $g^{(2)}$ are put in correspondence with Boolean functions defined by vectors (10), (1011), and (1000) respectively:

(1) $\mathfrak{A} = f_2^{(2)}(f_1^{(1)}(g^{(2)}(x, f_1^{(1)}(y))), y)$;

(2) $\mathfrak{A} = g^{(2)}(f_1^{(1)}(f_2^{(2)}(x, y)), g^{(2)}(x, f_1^{(1)}(y)))$;

(3) $\mathfrak{A} = f_1^{(1)}(f_2^{(2)}(x, g^{(2)}(f_2^{(2)}(x, y), f_1^{(1)}(y))))$.

1.2.16. Compile the tables of functions represented by the following formulas (generated by the set of connectives $\mathfrak{S}$):

(1) $(x \to y) \oplus ((y \to z) \oplus (z \to x))$;

(2) $\overline{(\bar{x} \vee y) \vee (x \cdot \bar{z})} \downarrow (x \sim y)$;

(3) $\bar{x} \to (\bar{z} \sim (y \oplus xz))$;

(4) $(((x \mid y) \downarrow z) \mid y) \downarrow z$.

1.2.17. A formula generated by a set $\mathfrak{S}$ is called *identically truth* (*identically false*) if the function represented by it is equal to 1 (resp. zero) on any tuple of values of variables. Find out which of the following formulas are identically truth and which are identically false:

(1) $(x \to y) \to ((x \vee z) \to (y \vee z))$;

(2) $((x \oplus y) \sim z)(x \to yz)$;

(3) $((\bar{x} \vee \bar{y}) \downarrow (x \oplus \bar{y})) \oplus \overline{(x \to \bar{y}} \to (\bar{x} \vee y))$;

(4) $((x \vee \bar{y}) z \to ((x \sim z) \oplus y)) \quad (x(yz))$.

1.2.18. Find out if the following formulas $\mathfrak{A}$ and $\mathfrak{B}$ are equivalent:

(1) $\mathfrak{A} = ((x \vee y) \vee z) \to ((x \vee y) \quad (x \vee z))$, $\mathfrak{B} = x \sim z$;

(2) $\mathfrak{A} = (x \to y) \to z$, $\mathfrak{B} = x \to (y \to z)$;

(3) $\mathfrak{A} = ((x \oplus y) \to (x \vee y))((\bar{x} \to y) \to (x \oplus y))$, $\mathfrak{B} = x \mid y$;

(4) $\mathfrak{A} = \overline{(x \to y) \vee ((x \to z)\, y)}$, $\mathfrak{B} = (x\bar{y}) \quad (\bar{y} \to x\bar{z})$.

1.2.19. Find all functions which depend only on variables of the set $\{x, y\}$ and are superpositions generated by the set P:

(1) $P = \{u_1 \oplus u_2, 1\}$;

(2) $P = \{u_1 u_2 \oplus (u_1 \vee u_2)\}$;

(3) $P = \{f(u_1, u_2) = (1101),\ g(u_1, u_2, u_3) = (10010110)\}$[2].

1.2.20. *Depth of a formula* $\mathfrak{A}$ *generated by the set* Φ (notation: $\text{dep}_\Phi(\mathfrak{A})$) is determined by induction:

(I) if $\mathfrak{A}$ is the symbol of a variable or a functional symbol of 0 arguments, then $\text{dep}_\Phi\, \mathfrak{A} = 0$;

---

[2] Here and below, while presenting a function in the vector form we shall use the notation $f = \tilde{\beta}$ instead of $\tilde{\alpha} = \tilde{\beta}$.

(II) if $\mathfrak{A} = f^{(n)}(\mathfrak{A}_1, \ldots, \mathfrak{A}_n)$, where $f^{(n)} \in \Phi$, then

$$\mathrm{dep}_\Phi(\mathfrak{A}) = \max_{1 \leqslant i \leqslant n} \mathrm{dep}_\Phi(\mathfrak{A}_i) + 1.$$

Find the formula $\mathfrak{A}$ generated by the set $\{\downarrow\}$, having a minimal depth and representing the function

(1) $f = x \vee y$; (2) $f = x \oplus y$; (3) $f = xy$.

1.2.21. Find out if a function f can be represented by a formula $\mathfrak{A}$ of depth k generated by a set of connectives S if

(1) $f = xy$, $k = 2$, $S = \{|\}$;
(2) $f = x \to y$, $k = 3$, $S = \{\vee, \sim\}$;
(3) $f = x \oplus y \oplus z$, $k = 2$, $S = \{\to, \&\}$.

1.2.22. Prove that a function f cannot be represented by a formula generated by a set of connectives S if

(1) $f = x \oplus y$, $S = \{\&\}$;
(2) $f = xy$, $S = \{\to\}$;
(3) $f = x \vee y$, $S = \{\sim\}$.

1.2.23. Is it possible to represent a function f by a formula of depth $k + 1$ generated by a set S if it can be represented by a formula of depth k generated by the same set?

1.2.24. A function f in P_2 can be represented by a formula of depth k generated by a set S. Show that the function f can be represented by a certain formula of depth more than k generated by the same set S.

In operations involving Boolean functions, it is often expedient to use the following equivalences (we shall call them *basic equivalences* in the following):

$x \circ y = y \circ x$ (commutativity of a connective $\circ$, where the symbol $\circ$ is a common notation for the connectives $\&$, $\vee$, $\oplus$, $\sim$, $|$, $\downarrow$;

$(x \circ y) \circ z = x \circ (y \circ z)$ (associativity of the connective $\circ$, where $\circ$ is the common notation for $\&$, $\vee$, $\oplus$, $\sim$);

$\overline{x \& y} = \bar{x} \vee \bar{y}$ and $\overline{x \vee y} = \bar{x} \& \bar{y}$ (De Morgan's laws);

$x \vee (x \& y) = x$ and $x \& (x \vee y) = x$ (absorption laws);

$x \vee (\bar{x} \& y) = x \vee y$ and $x \& (\bar{x} \vee y) = x \& y$;

$x \& (y \vee z) = (x \& y) \vee (x \& z)$ (distributivity of conjunction relative to disjunction);

$x \vee (y \& z) = (x \vee y) \& (x \vee z)$ (distributivity of disjunction relative to conjunction);

$x \,\&\, (y \oplus z) = (x \,\&\, y) \oplus (x \,\&\, z)$ (distributivity of conjunction relative to exclusive sum);

$0 = x \,\&\, \bar{x} = x \,\&\, 0 = x \oplus x;$

$1 = x \vee \bar{x} = x \vee 1 = x \sim x;$

$x = \bar{\bar{x}} = x \vee x = x \,\&\, x = x \,\&\, 1 = x \vee 0;$

$\bar{x} = x \oplus 1, \quad x \sim y = (x \oplus y) \oplus 1;$

$x \oplus y = (x \,\&\, \bar{y}) \vee (\bar{x} \,\&\, y), \quad x \to y = ((x \,\&\, y) \oplus x) \oplus 1.$

1.2.25. Verify if the following relations are valid:
(1) $x \vee (y \sim z) = (x \vee y) \sim (x \vee z);$
(2) $x \to (y \sim z) = (x \to y) \sim (x \to z);$
(3) $x \,\&\, (y \sim z) = (x \,\&\, y) \sim (x \,\&\, z);$
(4) $x \to (y \vee z) = (x \to y) \vee (x \to z);$
(5) $x \to (y \,\&\, z) = (x \to y) \,\&\, (x \to z);$
(6) $x \oplus (y \to z) = (x \oplus y) \to (x \oplus z);$
(7) $x \to (y \to z) = (x \to y) \to (x \to z).$

1.2.26. Represent the function f by a formula generated by the set of connectives S if
(1) $f = x \to y, \quad S = \{\neg, \vee\};$
(2) $f = x \vee y, \quad S = \{\to\};$
(3) $f = x \sim y, \quad S = \{\&, \to\};$
(4) $f = x \mid y, \quad S = \{\downarrow\}.$

1.2.27. Using the basic equivalences, prove the equivalence of formulas $\mathfrak{A}$ and $\mathfrak{B}$ if

(1) $\mathfrak{A} = \overline{(\bar{x} \,\&\, \bar{z}) \vee (x \,\&\, y)} \vee (x \,\&\, \bar{z}),$
$\mathfrak{B} = x \,\&\, \overline{y \,\&\, z} \vee \bar{x} \,\&\, z;$

(2) $\mathfrak{A} = (x \to y) \to ((x \,\&\, \bar{y}) \oplus (x \sim \bar{y})),$
$\mathfrak{B} = (x \vee y) \,\&\, (\bar{x} \vee \bar{y});$

(3) $\mathfrak{A} = x \to (x \,\&\, y \to ((x \to y) \to y) \,\&\, z),$
$\mathfrak{B} = y \to (x \to z).$

1.2.28. Let $\mathfrak{A}$ and $\mathfrak{B}$ be formulas generated by the set $\{\neg, \&, \vee, \oplus, \sim, \to, \mid, \downarrow\}$ and $\mathfrak{A}_1 = (\mathfrak{A} \to \mathfrak{B}), \to \mathfrak{B}$, $\mathfrak{A}_2 = \mathfrak{A} \to (\mathfrak{B} \to \neg(\mathfrak{A} \to \mathfrak{A}))$. Prove that $\mathfrak{A} \sim \mathfrak{B} = 1$ if and only if $\neg(\mathfrak{A}_1 \sim \mathfrak{A}_2) = 1$.

1.2.29. Prove that formula $\mathfrak{A}$ containing only the connective $\sim$ is identically truth if and only if any variable from formula $\mathfrak{A}$ appears in it an even number of times.

1.2.30. By $S^{x}_{\mathfrak{C}}\mathfrak{A} \mid$ we denote the formula obtained from formula $\mathfrak{A}$ as a result of a simultaneous replacement of each occurrence of variable x by formula $\mathfrak{S}$. If formula $\mathfrak{A}$ does not contain x, then by definition we put $S^x_{\mathfrak{S}}\mathfrak{A}| = \mathfrak{A}$.

(1) Prove that $\mathfrak{A}$ is identically truth if and only if the formula $S^x_{y_1 \to y_2}\ \mathfrak{A}\ |$ is identically truth, y_1 and y_2 being variables not appearing in formula $\mathfrak{A}$.

(2) Can the condition that y_1 and y_2 do not appear in formula $\mathfrak{A}$ be neglected in (1)?

1.2.31. (1) Suppose that the function $f(x, y)$ from P_2 satisfies the relation

$$f(f(x, f(y, z)), f(f, (x, y), f(x, z))) \equiv 1.$$

Prove that the following equivalences are truth in this case:

(a) $f(x, x) \equiv 1$;
(b) $f(x, f(y, x)) \equiv 1$;
(c) $f(f(f(x, y), f(x, z)), f(x, f(y, z))) \equiv 1$;
(d) $f(f(x, y), f(f(x, f(y, z)), f(x, z))) \equiv 1$;
(e) $f(f(x, f(y, z)), f(y, f(x, z))) \equiv 1$.

(2) Do the equivalences (b), (c) and (d) follow from (e)?

1.3. Special Forms of Formulas. Disjunctive and Conjunctive Normal Forms. Polynomials

The formula $x_{i_1}^{\sigma_1} \& x_{i_2}^{\sigma_2} \& \ldots \& x_{i_r}^{\sigma_r}$ (formula $x_{i_1}^{\sigma_1} \vee x_{i_2}^{\sigma_2} \vee \ldots \vee x_{i_r}^{\sigma_r}$), where $\sigma_k \in \{0, 1\}$, $x_{i_k}^0 = \bar{x}_{i_k}$, $x_{i_k}^1 = x_{i_k}$, $i_k \in \{1, 2, \ldots, n\}$ for all $k = \overline{1, r}$, is called a *conjunction* (resp. *disjunction*) *generated by the set of variables* $X^n = \{x_1, x_2, \ldots, x_n\}$. A conjunction (disjunction) is called *elementary* (abbreviated as *e.c.* and *e.d.* respectively) if $x_{i_j} \neq x_{i_k}$ for $j \neq k$. For the sake of brevity, the symbol & in an e.c. will be omitted. Expressions of the type $x_{i_k}^{\sigma_k}$ will be called *characters*. The number of characters in an e.c. (e.d.) is called the *rank* of the e.c. (e.d.). The constant 1 will be assumed to be the e.c. of the rank zero, and 0 will be called the e.d. of the rank zero.

Formulas of the type

$$\mathscr{D} = K_1 \vee K_2 \vee \ldots \vee K_s \quad (1)$$

(concise notation $\bigvee_{i=1}^{s} K_i$), where K_i $(i = \overline{1, s})$ are conjunctions, are called the *disjunctive normal forms* (abbreviated as *d.n.f.*).

Formulas of the type

$$\mathscr{K} = D_1 \& D_2 \& \ldots \& D_s \quad (2)$$

(concise notation $\underset{i=1}{\overset{s}{\&}} D_i$), where D_i $(i = \overline{1, s})$ are disjunctions, are called *conjunctive normal forms* (abbreviated as *c.n.f.*). The number s is called the *length of the d.n.f.* (*c.n.f.*). The sum of the ranks of conjunctions (disjunctions) is called the *complexity of d.n.f.* (*c.n.f.*). The disjunctive (conjunctive) normal form generated by the set $X^n = \{x_1, x_2, \ldots, x_n\}$ of variables is called *perfect* if it is formed by pairwise different elementary conjunctions (disjunctions) of the rank n.

Let $f(\tilde{x}^n)$ be a Boolean function and let $1 \leqslant i_1 < i_2 < \ldots < i_k \leqslant n$. By $f_{\sigma_1\sigma_2\ldots\sigma_k}^{i_1, i_2, \ldots, i_k}(\tilde{x}^n)$ (or sometimes by $S_{\sigma_1\sigma_2\ldots\sigma_k}^{i_1, i_2, \ldots, i_k} f(\tilde{x}^n)$) we shall denote the function obtained from $f(\tilde{x}^n)$ by substituting the constants $\sigma_1, \sigma_2, \ldots, \sigma_k$ respectively for the variables $x_{i_1}, x_{i_2}, \ldots, x_{i_k}$. The function $f_{\sigma_1\sigma_2\ldots\sigma_k}^{i_1, i_2, \ldots, i_k}(\tilde{x}^n)$ is called the $x_{i_1}^{\sigma_1} x_{i_2}^{\sigma_2} \ldots x_{i_k}^{\sigma_k}$-*component of the function* $f(\tilde{x}^n)$ or a *subfunction* of $f(\tilde{x}^n)$. The subfunction $f_{\sigma_1\sigma_2\ldots\sigma_k}^{i_1, i_2, \ldots, i_k}(\tilde{x}^n)$ is called *proper* if $k \neq n$, $k \neq 0$. The subfunctions of $f(\tilde{x}^n)$ are *different* if they differ as functions of the variables $x_1, x_2, \ldots, x_n$.

Let $1 \leqslant i_1 < i_2 < \ldots < i_k \leqslant n$. In this case, the following representation is valid:

$$f(\tilde{x}^n) = \bigvee_{(\sigma_1, \sigma_2, \ldots, \sigma_k)} x_{i_1}^{\sigma_1} x_{i_2}^{\sigma_2} \ldots x_{i_k}^{\sigma_k} f_{\sigma_1\sigma_2\ldots\sigma_k}^{i_1, i_2, \ldots, i_k}(\tilde{x}^n), \quad (3)$$

where the disjunction includes all vectors $(\sigma_1, \sigma_2 \ldots, \sigma_k)$ in B^k. For $k = n$ this representation has the form

$$f(\tilde{x}^n) = \bigvee_{(\sigma_1, \sigma_2, \ldots, \sigma_n)} x_1^{\sigma_1} x_2^{\sigma_2} \ldots x_n^{\sigma_n} f(\sigma_1, \sigma_2, \ldots, \sigma_n). \quad (4)$$

Representation (4) can be written in the form

$$f(\tilde{x}^n) = \bigvee_{\tilde{\sigma}:\ f(\tilde{\sigma})=1} x_1^{\sigma_1} x_2^{\sigma_2} \ldots x_n^{\sigma_n}. \qquad (5)$$

The right-hand side of this formula is the perfect d.n.f. of $f(\tilde{x}^n)$. Similarly, the following representations are valid:

$$f(\tilde{x}^n) = \mathop{\&}_{(\sigma_1, \sigma_2, \ldots, \sigma_k)} (x_{i_1}^{\bar{\sigma}_1} \vee x_{i_2}^{\bar{\sigma}_2} \vee \ldots \vee x_{i_k}^{\bar{\sigma}_k} \vee f_{\sigma_1\sigma_2 \ldots \sigma_k}^{i_1, i_2, \ldots, i_k}(\tilde{x}^n)), \qquad (6)$$

$$f(\tilde{x}^n) = \mathop{\&}_{\tilde{\sigma}^n:\ f(\tilde{\sigma}^n)=0} (x_1^{\bar{\sigma}_1} \vee x_2^{\bar{\sigma}_2} \vee \ldots \vee x_n^{\bar{\sigma}_n}). \qquad (7)$$

An elementary conjunction is called *monotonic* if it does not contain negations of variables. The formula

$$P(\tilde{x}^n) = K_1 \oplus K_2 \oplus \ldots \oplus K_s, \qquad (8)$$

where K_i ($i = \overline{1, s}$) are pairwise different elementary monotonic conjunctions generated by the set X^n, is called *Zhegalkin's polynomial* or *mod 2 polynomial*. The highest rank of elementary conjunctions appearing in a polynomial is called the *degree of this polynomial*. The number s is called the *length of polynomial* (8). For $s = 0$, we assume that $P(\tilde{x}^n) = 0$.

1.3.1. Using equivalent transformations, reduce the following formulas to d.n.f.:

(1) $F = (x_1 \vee x_2\bar{x}_3)(x_1 \vee x_3)$;

(2) $F = ((x_1 \vee x_2\bar{x}_3x_4)((\bar{x}_2 \vee x_4) \to x_1\bar{x}_3\bar{x}_4) \vee x_2x_3) \vee (\bar{x}_1 \vee x_4)$;

(3) $F = ((x_1 \to x_2x_3)(x_2x_4 \oplus x_3) \to x_1\bar{x}_4) \vee \bar{x}_1$.

1.3.2. Present the following functions as a perfect d.n.f.:

(1) $f(\tilde{x}^3) = (x_1 \oplus x_2) \to x_2x_3$;

(2) $f(\tilde{x}^3) = (01101100)$;

(3) $f(\tilde{x}^3) = (10001110)$.

1.3.3. Using transformations of the type $A = Ax \vee A\bar{x}$, $A \vee A = A$, reduce a given d.n.f. $D(\tilde{x}^3)$ to a perfect d.n.f. if

(1) $D(\tilde{x}^3) = x_1 \vee \bar{x}_2 x_3$;

(2) $D(\tilde{x}^3) = x_1\bar{x}_2 \vee \bar{x}_1 x_3$;

(3) $D(\tilde{x}^3) = x_1 \vee \bar{x}_1 x_2 \vee \bar{x}_2 x_3$.

1.3.4. Using relations of the type $x \vee yz = (x \vee y) \times (x \vee z)$, transform the d.n.f. in the previous problem to a c.n.f.

1.3.5. Construct a perfect c.n.f. for each of the functions in problem 1.3.3.

1.3.6. Count the number of functions $f(\tilde{x}^n)$ for which a perfect c.n.f. is simultaneously a d.n.f.

1.3.7. Find the length of the perfect d.n.f. of the function $f(\tilde{x}^n)$:

(1) $f(\tilde{x}^n) = x_1 \oplus x_2 \oplus \ldots \oplus x_n$, $n \geqslant 1$;

(2) $f(\tilde{x}^n) = (x_1 \vee x_2 \vee \ldots \vee x_n)(\bar{x}_1 \vee \bar{x}_2 \vee \ldots \vee \bar{x}_n)$, $n \geqslant 2$;

(3) $f(\tilde{x}^n) = (x_1 \vee x_2 \vee x_3)(\bar{x}_1 \vee \bar{x}_2 \vee \bar{x}_3) \oplus x_4 \oplus x_5 \oplus \ldots \oplus x_n$, $n \geqslant 4$.

1.3.8. Let us suppose that the sets $X^n = \{x_1, x_2, \ldots, x_n\}$ and $Y^m = \{y_1, y_2, \ldots, y_m\}$ do not intersect. Assuming that the perfect d.n.f.s of functions $f(\tilde{x}^n)$ and $g(\tilde{y}^m)$ have k and l terms respectively, find the length of the perfect d.n.f. of the following functions:

(1) $f(\tilde{x}^n) \,\&\, g(\tilde{y}^m)$; (2) $f(\tilde{x}^n) \vee g(\tilde{y}^m)$; (3) $f(\tilde{x}^n) \oplus g(\tilde{y}^m)$.

1.3.9. Present the x_1-, $\bar{x}_2$- and $\bar{x}_1 x_3$-components of the function $f(\tilde{x}^3)$ by using the minimal complexity d.n.f. for

(1) $f(\tilde{x}^3) = (01101101)$;

(2) $f(\tilde{x}^3) = (x_1 \to x_2) \oplus x_2\bar{x}_3$;

(3) $f(\tilde{x}^3) = (x_1 \to x_2\bar{x}_3) \oplus x_2$.

1.3.10. Find out which of the functions depending on variables x_1 and x_2 have largest number of pairwise different subfunctions.

1.3.11. Find the number of Boolean functions $f(\widetilde{x}^n)$ which are transformed into themselves upon a commutation of x_1 and x_2.

1.3.12. Two functions $f(\widetilde{x}^n)$ and $g(\widetilde{x}^n)$ are *commutatively equivalent* if there exists a permutation π of numbers $1, \ldots n$ such that $f(x_1, \ldots, x_n) = g(x_{\pi(1)}, \ldots, x_{\pi(n)})$. Find the number of classes of commutatively equivalent functions in $P_2(X^2)$.

1.3.13*. The function $f_n(\widetilde{x}^n)$ is defined by the following recurrence relations:

$$f_4(\widetilde{x}^4) = \overline{x}_1 x_2 (x_3 \vee x_4) \vee \overline{x}_2 (\overline{x}_1 \overline{x}_3 \vee x_1 x_4) \vee x_1 x_2 \overline{x}_3 \overline{x}_4,$$

$$f_{n+1}(\widetilde{x}^{n+1}) = f_n(\overline{x}_1, \overline{x}_2, \ldots, \overline{x}_n)\, \overline{x}_{n+1} \vee x_1 x_2 \ldots x_n x_{n+1} \quad (n \geqslant 4).$$

For each function in the sequence $\{f_n\}$ find the number of different subfunctions of the type $f_\sigma^i(\widetilde{x}^n)$, $i = \overline{1, n}$, $\sigma \in \{0, 1\}$, such that no two functions are commutatively equivalent.

1.3.14. Prove that the number of different functions $f(\widetilde{x}^n)$ for which a given function $g(\widetilde{x}^k)$ is a subfunction is not less than

$$2^{2^n} - (2^{2^k} - 1)^{2^{n-k}} \quad (k \leqslant n).$$

1.3.15. Let $f_0^i(\widetilde{x}^n) = f_1^i(\widetilde{x}^n)$ for any i $(1 \leqslant i \leqslant n)$. Prove that $f(\widetilde{x}^n)$ is a constant.

1.3.16. Find the number of functions $f(\widetilde{x}^n)$ such that $f_{00}^{i,j}(\widetilde{x}^n) = f_{11}^{i,j}(\widetilde{x}^n)$ for all $1 \leqslant i < j \leqslant n$ $(n \geqslant 3)$.

1.3.17. Let $h(\widetilde{x}^n) = f(g_1(\widetilde{x}^n), \ldots, g_{i-1}(\widetilde{x}^n), g_i(\widetilde{x}^n), g_{i+1}(\widetilde{x}^n), \ldots, g_n(\widetilde{x}^n))$, $n \geqslant 2$, and the following relation is satisfied for any $i = 1, 2, \ldots, n$ and $\sigma \in \{0, 1\}$:

$$S_\sigma^i h(\widetilde{x}^n) \,| = f(S_{\overline{\sigma}}^i g_1(\widetilde{x}^n) \,|, \ldots, S_{\overline{\sigma}}^i g_{i-1}(\widetilde{x}^n) \,|, \overline{\sigma}, S_{\overline{\sigma}}^i g_{i+1}(\widetilde{x}^n) \,|, \ldots, S_{\overline{\sigma}}^i g_n(\widetilde{x}^n) \,|),$$

in other words, the x_i^σ-component of the superposition of the functions $f, g_1, \ldots, g_{i-1}, g_i, g_{i+1}, \ldots, g_n$ is equal to the superposition of the $x_i^{\overline{\sigma}}$-components of the functions f, g_1, ..., g_{i-1}, g_{i+1}, ..., g_n. Moreover, let the $x_j^{\sigma_j} x_k^{\sigma_k}$-component of the function $g_i(\tilde{x}^n)$ coincide with the $x_j^{\bar{\sigma}_j} x_k^{\bar{\sigma}_k}$-component of the same function $g_i(\tilde{x}^n)$, $i = \overline{1, n}$, for all j, k $(1 \leqslant j < k \leqslant n)$ and σ_j, σ_k from $\{0, 1\}$. Prove that $h(\tilde{x}^n)$ is a constant.

1.3.18. Find the number of monotonic elementary conjunctions of the rank r generated by the set X^n.

1.3.19. Find the number of polynomials of power r generated by the set of variables X^n.

1.3.20. Find the number of different polynomials of length k generated by the set X^n, vanishing at the tuples $\tilde{0}$ and $\tilde{1}$. (Polynomials are considered to be different if they differ in the composition of e.c..)

We adopt the following numeration of monotonic elementary conjunctions generated by the set $X^n = \{x_1, x_2, \ldots, x_n\}$ of variables. We correspond each monotonic e.c. K with a vector $\tilde{\sigma}(K) = (\sigma_1, \sigma_2, \ldots, \sigma_n)$ in B^n, in which $\sigma_i = 1$ if and only if x_i appears in K. The number K of an e.c. is defined as $\nu(\tilde{\sigma}(\tilde{K})) = \sum_{i=1}^{n} \sigma_i 2^{n-i}$. The constant 1 will have the number zero in this numeration. Thus, each polynomial $P(\tilde{x}^n)$ can be presented in the form

$$P(\tilde{x}^n) = \beta_0 \cdot 1 \oplus \beta_1 K_1 \oplus \beta_2 K_2 \oplus \ldots \oplus \beta_{2^n-1} K_{2^n-1}, \quad (9)$$

where K_i is an e.c. with number i $(i = \overline{0, 2^n - 1})$. The vector $\tilde{\beta}_P = (\beta_0, \beta_1, \ldots, \beta_{2^n-1})$ is called the *vector of coefficients of polynomial* $P(\tilde{x}^n)$.

The *method of indeterminate coefficients* for constructing Zhegalkin's polynomial representing the function $f(\tilde{x}^n)$ can be described as follows. We consider a polynomial in the form (9) and for each $\tilde{\alpha} \in B^n$ we compose an equation $f(\tilde{\alpha}) = P(\tilde{\alpha})$. The solution of these equations gives the coefficients of the polynomial $P(\tilde{x}^n)$.

Example. $f(\widetilde{x}^2) = x_1 \to x_2$, $P(\widetilde{x}^2) = \beta_0 \oplus \beta_1 x_2 \oplus \beta_2 x_1 \oplus \beta_3 x_1 x_2$.

$$f(0, 0) = 1 = \beta_0 \oplus \beta_1 \cdot 0 \oplus \beta_2 \cdot 0 \oplus \beta_3 \cdot 0;$$
$$f(0, 1) = 1 = \beta_0 \oplus \beta_1 \cdot 1 \oplus \beta_2 \cdot 0 \oplus \beta_3 \cdot 0;$$
$$f(1, 0) = 0 = \beta_0 \oplus \beta_1 \cdot 0 \oplus \beta_2 \cdot 1 \oplus \beta_3 \cdot 0;$$
$$f(1, 1) = 1 = \beta_0 \oplus \beta_1 \cdot 1 \oplus \beta_2 \cdot 1 \oplus \beta_3 \cdot 1.$$

We obtain $\beta_0 = \beta_2 = \beta_3 = 1$, $\beta_1 = 0$. Hence $x_1 \to x_2 = 1 \oplus x_1 \oplus x_1 x_2$.

1.3.21. Using the method of indeterminate coefficients, find Zhegalkin's polynomials for the following functions:

(1) $f(\widetilde{x}^2) = (1001)$;

(2) $f(\widetilde{x}^3) = (01101000)$;

(3) $f(\widetilde{x}^3) = (11111000)$.

1.3.22. We introduce the operation T generated by vectors in B^{2^n}. If $n = 1$ and $\widetilde{\alpha} = (\alpha_0, \alpha_1)$, then $T(\widetilde{\alpha}) = (\alpha_0, \alpha_0 \oplus \alpha_1)$. Suppose that for each $\overline{\sigma} \in B^{2^n}$ the vector $T(\overline{\sigma})$ is defined and the vector $\widetilde{\alpha}$ in $B^{2^{n+1}}$ has the form $\widetilde{\alpha} = (\beta_0, \beta_1, \ldots, \beta_{2^n-1}, \gamma_0, \gamma_1, \ldots, \gamma_{2^n-1})$. Let

$$T(\beta_0, \beta_1, \ldots, \beta_{2^n-1}) = (\delta_0, \delta_1, \ldots, \delta_{2^n-1}),$$
$$T(\gamma_0, \gamma_1, \ldots, \gamma_{2^n-1}) = (\varepsilon_0, \varepsilon_1, \ldots, \varepsilon_{2^n-1}).$$

In this case,

$$T(\widetilde{\alpha}) = (\delta_0, \delta_1, \ldots, \delta_{2^n-1}, \delta_0 \oplus \varepsilon_0, \delta_1 \oplus \varepsilon_1, \ldots, \delta_{2^n-1} \oplus \varepsilon_{2^n-1}).$$

For example, if $\widetilde{\alpha} = (1011)$, we obtain $T(\widetilde{\alpha}) = (1101)$. Show that the vector $\widetilde{\alpha}_f$ of the values of the function $f(\widetilde{x}^n)$ is related as follows to the vector $\widetilde{\beta}_P$ of the coefficients of the polynomial $P(\widetilde{x}^n)$ representing the function $f(\widetilde{x}^n)$: $\widetilde{\alpha}_f = T(\widetilde{\beta}_P)$, $\widetilde{\beta}_P = T(\widetilde{\alpha}_f)$.

1.3.23. For a function $f(\widetilde{x}^4)$ such that $\widetilde{\alpha}_f = (1011001000101101)$ find the vector $\widetilde{\beta}_P$ of Zhegalkin's polynomial coefficients.

One of the methods of constructing Zhegalkin's polynomial from formula F involves the construction of an equivalent formula generated by the set of connectives $\{\&, -\}$, followed by the replacement of $\bar{x}$ by $x \oplus 1$ throughout, the opening of the brackets, the application of the distributive law $(x \oplus y) z = xz \oplus yz$ and the collection of terms.

Example. $x \to x_2 = \overline{x_1 \bar{x}_2} = x_1 (x_2 \oplus 1) \oplus 1 = x_1 x_2 \oplus x_1 \oplus 1$.

1.3.24. Construct the polynomials for the following functions:

(1) $f(\tilde{x}^3) = (x_1 \mid x_2) \downarrow x_3$;

(2) $f(\tilde{x}^3) = (x_1 \to x_2)(x_2 \downarrow x_3)$;

(3) $f(\tilde{x}^3) = ((x_1 \to x_2) \vee \bar{x}_3) \mid x_1$.

1.3.25. Any Boolean function f can be written in the form of a polynomial by using conventional arithmetic operations of multiplication, addition and subtraction. For this purpose, it is sufficient first to express f in terms of a conjunction and a negation, and then replace subformulas of the type $\bar{A}$ by $1 - A$ and open the brackets. Using the arithmetic operations, find the expressions for the following functions:

(1) $f(\tilde{x}^2) = x_1 \oplus x_2$;

(2) $f(\tilde{x}^3) = (x_1 \to x_2) \to x_3$;

(3) $f(\tilde{x}^3) = (10000001)$.

1.3.26. Find the function $f(\tilde{x}^n)$ whose polynomial has a length 2^n times the length of its perfect d.n.f.

1.3.27. Prove the validity of the following formula for expansion in k variables:

$$f(\tilde{x}^n) = \sum_{(\sigma_1, \sigma_2, \ldots, \sigma_k) \in B^k} x_1^{\sigma_1} x_2^{\sigma_2} \ldots x_k^{\sigma_k} f(\sigma_1, \sigma_2, \ldots, \sigma_k, x_{k+1}, \ldots, x_n).$$

1.3.28. Prove that

(1) $f(\tilde{x}^n) = x_1 (f_0^1(\tilde{x}^n) \oplus f_1^1(\tilde{x}^n)) \oplus f_0^1(\tilde{x}^n)$;

(2) $f(\tilde{x}^n) = (x_1 \vee (f_0^1(\tilde{x}^n) \sim f_1^1(\tilde{x}^n))) \sim f_1^1(\tilde{x}^n)$.

1.3.29. Show that the function $f(\tilde{x}^n)$ represented by a polynomial of degree $k > 0$ becomes equal to 1 on at least 2^{n-k} vectors in B^n.

1.3.30. Determine the number of tuples in B^n on which the polynomial $P(\widetilde{x}^n)$ becomes equal to unity:

(1) $P(\widetilde{x}^n) = x_1 \ldots x_k \oplus x_{k+1} \ldots x_n \; (1 \leqslant k < n)$;

(2) $P(\widetilde{x}^n) = 1 \oplus x_1 \oplus x_1 x_2 \oplus \ldots \oplus x_1 x_2 \ldots x_n = 1 \oplus \sum\limits_{i=1}^{n} x_1 \ldots x_i$.

1.3.31. Show that for any l $(l \leqslant 2^n)$ there exists a polynomial $P(\widetilde{x}^n)$ of length not exceeding n, so that $|N_{P(\widetilde{x}^n)}| = l$.

1.3.32. Prove that any function $f(\widetilde{x}^n)$ other than 0 at $n = 1$ can be presented in the form $f(\widetilde{x}^n) = \sum\limits_{i=1}^{s} K_i$, where K_i $(i = \overline{1, s})$ are elementary conjunctions containing not more than one negation of the variable, and $s \leqslant 2^{n-1}$.

1.3.33. Show that if the symbol $\vee$ in a perfect d.n.f. is replaced by $\oplus$ everywhere, a formula equivalent to the initial one is obtained. Is this statement true for an arbitrary d.n.f.?

The *derivative of a Boolean function* $f(\widetilde{x}^n)$ *with respect to the variables* $x_{i_1}, x_{i_2}, \ldots, x_{i_k}$ (or the *Boolean difference*) is defined as the function

$$\frac{\partial f(\widetilde{x}^n)}{\partial (x_{i_1}, x_{i_2}, \ldots, x_{i_k})} = f(x_1, \ldots, \bar{x}_{i_1}, \ldots, \bar{x}_{i_k}, \ldots, x_n) \oplus f(x_1, \ldots, x_{i_1}, \ldots, x_{i_k}, \ldots, x_n).$$

(For $k = 1$, the notation $\dfrac{\partial f(\widetilde{x}^n)}{\partial x_{i_1}}$ is used.)

1.3.34. Prove the following properties of a derivative:

(1) $\dfrac{d}{dx_j}\left(\dfrac{df(\widetilde{x}^n)}{dx_i}\right) = \dfrac{d}{dx_i}\left(\dfrac{df(\widetilde{x}^n)}{dx_j}\right)$;

(2) $\dfrac{\partial f(\widetilde{x}^n)}{\partial (x_{i_1}, \ldots, x_{i_k})} = \dfrac{\partial \bar{f}(\widetilde{x}^n)}{\partial (x_{i_1}, \ldots, x_{i_k})}$;

(3) $\dfrac{\partial (f(\widetilde{x}^n) \oplus g(\widetilde{x}^n))}{\partial (x_{i_1}, \ldots, x_{i_k})} = \dfrac{\partial f(\widetilde{x}^n)}{\partial (x_{i_1}, \ldots, x_{i_k})} \oplus \dfrac{\partial g(\widetilde{x}^n)}{\partial (x_{i_1}, \ldots, x_{i_k})}$;

(4) $$\frac{d(f(\tilde{x}^n) \vee g(\tilde{x}^n))}{dx_i} = \bar{f}(\tilde{x}^n)\frac{dg(\tilde{x}^n)}{dx_i} \oplus \bar{g}(\tilde{x}^n)\frac{df(\tilde{x}^n)}{dx_i} \oplus \frac{df(\tilde{x}^n)}{dx_i}\frac{dg(\tilde{x}^n)}{dx_i};$$

(5) $$\frac{d(f(\tilde{x}^n)g(\tilde{x}^n))}{dx_i} = f(\tilde{x}^n)\frac{dg(\tilde{x}^n)}{dx_i} \oplus g(\tilde{x}^n)\frac{df(\tilde{x}^n)}{dx_i} \oplus \frac{df(\tilde{x}^n)}{dx_i}\frac{dg(\tilde{x}^n)}{dx_i};$$

(6) $\frac{df(\tilde{x}^n)}{dx_i} = 0$ if and only if x_i does not appear explicitly in the Zhegalkin polynomial of the function $f(\tilde{x}^n)$;

(7) if $f(\tilde{x}^n) = x_1 g(x_2, x_3, \ldots, x_n) \oplus h(x_2, x_3, \ldots, x_n)$, then $\frac{df(\tilde{x}^n)}{dx_1} = g(x_2, x_3, \ldots, x_n)$.

1.3.35. If $g(x_1, \ldots, x_m)$ and $h(x_{m+1}, \ldots, x_n)$ are Boolean functions and $1 \leqslant i_j \leqslant m$ for all $j = \overline{1, k}$, then

(1) $$\frac{\partial(g \oplus h)}{\partial(x_{i_1}, \ldots, x_{i_k})} = \frac{\partial g}{\partial(x_{i_1}, \ldots, x_{i_k})};$$

(2) $$\frac{\partial(g \,\&\, h)}{\partial(x_{i_1}, \ldots, x_{i_k})} = h\frac{\partial g}{\partial(x_{i_1}, \ldots, \partial x_{i_k})};$$

(3) $$\frac{\partial(g \vee h)}{\partial(x_{i_1}, \ldots, x_{i_k})} = \bar{h}\frac{\partial g}{\partial(x_{i_1}, \ldots, \partial x_{i_k})}.$$

1.4. Minimization of Boolean Functions

The admissible conjunction or *implicant* of the function $f(\tilde{x}^n)$ is an elementary conjunction K of the set of variables $\{x_1, x_2, \ldots, x_n\}$, such that $K \vee f(\tilde{x}^n) = f(\tilde{x}^n)$. The *implicant* K of a function f is called *prime* if the rejection of any character from K leads to an elementary conjunction that is not an implicant of the function f. The disjunction of all prime implicants of the function f is called a *contracted d.n.f.* of the function f.

A disjunctive normal form is called

minimal if it has the smallest number of characters among all equivalent d.n.f.s;

shortest if it has the smallest length among all equivalent d.n.f.s;

terminal (*irredundant*) if the omission of any term or character leads to a non-equivalent d.n.f.; and

a *d.n.f. of the function* f if it represents the function f.

If an elementary conjunction K is an implicant of the function $f(\tilde{x}^n)$, the set N_K of such vectors $\tilde{\alpha}$ in B^n for which $K(\tilde{\alpha}) = 1$ forms a face belonging to the set N_f. This face is called the *interval of the function* $f(\tilde{x}^n)$ *corresponding to the implicant* K. The interval of the function f which is not included in any other interval of the function f is called the *maximal interval*. Maximal intervals correspond to the prime implicants of the function f.

1.4.1. Isolate prime implicants of the function $f(\tilde{x}^n)$ from a given set of elementary conjunctions $\mathscr{K}$ if

(1) $\mathscr{K} = \{x_1, \bar{x}_3, x_1x_2, x_2\bar{x}_3\}$, $f(\tilde{x}^3) = (00101111)$;

(2) $\mathscr{K} = \{x_1\bar{x}_2, x_2x_3, x_1, x_1x_2x_3\}$, $f(\tilde{x}^3) = (01111110)$;

(3) $\mathscr{K} = \{x_1, \bar{x}_4, x_2\bar{x}_3, \bar{x}_1\bar{x}_2x_3\}$,

$$f(\tilde{x}^4) = (1010111001011110).$$

Blake's method of obtaining a contracted d.n.f. from an arbitrary d.n.f. involves the application of the following rules: $xK_1 \vee \bar{x}K_2 = xK_1 \vee \bar{x}K_2 \vee K_1K_2$ (generalized pasting) and $K_1 \vee K_1K_2 = K_1$ (absorption). It is assumed that these rules are applied from left to right. At the first stage, the operations of generalized pasting are continued as long as possible. At the next stage, the absorption operation is carried out.

Example. Obtain the contracted d.n.f. for $\mathscr{D}(\tilde{x}^3) = x_1x_2 \vee \bar{x}_1x_3 \vee \bar{x}_2x_3$.

After the first stage, we obtain

$$\mathscr{D}_1 = x_1x_2 \vee \bar{x}_1x_3 \vee x_2x_3 \vee \bar{x}_2x_3 \vee x_3 \vee x_1x_3.$$

After the second stage, we get

$$\mathscr{D}_2 = x_1x_2 \vee x_3.$$

1.4.2. Using Blake's method, construct the contracted d.n.f. for a given d.n.f. $\mathscr{D}$:

(1) $\mathscr{D} = \bar{x}_1\bar{x}_2 \vee x_1\bar{x}_3x_4 \vee x_2\bar{x}_3x_4$;

(2) $\mathscr{D} = x_1\bar{x}_2x_3 \vee x_1x_2\bar{x}_4 \vee x_3x_4$;

(3) $\mathscr{D} = x_1x_2 \vee \bar{x}_1x_3 \vee x_1\bar{x}_2x_3x_4 \vee \bar{x}_1x_2x_3x_4$.

The contracted d.n.f. of the function $f(\tilde{x}^n)$ specified in the form of a c.n.f. can be obtained as follows. First the parentheses are opened by using the distributivity law. After this, characters and terms are cancelled from the obtained d.n.f. by using the relations $x\bar{x} = 0$, $xx = x$, $x \vee x = x$, $K_1 \vee K_1K_2 = K_1$.

Example. Find the contracted d.n.f. for

$$f(\tilde{x}^3) = (x_1 \vee x_2)(\bar{x}_1 \vee x_2 \vee x_3).$$

After opening the parentheses, we get

$$\mathscr{D}_1 = x_1\bar{x}_1 \vee x_1x_2 \vee x_1x_3 \vee x_2\bar{x}_1 \vee x_2x_2 \vee x_2x_3.$$

Applying the above-mentioned rules, we obtain

$$\mathscr{D} = x_1x_3 \vee x_2.$$

1.4.3. Construct the contracted d.n.f. by using the following given c.n.f.s:

(1) $(x_1 \vee x_2 \vee \bar{x}_3)(\bar{x}_1 \vee x_2 \vee x_3)(\bar{x}_2 \vee \bar{x}_3)$;

(2) $(x_1 \vee x_4)(x_2 \vee \bar{x}_3 \vee x_4)(\bar{x}_1 \vee x_2 \vee \bar{x}_3)$;

(3) $(x_1 \vee \bar{x}_2 \vee x_3)(\bar{x}_1 \vee x_4)(x_2 \vee x_3 \vee \bar{x}_4)$.

1.4.4. Let N_f be a set of tuples $\tilde{\alpha} \in B^n$ such that $f(\tilde{\alpha}) = 1$ and $l^c(f)$ is the length of the contracted d.n.f. of the function f. Show that $l^c(f) \leqslant \frac{1}{2}|N_f|(|N_f| + 1)$.

1.4.5. Find the length of the contracted d.n.f. of the following functions:

(1) $x_1 \oplus x_2 \oplus \ldots \oplus x_n$;

(2) $(x_1 \vee x_2 \vee x_3)(\bar{x}_1 \vee \bar{x}_2 \vee \bar{x}_3) \oplus x_4 \oplus x_5 \oplus \ldots \oplus x_n$;

(3) $(x_1 \vee x_2 \vee x_3)(\bar{x}_1 \vee \bar{x}_2 \vee \bar{x}_3)(x_4 \oplus x_5 \oplus \ldots \oplus x_n)$;

(4) $(x_1 \oplus \ldots \oplus x_k)(x_{k+1} \oplus \ldots \oplus x_n)$, $1 \leqslant k \leqslant n$;

(5) $(x_1 \vee \ldots \vee x_n)(x_1 \vee \ldots \vee x_k \vee \bar{x}_{k+1} \vee \ldots \vee \bar{x}_n)$, $1 \leqslant k \leqslant n$.

1.4.6. Let $f(\tilde{x}^n)$ be such that $N_f = \{\tilde{\alpha} : k \leqslant \| \tilde{\alpha} \| \leqslant k + m\}$, and $k \leqslant l \leqslant k + m$, $\tilde{\alpha} \in B_l^n$.

(1) Find the number of terms in the contracted d.n.f. of the function f, which become equal to unity on the tuple $\tilde{\alpha}$.

(2) Show that the length of the contracted d.n.f. of the function $f(\tilde{x}^n)$ is equal to $\binom{n}{k}\binom{n-k}{m}$.

1.4.7. Suppose that the functions $f(\tilde{x}^n)$ and $g(\tilde{y}^m)$ do not have common variables, K is a prime implicant of the function f, and L a prime implicant of the function g. Show that $K \& L$ is the prime implicant of the function $f \& g$.

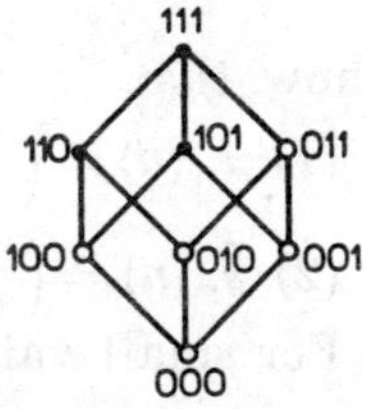

Fig. 2

Each elementary conjunction generated by the set of variables $\{x_1, x_2, \ldots, x_n\}$ has a one-to-one correspondence with the face of the cube B^n formed by the vertices $\tilde{\alpha}$ at which the e.c. becomes equal to unity. This allows us to construct a contracted d.n.f. from the geometrical representation of a Boolean function. Vertices of the set N_f of the function f are marked on the cube B^n. The faces contained in N_f and not contained in any other face formed by vertices from the set N_f are written down. Each such face is assigned a prime implicant.

Example. Let the function $f(\tilde{x}^3)$ be defined by the vector $\tilde{\alpha}_f = (11111000)$. Find its contracted d.n.f..

Solution. The set N_f is formed by $\{(000), (001), (010), (011), (100)\}$. The faces have the form $g_1 = \{(000), (001), (010), (011)\}$, $g_2 = \{(000), (100)\}$. The faces g_1 correspond to the e.c. $\bar{x}_1$, and g_2 to $\bar{x}_2\bar{x}_3$. The contracted d.n.f. is given by $\bar{x}_1 \vee \bar{x}_2\bar{x}_3$ (see Fig. 2).

1.4.8. Construct the contracted d.n.f. of the function $f(\tilde{x}^n)$:

(1) $f(\tilde{x}^4) = (1111100001001100)$;

(2) $f(\tilde{x}^4) = (0000001111111101)$;

(3) $f(\tilde{x}^4) = (0001101111011011)$.

1.4.9. Calculate the number of functions $f(\tilde{x}^n)$ for which a given elementary conjunction of rank r is:
(1) an implicant;
(2) a prime implicant.

1.4.10. Let $i_r(f)$ be the number of implicants of rank r for the function $f(\tilde{x}^n)$, and $s_r(f)$ be the number of prime implicants of rank r. Let P_n be the set of Boolean functions of the variables $x_1, x_2, \ldots, x_n$; and

$$\bar{i}_r(n) = \frac{1}{2^{2^n}} \sum_{f \in P_n} i_r(f);$$

$$\bar{s}_r(n) = \frac{1}{2^{2^n}} \sum_{f \in P_n} s_r(f).$$

Show that

(1) $\bar{i}_r(n) = \binom{n}{r} 2^{r-2^{n-r}}$;

(2) $\bar{s}_r(n) = \binom{n}{r} 2^{r-2^{n-r}} (1-2^{-2^{n-r}})^r$.

For small values of n, the contracted d.n.f. of the function $f(\tilde{x}^n)$ can be found with the help of the rectangular table (*minimizing chart* or a *Karnaugh map*). For example, suppose that the function $f(\tilde{x}^4)$ is defined with the help of Table 5. Combining the cells corresponding to

Table 5

	x_3	0	0	1	1
x_1	x_2 \ x_4	0	1	1	0
0	0	1	1	0	1
0	1	0	1	1	0
1	1	1	1	1	0
1	0	0	1	0	0

unit values of the function f into the maximal intervals as shown in Table 5 and comparing them with the e.c., we obtain the contracted d.n.f.:

$$x_1 x_2 \bar{x}_3 \vee \bar{x}_3 x_4 \vee x_2 x_4 \vee \bar{x}_1 \bar{x}_2 \bar{x}_3 \vee \bar{x}_1 \bar{x}_2 \bar{x}_4.$$

1.4.11. Find the contracted d.n.f. for functions defined by the following tables:

(1)

$x_1 x_2$ \ x_3 x_4	0 0	0 1	1 1	1 0
0 0	1	1	1	0
0 1	0	1	1	1
1 1	1	0	1	1
1 0	1	1	0	1

(2)

$x_1 x_2$ \ x_3 x_4	0 0	0 1	1 1	1 0
0 0	1	1	0	0
0 1	1	1	1	0
1 1	0	1	1	1
1 0	1	0	1	1

(3)

$x_1 x_2$ \ x_3 x_4	0 0	0 1	1 1	1 0
0 0	0	1	1	0
0 1	1	1	0	1
1 1	1	0	1	1
1 0	0	1	1	0

A prime implicant is called a *core* implicant if its removal from a contracted d.n.f. leads to a d.n.f. that is not equivalent to the initial one. For each core implicant K there exists a tuple of variables which turns K to unity and the remaining terms of the contracted d.n.f. to zero. Such a tuple is called a *proper tuple* of the core implicant.

1.4.12. Isolate the core implicants from the contracted d.n.f.:

(1) $\bar{x}_1 x_3 \vee x_2 x_3 \vee x_1 x_2 \vee \bar{x}_1 \bar{x}_2 x_4 \vee \bar{x}_2 x_3 x_4 \vee x_1 x_3 x_4$;

(2) $x_1 x_2 \bar{x}_3 \vee x_1 \bar{x}_2 x_3 \vee \bar{x}_3 x_4 \vee \bar{x}_2 x_4 \vee \bar{x}_1 x_2 x_3 \vee \bar{x}_2 x_4 \vee \bar{x}_1 x_4$;

(3) $x_2 \bar{x}_3 \vee \bar{x}_2 x_3 \vee x_1 x_2 \vee x_1 x_3 \vee \bar{x}_3 x_4 \vee \bar{x}_2 x_4 \vee x_1 x_4$.

1.4.13. Show that the number of core implicants of an arbitrary function $f(\widetilde{x}^n)$ does not exceed 2^{n-1}.

1.4.14. Find the number of core implicants of the functions in Problem 1.4.5.

1.4.15. Find the number of Boolean functions $f(\widetilde{x}^n)$ for which the e.c. $x_1 \ldots x_r$ is a core implicant.

1.4.16. Let K, K_1 and K_2 be conjunctions from the contracted d.n.f., and r, r_1 and r_2 be the ranks of these conjunctions. Let $K \vee K_1 \vee K_2 = K_1 \vee K_2$. Show that $r_1 + r_2 \geqslant r + 2$.

1.4.17. Construct all terminal d.n.f.s of the following functions:

(1) $f(\widetilde{x}^3) = (01111110)$;

(2) $f(\widetilde{x}^4) = (1110011000010101)$;

(3) $f(\widetilde{x}^4) = (0110101111011110)$.

1.4.18. Find the number of terminal and minimal d.n.f.s for the functions appearing in Problem 1.4.5.

1.4.19. Show that the number of terminal d.n.f.s for an arbitrary Boolean function $f(\widetilde{x}^n)$ does not exceed $\binom{3^n}{2^n}$.

1.4.20*. How many terminal d.n.f.s exist for a function having 2^{n-1} core implicants?

1.4.21. Find out if the following d.n.f.s are terminal or shortest, or minimal:

(1) $\mathscr{D} = x_1x_2 \vee \overline{x}_2$;

(2) $\mathscr{D} = \overline{x}_3\overline{x}_4 \vee x_1x_2x_4 \vee \overline{x}_1\overline{x}_2x_3$;

(3) $\mathscr{D} = x_1x_2 \vee \overline{x}_1x_3 \vee \overline{x}_2x_3x_4 \vee x_2x_3$.

1.4.22. Let $L(f)$ be the complexity of the minimal, and $l(f)$ the length of the shortest d.n.f. of the function f. Show that $L(f(\widetilde{x}^n)) \leqslant nl(f(\widetilde{x}^n))$ for an arbitrary function $f(\widetilde{x}^n)$.

1.4.23. Show that $l(f(\widetilde{x}^n)) \leqslant 2^{n-1}$, $L(f(\widetilde{x}^n)) \leqslant n2^{n-1}$ for any function $f(\widetilde{x}^n)$.

1.4.24. For how many functions $f(\widetilde{x}^n)$ are the following relations valid:

(1) $L(f(\widetilde{x}^n)) = n2^{n-1}$; (2) $L(f(\widetilde{x}^n)) = n2^{n-1} - n$?

1.4.25. Give an example of a number k $(0 < k \leqslant n2^n)$ such that there is no $f(\widetilde{x}^n)$ having a minimal d.n.f. of complexity k.

1.4.26. For the functions of Problem 1.4.5. find the

complexity of the minimal and length of the shortest d.n.f.

1.4.27. Let us consider a family of belt functions, i.e. the functions $f(\tilde{x}^n)$ for which there exist numbers k and m such that

$$N_f = \{\tilde{\alpha}: k \leqslant \|\tilde{\alpha}\| \leqslant k+m\}.$$

(1) Find the number of core implicants of the belt function $f(\tilde{x}^n)$ for different values of k and m.

(2) How many belt functions $f(\tilde{x}^n)$ have the maximum number of core implicants?

1.4.28. The function $f(\tilde{x}^n)$ is called a *chain* (*cyclic*) function if the set N_f can be arranged in a sequence that is a 2-chain (2-cycle).

(1) Find the number of terminal and minimal d.n.f.s of a chain function $f(\tilde{x}^n)$ if $|N_f| = l$.

(2) Find the same for a cyclic function $f(\tilde{x}^n)$ such that $|N_f| = 2m$ $(m > 2)$.

1.5. Essential and Apparent Variables

The variable x_i of the function $f(x_1, x_2, \ldots, x_n)$ is called *essential* if there exist tuples $\tilde{\alpha}$ and $\tilde{\beta}$ such that $\tilde{\alpha} = (\alpha_1, \ldots, \alpha_{i-1}, 1, \alpha_{i+1}, \ldots, \alpha_n)$, $\tilde{\beta} = (\alpha_1, \ldots, \alpha_{i-1}, 0, \alpha_{i+1}, \ldots, \alpha_n)$ and $f(\tilde{\alpha}) \neq f(\tilde{\beta})$. Otherwise, the variable x_i is called an *apparent* variable of the function $f(\tilde{x}^n)$. Two functions $f(\tilde{x}^n)$ and $g(\tilde{x}^n)$ are called *equal* if the sets of their essential variables coincide and on any two tuples, differing perhaps only in the values of apparent variables, the values of the functions are identical. Let $1 \leqslant i_1 < i_2 < \ldots < i_k \leqslant n$. We shall say that the function $\varphi(x_1, \ldots, x_{i_1-1}, x, x_{i_1+1}, \ldots, x_{i_2-1}, x_{i_2+1}, \ldots, x_{i_k-1}, x_{i_k+1}, \ldots, x_n)$ *is obtained from* $f(\tilde{x}^n)$ *by identification of variables* $x_{i_1}, x_{i_2}, \ldots, x_{i_k}$ if φ is obtained from f by substitution of x in place of variables $x_{i_1}, x_{i_2}, \ldots, x_{i_k}$. For x we can take any variable not belonging to the set $X^n \setminus \{x_{i_1}, x_{i_2}, \ldots, x_{i_k}\}$.

1.5.1. Show that the statement "x_i is an essential variable of the function $f(\tilde{x}^n)$" is equivalent to each of the following statements:

(1) $f_0^i(\tilde{x}^n) \neq f_1^i(\tilde{x}^n)$;

(2) there are variables $x_{i_1}, \ldots, x_{i_k}$ $(i_j \neq i, j = \overline{1, k})$ and constants $\sigma_1, \ldots, \sigma_k$ such that the function $f_{\sigma_1 \ldots \sigma_k}^{i_1, \ldots, i_k}(\tilde{x}^n)$ depends essentially on x_i.

1.5.2. Enumerate the essential variables of the following functions:

(1) $f(\tilde{x}^3) = (x_1 \to (x_1 \vee x_2)) \to x_3$;

(2) $f(\tilde{x}^2) = (x_1 \vee x_2) \to x_2$;

(3) $f(\tilde{x}^4) = (x_1 \vee x_2 \vee \bar{x}_2 x_3 \vee \bar{x}_1 \bar{x}_2 \bar{x}_3)\, x_4$;

(4) $f(\tilde{x}^3) = (\bar{x}_1 \vee x_2 \vee x_3)(\bar{x}_1 \vee \bar{x}_2 \vee x_3)(\bar{x}_1 \vee x_2 \vee \bar{x}_3)(\bar{x}_1 \vee \bar{x}_2 \vee \bar{x}_3)$.

1.5.3. Show that x_1 is an apparent variable of the function f (express f through a formula that does not contain x_1 explicitly):

(1) $f(\tilde{x}^2) = (x_1 \oplus x_2)(x_1 \downarrow x_2)$;

(2) $f(\tilde{x}^3) = (((x_3 \to x_2) \vee x_1)(x_2 \to x_1)\, x_3 \bar{x}_1) \oplus x_3$;

(3) $f(\tilde{x}^3) = ((x_1 \vee x_2)(x_1 \vee \bar{x}_3) \to (\bar{x}_1 \to x_2 x_3))\, x_2$.

1.5.4. Indicate the apparent variables of the function f:

(1) $f(\tilde{x}^3) = (11110000)$;

(2) $f(\tilde{x}^3) = (00110011)$;

(3) $f(\tilde{x}^3) = (00111100)$.

1.5.5. Let the function $f(\tilde{x}^n)$ be defined by the vector $\tilde{\alpha}_f = (\alpha_0, \alpha_1, \ldots, \alpha_{2^n-1})$. Show that if x_k is an apparent variable, $\alpha_i = \alpha_{2^{n-k}+i}$ for all i on the segment $[s2^{n-k+1}, (2s+1)\,2^{n-k} - 1]$, $s = \overline{0, 2^{k-1} - 1}$.

1.5.6. Show that if there are apparent variables among the variables of the function $f(\tilde{x}^n)$, $n \geqslant 1$, the function assumes the value 1 on an even number of tuples. Is the converse true?

1.5.7. Let the function $f(\tilde{x}^n)$ be such that $|N_f| = 2^m(2l - 1)$. What is the maximum possible number of apparent variables of the function f?

1.5.8. Using the results of Problems 1.5.5. and 1.5.6.,

find the variables on which the function f depends essentially:

(1) $f(\tilde{x}^4) = (1011100111001010)$;

(2) $f(\tilde{x}^4) = (0011110011000011)$;

(3) $f(\tilde{x}^4) = (0111011101110111)$;

(4) $f(\tilde{x}^4) = (0101111100001010)$.

1.5.9. For each function in Problem 1.5.8. construct an equal function that depends essentially on all its variables.

1.5.10. Find the values of n $(n \geqslant 2)$ for which the following functions depend essentially on all their variables:

(1) $f(\tilde{x}^n) = (x_1 \vee x_2) \oplus (x_2 \vee x_3) \oplus \ldots \oplus (x_{n-1} \vee x_n) \oplus (x_n \vee x_1)$;

(2) $f(\tilde{x}^n) = (x_1 \to x_2)(x_2 \to x_1) \oplus (x_2 \to x_3)(x_3 \to x_2) \ldots \oplus (x_n \to x_1)(x_1 \to x_n)$;

(3) $f(\tilde{x}^n) = (\ldots((x_1 \downarrow x_2) \downarrow x_3) \downarrow \ldots \downarrow x_n) \to (x_1 \mid (x_2 \mid (x_3 \mid \ldots \mid x_n) \ldots))$;

(4) $f(\tilde{x}^n) = (x_1 \to x_2)(x_2 \to x_3) \ldots (x_{n-1} \to x_n)(x_n \to x_1) \to (x_1 \oplus x_2 \oplus \ldots \oplus x_n \oplus 1)$;

(5) $f(\tilde{x}^n) = \left(\bigvee_{1 \leqslant i_1 < i_2 < \ldots < i_{[n/2]} \leqslant n} x_{i_1} x_{i_2} \ldots x_{i_{[n/2]}} \right) \& \left(\bigvee_{1 \leqslant i_1 < i_2 < \ldots < i_{[n/2]} \leqslant n} \bar{x}_{i_1} \bar{x}_{i_2} \ldots \bar{x}_{i_{[n/2]}} \right) \to (x_1 \oplus x_2 \oplus \ldots \oplus x_n)$.

1.5.11. Let the functions $f(\tilde{x}^n)$ and $g(\tilde{y}^m)$ depend essentially on all their variables and let the variables $x_1, \ldots, x_n, y_1, \ldots, y_m$ be pairwise different. Show that the function $f(x_1, \ldots, x_{n-1}, g(y_1, \ldots, y_m))$ depends essentially on all its variables.

1.5.12. Let $P^c(X^n)$ be a set of all Boolean functions depending, and that too essentially, on the variables $x_1, x_2, \ldots, x_n$.

(1) Enumerate all functions in $P^c(X^2)$.

(2) Find the number $|P^c(X^3)|$.

(3) Show that $|P^c(X^n)| = \sum_{k=0}^{n} (-1)^k \binom{n}{k} 2^{2^{n-k}}$.

(4) Show that $\lim_{n\to\infty} 2^{-2^n} |P^c(X^n)| = 1$.

1.5.13. Let $\tilde{\alpha}$, $\tilde{\beta}$ and $\tilde{\gamma}$ be such tuples in B^n that $\tilde{\alpha} \leqslant \tilde{\beta} \leqslant \tilde{\gamma}$. Let the function $f(\tilde{x}^n)$ be such that $f(\tilde{\alpha}) = f(\tilde{\gamma}) \neq f(\tilde{\beta})$. Show that $f(\tilde{x}^n)$ depends essentially on at least two variables.

1.5.14*. Show that x_i is an essential variable of the function f if and only if this variable appears explicitly in the contracted d.n.f. of the function f.

1.5.15. Show that x_i is an essential variable of the function f if and only if x_i appears explicitly in the Zhegalkin polynomial of the function f.

1.5.16. Let $\frac{\partial f(\tilde{x}^n)}{\partial (x_{i_1}, \ldots, x_{i_k}, x_j)} = 0$ for any non-empty set $\{x_{i_1}, \ldots, x_{i_k}\}$ of variables defferent from x_j. Does $f\{\tilde{x}^n\}$ have an apparent dependent on x_j?

1.5.17. Show that any symmetric function $f(\tilde{x}^n)$ other than a constant depends essentially on all its variables.

1.5.18. Suppose that the function $f(\tilde{x}^n)$ changes its value m times at the vertices of the chain $\tilde{\alpha}$, $\tilde{\beta}_1, \ldots, \tilde{\beta}_{k-1}$, $\tilde{\gamma}$ connecting these vertices $\tilde{\alpha}$, $\tilde{\gamma}$ of the cube B^n, for which $\rho(\tilde{\alpha}, \tilde{\gamma}) = k$. Show that $f(\tilde{x}^n)$ depends essentially on at least m variables.

1.5.19. Let $f(\tilde{x}^n)$ depend essentially on at least two variables. Show that there are three vertices $\tilde{\alpha}$, $\tilde{\beta}$, $\tilde{\gamma}$ of the cube B^n satisfying the condition

$$\rho(\tilde{\alpha}, \tilde{\beta}) = \rho(\tilde{\beta}, \tilde{\gamma}) = 1, \quad \tilde{\alpha} \neq \tilde{\gamma}, \quad f(\tilde{\alpha}) = f(\tilde{\gamma}) \neq f(\tilde{\beta}).$$

1.5.20. Let $f(\tilde{x}^n)$ depend essentially on all its variables. Prove that for any i $(1 \leqslant i \leqslant n)$ there exists a j such that a certain substitution of constants in place of variables other than x_i and x_j leads to a function depending essentially on x_i and x_j.

1.5.21. Show that for any function $f(\widetilde{x^n})$ that depends essentially on n variables there exists a variable x_i and a constant α such that the function $f_\alpha^i(\widetilde{x^n}) = f(\widetilde{x_1, \ldots}, x_{i-1}, \alpha, x_{i+1}, \ldots, x_n)$ depends essentially on $n-1$ variables.

1.5.22. Let $f(\widetilde{x^n})$ depend essentially on all its variables. Check the validity of the following statements:

(1) there exists an i such that for any j there exist constants whose substitution into $f(\widetilde{x^n})$ for variables other than x_i and x_j leads to a function depending essentially on x_i and x_j;

(2) for any two variables x_i and x_j there exist constants whose substitution into $f(\widetilde{x^n})$ for variables other than x_i and x_j leads to a function depending essentially on x_i and x_j.

1.5.23. Enumerate the functions in $P_2(X^2)$ that can be obtained by identifying the variables of the following functions:

(1) $f(\widetilde{x^3}) = (10010110)$; (2) $f(\widetilde{x^3}) = (11111101)$;

(3) $f(\widetilde{x^3}) = x_1x_2 \vee x_2x_3 \vee x_3x_1$;

(4) $f(\widetilde{x^3}) = x_1x_2x_3 \oplus x_2x_3 \oplus x_3x_1 \oplus x_2 \oplus 1$.

1.5.24. Show that the identity operation on the function $f(\widetilde{x^n})$ can lead to a constant if and only if $f(\widetilde{0}) = f(\widetilde{1})$.

1.5.25. Find the number of functions $f(\widetilde{x^2})$ for which the identification of variables cannot lead to a function depending essentially on one variable.

1.5.26. Can the identity operation on the symmetric function $f(\widetilde{x^n})$ lead to a function that depends essentially on all its variables and that is not symmetric?

1.5.27. Let $\widetilde{\alpha}, \widetilde{\beta}, \widetilde{\gamma}$ be three vertices in B^n such that $\widetilde{\alpha} \leqslant \widetilde{\beta} \leqslant \widetilde{\gamma}$, and let $f(\widetilde{x^n})$ be such that $f(\widetilde{\alpha}) = f(\widetilde{\gamma}) \neq f(\widetilde{\beta})$. Show that it is possible to identify certain variables of the function f in such a way that the function depends essentially on at least two and at the most three variables.

1.5.28*. Show that for the function $f(\widetilde{x^n})$, $n \geqslant 4$, reresented by Zhegalkin's polynomial of power not less than 2, there exist two variables whose identification decreases the number of essential variables by one.

1.5.29. Enumerate all the functions $f(\widetilde{x}^3)$ that depend essentially on three variables and in which the identification of any two variables leads to a function depending essentially on exactly one variable.

1.5.30. Let the function $f(\widetilde{x}^n)$ be such that $|N_f| > 2^{n-1}$. Show that the identification of any two variables of the function leads to a function that is not identically equal to zero.

1.5.31. Show that the number of functions $f(x_1, x_2, \ldots, x_{n+1})$ which lead to a certain function $g(x_1, x_2, \ldots, x_n)$ as a result of identification, is asymptotically equal to $\binom{n+1}{2} 2^{2n}$ as $n \to \infty$.

1.5.32. Show that if $f(\widetilde{x}^n)$ depends apparently on $\widetilde{x}_i$, the identification of this variable to any other variable leads to a function that depends essentially on the same variables as the function $f(\widetilde{x}^n)$.

1.5.33. Let $n \geqslant 1$ and let the functions $f(\widetilde{x}^n)$ and $g(\widetilde{x}^n)$ be such that $|N_{f\oplus g}| = 1$. Show that for any $i = \overline{1, n}$, at least one of the functions f or g depends essentially on x_i.

1.5.34. Show that if $|N_{f_{00}^{ij}(\widetilde{x}^n)\oplus f_{11}^{ij}(\widetilde{x}^n)}|$ is odd, the function φ obtained from $f(\widetilde{x}^n)$ by identifying the variables x_i and x_j depends essentially on $n - 1$ variables ($n \geqslant 3$).

1.5.35*. Let the function $f(\widetilde{x}^n)$ depend essentially on n variables. Let $v_f(\widetilde{\alpha})$ be the number of vertices $\widetilde{\beta}$ for which $f(\widetilde{\alpha}) \neq f(\widetilde{\beta})$ and $\rho(\widetilde{\alpha}, \widetilde{\beta}) = 1$. Let $v(f) = \max\limits_{\widetilde{\alpha}\in B^n} v_f(\widetilde{\alpha})$.

Find $v(f)$ for the following functions f:

(1) $f(\widetilde{x}^n) = x_1 \oplus x_2 \oplus \ldots \oplus x_n$;

(2) $f(\widetilde{x}^{k[n/k]}) = (x_1 \vee \ldots \vee x_k) \& (x_{k+1} \vee \ldots \vee x_{2k}) \& \ldots \& (x_{k([n/k]-1)+1} \vee \ldots \vee x_{k[n/k]})$, $1 \leqslant k \leqslant n$;

(3) $f(\widetilde{x}^{k+2^k}) = \alpha_{k+1+\nu(\alpha_1, \ldots, \alpha_k)}$ if $\widetilde{x}^{k+2^k} = (\alpha_1, \alpha_2, \ldots, \alpha_k, \alpha_{k+1}, \ldots, \alpha_{k+2^k})$, where $\nu(\alpha_1, \ldots, \alpha_k)$ is the number of the tuple $(\alpha_1, \ldots, \alpha_k)$.

Chapter Two

Closed Classes and Completeness

2.1. Closure Operation. Closed Classes

Let M be a certain set of Boolean functions. The *closure* $[M]$ of the set M is defined as the set of all functions from P_2 that are superpositions of functions in the set M. The operation of obtaining the set $[M]$ from M is called the *closure operation*. The set M is called a *functionally closed class* (in short, *closed class*) if $[M] = M$.

Let M be a closed class in P_2. The subset A in M is called a *functionally complete system* (in short, *complete system*) in M if $[A] = M$. The set A of Boolean functions is called an *irreducible system* if the closure of any proper subset A' in A is different from the closure of the entire set A, i.e. $[A'] \subset [A]$ and $[A'] \neq [A]$. An irreducible complete system in the closed class M is called the *basis* of the class M. The set M' contained in the closed class M (among other things, in the entire set P_2) is called a *precomplete class in M* if it is not a complete set in M, but the equality $[M' \cup \{f\}] = M$ is satisfied for any function $f \in M \setminus M'$.

Functions f_1 and f_2 will be called *congruent* if one of them can be obtained from the other by a change of variables (without identification). For example, the functions $x \cdot \bar{y}$ and $y \cdot \bar{z}$ are congruent while the functions $x \cdot y$ and $z \cdot z$ are not. While considering questions concerning closed classes, it is convenient to indicate one representative each from the set of pairwise congruent functions. For example, the class $\{x, y, z, \ldots, x_1, x_2, \ldots\}$ formed by all identity functions will be denoted by $\{x\}$.

If M is a certain set of functions, then $M(X^n)$ (or M^n) will denote a subset of all functions in M depending *only* on the variables $x_1, x_2, \ldots, x_n$.

2.1.1. Justify the following closure properties:
(1) $[[M]] = [M]$;
(2) if $M_1 \subseteq M_2$, $[M_1] \subseteq [M_2]$;
(3) $[M_1 \cup M_2] \supseteq [M_1] \cup [M_2]$;
(4) $[\varnothing] = \varnothing$.

2.1.2. Does relation (4) in 2.1.1 follow from relations (1)-(3)?

2.1.3. Is the set A a closed class? Assume that together with each function f in A, the set A also contains all functions in P_2 that are congruent to f:
(1) $A = \{0, 1\}$; (2) $A = \{\bar{x}\}$; (3) $A = \{1, \bar{x}\}$;
(4) $A = \{x_1, x_1 \cdot x_2, x_1 \cdot x_2 \cdot x_3, \ldots, x_1 \cdot x_2 \cdot \ldots \times x_n, \ldots\}$;
(5) $A = \{0, x_1 \vee x_2 \vee \ldots \vee x_n, \quad n \geqslant 1\}$;
(6) $A = \{x_1 \oplus x_2 \oplus \ldots \oplus x_n, \quad n \geqslant 1\}$;
(7) $A = \{0, x_1 \oplus x_2 \oplus \ldots \oplus x_{2n}, \quad n \geqslant 1\}$.

2.1.4. Write down all functions from the closure of the set A which depend only on variables x_1, x_2, x_3 and are pairwise non-congruent:
(1) $A = \{x \rightarrow 1\}$;
(2) $A = \{0, \bar{x}\}$;
(3) $A = \{x \vee y\}$;
(4) $A = \{x^y\}$;
(5) $A = \{x \oplus y \oplus z\}$;
(6) $A = \{00000001\}$.

2.1.5. Prove that if a closed class in P_2 contains a function depending essentially on $n \geqslant 2$ variables, it contains an infinitely large number of pairwise non-congruent functions.

2.1.6. Enumerate all closed classes in P_2 which contain only a finite number of pairwise non-congruent functions.

2.1.7. In P_2,
(1) is the intersection of closed classes always a closed class?
(2) is the difference between closed classes always a closed class?
(3) can the completion of a closed class never be a closed class?

2.1.8. Through reduction to a priori complete set in P_2, show that the set A is complete in P_2 if:
(1) $A = \{x \downarrow y\}$;
(2) $A = \{x \cdot y \oplus z, (x \sim y) \oplus z\}$;

(3) $A = \{x \rightarrow y, \overline{x \oplus y \oplus z}\}$;
(4) $A = \{x \rightarrow y, (1100001100111100)\}$;
(5) $A = \{0, m(x, y, z), x^y \oplus z\}$[1];
(6) $A = \{(1011), (1111110011000000)\}$.

2.1.9. Which of the relations $\supset$, $\subset$, $\supseteq$, $\subseteq$, $=$, $\not\supseteq\!\!\!\!\not\subseteq$ is satisfied for the sets[2] K_1 and K_2 (the relation $\not\supseteq\!\!\!\!\not\subseteq$ means that none of the relations $\supset$, $\subset$, $\supseteq$, $\subseteq$, $=$ is satisfied)?
(1) $K_1 = [M_1 \cap M_2]$, $K_2 = [M_1] \cap [M_2]$;
(2) $K_1 = [M_1 \setminus M_2]$, $K_2 = [M_1] \setminus [M_2]$;
(3) $K_1 = [M_1 \cup (M_2 \cap M_3)]$, $K_2 = [M_1 \cup M_2] \cap [M_1 \cup M_3]$;
(4) $K_1 = [M_1 \cap (M_2 \cup M_3)]$, $K_2 = [M_1 \cap M_2] \cup [M_1 \cap M_3]$;
(5) $K_1 = [M_1 \setminus (M_1 \cap M_2)]$, $K_2 = [M_1] \setminus [M_1 \cap M_2]$.

2.1.10. Let M_1 and M_2 be closed classes in P_2, such that $M_1 \setminus M_2 \neq \varnothing$. Give examples of concrete classes M_1 and M_2 that also satisfy the following conditions:
(1) $M_1 \cap M_2 = \varnothing$, $M_2 \setminus M_1 \neq \varnothing$, $[M_1 \cup M_2] = M_1 \cup M_2$;
(2) $M_1 \cap M_2 \neq \varnothing$, $M_2 \setminus M_1 \neq \varnothing$, $[M_1 \cup M_2] = M_1 \cup M_2$;
(3) $M_1 \supset M_2$, $[M_1 \setminus M_2] \neq M_1 \setminus M_2$;
(4) $M_1 \cap M_2 \neq \varnothing$, $M_2 \setminus M_1 \neq \varnothing$, $[M_1 \setminus M_2] = M_1 \setminus M_2$;
(5) $M_1 \cap M_2 \neq \varnothing$, $M_2 \setminus M_1 \neq \varnothing$, $[M_1 \oplus M_2] = M_1 \oplus M_2$.

2.1.11. Isolate the basis from the set A that is complete for the closed class $M = [A]$.
(1) $A = \{0, 1, x, \bar{x}\}$;
(2) $A = \{1, x \oplus y \oplus z \oplus 1\}$;
(3) $A = \{x \vee y, x \cdot y \cdot z, x \vee y \cdot z, (x \vee y) \cdot z\}$;
(4*) $A = \{x \oplus 1, x \oplus y \oplus z, m(x, y, z)\}$;
(5) $A = \{x \vee y \vee z, x \cdot y \cdot z, (x \rightarrow y) \rightarrow z, (x \vee y) \rightarrow z\}$;
(6) $A = \{(x \rightarrow y) \rightarrow (y \rightarrow z), x \vee y \vee (y \oplus z)\}$;
(7) $A = \{x \cdot y, x \vee y, x \rightarrow y, x \oplus y \oplus z \oplus u\}$.

2.1.12. Show that any precomplete class in P_2 is a closed class.

---

[1] By $m(x, y, z)$ (or $h_2(x, y, z)$) we denote the function $xy \vee xz \vee yz$ called the *median* (or *majority function*).
[2] The sets are taken in P_2.

2.1.13. Let M_1 and M_2 be different precomplete classes in the same closed class M[3]. Show that if $M_1^1 \neq M^1$, then $M_1^1 \neq M_2^1$ (i.e. classes M_1 and M_2 "differ" even on a set of functions that depend on only one variable).

2.1.14. Enumerate all precomplete classes in the closed class M.

(1) $M = [0, \bar{x}]$; (4) $M = [0, x \vee y]$;
(2) $M = [0, 1]$; (5) $M = [0, x \cdot y \cdot z]$.
(3) $M = [x \cdot y]$;

2.1.15. Verify if the following sets form closed classes in P_2:

(1) the set of all symmetric functions;

(2) the set of all functions $f(\tilde{x}^n)$, $n \geqslant 0$, satisfying the condition $f(\tilde{0}^n) = f(\tilde{1}^n) = 0$[4];

(3) the set of all functions $f(\tilde{x}^n)$, $n \geqslant 1$ for which $|N_f| = 2^{n-1}$.

2.1.16. Show that if M is a closed class in P_2, then $[M \cup \{x\}] = M \cup \{x\}$.

2.1.17. Prove that the set P_2 of all Boolean functions cannot be presented as a union $\bigcup_{i=1}^{s} M_i$ $(s \geqslant 2)$ on pairwise non-intersecting closed classes in P_2.

2.1.18. Prove that any closed class in P_2 containing a function other than a constant also contains the function x.

2.1.19. Prove that if a closed class in P_2 has a finite basis, then each basis of this class is finite.

2.1.20. Majorize the power of the set of all closed classes in P_2 containing finite complete sets.

2.1.21. Prove that if a non-empty closed class in P_2 differs from the sets $\{0\}$, $\{1\}$ and $\{0, 1\}$, it cannot be extended to a basis in P_2.

2.1.22. Let M be a closed class in P_2 containing a finite number of precomplete classes (in M). It is assumed that any closed class in M can be extended to a precomplete class in M. Prove that the number of functions in any basis of class M does not exceed the number of precomplete classes (in M).

---

[3] We assume that $M \subseteq P_2$.

[4] If $n = 0$, then f is a constant equal to zero, considered as a function of zero arguments.

2.1.23. Prove that a closed class $[x \to y]$ contains only such functions in P_2 which can be presented (barring the notation of the variables) in the form $x_i \vee f(x_1, x_2, \ldots, x_n)$, where $f(\tilde{x}^n) \in P_2$.

2.1.24. Let the function $f(\tilde{x}^n)$ belong to the closed class $[x \to y]$, and depend essentially on at least two variables. Prove that $|N_f| > 2^{n-1}$.

2.1.25. Prove (without the help of Problem 2.1.18) that each precomplete class in P_2 contains an identity function.

2.2. Duality and the Class of Self-Dual Functions

The function $g(x_1, x_2, \ldots, x_n)$ is called *dual to the function* $f(x_1, x_2, \ldots, x_n)$ if $g(x_1, x_2, \ldots, x_n) = \overline{f}(\overline{x}_1, \overline{x}_2, \ldots, \overline{x}_n)$. By definition, the dual function for constant 0 is constant 1 and, conversely, constant 0 is a function dual to constant 1. The function dual to $f(x_1, x_2, \ldots, x_n)$ is denoted by $f^*(x_1, x_2, \ldots, x_n)$.

The following statement, called the *duality principle*, is true: if $\Phi(x_1, x_2, \ldots, x_n) = f(f_1(x_1, x_2, \ldots, x_n), \ldots, f_m(x_1, x_2, \ldots, x_n))$, then $\Phi^*(x_1, x_2, \ldots, x_n) = f^*(f_1^*(x_1, x_2, \ldots, x_n), \ldots, f_m^*(x_1, x_2, \ldots, x_n))$.

Let M be a set of Boolean functions. By M^* we shall denote the set of all functions that are dual to the functions in the set M. The set M^* will be called *dual to the set* M. If $M^* = M$, the set M is called *self-dual*.

The function $f(x_1, x_2, \ldots, x_n)$ is called *self-dual* if $f^*(x_1, x_2, \ldots, x_n) = f(x_1, x_2, \ldots, x_n)$. The set of all self-dual functions is denoted by S.

It follows from the definition of self-dual functions that a function is self-dual if and only if it assumes opposite values on any two opposite tuples of values of variables.

The following statement, called the *lemma of non-self-dual function*, is valid: if the function $f(\tilde{x}^n)$ is non-self-dual, the substitution of the functions x and $\overline{x}$ for its variables will lead to a constant,

2.2.1. Is the function g dual to the function f if
(1) $f = x \oplus y, \quad g = x \sim y$;
(2) $f = x \to y, \quad g = y \to x$;
(3) $f = xy \vee xz \vee yz, \quad g = xy \oplus xz \oplus yz$;
(4) $f = x \oplus y \oplus z, \quad g = x \oplus y \oplus z$;
(5) $f = x\overline{y}z \vee x(y \sim z), \quad g(x, y, z) = (01101101)$?

2.2.2. Let the function $f(\tilde{x}^n)$ be defined by the vector $\tilde{\alpha}_f = (\alpha_0, \alpha_1, \ldots, \alpha_{2n-1})$. Prove that the function $f^*(\tilde{x}^n)$ is defined by the vector $(\overline{\alpha}_{2n-1}, \ldots, \overline{\alpha}_1, \overline{\alpha}_0)$.

2.2.3. Using the duality principle, derive a formula for representing a function dual to f. Simplify the obtained expression (by presenting it in a disjunctive normal form or in the form of a Zhegalkin polynomial).
(1) $f = xy \vee yz \vee xt \vee zt$;
(2) $f = x \cdot 1 \vee y(zt \vee 0) \vee x\overline{y}z$;
(3) $f = (x \to y) \oplus ((x \downarrow y) \mid \overline{(x \sim yz)})$;
(4) $f = (\overline{x} \vee y \vee (y\overline{z} \oplus 1)) \to 1$.

2.2.4. Suppose that the function $f(\tilde{x}^n)$ is represented by the formula $\mathfrak{A}$ generated by the set $\{0, 1, \neg, \&, \vee\}$. Prove that the function $f^*(\tilde{x}^n)$ is represented by the formula $\mathfrak{A}^*$, called the *formula dual to* $\mathfrak{A}$ and obtained from it by replacing each symbol & by $\vee$, $\vee$ by &, 0 by 1 and 1 by 0.

2.2.5. Prove that if formulas $\mathfrak{A}$ and $\mathfrak{B}$ generated by the set $\{0, 1, \neg, \&, \vee\}$ are equivalent, formulas $\mathfrak{A}^*$ and $\mathfrak{B}^*$ are also equivalent.

2.2.6. Prove that if the function $f(\tilde{x}^n)$ depends essentially on the variable x_i $(1 \leqslant i \leqslant n)$, the function $f^*(\tilde{x}^n)$ also depends essentially on x_i.

2.2.7. Prove that
(1) the set dual to M^* is identical to M;
(2) the set M is a closed class if and only if the set M^* is a closed class;
(3) if the set M_1 is a complete system (or basis) in the closed class M, the dual set M_1^* forms a complete system (resp. basis) in the closed class M^*;
(4) if $M_1 \supseteq M_2$, then $M_1^* \supseteq M_2^*$.

2.2.8. Count the number of functions depending on variables $x_1, x_2, \ldots, x_n$ in the set $[M^*] \setminus [M]$.

(1) $M = \{0, \overline{x}\}$; (2) $M = \{x \oplus y\}$; (3) $M = \{xy, x \vee y, 1\}$.

2.2.9. Is the function f self-dual in the following case:
(1) $f = \overline{m(x, y, z)}$;
(2) $f = (x \to y) \to xz \to (y \to z)$;
(3) $f = (\overline{x} \vee y \vee \overline{z})\, t \vee \overline{x}\overline{y}z$;
(4) $f = (0001001001100111?)$;
(5) $f = x_1 \oplus x_2 \oplus \ldots \oplus x_{2m+1} \oplus \sigma$, where $\sigma \in \{0, 1\}$?

2.2.10. Prove that the function $f(\widetilde{x}^n)$ is self-dual if and only if its x_1-component $f_1^1(\widetilde{x}^n)$ is dual to its $\overline{x}_1$-component $f_0^1(\widetilde{x}^n)$.

2.2.11. Prove that if $f(\widetilde{x}^n)$ is a self-dual function, then $|N_f| = 2^{n-1}$.

2.2.12. Show that there are no self-dual functions depending essentially on two variables.

2.2.13. Determine the number of self-dual functions depending essentially on variables $x_1, x_2, \ldots, x_n$.

2.2.14. Enumerate all self-dual functions that depend essentially on variables x, y, z. Show that each of these functions can be presented in the form $m(x^\alpha, y^\beta, z^\gamma)$ or $x \oplus y \oplus z \oplus \sigma$, where $\alpha, \beta, \gamma, \sigma$ belong to the set $\{0, 1\}$.

2.2.15. Determine the values $n \geqslant 2$ which make the function $f(\widetilde{x}^n)$ self-dual:

$$(1)\ f(\widetilde{x}^n) = x_1(x_2 \vee x_3 \vee \ldots \vee x_n) \vee x_2(x_3 \vee \ldots \vee x_n) \vee \ldots \vee x_{n-2}(x_{n-1} \vee x_n) \vee x_{n-1}x_n;$$

$$(2)\ f(\widetilde{x}^n) = x_1(x_2 \oplus x_3 \oplus \ldots \oplus x_n) \oplus x_2(x_3 \oplus \ldots \oplus x_n) \oplus \ldots \oplus x_{n-2}(x_{n-1} \oplus x_n) \oplus x_{n-1}x_n;$$

$$(3)\ f(\widetilde{x}^n) = (x_1 \to x_2) \oplus (x_2 \to x_3) \oplus \ldots \oplus (x_{n-1} \to x_n) \oplus (x_n \to x_1);$$

$$(4)\ f(\widetilde{x}^n) = \bigvee_{1 \leqslant i_1 < i_2 < \ldots < i_{]n/2[} \leqslant n} x_{i_1} x_{i_2} \ldots x_{i_{]n/2[}}.$$

Here the disjunction is taken over all monotonic conjunctions of length $]n/2[$ formed by variables $x_1, x_2, \ldots, x_n$.

2.2.16. Let $f(\widetilde{x}^n) = x_1 \oplus x_2 \oplus \ldots \oplus x_n$, $g_i^*(\widetilde{x}^m) = g_{n-i+1}(\widetilde{x}^m)$, $i = 1, 2, \ldots, n$. Is the function $f(g_1(\widetilde{x}^m), \ldots, g_n(\widetilde{x}^m))$ self-dual?

2.2.17. Prove that if

(1) $f(x_1, x_2, \ldots, x_n) \in S$, then $f(x, x, \ldots, x) \in \{x, \bar{x}\}$;

(2) $f \in S$, then $m(x_1, \bar{f}, x_2 \oplus x_3 \oplus f) \oplus x_3 \oplus x_4 \in S$;

(3) $f \in S$, then $\bar{x}_1 f \oplus x_2 x_3 \oplus \bar{x}_1 x_3 \oplus x_2 \bar{f} \oplus x_4 \oplus x_5 \in S$.

2.2.18. Let the functions $f(\tilde{x}^n)$ and $f^*(\tilde{x}^n)$, $n \geqslant 1$, satisfy the condition $|N_f| = |N_{f^*}|$. Prove that

(1) if $\bar{f} \vee f^* \equiv \text{const}$, then $f \in S$;

(2) if $f \oplus f \cdot f^* \equiv \text{const}$, then $f \in S$.

2.2.19. Substituting the functions x and $\bar{x}$ for the variables obtain a constant using the non-self-dual function f:

(1) $f = (00111001)$;

(2) $f = (x \vee \bar{y} \vee z)\, t \vee \bar{x} yz$;

(3) $f = (x \downarrow y) \rightarrow (x \oplus z)$;

(4) $f = xy \vee xz \vee yt \vee zt$.

2.2.20. Prove that if a non-self-dual function depends essentially on at least three variables, the identification of some of its variables can lead to a function that depends essentially on two variables.

2.2.21. Prove that if identification of variables in the function $f(\tilde{x}^n)$, $n \geqslant 3$, cannot lead to a function depending essentially on two variables, the function f is self-dual.

2.2.22. Let $f(x, y, z) = m(x^\alpha, y^\beta, z^\gamma)$, where α, β, γ belong to the set $\{0, 1\}$. Prove that for any $n \geqslant 4$, we can apply the superposition operation to the function f to obtain a function depending essentially on n variables.

2.2.23. Prove the following equivalences:

(1) $x \oplus y \oplus z = m(m(x, y, \bar{z}), m(x, \bar{y}, z), m(\bar{x}, y, z)) = m(x, m(\bar{x}, y, z), \overline{m(x, y, z)}) = m(m(x, \bar{y}, z), m(\bar{x}, y, z), \bar{z})$;

(2) $(x \vee y \vee z)\, t \vee xyz = m(m(y, z, t), x, t) = m(m(x, y, t), m(x, z, t), m(y, z, t))$;

(3) $m(x, y, z) = m(\overline{m(\bar{x}, y, z)}, y, z)$.

2.2.24. Let $f(\tilde{x}^n) \in P_2$ and $n \geqslant 3$. Prove the following relation:

$$f(x_1, x_2, x_3, x_4, \ldots, x_n) = f(x_1, m(x_1, x_2, x_3), m(x_1, x_2, x_3), x_4, \ldots, x_n)$$

$\oplus f(m(x_1, x_2, x_3), x_2, m(x_1, x_2, x_3), x_4, \ldots, x_n)$
$\oplus f(m(x_1, x_2, x_3), m(x_1, x_2, x_3), x_3, x_4, \ldots, x_n)$.

2.2.25. (1) Using the problems 2.2.12., 2.2.14., 2.2.23. (item 1) and 2.2.24., prove that $[\overline{m(x, y, z)}] = [m(\overline{x}, \overline{y}, \overline{z})] = S$.

(2) Prove that any basis of the class S of all self-dual functions contains not more than two functions.

2.2.26. Can the superposition operation be used to obtain the function g using the function f if:

(1) $f = (10110010)$, $g = (1000)$;
(2) $f = (1111011100010000)$, $g = (00010111)$;
(3) $f = (11001100)$, $g = (00110011)$?

2.2.27. The function $f(\tilde{x}^n)$, $n \geqslant 2$, has the following properties: $f(\tilde{x}^n) \notin S$, and the identification of any $2[\sqrt{n}]$ variables in it leads to a function from class S. What is the largest number of essential variables in the function $f(\tilde{x}^n)$?

2.2.28. Enumerate all functions depending essentially on the variables x_1, x_2, x_3, x_4, any $x_i^{\sigma_i}$-component of which is a self-dual function.

2.2.29. Is the set M self-dual in the following case:

(1) $M = \{x \oplus y \oplus z,\ m(x \oplus y,\ x \sim z,\ y \sim \overline{z})\}$;
(2) $M = \{x \cdot y,\ x \vee y,\ x \oplus y \oplus m(x, y, z)\}$;
(3) $M = \{(x \to y) \to y,\ (x \vee y) \oplus x \oplus y,\ (x \vee y \vee z) t \vee \overline{x}yz\}$;
(4) $M = S \setminus \{x \oplus y \oplus \overline{z},\ m(x, \overline{y}, \overline{z})\}$?

2.3. Linearity and the Class of Linear Functions

The function $f(\tilde{x}^n)$ is called *linear* if it can be presented in the form

$$f(\tilde{x}^n) = \alpha_0 \oplus \alpha_1 x_1 \oplus \alpha_2 x_2 \oplus \ldots \oplus \alpha_n x_n,$$

where $\alpha_i \in \{0, 1\}$, $0 \leqslant i \leqslant n$. The set of all linear functions is denoted by L, while the set of all linear functions depending on the variables $x_1, x_2, \ldots, x_n$ is denoted by L^n. The set L is a closed and precomplete class in P_2. The following statement (*lemma on non-linearity of a function*) is valid:

If $f \notin L$, the substitution of the functions $0, 1, x, y,$

$\bar{x}$, $\bar{y}$ for the variables can lead either to xy or to $\overline{xy}$.

If $f \notin L$, f is called *nonlinear*.

2.3.1. Expanding the function f into a Zhegalkin polynomial, find if it is linear:

(1) $f(\tilde{x}^3) = (x_1x_2 \vee \bar{x}_1\bar{x}_2) \oplus x_3$;

(2) $f(\tilde{x}^2) = x_1x_2(x_1 \oplus x_2)$;

(3) $f(\tilde{x}^4) = \bar{x}_1x_2 \vee \bar{x}_2x_3 \vee \bar{x}_3x_4 \vee \bar{x}_4x_1$;

(4) $f(\tilde{x}^3) = (x_1 \to x_2)(x_2 \to x_1) \sim x_3$.

2.3.2. Prove that if the function $f(\tilde{x}^n)$ assumes opposite values on any two adjacent vertices $\tilde{\alpha}$ and $\tilde{\beta}$ in B^n, it is a linear function. Is the converse true?

2.3.3. Prove that if $f(\tilde{x}^n)$ is a linear function other than a constant, $|N_f| = 2^{n-1}$. Is the converse true?

2.3.4. Find out if the function f is linear:

(1) $f(\tilde{x}^4) = (1010\ \ 1010\ \ 0110\ \ 1000)$;

(2) $f(\tilde{x}^4) = (1001\ \ 0110\ \ 1001\ \ 0110)$;

(3) $f(\tilde{x}^4) = (1001\ \ 0110\ \ 0110\ \ 1001)$;

(4) $f(\tilde{x}^4) = (0110\ \ 1001\ \ 1010\ \ 0101)$.

2.3.5. Show that the number of linear functions $f(\tilde{x}^n)$ depending essentially on exactly k variables of the set $\{x_1, x_2, \ldots, x_n\}$ is equal to $2\binom{n}{k}$.

2.3.6. Find the number of self-dual functions belonging to the set L^n.

2.3.7. Show that the function $x \to y$ cannot be obtained from the functions $x \oplus y \oplus z$, $x \oplus 1$, $x \oplus y$ with the help of the superposition operation.

2.3.8. Can the function f lead to xy if the functions $0, 1, x, y, \bar{x}, \bar{y}$ are substituted for its variables?

(1) $f(\tilde{x}^3) = (1110\ \ 1000)$;

(2) $f(\tilde{x}^3) = (0111\ \ 1111)$;

(3) $f(\tilde{x}^3) = (1001\ \ 1001)$;

(4) $f(\tilde{x}^n) = (x_1 \to x_2)(x_2 \to x_1) \oplus (x_2 \to x_3)(x_3 \to x_2) \oplus \ldots \oplus (x_{n-1} \to x_n)(x_n \to x_{n-1})$.

2.3.9. Find out if the function $x \to y$ can be represented by a formula generated by the set Φ, where

(1) $\Phi = L \cup \{xy \vee xz \vee zy\}$;
(2) $\Phi = L \setminus S$;
(3) $\Phi = (L \cup \{xy \vee yz \vee zx\}) \setminus S$;
(4) $\Phi = (L \cup \{xy\}) \setminus S$.

2.3.10*. Prove that if the function $f(\tilde{x}^n)$, depending essentially on all its variables, is linear if and only if the substitution of any subset of constants for any subset of variables leads to a function that depends essentially on all the remaining variables.

2.3.11*. Prove that a polynomial of degree $k \geqslant 3$ can lead to a polynomial of degree $k - 1$ as a result of identification of variables.

2.3.12*. Show that by identifying the variables in the function $f(\tilde{x}^n)$, $n \geqslant 4$, we can obtain a nonlinear function depending on not more than three variables. Enumerate all nonlinear functions $f(\tilde{x}^3)$ from which a nonlinear function cannot be obtained as a result of identification of variables.

2.3.13. Show that the identification of variables in a non-linear function can lead to a function that is congruent either to $xy \oplus l(x, y)$, or to $xy \oplus yz \oplus zx \oplus l(x, y, z)$, where $l(x, y)$ and $l(x, y, z)$ are linear functions.

2.3.14. Show that if $f(\tilde{x}^n) \notin L$, there exists a two-dimensional face in B^n on exactly three of whose vertices the function $f(\tilde{x}^n)$ assumes the same value.

2.3.15. Let the function $f(\tilde{x}^n)$, $n \geqslant 3$, be such that for all i, j $(1 \leqslant i < j \leqslant n)$ $f_{00}^{ij}(\tilde{x}_n) = f_{11}^{ij}(\tilde{x}^n)$. Show that $f(\tilde{x}^n) \in L$.

2.3.16. Which of the following sets form a basis in L:
(1) $\{1, x \oplus y\}$;
(2) $\{x \sim y, x \oplus y \oplus z\}$;
(3) $\{x \sim y, x \oplus y, 0\}$;
(4) $\{0, x \oplus y \oplus z \oplus 1\}$;
(5) $\{1 \oplus x \oplus y \oplus z, x \oplus y \oplus z\}$?

2.3.17. Single out all bases in the following sets complete in L:
(1) $\{0, 1, x \oplus 1, x \sim y, x \oplus y \oplus z\}$;
(2) $\{x \sim y, (x \sim y) \sim z, x \oplus y \oplus z, x \oplus y \oplus z \oplus t\}$;
(3) $\{0, (x \sim y) \sim z, x \oplus 1, x \oplus y\}$.

2.3.18. Prove that any set complete in L contains at least two functions.

2.3.19. Prove that any basis in L contains not more than three functions.

2.3.20. Prove that only a finite number of closed classes can be formed by linear functions alone. List all such classes.

2.3.21. Show that any closed class formed by a finite number of pairwise non-congruent functions is contained in L.

2.3.22. The function $f(\tilde{x}^n)$, $n \geqslant 1$, satisfies the conditions:

(1) $|N_f| = 2^{n-1}$, (2) $f \oplus f^*$ is a constant.

Show that $f(\tilde{x}^n) \in L \cup S$ for $n \leqslant 3$. Give an example when the function f satisfying the conditions (1) and (2) does not belong to the set $L \cup S$.

2.3.23. Prove that $L \cap S = [x \oplus y \oplus z \oplus 1]$.

2.3.24. Show that $L^* = L$.

2.3.25. Enumerate all pairwise non-congruent functions f satisfying the following conditions:

(1) $f \notin L$,

(2) any proper subfunction of f is linear.

2.3.26. Determine the number of self-dual linear functions $f(\tilde{x}^n)$ that depend essentially on all their variables.

2.3.27. In how many ways can brackets be arranged in the expression $x_1 \to x_2 \to x_3 \to x_4 \to x_1$ so that to give a linear function?

2.3.28. Let $f(\tilde{x}^n) \in L \cap S$, $f(0, 0, \ldots, 0) = f(0, 0, \ldots, 0, 1)$ and $f(\alpha_1, \alpha_2, \ldots, \alpha_n) = 0$. Find the value of the function $f(\bar{\alpha}_1, \bar{\alpha}_2, \ldots, \bar{\alpha}_{n-1}, \alpha_n)$.

2.3.29. For what values of n can the linear function $f(\tilde{x}^n)$ satisfy the condition $f(0, 0, \ldots, 0) \neq f(1, 1, \ldots, 1)$ and be symmetric?

2.3.30. Let $f(\tilde{x}^n)$ be such that

$$f_{1\,1\,\ldots\,1}^{3,\,4,\,\ldots,\,n}(\tilde{x}^n) = \bar{x}_1 x_2 \quad \text{and} \quad f_{0\,0\,\ldots\,0}^{3,\,4,\,\ldots,\,n}(\tilde{x}) = \bar{x}_2.$$

Show that $f(\tilde{x}^n) \notin L \cup S$.

2.3.31*. Enumerate all pairwise non-congruent nonlinear functions f such that any identification of variables leads to a function in L.

2.3.32*. Enumerate all pairwise non-congruent functions $f \notin L \cup S$ for which any identification of variables leads to a function in $L \cap S$.

2.3.33. Let $Q = \{0, \bar{x}, f_1, f_2, f_3\}$, where f_1, f_2, f_3 are pairwise different functions depending essentially on variables $x_1, x_2, \ldots, x_n$ $(n > 1)$. Prove that the system Q is complete in P_2.

2.4. Classes of Functions Preserving the Constants

The function $f(\tilde{x}^n)$ is said to *preserve the constant* 0 (or *constant* 1) if $f(0, 0, \ldots, 0) = 0$ (resp. if $f(1, 1, \ldots, 1) = 1$). The set of all Boolean functions preserving the constant 0 is denoted by T_0, while the set of functions preserving the constant 1 is denoted by T_1. The set of all Boolean functions that depend on variables $x_1, x_2, \ldots, x_n$ and preserve the constant 0 (constant 1) is denoted by T_0^n (resp. by T_1^n). Each of the sets T_0 and T_1 is a closed and precomplete class in P_2.

2.4.1. To which of the sets $T_0 \cup T_1$, $T_1 \setminus T_0$ do the following functions belong:

(1) $((x \vee y) \rightarrow (x \mid yz)) \downarrow ((y \sim z) \rightarrow x)$;

(2) $(xy \rightarrow z) \mid ((x \rightarrow y) \downarrow (z \oplus \bar{x}y))$;

(3) $(x \rightarrow y) \& (y \downarrow z) \vee (z \rightarrow y)$?

2.4.2. For what values of n does the function $f(\tilde{x}^n)$ belong to the set $T_1 \setminus T_0$:

(1) $f(\tilde{x}^n) = x_1 \oplus x_2 \oplus \ldots \oplus x_n \oplus 1$;

(2) $f(\tilde{x}^n) = (\ldots((x_1 \rightarrow x_2) \rightarrow x_3) \rightarrow \ldots \rightarrow x_n)$;

(3) $f(\tilde{x}^n) = (\ldots((x_1 \rightarrow x_2) \rightarrow x_3) \rightarrow \ldots \rightarrow x_n)$
$\oplus ((\ldots((x_2 \rightarrow x_3) \rightarrow x_4) \rightarrow \ldots \rightarrow x_n) \rightarrow x_1) \oplus \ldots$
$\oplus (\ldots((x_n \rightarrow x_1) \rightarrow x_2) \rightarrow \ldots \rightarrow x_{n-1})$;

(4) $f(\tilde{x}^n) = 1 \oplus \sum\limits_{1 \leqslant i_1 < i_2 < i_3 \leqslant n} x_{i_1} x_{i_2} x_{i_3}$?

2.4.3. For what values of n does the recurrently defined function $f(\tilde{x}^n)$ belong to the set $(T_0 \setminus T_1) \cup (T_1 \setminus T_0)$:

(1) $f_2(\widetilde{x^2}) = x_1 \oplus x_2$,

$f_n(\widetilde{x^n}) = (x_n \rightarrow f_{n-1}(\widetilde{x^{n-1}}))(x_n \vee f_{n-1}(\widetilde{x^{n-1}})), \quad n > 2;$

(2) $f_1(x_1) = x_1, \quad f_2(x_1, x_2) = x_1 \oplus x_2$,

$$f_n(\widetilde{x^n}) = x_{n-1}x_n \oplus f_{n-2}(\widetilde{x^{n-2}}), \quad n \geqslant 3;$$

(3) $f_1(x_1) = x_1, \quad f_2(x_1, x_2) = x_1 \vee x_2$,

$$f_n(\widetilde{x^n}) = f_{n-1}(\widetilde{x^{n-1}}) \oplus x_n f_{n-2}(\widetilde{x^{n-2}}), \quad n \geqslant 3?$$

2.4.4. In how many ways can the brackets be arranged in the expression $x_1 \rightarrow x_2 \rightarrow x_1 \rightarrow x_2 \rightarrow x_1$ in order to obtain a formula representing a function in T_0?

2.4.5. Count the number of functions depending on $x_1, x_2, \ldots, x_n$ in each of the following sets:

(1) $T_0 \cap T_1$; (2) $T_0 \cup T_1$; (3) $T_0 \cap L$; (4) $T_1 \cup L$; (5) $L \setminus (T_0 \cap T_1)$; (6) $L \setminus (T_0 \cup T_1)$; (7) $T_1 \cap S$; (8) $T_0 \setminus S$; (9) $S \cap (T_0 \cup T_1)$; (10) $S \cap (T_0 \setminus T_1)$; (11) $S \setminus (T_0 \cup T_1)$; (12) $(S \setminus T_0) \cap T_1$;

(13) $L \cap S \cap T_1$;
(14) $L \setminus (T_0 \cup (T_1 \cap S))$;
(15) $(L \cup S) \setminus (T_0 \cup T_1)$.

2.4.6. Find the function $f(x, x, \ldots, x)$ if:

(1) $f(x_1, x_2, \ldots, x_n) \in T_1 \setminus T_0$;
(2) $f(x_1, x_2, \ldots, x_n) \in L \setminus (T_1 \cap S)$;
(3) $f(x_1, x_2, \ldots, x_n) \in S \setminus T_0$.

2.4.7. Is it possible to obtain the function f as a result of the superposition operation generated by the set Φ if:

(1) $f = x \oplus y, \quad \Phi = \{x \rightarrow y\}$;
(2) $f = x \rightarrow y, \quad \Phi = \{xy, x \vee y\}$;
(3) $f = \bar{x} \vee \bar{y}, \quad \Phi = T_0 \cup (S \setminus (L \cup T_1))$;
(4) $f = x\bar{y}, \quad \Phi = (T_1 \setminus L) \cup \{x \oplus y\}$?

2.4.8. Prove that

(1) $T_0 = [xy, x \oplus y] = [x \vee y, x \oplus y]$;
(2) $T_1 = [x \vee y, x \sim y] = [xy, x \sim y]$;
(3) $T_0 \cap T_1 = [xy, x \oplus y \oplus z]$.

2.4.9*. Prove that any basis in T_0 contains not more than three functions. Give examples of bases of the class T_0 consisting of one, two or three functions.

2.4.10*. Prove that any basis in $T_0 \cap T_1$ contains not more than two functions. Give examples of a basis consisting of one function.

2.4.11. Does the class $T_0 \cap T_1$ contain a function depending on three variables and forming a basis in it?

2.4.12. Prove that

(1) $L \cap T_0 = [x \oplus y]$;

(2) $L \cap T_1 = [x \sim y]$;

(3*) $S \cap T_0 = [xy \vee yz \vee zx,\ x \oplus y \oplus z] = [x\bar{y} \vee yz \vee z\bar{x}]$.

2.4.13. The function $f(\tilde{x}^3)$, which is not defined everywhere, is equal to zero on the tuples (000), (001) and equal to unity on the tuples (011), (100), (110). Extend the definition of the function $f(\tilde{x}^3)$ on the remaining tuples so that the obtained function forms a basis in T_0.

2.4.14*. Prove that if the function f depends essentially on at least two variables and $f \notin T_0 \cup T_1$, then $L \cap S \subseteq [f]$.

2.4.15. Are the following sets bases in T_0:

(1) $\{xy \vee y\bar{z} \vee \bar{z}x,\ x \oplus y \oplus z\}$;

(2) $\{x\bar{y},\ x \oplus y \oplus z,\ x \vee y\}$;

(3) $\{xy \oplus z\}$;

(4) $\{x \oplus y \oplus z,\ xy,\ x \vee y \vee z\}$?

2.4.16. Give an example of a symmetric function $f(\tilde{x}^4)$ forming a complete system in T_0.

2.4.17. Prove that

$$L \cap T_0 \cap T_1 = L \cap S \cap T_0 = L \cap S \cap T_1 = L \cap S \cap T_0 \cap T_1.$$

2.4.18*. Prove that the classes $T_0 \cap L$, $T_0 \cap S$, $T_0 \cap T_1$ are precomplete in T_0.

2.4.19. Prove that the system of functions $\{1\} \cup (T_0 \setminus (T_1 \cup L \cup S))$ is complete in P_2.

2.4.20. Prove that

(1) $T_0^* = T_1$;

(2) $(T_0 \cup T_1)^* = T_0 \cup T_1$;

(3) $(T_0 \cup T_1 \cup S \cup L)^* = T_0 \cup T_1 \cup S \cup L$;

(4) $((T_0 \cap T_1) \cup S)^* = (T_0 \cap T_1) \cup S$;

(5) $(T_0 \setminus S)^* = T_1 \setminus S$;

(6) $(S \setminus T_0)^* = S \setminus T_1$.

2.4.21. Let $f(\tilde{x}^4) \in S \cap L \cap T_0$, $f(1, 1, 0, 1) = 1$, and let $f(\tilde{x}^4)$ depend essentially on at least two variables. Find $f(\tilde{x}^4)$.

2.4.22*. Prove that the classes $L \cap T_0$, $L \cap T_1$, $L \cap S$, $[0, \bar{x}]$ are the only precomplete classes in L.

2.4.23. Prove that $f \in (T_0 \cap T_1) \cup S$ if and only if not a single constant can be obtained from f by carrying out the superposition operation.

2.5. Monotonicity and the Class of Monotonic Functions

The Boolean function $f(\tilde{x}^n)$ is called *monotonic* if for any two tuples $\tilde{\alpha}$ and $\tilde{\beta}$ in B^n, such that $\tilde{\alpha} \leqslant \tilde{\beta}$, the inequality $f(\tilde{\alpha}) \leqslant f(\tilde{\beta})$ is satisfied. Otherwise, the function $f(\tilde{x}^n)$ is called *non-monotonic*. The set of all monotonic Boolean functions is denoted by M, while the set of all monotonic functions depending on variables $x_1, x_2, \ldots, x_n$ is denoted by M^n. The set M is a closed and precomplete class in P_2. The following statement (*lemma on a non-monotonic function*) is valid: if $f \notin M$, substitutions of the functions 0, 1, x for its variables can lead to the function $\bar{x}$.

The vertex $\tilde{\alpha}$ of the cube B^n is called the *lower unity* (*upper zero*) of the monotonic function $f(\tilde{x}^n)$ if $f(\tilde{\alpha}) = 1$ (resp. $f(\tilde{\alpha}) = 0$). For any vertex $\tilde{\beta}$, it follows from $\tilde{\beta} < \tilde{\alpha}$ that $f(\tilde{\beta}) = 0$ (resp. $f(\tilde{\beta}) = 1$ follows from the condition $\tilde{\alpha} < \tilde{\beta}$).

2.5.1. Which of the following functions are monotonic:

(1) $x \to (x \to y)$;
(2) $x \to (y \to x)$;
(3) $xy\,(x \oplus y)$;
(4) $xy \oplus yz \oplus zx \oplus z$;
(5) $f(\tilde{x}^3) = (00110111)$;
(6) $f(\tilde{x}^3) = (01100111)$;
(7) $f(\tilde{x}^4) = (0001010101010111)$;
(8) $f(\tilde{x}^4) = (0000000010111111)$?

2.5.2. For what values of n is the function $f(\widetilde{x^n})$ monotonic:

(1) $f(\widetilde{x^n}) = \sum\limits_{1 \leqslant i < j \leqslant n} x_i x_j$;

(2) $f(\widetilde{x^n}) = x_1 x_2 \ldots x_{n-1} \bar{x}_n \to (x_1 \oplus x_2 \oplus \ldots \oplus x_n)$;

(3) $f(\widetilde{x^n}) = x_1 x_2 \ldots x_n \oplus \sum\limits_{i=1}^{n} x_1 \ldots x_{i-1} x_{i+1} \ldots x_n$?

2.5.3. Prove that for the monotonic functions $f(\widetilde{x^n})$ the following expansion formulas are valid:

$$f(\widetilde{x^n}) = x_i f_1^i(\widetilde{x^n}) \vee f_0^i(x^n);\ f(\widetilde{x^n}) = (x_i \vee f_0^i(\widetilde{x^n})) \,\&\, f_1^i(\widetilde{x^n}).$$

2.5.4. Prove that for any monotonic function f other than a constant, there exist disjunctive normal forms and conjunctive normal forms that do not contain negations of variables and that represent f.

2.5.5. For how many monotonic functions $f(\widetilde{x^3})$ are the relations $f(0, 1, 1) = f(1, 0, 1) = 1$, $f(0, 0, 1) = 0$ valid? How many of these functions belong to the set $M \setminus S$? Do they include functions with apparent variables?

2.5.6. Let $f(\widetilde{x^4}) \in S \cap M$, $f(0, 0, 1, 1) = f(0, 1, 1, 0) = f(1, 0, 1, 0)$ and let $f(\widetilde{x^4})$ depend essentially on all its variables. Compile the table for the $f(\widetilde{x^4})$.

2.5.7. Prove that if f is not a constant and $f \vee f^*$ a constant, then $f \notin M \cup S$.

2.5.8. (1) Is it true that if $f(\widetilde{x^n})$ is monotonic, the condition

$$\tilde{\alpha}, \tilde{\beta} \in B^n,\ \nu(\tilde{\beta}) > \nu(\tilde{\alpha}),\ \|\tilde{\beta}\| > \|\tilde{\alpha}\|, f(\tilde{\alpha}) = 1,$$

leads to the equality $f(\tilde{\beta}) = 1$?

(2)* Suppose that for all k $(0 \leqslant k < n)$ the conditions $f(\widetilde{\alpha^n}) = 1$, $\nu(\widetilde{\alpha^n}) \leqslant 2^{n-1} - 2^k$, $\nu(\widetilde{\beta^n}) = \nu(\widetilde{\alpha^n}) + 2^k$ lead to the relation $f(\widetilde{\beta^n}) = 1$. Prove that $f(\widetilde{x^n}) \in M$.

2.5.9. Enumerate all functions $f(\widetilde{x^4}) \in M$ satisfying the following conditions:

(1) $f(1, 0, 0, 0) = 1$, $f(0, 1, 1, 1) = 0$;
(2) $f(1, 0, 0, 0) = 1$, $f(\tilde{x}^4) \in L$;
(3) $f(0, 1, 0, 0) \neq f(1, 0, 1, 1)$, $f(\tilde{x}^4)$ is symmetric;
(4) $f(1, 0, 0, 1) = 0$, $f \in S$.

2.5.10. Show that if $f(\tilde{x}^n)$ is non-monotonic, there exist two vectors $\tilde{\alpha}$, $\tilde{\beta}$ in B^n that differ exactly in one coordinate, and for which $\tilde{\alpha} < \tilde{\beta}$ but $f(\tilde{\alpha}) > f(\tilde{\beta})$.

2.5.11. Show that the function f, which depends essentially on at least two variables, is monotonic if and only if any (proper) subfunction of the function f is monotonic.

2.5.12. Give an example of a non-monotonic function $f(\tilde{x}^n)$ whose each subfunction of the type $f_\sigma^i(\tilde{x}^n)$, $i = \overline{1, n}$, $\sigma \in \{0, 1\}$, is monotonic. How many of such functions depend (not always) essentially on the variables of the set $\{x_1, x_2, \ldots, x_n\}$?

2.5.13. Show that the function $f(\tilde{x}^n)$ is monotonic if and only if for any k ($k = \overline{1, n-1}$), for any non-empty subset $\{i_1, \ldots, i_k\} \subset \{1, \ldots, n\}$ and for any tuples $\tilde{\sigma} = (\sigma_1, \ldots, \sigma_k)$ and $\tilde{\tau} = (\tau_1, \ldots, \tau_k)$ with $\tilde{\sigma} \leqslant \tilde{\tau}$, the following relation is satisfied:

$$f_{\sigma_1 \ldots \sigma_k}^{i_1, \ldots, i_k}(\tilde{x}^n) \vee f_{\tau_1 \ldots \tau_k}^{i_1, \ldots, i_k}(\tilde{x}^n) = f_{\tau_1 \ldots \tau_k}^{i_1, \ldots, i_k}(\tilde{x}^n).$$

2.5.14. Prove that no simple implicant of a monotonic function contains a negation of variables.

2.5.15. The monotonic elementary conjunction K generated by the set of variables $x_1, x_2, \ldots, x_n$ is said to correspond to the vector $(\alpha_1, \alpha_2, \ldots, \alpha_n)$ in B^n when $\alpha_i = 1$ for any $i = \overline{1, n}$ if and only if x_i is included in K. Prove that if the vector $\tilde{\alpha}$ is the lower unity of the monotonic function $f(\tilde{x}^n)$, the Zhegalkin polynomial contains as one of the terms the conjunction K corresponding to the vector $\tilde{\alpha}$.

2.5.16*. Let $f(\tilde{x}^n)$ be a monotonic symmetric function, such that $N_f = \{\tilde{\alpha}\colon \|\tilde{\alpha}\| \geqslant k,\ \tilde{\alpha} \in B^n\}$. For what values of n and k does its Zhegalkin polynomial contain at least one elementary conjunction of rank $k+2$ as one of its terms?

2.5.17. Show that there are no monotonic self-dual functions with exactly two lower unities.

2.5.18. Can monotonic self-dual functions have three lower unities and depend essentially on:

(1) three variables, (2) more than three variables?

2.5.19*. Show that the maximum number of lower unities of the monotonic function $f(\widetilde{x}^n)$ is equal to $\binom{n}{[n/2]}$.

2.5.20. Show that if the values of a monotonic function are arbitrarily replaced by zeros at some of its lower unities, the resulting function will also be monotonic.

2.5.21. Show that $|M^n| \geqslant 2^{\binom{n}{[n/2]}}$.

2.5.22. Find the number of functions in each of the following sets:

(1) $M^n \diagdown (T_1^n \cap T_0^n)$; (4) $M^n \cap L^n \cap S^n$;

(2) $M^n \diagdown (T_1^n \cup T_0^n)$; (5) $L^n \diagdown (M^n \cup S^n)$.

(3) $M^n \cap L^n$;

2.5.23*. Show that

(1) $|S^n \cap M^n| < |M^{n-1}|$ for $n \geqslant 1$;

(2) $|M^n| < |M^{n-1}|^2$ for $n \geqslant 1$;

(3) $|M^n| \leqslant |M^{n-2}|^2 \, 2^{2n-2}$ for $n \geqslant 2$.

2.5.24. Let $m(n)$ be the number of monotonic functions depending on the variables $x_1, x_2, \ldots, x_n$. Show that $m(1) = 3$, $m(2) = 6$, $m(3) = 20$, $m(4) = 168$.

2.5.25. Count the number $m_e(n)$ of monotonic functions $f(\widetilde{x}^n)$ depending essentially on n variables for $n = \overline{1, 4}$.

2.5.26*. Prove that $|M^n| < |S^n|$ for $n \geqslant 4$.

2.5.27. What is the number of monotonic self-dual functions $f(\widetilde{x}^4)$ that depend essentially on all their variables?

2.5.28*. Using the fact (see problem 1.1.18) that a cube B^n can be divided into $\binom{n}{[n/2]}$ non-intersecting ascending chains, prove that

(1) $|M^n| \leqslant (n+1)^{\binom{n}{[n/2]}}$;

(2) $|M^n| \leqslant (n-1)^{\binom{n}{[n/2]}} + 2$.

2.5.29*. With the help of the problem 1.1.18, show that $|M^n| \leqslant 3^{\binom{n}{[n/2]}}$.

2.5.30. Prove that the number of monotonic functions $f(\tilde{x}^n)$ each of whose lower unities has a weight not exceeding k $(0 \leqslant k \leqslant n/2)$ is not more than $1 + k^{\binom{n}{k}}$.

2.5.31. Let $(A, \leqslant)$ be a partially ordered set. The function f, defined on the set A and assuming its values from the set $\{0, 1\}$, is called *monotonic* if for any α and β in A, such that $\alpha \leqslant \beta$, the relation $f(\alpha) \leqslant f(\beta)$ is satisfied. Let $m(A, \leqslant)$ be the number of different monotonic functions defined on A. Find $\min\limits_{|A|=n} m(A, \leqslant)$, $\max\limits_{|A|=n} m(A, \leqslant)$ and find the partially ordered sets with n elements on which these minimum and maximum values are attained.

2.5.32*. Give an example of a sequence of monotonic functions $f_n(\tilde{x}^{2n})$, $n = 1, 2, \ldots$, such that the number of lower unities of the function $f_n(\tilde{x}^{2n})$ exceeds the number of upper zeros by a factor of $2^n/n$.

2.5.33. Prove that if the number of lower unities of the monotonic function $f(\tilde{x}^n)$ is not less than 2, the function depends essentially on at least two variables.

2.5.34*. Let $t(f)$ be the number of lower unities of the monotonic function f and let $p(f)$ be the number of its essential variables. Prove that $p(f) \geqslant \log_2 t(f) - \log_2 \log_2 t(f)$.

2.5.35. Let $f \in M^n$, and let $m_k(f)$ be the number of vectors $\tilde{\alpha}$ in B_k^n, such that $f(\tilde{\alpha}) = 1$, $q_k(f) = m_k(f) \Big/ \binom{n}{k}$. Show that $q_{k-1}(f) \leqslant q_k(f)$, $k = \overline{1, n}$.

2.5.36*. The function $\varphi(\tilde{x}^n)$, defined on B^n and assuming arbitrary real values, is called a *generalized monotonic function* if it follows from $\tilde{\alpha} \leqslant \tilde{\beta}$ that $\varphi(\tilde{\alpha}) \leqslant \varphi(\tilde{\beta})$. Prove that a generalized monotonic function can be presented as a linear combination of monotonic Boolean functions of the type

$$\varphi(\tilde{x}^n) = c + \sum_{f(\tilde{x}^n) \in M \cap T_0} a_f \cdot f(\tilde{x}^n),$$

where c is real and $a_f \geqslant 0$.

2.5.37*. Let $\varphi(\tilde{x}^n)$ be a generalized monotonic function and $q_k(\varphi) = \binom{n}{k}^{-1} \sum_{\tilde{\alpha} \in B_k^n} \varphi(\tilde{\alpha})$. Show that $g_{k-1}(\varphi) < q_k(\varphi)$, $k = \overline{1, n}$.

2.5.38*. Show that if $f(\tilde{x}^n) \notin M$ $(n \geqslant 4)$, an identification of variables in it can lead to a non-monotonic function depending on not more than three variables.

2.5.39. Is it possible to obtain $\bar{x}$ from $x\bar{y}\bar{z} \vee t\ (xy \to z)$:

(1) by identification of variables;

(2) by identification of variables and substitution of the constant 0 for certain variables;

(3) by identification of variables and substitution of constants?

2.5.40*. Show that if $f(\tilde{x}^n) \notin M \cup S$ $(n \geqslant 3)$, the identification of variables in it can lead to a non-monotonic non-self-dual function that depends essentially on two variables.

2.5.41. Show that if $f \in M$, $f^* \in M$.

2.5.42. Show that any monotonic function is contained in not more than two classes in T_0, T_1 and L.

2.5.43. Show that none of the classes T_0, T_1, S or L contains M.

2.5.44. Can the function $xy \vee yz \vee zx$ lead to the function xy with the help of the superposition operation?

2.5.45. Is it possible to obtain 0 from the functions xy, $x \vee y$ and 1?

2.5.46. Show that the set $\{0, 1, xy, x \vee y\}$ forms a basis in M.

2.5.47. Isolate all subsets from the set $\{0, 1, xy, x \vee y, xy \vee z, xy \vee yz \vee zx\}$ which are bases in M.

2.5.48*. Show that any basis in M contains not more than four and not less than three functions.

2.5.49. Show that any basis in M consisting of three functions contains a function that depends essentially on three or more variables.

2.5.50. Give examples of bases in the following classes: (1) $T_0 \cap M$; (2) $T_1 \cap M$; (3) $L \cap M$.

2.5.51. Let $f(\tilde{x}^n) \in M$ $(n \geqslant 3)$. Prove the validity of the representation

$$f(\tilde{x}^n) = m(f(x_1, x_1, x_3, x_4, \ldots, x_n), f(x_1, x_2, x_2, x_4, \ldots, x_n), f(x_3, x_2, x_3, x_4, \ldots, x_n)),$$

where $m(x, y, z) = xy \vee yz \vee zx$.

2.5.52. Show that

(1) $xy \vee yz \vee zx$ forms a basis in $M \cap S$,

(2) any function from $M \cap S$, which depends essentially on more than one variable, forms a basis in $M \cap S$.

2.5.53*. Let $\mathscr{D}$ be a set consisting of all monotonic elementary disjunctions, and let $\mathscr{K}$ be a set of all monotonic elementary conjunctions. Show that only the sets $\mathscr{D} \cup \{0, 1\}$, $\mathscr{K} \cup \{0, 1\}$, $M \cap T_0$, $M \cap T_1$ are precomplete in M.

2.5.54*. Let $f(\tilde{y}^{2^n}) = \mathop{\&}\limits_{k=0}^{n-1} \mathop{\&}\limits_{i=0}^{2^n-2^k-1} (y_i \to y_{i+2^k})$. Show that $|N_f| = |M^n|$.

2.5.55. Let $f(\tilde{x}^n)$ and $g(\tilde{x}^n)$ be monotonic functions and $h = f \,\&\, g$. Prove that

$$|N_h| \geqslant |N_f|\,|N_g|\,2^{-n}.$$

2.6. Completeness and Closed Classes

The following *completeness criterion* is valid in P_2:

Theorem (Post). *A system A is complete in P_2 if and only if it is not contained entirely in one of the classes T_0, T_1, L, S,* and *M.*

The function $f(\tilde{x}^n)$ is called *Sheffer's function* if it forms a basis in P_2.

Let the function $f(\tilde{x}^n)$ depend essentially on all its variables. By $\mathfrak{N}(f(\tilde{x}^n))$ we denote the set of all such functions which are obtained from the function $f(\tilde{x}^n)$ through identification of variables, the function f does not belong to the set $\mathfrak{N}(f)$. If $n < 2$, then by definition $\mathfrak{N}(f(\tilde{x}^n)) \neq \varnothing$. The set $\mathfrak{N}(f(\tilde{x}^n))$ is called the *hereditary system of the function* $f(\tilde{x}^n)$. The function f is called *irreducible* if $[\mathfrak{N}(f)] \neq [f]$. The basis $\mathfrak{B}$ of the closed class K is called *prime* if the substitution of its hereditary system for an arbitrary function f from $\mathfrak{B}$ leads to an incomplete system in K. The function f not belonging to the closed class K is called *a prime function with respect to* K if its hereditary system $\mathfrak{N}(f)$ is contained in K.

2.6.1. Show that P_2 does not contain precomplete classes other than the classes T_0, T_1, L, S, M.

2.6.2. Using the completeness criterion, find out if the system A is complete in the following cases:

(1) $A = \{x \to y,\ x \to \bar{y} \cdot z\}$;
(2) $A = \{x \cdot \bar{y},\ \bar{x} \sim yz\}$;
(3) $A = \{0, 1, x(y \sim z) \vee \bar{x}(y \oplus z)\}$;
(4) $A = \{(01101001),\ (10001101),\ (00011100)\}$;
(5) $A = \{(0010),\ (1010110111110011)\}$;
(6) $A = (S \setminus M) \cup (L \setminus (T_0 \cup T_1))$;
(7) $A = (S \cap M) \cup (L \setminus M) \cup (T_0 \setminus S)$;
(8) $A = (M \setminus (T_0 \cap T_1)) \cup (L \setminus S)$.

2.6.3. Prove that if the function f depends essentially on at least two variables and belongs to the class $S \cap M$, the system $\{0, \bar{f}\}$ is complete in P_2.

2.6.4. Is the system $A = \{f_1(\tilde{x}^n), f_2(\tilde{x}^n)\}$ complete if:

(1) $f_1 \in S \setminus M,\quad f_2 \notin L \cup S,\quad \bar{f}_1 \to f_2 \equiv 1$;
(2) $f_1 \notin T_0 \cup L,\quad f_2 \notin S,\quad f_1 \to \bar{f}_2 \equiv 1$;
(3) $f_1 \notin T_0 \cap T_1,\quad f_2 \in M \setminus T_1,\quad f_1 \to f_2 \equiv 1$?

2.6.5. Is the system $A = \{f_1(\tilde{x}^n), f_2(\tilde{x}^n), f_3(\tilde{x}^n)\}$ complete if $f_1 \notin L \cup (T_0 \cap T_1)$, $f_2 \in M \setminus L$, $f_1 \to f_2 \equiv 1$ and $f_1 \vee f_3 \equiv 1$?

2.6.6. From the complete system A in P_2, isolate all possible bases in the following cases:

(1) $A = \{(x \vee y)(\bar{x} \vee \bar{y}),\ xy \oplus z,\ (x \oplus y) \sim z,\ m(x, y, z)\}$;
(2) $A = \{1,\ \bar{x},\ xy(y \sim z),\ x \oplus y \oplus m(x, y, z)\}$;
(3) $A = \{0,\ x \oplus y,\ (x \to y) \downarrow (y \sim z),\ (x \mid (xy)) \to \bar{z}\}$;
(4) $A = \{x \vee (x \oplus y) \vee z,\ (x \sim y) \sim z,\ xy \oplus zu,\ m(x, \bar{y}, \bar{z})\}$.

2.6.7. Give three examples of bases in P_2 each containing one, two, three and four functions.

2.6.8. Enumerate all different bases in P_2 containing only such functions that depend essentially on two variables x and y (bases are considered to be different if one of these cannot be reduced to the other through a redesignation of variables).

2.6.9. Find out if the set $\mathfrak{A}$ can be extended to a basis in P_2 in the following cases:

(1) $\mathfrak{A} = \{x \oplus y, m(x, y, z)\}$; (3) $\mathfrak{A} = M \setminus (T_0 \cup T_1)$;

(2) $\mathfrak{A} = \{x \sim y, x \vee yz\}$; (4) $\mathfrak{A} = L \cap M$.

2.6.10*. Does P_2 contain a basis consisting of four functions f_1, f_2, f_3 and f_4, such that $f_1 \notin T_0$, $f_2 \notin T_1$, $f_3 \notin L$ and $f_4 \notin S$?

2.6.11. Using the operations in the theory of sets, express the closure of the set $\mathfrak{A}$ through known closed classes T_0, T_1, L, S, M and P_2 if:

(1) $\mathfrak{A} = P_2 \setminus (T_0 \cup T_1 \cup L \cup S \cup M)$;
(2) $\mathfrak{A} = M \setminus (T_0 \cup L)$; (6) $\mathfrak{A} = S \setminus (T_0 \setminus T_1)$;
(3) $\mathfrak{A} = M \setminus (T_0 \cap T_1)$; (7) $\mathfrak{A} = L \setminus (T_0 \cup T_1)$;
(4) $\mathfrak{A} = M \setminus L$; (8) $\mathfrak{A} = T_0 \setminus T_1$;
(5) $\mathfrak{A} = T_0 \cap (L \setminus S)$; (9) $\mathfrak{A} = (T_0 \cap T_1) \setminus M$.

2.6.12. Which of the relations $\supset, \subset, \supseteq, \subseteq, =, \not\gtrless$ is valid for classes K_1 and K_2 (the relation $\not\gtrless$ means that none of the other five relations is satisfied):

(1) $K_1 = [x \vee (x \oplus y) \vee z]$, $K_2 = [x \vee y, x \oplus y]$;
(2) $K_1 = [x \sim y, x \vee yz]$, $K_2 = [x \oplus yz]$;
(3) $K_1 = [xy, x \oplus y]$, $K_2 = [x \rightarrow y, x \sim \bar{y}]$;
(4) $K_1 = [1, x \vee y]$, $K_2 = [x \oplus y, x \vee \bar{y}z]$;
(5) $K_1 = [x \oplus y, x \sim yz]$, $K_2 = [m(y, x, z), x \oplus y \oplus z \oplus 1, xy \oplus z]$;
(6) $K_1 = [x \rightarrow y]$, $K_2 = [x \oplus y, m(x, y, z)]$?

2.6.13. (1) Prove that $P_2(X^2)$ contains exactly two Sheffer's functions.

(2) Find the number of Sheffer's functions in $P_2(X^3)$.

2.6.14. (1) Prove that if $f \notin T_0 \cup T_1 \cup S$, then f is a Sheffer function.

(2)* Find the number of Sheffer's functions in $P_2(X^n)$.

2.6.15. Prove that by identification of variables, we can obtain from Sheffer's functions depending essentially on at least three variables a Sheffer function depending essentially on two variables.

2.6.16. For what values of n ($n \geqslant 2$) is the function f a Sheffer function if:

(1) $f = 1 \oplus \sum_{1 \leqslant i < j \leqslant n} x_i x_j$;

(2) $f = 1 \oplus x_1 x_2 \oplus x_2 x_3 \oplus \ldots \oplus x_{n-1} x_n \oplus x_n x_1$;

(3) $f = (x_1 \to x_2) \oplus (x_2 \to x_3) \oplus \ldots \oplus (x_{n-1} \to x_n) \oplus (x_n \to x_1)$;

(4) $f = (x_1 | x_2) \oplus (x_2 | x_3) \oplus \ldots \oplus (x_{n-1} | x_n) \oplus (x_n | x_1)$;

(5) $f = 1 \oplus (x_1 \to x_2)(x_2 \to x_3) \ldots (x_{n-1} \to x_n)(x_n \to x_1)$;

(6) $f = \bigvee_{1 \leqslant i_1 < i_2 < \ldots < i_{]n/2[} \leqslant n} \bar{x}_{i_1} \bar{x}_{i_2} \ldots \bar{x}_{i_{]n/2[}}$?

2.6.17. The function f does not belong to the set $T_1 \cup M$ and has exactly one value equal to zero. Prove that it either is a Sheffer function, or depends essentially on one variable.

2.6.18. Give an example of a Sheffer function that depends essentially on the smallest possible number of variables and assumes the value unity on exactly half the tuples of values of variables.

2.6.19. Give an example of a function $f(\tilde{x}^n)$ such that for any k ($1 \leqslant k \leqslant n-2$) and for any subset i_1, $i_2, \ldots, i_k$ in $\{1, 2, \ldots, n\}$ the function

$$\frac{d}{dx_{i_1}}\left(\frac{d}{dx_{i_2}}\left(\cdots\left(\frac{d}{dx_{i_k}} f(\tilde{x}^n)\right)\cdots\right)\right)$$

is a Sheffer function.

2.6.20. Let the function f be monotonic with exactly two lower unities. Prove that $\bar{f}$ is a Sheffer function.

2.6.21. Prove that any self-dual function not belonging to the set $T_0 \cup T_1 \cup L \cup M$ forms a basis in the class S.

2.6.22. (1) Prove that any function in class L belongs to at least one of the classes T_0, T_1, S and M.

(2) Give examples of linear functions contained exactly in one of the classes T_0, T_1 and S.

2.6.23*. Give example of a function in class T_0 that does not form a basis in T_0 and does not belong to the set $T_1 \cup L \cup S \cup M$.

2.6.24. Give example of a nonlinear function $f(\tilde{x}^n)$ that depends essentially on the smallest possible number of variables and satisfies the following condition: for any number i $(1 \leqslant i \leqslant n)$, each of the functions $f_0^i(\tilde{x}^n)$ and $f_1^i(\tilde{x}^n)$ assumes the value unity at exactly 2^{n-2} tuples of values of variables.

2.6.25*. Prove that identification of variables in a nonlinear function depending essentially on $n \geqslant 4$ variables leads to a nonlinear function depending essentially on $n-1$ variables.

2.6.26*. Prove that identification of variables in a function f not belonging to the set $L \cup S$ and depending essentially on $n \geqslant 4$ variables leads to a non-self-dual nonlinear function depending essentially on $n-1$ variables. Is this statement valid for $n = 3$?

2.6.27*. Prove that identification of variables in a Sheffer function $f(\tilde{x}^n)$ depending essentially on $n \geqslant 3$ variables leads to a Sheffer function depending essentially on $n-1$ variables.

2.6.28. Find out if the following implications are true:

(1) $f \notin (T_0 \cup T_1) \setminus S \Rightarrow f \in L \cup M$;
(2) $f \notin T_0 \cup T_1 \cup M \Rightarrow f$ is a Sheffer function;
(3) $f \notin T_0 \cup S \cup M \Rightarrow f \notin (L \setminus T_1) \cap (S \setminus M)$;
(4) $f \notin L \cup S \cup M \Rightarrow f$ is a Sheffer function.

2.6.29. Let the subset $\mathfrak{A}$ from $P_2(X^n)$ contain more than 2^{2^n-1} functions. Prove that $\mathfrak{A}$ is a complete system in P_2 for $n \geqslant 2$.

2.6.30. Prove that each function from a prime basis in P_2 is a prime function with respect to precomplete class in P_2.

2.6.31. Find all pairwise non-congruent functions that are prime functions with respect to the class K:

(1) $K = T_0$; (3) $K = [0, 1, x]$;
(2)* $K = L \cap S$; (4) $K = [0, 1, x, \bar{x}]$.

2.6.32. Prove that prime functions with respect to the class K that are pairwise non-congruent, are exhausted by functions of the set $\mathfrak{A}$.

(1) $K = T_0 \cap T_1$, $\mathfrak{A} = \{0, 1, \bar{x}\}$;
(2) $K = T_0 \cap L \cap S$, $\mathfrak{A} = \{0, 1, \bar{x}, xy, x \vee y, m(x, y, z), m(x, y, \bar{z})\}$.

2.6.33. Prove that P_2 contains:

(1) only two prime bases $\{x \mid y\}$ and $\{x \downarrow y\}$ consisting of one function;

(2)* only three prime bases $\{0, 1, x \oplus y \oplus z, xy\}$, $\{0, 1, x \oplus y \oplus z, x \vee y\}$ and $\{0, 1, x \oplus y \oplus z, m(x, y, z)\}$ consisting of four functions.

Chapter Three

k-Valued Logics

3.1. Representation of Functions of k-Valued Logics Through Formulas

3.1.1. Elementary Functions of k-Valued Logics and Relations Between Them

Throughout this chapter, we shall assume that k is a natural number larger than 2. By E_k we shall denote the set $\{0, 1, \ldots, k-1\}$. The function $f(\tilde{x}^n) = f(x_1, x_2, \ldots, x_n)$ is called a *function of the k-valued logic* if on any tuple $\tilde{\alpha} = (\alpha_1, \alpha_2, \ldots, \alpha_n)$ of values of the variables $x_1, x_2, \ldots, x_n$, where $\alpha_i \in E_k$, the value $f(\tilde{\alpha})$ also belongs to the set E_k. The *set of all functions of the k-valued logic* is denoted by P_k. The concept of fictitious and essential variables, equal functions, formulas generated by a set of functions (and connectives), superposition and closure operations, closed class, basis, etc. are defined in k-valued logics in the same way as in a Boolean algebra. Hence we shall give only the definitions of such concepts that differ essentially from the corresponding concepts in P_2.

The following functions in a k-valued logic are assumed to be *elementary functions*:

constants $0, 1, \ldots, k-1$; these functions will be treated as functions depending on an arbitrary finite number of variables (including zero variables);

Post's negation: $x+1 \pmod{k}$, denoted as $\bar{x}$;

Lukasiewicz's negation: $(k-1)-x$, denoted as $\sim x$ or Nx;

characteristic function (of the first kind) of a number i:

$$j_i(x),\ i=0, 1, \ldots, k-1; \quad j_i(x) = \begin{cases} 1 & \text{if } x = i, \\ 0 & \text{if } x \neq i; \end{cases}$$

characteristic function of the second kind of a number i:

$$J_i(x), \quad i=0, 1, \ldots, k-1; \quad J_i(x) = \begin{cases} k-1 & \text{if } x = i, \\ 0 & \text{if } x \neq i; \end{cases}$$

the smaller of x or y: $\min(x, y)$ (alternatively, xy or $x \,\&\, y$);

the larger of x or y: $\max(x, y)$ (alternatively, $x \vee y$);

mod k sum: $x + y \pmod{k}$ (read as "*mod k x plus y*")[1];
mod k product: $x \cdot y \pmod{k}$ (read as "*mod k product of x and y*"[1]);
truncated difference:

$$x \dot{-} y = \begin{cases} 0 & \text{if } 0 \leqslant x < y \leqslant k-1, \\ x-y & \text{if } 0 \leqslant y \leqslant x \leqslant k-1; \end{cases}$$

implication:

$$x \supset y = \begin{cases} k-1 & \text{if } 0 \leqslant x < y \leqslant k-1, \\ (k-1)-x+y & \text{if } 0 \leqslant y \leqslant x \leqslant k-1; \end{cases}$$

joint denial: $\max(x, y)+1 \pmod{k}$, denoted by $v_k(x, y)$;
mod k difference:

$$x - y = \begin{cases} x-y & \text{if } 0 \leqslant y \leqslant x \leqslant k-1, \\ k-(y-x) & \text{if } 0 \leqslant x < y \leqslant k-1. \end{cases}$$

The functions (operations) min, max, + and · are commutative and associative. Moreover, the following relations are valid:

$$(x + y) \cdot z = (x \cdot z) + (y \cdot z)$$

called distributivity of multiplication with respect to addition;

$$\max(\min(x, y), z) = \min(\max(x, z), \max(y, z))$$

called the distributivity of the operation max with respect to the operation min;

$$\min(\max(x, y), z) = \max(\min(x, z), \min(y, z))$$

called the distributivity of the operation min with respect to the operation max;

$$\max(x, x) = x, \quad \min(x, x) = x$$

called idempotency of the operations min and max; and

$$\min(\sim x, \sim y) = \sim\max(x, y), \quad \max(\sim x, \sim y) = \sim\min(x, y)$$

called the analogs of De Morgan's rules (laws) in P_2.

---

[1] Unless otherwise stated, the symbols + and · in this chapter will denote mod k sum and product.

The following equalities are introduced by definition:

$\max(x_1, x_2, \ldots, x_{n-1}, x_n) = \max(\max(x_1, x_2, \ldots, x_{n-1}), x_n), \quad n \geqslant 3;$

$\min(x_1, x_2, \ldots, x_{n-1}, x_n) = \min(\min(x_1, x_2, \ldots, x_{n-1}), x_n), \quad n \geqslant 3;$

$$-x = \begin{cases} 0 & \text{if} \quad x = 0, \\ k - x & \text{if} \quad x \neq 0. \end{cases}$$

In view of the associative nature of mod k product, the product $x \cdot x \cdot x \cdot \ldots \cdot x$ (l cofactors, $l \geqslant 1$) is often written in the power form x^l.

3.1.1. Prove the validity of the following equalities:

(1) $-(\bar{x}) = \sim x;$

(2) $x \supset y = \sim (x \dotminus y);$

(3) $x \dotminus (x \dotminus y) = \min(x, y);$

(4) $(x \supset y) \supset y = \max(x, y);$

(5) $(x \supset y) + \bar{x} = \min(x, y);$

(6) $x \dotminus y = x - \min(x, y);$

(7) $x \dotminus y = \max(x, y) - y;$

(8) $(\sim x) \dotminus (y \dotminus x) = \sim \max(x, y);$

(9) $(\sim x) \dotminus (\sim y) = y \dotminus x;$

(10) $\sim(\bar{x} + y) = (\sim x) + (\sim y);$

(11) $\overline{\sim(\bar{x} \cdot \bar{y})} = (\sim x) \cdot \bar{y};$

(12) $\max((x + 2) \dotminus 1, J_{k-2}(x)) = \bar{x};$

(13) $\min(\sim J_{k-1}(x), (k - 2) \supset x) = \bar{x};$

(14) $\bar{x} \dotminus \bar{y} = (x \dotminus y) + \bar{x} \cdot j_{k-1}(y) + \bar{y} \cdot j_{k-1}(x);$

(15) $v_k(x, y) + \bar{x} \cdot j_{k-1}(y) + \bar{y} \cdot j_{k-1}(x) = \max(\bar{x}, \bar{y});$

(16) $\overline{\max(x, y)} + j_0(y \dotminus x) + j_{k-1}(x) \cdot y = \max(x, \bar{y});$

(17) $\overline{\min(x, y)} + J_0(y \dotminus x) - j_{k-1}(x) \cdot y = \min(\bar{x}, y);$

(18) $J_0(\max(J_0(x), J_1(x), \ldots, J_{k-2}(x))) = J_{k-1}(x);$

(19) $J_1(\max(x, 1, J_1(x), J_2(x), \ldots, J_{k-2}(x))) = J_0(x);$

(20) $x \cdot j_0(j_1(x)) + j_0(x) \cdot \overline{j_1(x)} = x + j_0(x) - j_1(x);$

(21) $J_0(x \dotminus i) \dotminus J_0(x \dotminus (i - 1)) = J_i(x), \quad i = 1, 2, \ldots, k - 1;$

(22) $(\sim((\sim x) \dotminus 1)) \dotminus (\ldots(((k-1) \dotminus \underbrace{(\sim x)) \dotminus (\sim x)) \dotminus \ldots \dotminus (\sim x))}_{k-1 \text{ times}} = \bar{x};$

(23) $(\ldots(((k - 1) \dotminus j_0(x)) \dotminus j_0(x \dotminus 1)) \dotminus \ldots \dotminus j_0(x \dotminus (k - 3))) \dotminus ((k - 1) j_0(\sim x)) = \bar{x}.$

3.1.2. Prove that the function f in P_k is generated by the set of functions A ($A \subset P_k$) as a result of superposition operation.

(1) $f = J_1(x)$, $A = \{J_0(x),\ J_2(x),\ \max(x, y)\}$, $k = 3$;

(2) $f = \sim x$, $A = \{J_0(x), J_1(x), \min(x, y), \max(x,y)\}$, $k = 3$;

(3) $f = \bar{x}$, $A = \{1, x^2, J_1(x), \max(x, y)\}$, $k = 3$;

(4) $f = j_0(x)$, $A = \{x - 1,\ x^2\}$, $k = 3,\ 5$;

(5) $f = j_1(x)$, $A = \{x \cdot y + x - y^2 + 1\}$, $k = 3, 5$;

(6) $f = \sim x$, $A = \{1,\ x \cdot \bar{y}\}$, $k = 3,\ 5$;

(7) $f = \bar{x}$, $A = \{3,\ j_0(x),\ x \dot{-} y\}$, $k = 4$;

(8) $f = \sim x$, $A = \{x + 2, J_0(x), J_1(x), \max(x, y), x \cdot y\}$, $k = 4$;

(9) $f = j_4(x)$, $A = \{x \dot{-} 1,\ J_2(x)\}$, $k = 6$;

(10) $f = j_5(x)$, $A = \{x + 2,\ x^2,\ J_3(x)\}$, $k = 6$;

(11) $f = j_1(x)$, $A = \{\bar{x},\ -x,\ J_{k-1}(x)\}$;

(12) $f = J_{k-1}(x)$, $A = \{\sim x,\ x \dot{-} y\}$;

(13) $f = J_{k-2}(x)$, $A = \{k - 1,\ x + 2,\ x \dot{-} y\}$;

(14) $f = j_0(x)$, $A = \{1,\ \sim x,\ x \dot{-} 2y\}$;

(15) $f = \bar{x}$, $A = \{1,\ \sim x,\ x \dot{-} y\}$.

3.1.3. Prove that if α belongs to E_k and is coprime to k, each function $J_i(x)$, $0 \leqslant i \leqslant k - 2$ can be presented as a superposition generated by the set $\{x + \alpha,\ J_{k-1}(x)\}$.

3.1.4. Show that the function φ from P_k can be represented by a formula generated by the set $\{0, 1, \ldots, k-1, x \dot{-} 2y\}$ if

(1) $\varphi = j_{k-1}(x)$, $k = 2m$ $(m \geqslant 2)$;

(2) $\varphi = j_0(x)$, $k = 2m + 1$ $(m \geqslant 1)$.

3.1.5. Let $h_1(x) = \sim x$, $h_{i+1}(x) = x \supset h_i(x)$, $i \geqslant 1$. Prove that $\sim h_{k-1}(x) = J_{k-1}(x)$.

3.1.6. For what values of k $(k \geqslant 3)$ are the functions x^2, x^3 and x^4 pairwise different?

3.1.7. Let $k = 3, 4, \ldots, 9, 10$. How many different functions in P_k, depending only on the variable x, can be presented in the form x^l ($l \geqslant 1$ and the power is taken over mod k) for each k?

3.1.8. Prove that each function $f(x)$ in P_k can be presented as a superposition generated by the set $\{1, J_{k-1}(x), x+y\}$.

3.1.9. The functions $f_1(x)$ and $f_2(x)$ in P_3 satisfy the following conditions: $f_1(x) \not\equiv \text{const}$, $f_1(E_3) \neq E_3$ and $f_2(E_3) = E_3$.

Prove that the function $g(x) = f_1(x) + f_2(x)$, where mod 3 sum is taken, does not assume at least one value in E_3, i.e. $g(E_3) \neq E_3$.

3.1.10. Let the function $f(x)$ in P_3 be presented in the form $a_0x^2 + a_1x + a_2$ (mod 3 sum and product are taken, and a_0, a_1, a_2 belong to E_3). What values can be assumed by the coefficients a_0, a_1 and a_2 if the function $f(x)$ is known not to assume at least one value in E_3, i.e. $f(E_3) \neq E_3$.

3.1.2. Decomposition of Functions of k-Valued Logics into First and Second Forms

Any function $f(x_1, x_2, \ldots, x_n)$ in P_k, $n \geqslant 1$, can be presented in the *first form*, which is an analog of the perfect disjunctive normal form for Boolean functions:

$$f(x_1, x_2, \ldots, x_n) = \max_{\tilde{\sigma}} \{\min (f(\sigma_1, \sigma_2, \ldots, \sigma_n), J_{\sigma_1}(x_1), J_{\sigma_2}(x_2), \ldots, J_{\sigma_n}(x_n))\}.$$

Here the maximum is taken over all tuples $\tilde{\sigma} = (\sigma_1, \sigma_2, \ldots, \sigma_n)$ of values of the variables $x_1, x_2, \ldots, x_n$.

Another representation of the functions of k-valued logic, called the *second form*, is also possible:

$$f(\tilde{x}^n) = \sum_{\tilde{\sigma}} f(\tilde{\sigma}) \cdot j_{\sigma_1}(x_1) \cdot \ldots \cdot j_{\sigma_n}(x_n),$$

where the sum is taken over all tuples $\tilde{\sigma} = (\sigma_1, \ldots, \sigma_n)$ of values of variables $x_1, x_2, \ldots, x_n$ (mod k sum and product are taken).

Example. Present the function $f(x, y) = \max (j_0(x) \times j_0(y), x \cdot (j_1(y) + 2j_2(y)))$ in P_3 in the first and second forms.

Solution. First of all, we compile the table of values of the function f:

y \ x	0	1	2
0	1	0	0
1	0	1	2
2	0	2	1

With the help of this table, we obtain the first and second forms of the function f.

$f(x, y) = \max\{\min(1, J_0(x), J_0(y)), \min(0, J_0(x), J_1(y)), \min(0, J_0(x), J_2(y)), \min(0, J_1(x), J_0(y)), \min(1, J_1(x), J_1(y)), \min(2, J_1(x), J_2(y)), \min(0, J_2(x), J_0(y)), \min(2, J_2(x), J_1(y)), \min(1, J_2(x), J_2(y))\}$.

Using simple manipulations, we get

$f(x, y) = \max\{\min(1, J_0(x), J_0(y)), \min(1, J_1(x), J_1(y)), \min(J_1(x), J_2(y)), \min(J_2(x), J_1(y)), \min(1, J_2(x), J_2(y))\}$.

$f(x, y) = 1 \cdot j_0(x) \cdot j_0(y) + 0 \cdot j_0(x) \cdot j_1(y) + 0 \cdot j_0(x) \times j_2(y) + 0 \cdot j_1(x) \cdot j_0(y) + 1 \cdot j_1(x) \cdot j_1(y) + 2 \cdot j_1(x) \cdot j_2(y) + 0 \cdot j_2(x) \cdot j_0(y) + 2 \cdot j_2(x) \cdot j_1(y) + 1 \cdot j_2(x) \cdot j_2(y) = j_0(x) \times j_0(y) + j_1(x) \cdot j_1(y) + 2 \cdot j_1(x) \cdot j_2(y) + 2 \cdot j_2(x) \cdot j_1(y) + j_2(x) \cdot j_2(y)$.

3.1.11. For a given k, present the function f in the first and second forms (simplify the obtained expressions).

(1) $f = \bar{x}$, $k = 3$;
(2) $f = \sim x$, $k = 4$;
(3) $f = -j_0(x)$, $k = 5$;
(4) $f = 2 \cdot J_1(x)$, $k = 6$;
(5) $f = J_2(x^2 + x)$, $k = 5$;
(6) $f = (\sim x)^2 + x$, $k = 4$;
(7) $f = 3 \cdot j_1(x) - j_3(x)$, $k = 4$;
(8) $f = x + 2y$, $k = 3$;
(9) $f = \max(x, y)$, $k = 3$;
(10) $f = x \dot{-} y^2$, $k = 3$;
(11) $f = x^2 \cdot y$, $k = 3$;
(12) $f = x \cdot \bar{y}$, $k = 4$.

3.1.12. Prove the validity of the following relation, which is the analog of the perfect conjunctive normal form of the functions in P_k:

$$f(\tilde{x}^n) = \min_{\tilde{\sigma}} \{\max (f(\sigma_1, \sigma_2, \ldots, \sigma_n), \sim J_{\sigma_1}(x_1), \sim J_{\sigma_2}(x_2), \ldots, \sim J_{\sigma n}(x_n))\}$$

(here, $n \geqslant 1$, and the minimum is taken over all tuples $\tilde{\sigma} = (\sigma_1, \sigma_2, \ldots, \sigma_n)$ of values of the variables $x_1, x_2, \ldots, x_n$).

3.2. Closed Classes and Completeness in k-Valued Logics

3.2.1. Some Closed Classes of k-Valued Logics. Representation of Functions in P_k Through Mod k Polynomials

Let $\mathscr{E}$ be a subset of the set E_k. The function $f(\tilde{x}^n)$ in P_k $(n \geqslant 1)$ is said to *preserve the set* $\mathscr{E}$ if it assumes a value $f(\tilde{\alpha}^n)$, also belonging to $\mathscr{E}$, on any tuple $\tilde{\alpha}^n = (\alpha_1, \alpha_2, \ldots, \alpha_n)$ such that $\alpha_i \in \mathscr{E}$ $(i = 1, 2, \ldots, n)$. For $n = 0$, it is assumed by definition that the function $f \equiv a$ $(a \in E_k)$ preserves the set $\mathscr{E}$ only if $a \in \mathscr{E}$. The *set of all functions in* P_k *preserving the set* $\mathscr{E}$ *is a closed class*, denoted by $T(\mathscr{E})$ and called the *class preserving the set* $\mathscr{E}$. If $\mathscr{E}$ is a proper subset of the set E_k (i.e. $\varnothing \subsetneqq \mathscr{E} \subsetneqq E_k$), $T(\mathscr{E}) \neq P_k$ (see Problem 3.2.3).

Let $D = \{\mathscr{E}_1, \mathscr{E}_2, \ldots, \mathscr{E}_s\}$ denote the partition of a set E_k, i.e. $E_k = \bigcup_{i=1}^{s} \mathscr{E}_i$, $\mathscr{E}_i \neq \varnothing$ for $i = 1, 2, \ldots, s$ and $\mathscr{E}_i \cap \mathscr{E}_j = \varnothing$ for $i \neq j$. *Elements* a *and* b are said to be *equivalent with respect to the partition* D (notation $a \sim b \pmod D$) if a and b belong to a certain (the same!) subset $\mathscr{E}_j$ of the partition D. *Two tuples* $\tilde{\alpha}^n$ *and* $\tilde{\beta}^n$ are called *equivalent with respect to the partition* D (notation $\tilde{\alpha}^n \sim \tilde{\beta}^n \pmod D$) if $\alpha_i \sim \beta_i \pmod D$ for $i = 1, 2, \ldots, n$. The function $f(\tilde{x}^n)$ in P_k $(n \geqslant 1)$ is said to *preserve the partition* D if for any tuples $\tilde{\alpha}^n$ and $\tilde{\beta}^n$ the equivalence $\tilde{\alpha}^n \sim \tilde{\beta}^n \pmod D$ leads to the equivalence

$f(\tilde{\alpha}^n) \sim (\tilde{\beta}^n) \pmod D$. It is assumed by definition that any constant function (i.e., a function depending on zero variables of the type $f = a$, $a \in E_k$) preserves any partition D. The *set of all functions in* P_k *preserving the partition* $D = \{\mathscr{E}_1, \mathscr{E}_2, \ldots, \mathscr{E}_s\}$ is a *closed class*, denoted by $U(D)$ or $U(\mathscr{E}_1, \mathscr{E}_2, \ldots, \mathscr{E}_s)$ and called the *class preserving the partition* D. If $s \neq 1$ and $s \neq k$, the class $U(D)$ is different from P_k (see Problem 3.2.4).

The function $f(\tilde{x}^n)$ in P_k $(n \geqslant 0)$ is called *linear* if it can be presented in the form $a_0 + a_1 \cdot x_1 + \ldots + a_n \cdot x_n$, where $a_j \in E_k$, $j = 0, 1, \ldots, n$, and the sum and products are taken over mod k. The *set of all linear functions in* P_k forms a *closed class of linear functions*, denoted by L_k (or L). The class L_k differs from P_k for all $k \geqslant 3$. A *mod k polynomial* of variables $x_1, x_2, \ldots, x_n$ is an expression of the type $a_0 + a_1 \cdot X_1 + \ldots + a_m \cdot X_m$, where the coefficients a_i belong to the set E_k, and X_j is either a variable in $\{x_1, x_2, \ldots, x_n\}$ or a product of variables in this set $(j = 1, \ldots, m)$. A certain function from P_k can *be presented by a mod k polynomial* if there exists such a polynomial equal to this formula. The set of all functions in P_k that can be presented by mod k polynomials (or, in short, the set of all mod k polynomials) forms a closed class in P_k.

Theorem. *Each function in* P_k *can be represented by a mod k polynomial if and only if k is a prime number* (in other words, the *set of mod k polynomials in* P_k *is complete in* P_k *if and only if k is a prime number*).

If k is a composite number, then P_k contains functions that can be represented by polynomials, as well as functions that cannot (for example, the constant functions $0, 1, \ldots, k-1$ and "polynomial" functions x, x^2, $x \cdot y$, $x + y$ can be represented by polynomials, while the functions $j_0(x)$, $\max(x, y)$, $\min(x, y)$, $x \dot{-} y$ cannot).

Example 1. Represent the function $f(x) = x^2 \dot{-} x$ in P_5 by a mod 5 polynomial.

Solution. Let us first represent the function $f(x)$ in the second form:

$$f(x) = \sum_{\sigma} f(\sigma) \cdot j_\sigma(x) = f(0) \cdot j_0(x) + f(1) \cdot j_1(x) + f(2) \cdot j_2(x)$$
$$+ f(3) \cdot j_3(x) + f(4) \cdot j_4(x) = 0 \cdot j_0(x) + 0 \cdot j_1(x) + 2 \cdot j_2(x)$$
$$+ 1 \cdot j_3(x) + 0 \cdot j_4(x) = 2 \cdot j_2(x) + j_3(x)$$

Then we shall use the fact that $j_0(x) = 1 - x^{k-1}$ if k is a prime number ($k \geqslant 3$) and that $j_i(x) = j_0(x - i)$, $i = 1, \ldots, k - 1$. (Here, as usual, we take mod k difference and mod k power). We have

$$j_2(x) = 1 - (x - 2)^4 = -x^4 - 2x^3 + x^2 + 2x,$$
$$j_3(x) = 1 - (x - 3)^4 = 1 - (x + 2)^4 = -x^4 + 2x^3 + x^2 - 2x.$$

Consequently, $x^2 \dot{-} x = 2x^4 - 2x^3 - 2x^2 + 2x$. (This expression can be written in a more compact form $2x \cdot (x - 1)^2 \cdot (x + 1)$.)

The polynomial representing the function $f(x)$ can also be determined by using the *method of indeterminate coefficients*. Let $f(x) = a_0 + a_1 \cdot x + a_2 \cdot x^2 + a_3 \cdot x^3 + a_4 \cdot x^4$ (higher powers need not be considered since $x^{4+l} = x^l$ (mod 5)). We compile the following set of equations:

$$\begin{cases} a_0 = f(0) = 0, \\ a_0 + a_1 + a_2 + a_3 + a_4 = f(1) = 0, \\ a_0 + 2a_1 + 4a_2 + 3a_3 + a_4 = f(2) = 2, \\ a_0 + 3a_1 + 4a_2 + 2a_3 + a_4 = f(3) = 1, \\ a_0 + 4a_1 + a_2 + 4a_3 + a_4 = f(4) = 0. \end{cases}$$

Solving this set of equations, for example, by the method of elimination, we obtain $a_0 = 0$, $a_1 = 2$, $a_2 = 3$, $a_3 = 3$, $a_4 = 2$.

Example 2. Prove that the function $f(x, y) = x^2 \dot{-} y$ in P_4 cannot be represented by a mod 4 polynomial.

Solution. We assume that the converse is true, i.e. that $f(x, y)$ can be represented by a mod 4 polynomial, and write this hypothetical polynomial in the general form (with indeterminate coefficients): $f(x, y) = (a_{00} + a_{10}x + a_{20}x^2 + a_{30}x^3) + (a_{01}y + a_{11}xy + a_{21}x^2y + a_{31}x^3y) + (a_{02}y^2 + a_{12}xy^2 + a_{22}x^2y^2 + a_{32}x^3y^2) + (a_{03}y^3 + a_{13}xy^3 + a_{23}x^2y^3 + a_{33}x^3y^3)$.

(Powers higher than 3 need not be considered since for $l \geqslant 1$ the relations $x^{2l+2} = x^2$ (mod 4) and $x^{2l+3} = x^3$ (mod 4) are valid.) We have:

$$\begin{cases} f(1, 0) = a_{00} + a_{10} + a_{20} + a_{30} = 1, \\ f(1, 2) = a_{00} + a_{10} + a_{20} + a_{30} + 2(a_{01} + a_{11} + a_{21} + a_{31}) = 0. \end{cases} \quad (*)$$

Since the equation $1 + 2a = 0 \pmod 4$ has no solutions (this can be proved directly by successively putting $a = 0, 1, 2$ and 3), the set (*) has no solution either. Consequently, we cannot choose appropriate coefficients a_{ij} which would ensure the representation of the function f in the form of a mod 4 polynomial. Therefore, the function f cannot be represented by a mod 4 polynomial.

Remark. In the problem considered above, it would suffice to write two relations between the coefficients. However, a more complete set of equations should be often considered. In the following subsection, we shall demonstrate other methods to prove that the functions in P_k cannot be represented by mod k polynomials.

3.2.1. (1) Which of the classes, $T(\{0, 2\})$ or $U(\{0, 1\}, \{2\})$, do the following functions in P_3 belong to:

(a) $\sim x$; (b) $j_1(x)$; (c) $J_2(x)$; (d) $x \dot{-} y$; (e) $x + y$; (f) $\min(x, y)$.

(2) Which of the classes, $T(\{1, 3\})$, $U(\{0, 1\}, \{2\}, \{3\})$ and/ or $U(\{0, 3\}, \{1, 2\})$ do the following functions in P_4 belong to:

(a) $\bar{x}$; (b) $\sim x$; (c) $j_0(x)$; (d) $x + 2y$; (e) $\max(x, y)$; (f) $x^2 \cdot y$.

3.2.2. For a function f in P_k and for a given k, choose the classes of the types $T(\mathscr{E})$ or/and $U(D)$ to which this function belongs. Here $\mathscr{E}$ must be a proper subset of the set E_k (i.e. $\mathscr{E} \neq \varnothing$ and $\mathscr{E} \neq E_k$), and D must be a partition $\{\mathscr{E}_1, \ldots, \mathscr{E}_s\}$ of the set E_k, such that the following inequalities are satisfied: $1 < s < k$.

(1) $k = 3$, $f = x^2 + 1$;
(2) $k = 3$, $f = J_0(x^2 + 2x)$;
(3) $k = 3$, $f = (x^2 \dot{-} y^2) + 1$;
(4) $k = 3$, $f = x \cdot \bar{y} - y + 1$;
(5) $k = 4$, $f = j_2(x - x^2)$;
(6) $k = 4$, $f = J_3(2x + x^2)$;
(7) $k = 4$, $f = x \cdot y - y + 3$;
(8) $k = 5$, $f = \min(x^2, y)$;
(9) $k = 5$, $f = (2x^2 \dot{-} y) - 1$;
(10) $k = 6$, $f = x^3 \cdot y + 2$;
(11) k is an arbitrary integer not less than 3, $f = j_1(2x \dot{-} x^2)$;
(12) k is an arbitrary integer not less than 3, $f = J_{k-1}(x \cdot y - 1)$.

3.2.3. (1) Let $\mathscr{E} \subseteq E_k$. Prove that $T(\mathscr{E}) \neq P_k$ if and only if $\mathscr{E}$ is a proper subset of the set E_k (i.e. $\mathscr{E} \neq \varnothing$ and $\mathscr{E} \neq E_k$).

(2) Determine the number of different closed classes in P_k which preserve the sets?

(3) Let $\mathscr{E} \subseteq E_k$. Calculate the number of functions in P_k contained in the class $T(\mathscr{E})$ and depending on the variables $x_1, x_2, \ldots, x_n$ $(n \geqslant 0)$.

3.2.4. (1) Let $D = \{\mathscr{E}_1, \ldots, \mathscr{E}_s\}$ be a partition of the set E_k. Prove that $U(D) \neq P_k$ if and only if $1 < s < k$.

(2) For $k = 3, 4$ and 5, calculate the number of different closed classes in P_k which preserve the partitions.

(3) Let $D = \{\mathscr{E}_1, \ldots, \mathscr{E}_s\}$ be a partition of the set E_k. Calculate the number of functions in P_k contained in the class $U(D)$ and depending on the variables $x_1, x_2, \ldots, x_n$ $(n \geqslant 0)$.

3.2.5. Let $\mathscr{E}$ be non-empty subset in E_k other than the entire E_k and let $D = \{\mathscr{E}, E_k \setminus \mathscr{E}\}$. Calculate the number of functions in P_k which depend on the variables $x_1, x_2, \ldots, x_n$ $(n \geqslant 0)$ and are contained in the set: (1) $T(\mathscr{E}) \setminus U(D)$; (2) $U(D) \setminus T(\mathscr{E})$; (3) $T(\mathscr{E}) \cup U(D)$.

3.2.6. By S_k we denote the *set of all different-valued functions in P_k which depend on a single variable* (i. e. $g(x)$ belongs to S_k if and only if $g(E_k) = E_k$). Let $P_k^{(1)}$ be the *set of all functions of the k-valued logic P_k, which depend on a single variable, and $CS_k = P_k^{(1)} \setminus S_k$.*

(1) Prove that the sets S_k and CS_k are closed classes.

(2) Calculate the number of functions depending on the variable x and belonging to the class $S_k \cap U(\{0, k-2\}, \{1, \ldots, k-3\}, \{k-1\})$. (For $k = 3$, we assume that $\{1, \ldots, k-3\} = \varnothing$.)

3.2.7. Expand a function f in P_k into a mod k polynomial:

(1) $f = 2\cdot x \dot{-} x^2$, $k = 5$;
(2) $f = \min(x^2, x^3)$, $k = 5$;
(3) $f = \max(2\cdot x \dot{-} 1, x^2)$, $k = 5$;
(4) $f = 3\cdot x \dot{-} (x \dot{-} 2\cdot x)$, $k = 7$;
(5) $f = \max((x \dot{-} 1)^2, x^3)$, $k = 7$;
(6) $f = \min(x^2, y)$, $k = 3$;
(7) $f = \max(2\cdot x \dot{-} y, x\cdot y)$, $k = 3$;
(8) $f = x \dot{-} y$, $k = 3$;

(9) $f = J_{k-2}(x)$, k is an arbitrary prime number;

(10) $f = j_2(x - x^2)$, k is an arbitrary prime number.

3.2.8. (1) Prove that the function $2j_i(x)$ in P_4 can be represented by a mod 4 polynomial for any $i = 0, 1, 2$ and 3.

(2) Prove that if a function in P_4 depending on a single variable assumes the values either from the set $\{0, 2\}$ or in the set $\{1, 3\}$, it can be represented by a mod 4 polynomial.

(3) Using the function $f(x, y) = 2\cdot j_0(x)\cdot j_0(y)$ (in P_4), prove that statement (2) cannot be generalized to functions depending on more than one variable.

3.2.9. Prove that if a function $f(x)$ in P_4 cannot be represented by a mod 4 polynomial, then for any integer $m \geqslant 2$ the function $(f(x))^m$ cannot be represented by a mod 4 polynomial either.

3.2.10. Calculate the number of functions in P_4 which depend only on a variable x and can be represented by mod 4 polynomials.

3.2.11. (1) Let a function $f(x)$ in P_6 be represented by a mod 6 polynomial. Prove that it can be represented by a mod 6 polynomial of the form $a_0 + a_1x + a_2x^2$.

(2) Prove that the number of functions in P_6, which depend on a variable x and can be presented by mod 6 polynomials, is equal to 108.

(3) Enumerate all the functions $f(x)$ in P_6 which can be represented in the form $a + b\cdot j_0(x)$ (a and b belong to E_6), cannot be represented by mod 6 polynomials, and satisfy the condition that the function $(f(x))^2$ can be represented by a mod 6 polynomial.

3.2.12. Find out whether a function f in P_k can be represented by a mod k polynomial:

(1) $f = 3\cdot x \dot{-} 2\cdot x^2, \quad k = 4;$

(2) $f = 3\cdot j_0(x), \quad k = 6;$

(3) $f = 2\cdot(J_1(x) + J_4(x)), \quad k = 6;$

(4) $f = (x \dot{-} y) \dot{-} y, \quad k = 4;$

(5) $f = (\max(x, y) - \min(x, y))^2, \quad k = 4.$

3.2.2. Completeness Test for Sets of Functions of k-Valued Logic

In k-valued logics, the completeness testing of an arbitrary set of functions is fraught with serious technical difficulties because the application of the completeness criterion based on an analysis of the set of all precomplete classes in P_k involves the verification of a large number of conditions even for $k = 3$ and 4 (since there are exactly 18 precomplete classes in P_3 and 82 in P_4). The completeness of concrete sets in P_k is normally proved by reducing them to sets known to be complete beforehand (like the *Rosser-Turquette set* $\{0, 1, \ldots, k-1, J_0(x), J_1(x), \ldots, J_{k-1}(x), \min(x, y), \max(x, y)\}$ or *Post's set* $\{\bar{x}, \max(x, y)\}$). Besides, there exists a number of completeness criteria in which we consider sets of functions containing some functions of a single variable and only one function which essentially depends on at least two variables. Let us formulate the most important criteria of this kind. It should be recalled that

S_k is a set of all different-valued functions in P_k, which depend on a single variable,

$P_k^{(1)}$ is a set of all functions of one variable in P_k and $CS_k = P_k^{(1)} \setminus S_k$. The function $f(\tilde{x}) \in P_k$ is referred to as *essential* if it essentially depends on at least two variables and assumes all k values in the set E_k.

Theorem 1 (Słupecki's criterion). *A set* $P_k^{(1)} \cup \{f(\tilde{x})\}$ *is complete in* P_k *(for* $k \geqslant 3$*) if and only if* $f(\tilde{x})$ *is an essential function.*

Theorem 2 (S. V. Yablonsky's criterion). *A set* $CS_k \cup \{f(\tilde{x})\}$ *is complete in* P_k *(for* $k \geqslant 3$*) if and only if* $f(\tilde{x})$ *is an essential function.*

Theorem 3 (A. Sálomaa's criterion). *A set* $S_k \cup \{f(\tilde{x})\}$ *is complete in* P_k *(for* $k \geqslant 5$*) if and only if* $f(\tilde{x})$ *is an essential function.*

The statements containing various completeness criteria for sets of functions in the sets $P_k^{(1)}$, S_k and CS_k can be used along with these theorems. Let us consider an example of application of such criteria. Let a function

$h_{ij}(x)$ where $0 \leqslant i < j \leqslant k-1$, be defined as follows:

$$h_{ij}(x) = \begin{cases} i & \text{if } x = j, \\ j & \text{if } x = i, \\ x & \text{otherwise.} \end{cases}$$

Theorem 4 (S. Piccard). *Each of the sets* $\{\bar{x}, h_{01}(x), x + j_0(x)\}$ *and* $\{h_{01}(x), h_{02}(x), \ldots, h_{0(k-1)}(x), x + j_0(x)\}$ *is complete in* $P_k^{(1)}$.

3.2.13. By choosing an appropriate class of the type $T(\mathfrak{E})$ or $U(D)$, prove that the set A is incomplete in P_k if:

(1) $A = \{\sim x, \min(x, y), x \cdot y^2\}$;
(2) $A = \{1, 2, \overline{x \dot{-} j_2(x)}, \max(x, y)\}$;
(3) $A = \{2, j_0(x), x + j_0(x) + J_1(x) + J_{k-1}(x), \min(x, y)\}$;
(4) $A = \{J_2(x), x + j_0(x), x + j_0(x) + J_1(x), \max(x, y)\}$;
(5) $A = \{k-1, J_0(x), x \dot{-} y, x \cdot y \cdot z\}$;
(6) $A = \{2 \cdot x^3, 2 \cdot x + y, x^2 \cdot y, x \cdot J_0(y), \bar{x}^2 + (\sim y)\}$;
(7) $A = \{x \cdot y, \max(x, y) - z + 1\}$;
(8) $A = \{-x^2, \max(x, y) + \bar{z}\}$;
(9) $A = \{1, -x \cdot y, \overline{x^2 \dot{-} y}\}$;
(10) $A = \{2, \max(x, y), x \dot{-} y\}$;
(11) $A = \{k-2, \sim j_{h-2}(x), \max(x, y), \bar{x} + \bar{y}\}$;
(12) $A = \{j_2(x), x + j_0(x) + J_1(x), x \cdot y, x \dot{-} y\}$;
(13) $A = \{1, J_0(x), \bar{x} + j_{k-1}(x), \min(x, y), \max(x, y)\}$;
(14) $A = \{1, 2, \ldots, k-1, x + j_0(x), j_2(x) \dot{-} 1, \max(x, y)\}$;
(15) $A = \{1, \sim x, \overline{x \dot{-} y}, \min(x, y)\}$;
(16) $A = \{\sim x, J_0(x), J_1(x), \ldots, J_{k-1}(x), \min(x, y) + (j_0(x) + j_0(y))(x + y)\}$;
(17) $A = \{1 - x, j_0(x), j_1(x), \ldots, j_{k-1}(x), x \cdot y, x \dot{-} y, \min(x, \bar{y})\}$.

3.2.14. It is known (see Sec. 3.1) that the set

$$A = \{0, 1, \ldots, k-1, j_0(x), j_1(x), \ldots, j_{k-1}(x), x + y, x \cdot y\}$$

is complete in P_k.

(1) Prove that it is possible to isolate from A a subset which is complete in P_k and consists of two functions.

(2) Prove that any subset of the set A, which consists of a single function, is incomplete in P_k.

3.2.15. The Rosser-Turquette set

$$A_1 = \{0, 1, \ldots, k-1, J_0(x), J_1(x), \ldots, J_{k-1}(x), \min(x, y), \max(x, y)\},$$

is known to be complete in P_k (see Sec. 3.1).

(1) Verify that by omitting in A_1 any constant other than 0 and $k-1$, we obtain a subset contained in a certain class of the type $T(\mathscr{E})$, where $\varnothing \neq \mathscr{E} \neq E_k$ (and hence incomplete in P_k).

(2) Isolate from the set A_1 a subset which is complete in P_k and consists of $2k-2$ functions.

3.2.16. For given k, test the completeness of the following subsets of the Rosser-Turquette set:

(1) $k = 3$, $\{1, J_0(x), J_2(x), \min(x, y), \max(x, y)\}$;
(2) $k = 3$, $\{1, 2, J_2(x), \min(x, y), \max(x, y)\}$;
(3) $k = 4$, $\{1, 2, J_0(x), J_1(x), \min(x, y), \max(x, y)\}$;
(4) $k = 4$, $\{1, 2, J_0(x), J_3(x), \min(x, y), \max(x, y)\}$.

3.2.17. Prove that each of the following sets is complete in S_k:

(1) $\{h_{01}(x), h_{02}(x), \ldots, h_{0(k-1)}(x)\}$;
(2) $\{h_{01}(x), h_{12}(x), \ldots, h_{i,(i+1)}(x), \ldots, h_{(k-2),(k-1)}(x)\}$;
(3) $\{\bar{x}, h_{01}(x)\}$.

3.2.18. Prove that the set $\{h_{01}(x), h_{02}(x), \ldots, h_{0(k-1)}(x), x + j_0(x)\}$ is complete in $P_k^{(1)}$.

3.2.19. Using the method of reduction to sets which are known beforehand to be complete, prove the completeness in P_k of the following sets:

(1) $\{J_0(x), J_1(x), \ldots, J_{k-1}(x), x^2, x \dot{-} y\}$;
(2) $\{k-1, x \dot{-} y, x+y\}$;
(3) $\{\sim x, x+2, x \dot{-} y\}$;
(4) $\{-x, 1-x^2, x \dot{-} y\}$;
(5) $\{x \dot{+} y, (\sim x) \dot{-} 2y\}$;
(6) $\{\bar{x}, \min(x, y)\}$;
(7) $\{\min(x, y) - 1\}$;
(8) $\{J_{k-1}(x), \bar{x}+y, x \cdot y\}$;
(9) $\{1, x^2+y, x^2 \dot{-} y\}$;
(10) $\{J_0(x), x+y, x \cdot y^2\}$;
(11) $\{k-2, x \cdot y+1, (\sim x) \dot{-} y\}$;

(12) $\{k-1,\ x^2-y,\ x^2 \dot{-} y\}$;
(13) $\{1,\ 2\cdot x+y,\ x \dot{-} y\}$;
(14) $\{1,\ x^2-y,\ \min(x,\ y)\}$;
(15) $\{1,\ x+y+2,\ x^2 \dot{-} y\}$;
(16) $\{\bar{x}\cdot j_0(y),\ \min(x,\ y)\}$.

3.2.20. Using the Słupecki criterion, prove the completeness in P_k of the following sets:

(1) $\{k-1,\ x-y+2,\ x^2 \dot{-} y\}$;
(2) $\{j_2(x),\ x+y^2,\ x\cdot y+1\}$;
(3) $\{x \dot{-} y,\ (\sim x)-y\}$;
(4) $\{j_1(x),\ \bar{x}-y,\ x^2-y\}$;
(5) $\{\bar{x},\ j_0(x),\ x\cdot y\}$;
(6) $\{x-1,\ (x+j_0(x))\cdot(1 \dot{-} y)+(1 \dot{-} x)\cdot(y-j_1(y))\}$;
(7) $\{(1 \dot{-} x)\cdot y+\bar{x}\cdot(1 \dot{-} y)\}$;
(8) $\{\bar{x}\cdot j_0(x-y)+(x-j_1(x))\cdot j_0(y)+y\cdot j_0(x)\}$;
(9) $\{j_0(x-y)+x\cdot j_0(y)+(\bar{x}-j_1(x))\cdot j_1(y)\}$;
(10) $\{x\cdot j_0(y)+j_0(x)\cdot(y+j_0(y)-j_1(y)) + j_1(x)\cdot(\overline{y}-j_0(y))\}$;
(11) $\{\bar{y}\cdot j_0(x)+j_1(x)\cdot(y+j_0(y))+j_1(y)\cdot(j_2(x)-j_1(x))\}$.

3.2.21. Test the completeness in P_k of the following sets:

(1) $\{k-1,\ x+2,\ \max(x,\ y)\}$;
(2) $\{1,\ 2,\ \overline{x \dot{-} y}\}$;
(3) $\{k-2,\ x+y,\ \min(x,\ y)\}$;
(4) $\{0,\ 1,\ \bar{x} \dot{-} (\sim y)\}$;
(5) $\{2,\ 2\cdot x+y,\ x^2 \dot{-} y\}$;
(6) $\{1,\ 2,\ \max(\bar{x},\ y)\}$;
(7) $\{2 \dot{-} x,\ x\cdot y,\ \max(x,\ y)\}$;
(8) $\{k-2,\ 2\cdot x+y,\ x \dot{-} y\}$;
(9) $\{\sim x,\ -x\cdot y,\ \min(x,\ y)\}$;
(10) $\{2,\ x+y,\ x \dot{-} y\}$;
(11) $\{\sim x,\ 2\cdot j_0(x),\ J_1(x),\ x \dot{-} y\}$;
(12) $\{1,\ \sim x,\ J_0(x)+J_1(x),\ \max(x,\ y)\}$;
(13) $\{0,\ 1,\ \sim x,\ 2-j_0(x)-2\cdot j_1(x),\ \min(x,\ y)\}$;
(14) $\left\{1,\ k-1,\ x \dot{-} \left[\frac{k}{2}\right],\ \min(x,\ y)\right\}$;
(15) $\{k-2,\ \sim x,\ x \dot{-} y\}$.

3.2.22. Prove that the following sets are complete in P_k if and only if k is a prime number:

(1) $\{1,\ x+y+x\cdot z\}$;
(2) $\{x-y+1,\ x^2-y,\ x\cdot y^2\}$;
(3) $\{x-1,\ x+y,\ x^2\cdot y\}$;
(4) $\{k-1,\ x\cdot y+x-y+z\}$;
(5) $\{k-2,\ x+2\cdot y,\ x\cdot y^2\}$;
(6) $\{\sim x,\ x-y,\ x^2\cdot y\}$;
(7) $\{x-2,\ x+2\cdot y+1,\ x\cdot y-x-y\}$;
(8) $\{1,\ 2\cdot x+y,\ x\cdot y^2-x+y\}$;
(9) $\{x+y+1,\ x\cdot y-x^2\}$;
(10) $\{x-2y,\ x\cdot y+x+1\}$;
(11) $\{1+x_1-x_2+x_1\cdot x_2\cdot\ldots\cdot x_k\}$.

3.2.23. Prove that the following functions in P_k cannot be represented by mod k polynomials in the case of a composite k:

(1) $j_i(x),\ 0\leqslant i\leqslant k-1$;
(2) $\max(x,\ y)$;
(3) $\min(x,\ y)$;
(4) $x\dot{-}y$;
(5) $x\supset y$;
(6) $(x\dot{-}y)\dot{-}z$;
(7) any Sheffer's function (i.e. a function forming a complete system in P_k).

3.2.24. Having chosen for a function $f(x,\ y)$ the polynomials $Q_0(x)$, $Q_1(x)$ and $Q_2(x)$ such that at least one of the functions $Q_0(f(Q_1(x),\ Q_2(y)))$ or $Q_0(f(Q_1(x),\ Q_2(x)))$ is known not to be expandable into a mod k polynomial, prove that for $k=4$ and 6 the following functions cannot be represented by mod k polynomials:

(1) $f=(2\dot{-}x^3)\cdot\bar{y}$;
(2) $f=((\sim\bar{x})+y)\dot{-}(\bar{x}\dot{-}1)$;
(3) $f=\min(\sim x,\ y)\dot{-}(1\dot{-}x)$;
(4) $f=\max(x,\ y)\dot{-}(x\dot{-}2)$.

3.2.25. Isolate the basis of the set A which is complete in P_k:

(1) $A=\{k-1,\ j_0(x),\ j_1(x),\ \ldots,\ j_{k-1}(x),\ x\cdot y,\ x\div y\}$;
(2) $A=\{x-2,\ J_0(x),\ \max(x,\ y),\ x\dot{-}y^2,\ x^2\cdot y\}$;
(3) $A=\{\sim x,\ \min(x,\ y),\ x\cdot y,\ x+y\}$;
(4) $A=\{k-1,\ x+2,\ \max(x,\ y),\ x\dot{-}y\}$;
(5) $A=\{2,\ j_0(x),\ x+y^2,\ x^2\dot{-}y,\ x\cdot y\cdot z\}$.

3.2.26. Let B be an arbitrary basis isolated from a Rosser-Turquette set. Prove that in this basis

(a) there exists at least one of the functions $J_i(x)$, $0 \leqslant i \leqslant k-2$,

(b) there does not exist the constant 0,

(c) if $J_0(x) \in B$, then $k-1 \notin B$.

3.2.27. Prove that if a closed class in P_k has a finite complete set, the set of all different bases of this class is not more than countable. (Two bases are assumed to be different if one of them cannot be obtained from the other by redesignating the variables without identifying them.)

3.2.28. (1) Let A be a non-empty set of functions of one variable from P_k, which differs from the entire set $P_k^{(1)}$ and satisfies the following conditions: there exists a precomplete class B in P_k, such that $B \cap P_k^{(1)} = A$. Prove that such a class B is unique.

(2) Prove that the number of precomplete classes in P_k, each of which does not contain the complete set $P_k^{(1)}$, is smaller than 2^{k^k}.

3.2.29. (1) Prove that the closed class $K_1 = [x^2 \cdot y^2]$ in P_3 contains neither constants nor identical function.

(2) Prove that the closed class $K_2 = [j_1(x) \cdot j_2(y)]$ in P_3 does not contain functions which essentially depend on a single variable.

3.2.30. Two functions in P_k are referred to as *congruent* if they can be obtained from each other by redesignating their variables without identifying them. Prove that the closed class $K_3 = [j_2(x) \cdot j_2(y)]$ in P_3 consists of a finite number of pairwise non-congruent functions which essentially depend on all their arguments.

3.2.31. Let us consider the closed class $K_4 = [f_1(x_1), f_2(x_1, x_2), f_3(x_1, x_2, x_3), \ldots, f_n(x_1, x_2, \ldots, x_n), \ldots]$, where $f_n(\tilde{x}^n) = j_2(x_1) \cdot j_2(x_2) \cdot \ldots \cdot j_2(x_n)$, $n = 1, 2, 3, \ldots$. It obviously contains an infinite number of pairwise-non-congruent functions. Prove that K_4 does not contain precomplete classes.

3.2.32. Calculate the number of essential functions in P_k, which depend on the variables $x_1, x_2, \ldots, x_n$ $(n \geqslant 2)$.

3.2.33. It is known that for $k \geqslant 3$, there exist in P_k closed classes having no bases and closed classes with countable infinite bases. Let us consider the class $A_k =$

$[f_2, \ldots, f_m, \ldots]$, where

$$f_m(x_1, \ldots, x_m) = \begin{cases} 1 & \text{for } x_1 = \ldots = x_{i-1} = x_{i+1} = \ldots = x_m = 2, \\ & x_i = 1 \ (i = 1, \ldots, m), \\ 0 & \text{otherwise} \end{cases}$$

$m \geqslant 2$. The set $\{f_2, \ldots, f_m, \ldots\}$ is a basis in A_k. Using the class A_k, prove that there exists in P_k ($k \geqslant 3$) a continual family $\{B_\gamma\}$ of closed classes which form a chain upon inclusion, i.e. only one of the inclusions $B_{\gamma_1} \subset B_{\gamma_2}$ or $B_{\gamma_2} \subset B_{\gamma_1}$ is valid for any two classes B_{γ_1} and B_{γ_2} in the family $\{B_\gamma\}$.

Chapter Four

Graphs and Networks

4.1. Basic Concepts in the Graph Theory[1]

Let V be a finite non-empty set and X be a tuple of pairs of elements in V. The tuple X may contain both pairs with identical elements and identical pairs. The set V and the tuple X define a *graph with loops and multiple edges* (to make it short, a *pseudograph*) $G = (V, X)$. The elements of the set V are called *vertices*, while the elements of the tuple X the *edges* of the pseudograph. The edges of the type (v, v) $(v \in V)$ are called *loops*. A pseudograph without loops is referred to as a *graph with multiple edges* (or a *multigraph*). If none of the pairs is contained in the tuple X more than once, the multigraph $G = (V, X)$ is called a *graph*.[2] If the pairs in the tuple X are ordered, the graph is referred to as *oriented*. The edges of an oriented graph are often called *arcs*. If the pairs in the tuple X are disordered, the graph is called *non-oriented* and is referred to simply as a *graph*. If $x = (u, v)$ is an edge of a graph, the vertices u and v are known as the *ends of the edge* x. If v is an end of the edge x, v and x are known to be *incident*. The vertices u and v of the graph G are called *adjacent* if there exists an edge of the graph G connecting them. Two edges are called *adjacent* if they have a vertex in common. The *degree of a vertex* v is defined as the number $d(v)$ of the edges of the graph which are incident with the vertex v. The degree of a vertex v in a pseudograph is equal to the total number of the edges incident with this vertex plus the

---

[1] The definitions given in this section coincide with, or are similar to, those presented in ref. [23] and in the chapter "Graphs and Networks" in ref. [6].

[2] Henceforth, all definitions are given for graphs. As a rule, these definitions can be extended naturally to multigraphs and pseudographs. When the differences in definitions are significant, definitions for pseudographs are formulated separately.

number of the loops incident with it. A vertex of a graph of degree 0 is referred to as an *isolated* vertex, and a vertex of degree 1 is known as a *pendant vertex.*

The sequence

$$v_1x_1v_2x_2v_3\ldots x_{n-1}v_n \quad (n \geqslant 2), \qquad (1)$$

in which vertices and edges alternate and such, that for each $i = \overline{1, n-1}$ the edge x_i has the form (v_i, v_{i+1}), is called the *walk* connecting the vertices v_1 and v_n. The number of edges in a walk is termed as the *length of walk.* A sequence consisting of a single vertex is called a *walk of length* 0. A walk in which all edges are pairwise different is known as a *chain.* A walk in which all vertices are pairwise different is called a *simple chain.* The walk (1) is called *closed* if $v_1 = v_n$. A closed walk all whose edges are pairwise different is called a *cycle.* A cycle in which all vertices, except the first and the last, are pairwise different is called a *simple cycle.* A graph is termed *connected* if for any of its two vertices there exists a chain connecting these vertices. The *distance* between vertices of a connected graph is defined as the length of the shortest chain connecting these vertices. The *diameter* of a connected graph is the distance between two vertices having the largest separation. The diameter of a graph G is denoted by $D(G)$. A *subgraph* of a graph G is a graph all whose vertices and edges are contained among the vertices and edges of the graph G. A subgraph is *proper* if it differs from the graph itself. A *connected component* of a graph G is its connected subgraph which is not a proper subgraph of any other connected subgraph of the graph G. A subgraph containing all the vertices of a graph is called a *spanning subgraph.* A *subgraph* of the graph $G = (V, X)$ *generated by the subset* $U \subseteq V$ is a graph $H = (U, Y)$ the set of whose edges consists of those and only those edges of the graph G whose both ends lie in U. The graphs (pseudographs) $G = (V, X)$ and $H = (U, Y)$ are *isomorphic* if there exist two one-to-one correspondences $\varphi: V \leftrightarrow U$ and $\psi: X \leftrightarrow Y$ such that for any edge $x = (u, v)$ in X we have $\psi(x) = (\varphi(u), \varphi(v))$. For graphs, we can formulate the following definition. *Graphs* $G = (V, X)$ and $H = (U, Y)$ *are isomorphic* if there exists a one-to-one mapping $\varphi: V \leftrightarrow U$ such that $(u, v) \in X$ if and only if

$(\varphi(u), \varphi(v)) \in Y$. Such a *mapping* φ is termed an *isomorphic mapping* (or *isomorphism*). An *automorphism* is an isomorphic mapping of a graph onto itself. The *operation of omitting a vertex* from a graph G consists in the removal of a certain vertex together with the edges incident with it. The *operation of omitting an edge* from a graph $G = (V, X)$ consists in the removal of the corresponding pair from X. Unless stipulated otherwise, all the vertices are preserved in this case. The *complement* $\overline{G}$ *of a graph* G is a graph in which two vertices are adjacent if and only if they are non-adjacent in G. The *operation of subpartition of the edge* (u, v) in the graph $G = (V, X)$ consists in the removal of the edge (u, v) from X, addition of a new vertex w to V, and addition of two edges (u, w) and (w, v) to $X \setminus \{(u, v)\}$. A graph G is called a *subpartition of a graph* H if G can be obtained from H by a consecutive application of the edge subpartition operation. Graphs G and H are *homeomorphic* if they have subpartitions which are isomorphic. Let $G = (V, X)$ and $H = (U, Y)$ be two graphs. We shall denote by $G \oplus H$ a graph known as the *symmetric difference of the graphs* G *and* H with a set of vertices $W = V \cup U$ and the set of edges $Z = X \oplus Y$ consisting of those and only those edges which are included exactly in one of the sets X or Y. We shall denote by $G \times H$ the *Cartesian product of the graphs* $G = (V, X)$ and $H = (U, Y)$ i.e. a graph whose vertices are the pairs of the form (v, u) $(v \in V, u \in U)$ and such that its vertices (v_1, u_1) and (v_2, u_2) are adjacent if and only if at least one of the pairs v_1, v_2 (in the graph G) or u_1, u_2 (in the graph H) is adjacent. The *union of the graphs* $G = (V, X)$ and $H = (U, Y)$ is the graph $E = (V \cup U, X \cup Y)$.

A *tree* is a connected graph containing no cycles. A graph without cycles is known as a *forest*. A graph is *complete* if each two of its different vertices are connected by an edge. A complete graph with n vertices is denoted by K_n. An *empty* (*null*, *completely disconnected*) *graph* is a graph without edges. A single-vertex graph without edges is referred to as *trivial*. A graph is *bipartite* if the set of its vertices can be divided into two subsets (two fractions) V_1 and V_2 so that each edge of the graph connects the vertices of different fractions. A bipartite graph with fractions V_1 and V_2 and the set of edges X

will be denoted by (V_1, V_2, X). If each vertex of V_1 is connected by an edge with each vertex of V_2, the graph is referred to as a *complete bipartite graph*. A complete bipartite graph (V_1, V_2, X) such that $|V_1| = n_1$, $|V_2| = n_2$ is denoted by K_{n_1, n_2}. A graph is assumed to be *k-connected* if the omission of any of its $k - 1$ vertices results in a connected graph differing from trivial. The vertex of a graph whose removal increases the number of connected components is known as a *separating vertex*, or a *cut vertex*. A graph is called a *regular graph of degree d* if all its vertices are of degree d. A regular graph of degree 1 is known as a *matching*. A regular graph of degree 3 is called a *cubic graph*. The *k-factor* of a graph is its spanning regular subgraph of degree k. The 1-factor is known as a *perfect matching*. The *maximum matching* of a graph G is the matching containing the maximum number of edges. The *Hamiltonian cycle* of a graph is a simple cycle containing all the vertices of the graph. The *Hamiltonian chain* is a simple chain containing all the vertices of a graph. A *unit n-dimensional cube* is a graph B^n whose vertices are Boolean vectors of length n, and the edges are one-dimensional faces (see Sec. 1.1).

4.1.1. Prove that for an arbitrary graph $G = (V, X)$, the equality $2|X| = \sum_{v \in V} d(v)$ is valid.

4.1.2. Let $i_k(G)$ be the number of vertices of degree k in a graph G. Determine the number of pairwise non-isomorphic graphs for which

(1) $i_2(G) = i_3(G) = i_4(G) = 2$, $i_k(G) = 0$ for $k \neq 2, 3, 4$;

(2) $i_2(G) = i_3(G) = i_4(G) = 3$, $i_k(G) = 0$ for $k \neq 2, 3, 4$.

4.1.3. Prove that in any graph without multiple edges and loops which has not less than two vertices, there exist two vertices with identical degrees.

4.1.4. Prove that for any tuple of non-negative integers $(k_0, k_1, \ldots)$, such that $\sum_i ik_i = 2m$, there exists a pseudograph with m edges, which has exactly k_i vertices of degree i for each $i = 0, 1, \ldots$.

4.1.5. Let $d_0(G)$ be the minimal of degrees of the vertices of an n-vertex graph G.

(1) Prove that if $d_0(G) \geqslant \frac{n-1}{2}$, the graph is connected.

(2) Can $\left[\frac{n-1}{2}\right]$ be substituted for $(n-1)/2$ in the previous statement?

4.1.6. Prove that any closed walk of an odd length contains a simple cycle. Is a similar statement valid for walks of an even length?

4.1.7. Prove that a connected graph with n vertices contains at least $n-1$ edges.

4.1.8. Prove that a connected graph with n vertices and c connected components has a number of edges not exceeding $\frac{1}{2}(n-c)(n-c+1)$.

4.1.9. Prove that any non-trivial connected graph contains a vertex which is not a separating vertex.

4.1.10. Prove that any two simple chains of maximum length in a connected graph have at least one vertex in common. Do they always have a common edge?

4.1.11. Prove that if an arbitrary edge contained in a simple cycle is removed from a connected graph, the latter will remain connected.

4.1.12. Show that if a graph with n vertices contains no cycles of an odd length and if the number of edges exceeds $(n-1)^2/4$, the graph is connected.

4.1.13. Determine the number $p_k(n)$ for which any graph with n vertices and $p_k(n)$ cycles of length k is connected.

4.1.14. Let graphs G and H be isomorphic. Prove that

(1) for every $d \geqslant 0$, the number of vertices of degree d in the graphs G and H is the same;

(2) for every l, the number of simple cycles of length l in graphs G and H is the same.

4.1.15. Prove that conditions (1) and (2) in Problem 4.1.14 are insufficient for the graph G and H to be isomorphic.

4.1.16. Indicate the pairs of isomorphic and non-isomorphic graphs among those shown in Figs 3-6. Justify your answer.

4.1.17. Let graphs G and H be 2-connected, have six vertices and eight edges each. The graph G has exactly two vertices of degree 2, while the graph H has four ver-

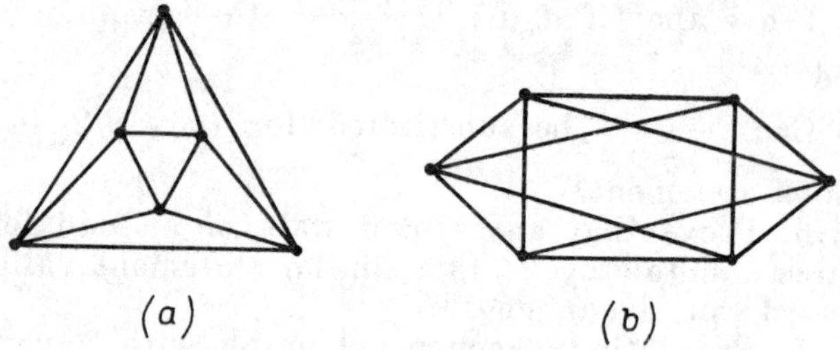

Fig. 3

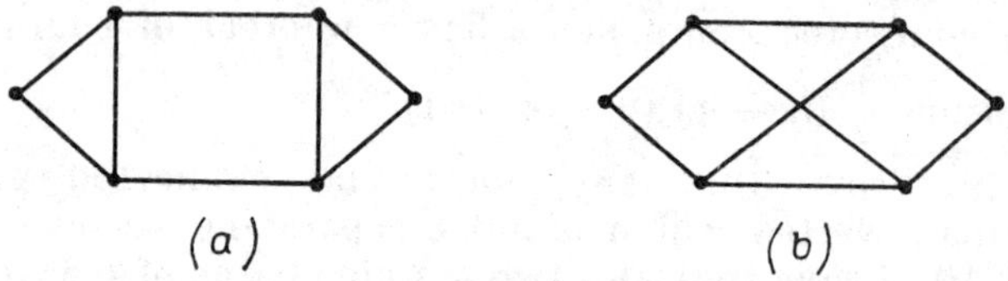

Fig. 4

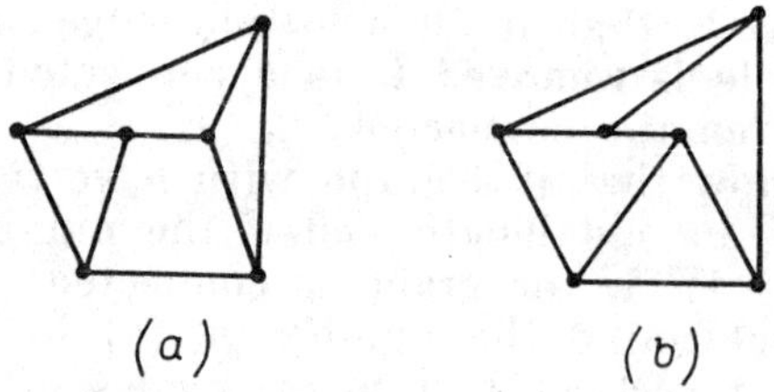

Fig. 5

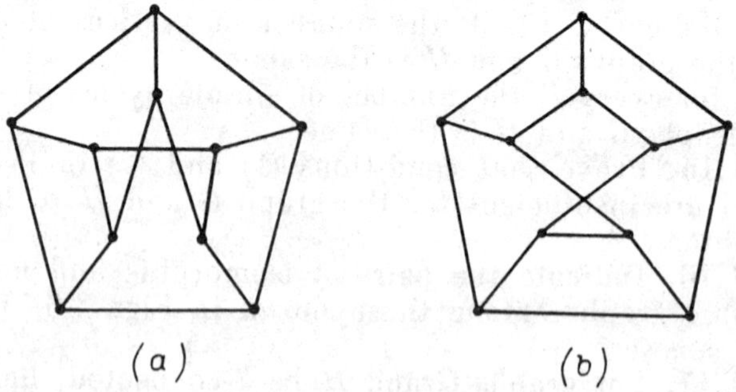

Fig. 6

tices of degree 3. Can we state that the graphs G and H are (1) isomorphic, (2) non-isomorphic?

4.1.18. Graphs G and H are 2-connected, have six vertices and ten edges each. One vertex in each graph is of degree d $(1 \leqslant d \leqslant 5)$, while the remaining vertices are of degree d_1 $(d_1 < d)$. Prove that graphs G and H are isomorphic.

4.1.19. Prove that in a graph having no non-trivial automorphisms

(1) the distance between any two vertices of degree 1 is larger than two;

(2) there exists a vertex of degree 3 or higher.

4.1.20. What is the number of automorphisms of a graph which is a cycle of length p?

4.1.21. Construct a graph without cycles which does not contain non-trivial automorphisms and has the minimum possible number of edges.

4.1.22. What is the smallest number n $(n > 1)$ of vertices in a graph having no non-trivial automorphisms?

4.1.23. Let the degrees of all six vertices in a 2-connected graph having nine edges be the same, and the number of simple cycles of length 3 be two. Reconstruct the graph and determine the number of its automorphisms.

4.1.24. Let $O(v)$ be the set of all vertices adjacent to v and $O'(v) = O(v) \cup \{v\}$. Let R_n be the set of all graphs G with n vertices, which possess the following properties: for any two non-adjacent vertices u and v, either $O(v) \subseteq O(u)$ or $O(u) \subseteq O(v)$, and for any adjacent vertices u and v, either $O'(v) \subseteq O'(u)$ or $O'(u) \subseteq O'(v)$. Let $G \in R_n$. Prove the following statements:

(1) all the vertices of the same degree in a graph G are either pairwise adjacent or pairwise non-adjacent;

(2) there is at least one vertex of degree $n - 1$;

(3) if for a certain d the vertices of degree d are pairwise adjacent, the vertices of a degree higher than d are also pairwise adjacent;

(4) a graph G is uniquely determined (within isomorphism) by specifying the degrees of its vertices;

(5) the removal of a vertex in G leads to a graph in R_{n-1}.

4.1.25. Let $n \geqslant 2$, and let the family $F(G) = \{H_1, H_2, \ldots, H_n\}$ of graphs be specified so that a certain graph H_i is obtained from an n-vertex graph G

by the removal of a vertex with a number i ($i = \overline{1, n}$). It should be noted that the vertices in graphs H_i are not labelled. Prove that using the family $F(G)$, it is possible

(1) to determine the number of edges of the graph G;

(2) to determine the degree of the vertex in each H_i whose removal from G leads to the graph H_i;

(3) to find out whether an arbitrary graph L having not more than $n - 1$ vertices is a subgraph of the graph G;

(4) to find out whether the graph G is connected; and

(5) to reconstruct the graph G if it is disconnected.

4.1.26. Let $D(G)$ be the diameter of a graph G and $\overline{G}$ be a complement to the graph G. Prove that $D(\overline{G}) \leqslant 3$ if the graph G is disconnected or if $D(G) \geqslant 3$.

4.1.27. A graph G is termed *self-complementary* if the graphs G and $\overline{G}$ are isomorphic.

(1) Find a self-complementary non-trivial graph with the smallest number of vertices.

(2) Prove that a self-complementary graph is connected.

(3) Prove that if G is a self-complementary graph, then $2 \leqslant D(G) \leqslant 3$.

4.1.28. Find the number of pairwise non-isomorphic graphs with 20 vertices and 188 edges.

4.1.29. Prove that if graphs G and H are homeomorphic, then

(1) for every $d \neq 2$, the number of vertices of degree d in both graphs is the same;

(2) there exists a one-to-one mapping of the set of simple cycles of the graph G onto the set of simple cycles of the graph H, for which the number of vertices of degree d in the corresponding cycles is the same for all $d \neq 2$.

4.1.30. Find out whether the graphs shown in Fig. 7 contain subgraphs homeomorphic to the graph G if:

(1) $G = K_4$ (see Fig. 8*a*);

(2) $G = K_5$ (see Fig. 8*b*);

(3) $G = K_{3,3}$ (see Fig. 8*c*).

4.1.31. The *superdivision operation* consists in the replacement of two adjacent edges (u, v) and (v, w), whose common vertex v is of degree 2, by a single edge (u, w). Applying the superdivision operation step by step, we can obtain from an arbitrary graph G containing vertices of degree 2 a pseudograph which does not contain vertices of degree 2. This pseudograph will be referred to as a *complete superdivision of the graph G*.

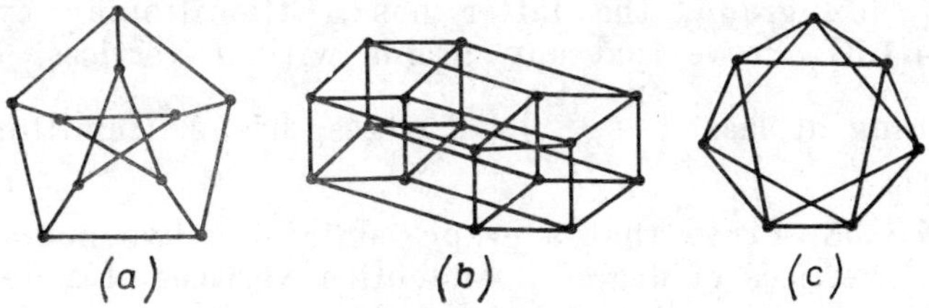

Fig. 7

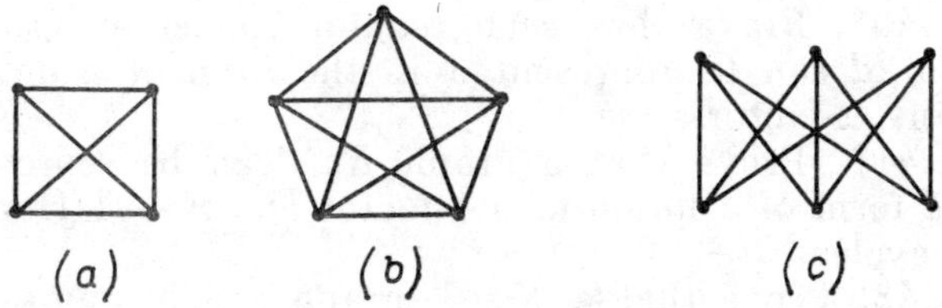

Fig. 8

(1) Prove that a complete superdivision of graph G does not depend on the order in which the superdivision operation has been applied to pairs of adjacent edges of the graph G.

(2) Prove that the graphs G and H are homeomorphic if and only if their complete superdivisions are isomorphic (as pseudographs).

4.1.32. Prove that the Petersen graph (Fig. 7*a*) has no Hamiltonian cycle, but the graph obtained from it by the removal of a vertex has a Hamiltonian cycle.

4.1.33. Prove that each of the graphs K_n, $K_{n,n}$, and B^n has a Hamiltonian cycle.

4.1.34. Let n be odd and B_k^n be a subset of vertices of a cube B^n, consisting of vertices of weight k. Let G be a subgraph of the cube B^n, generated by the set $B^n_{\frac{n-1}{2}} \cup B^n_{\frac{n+1}{2}}$. Does a perfect matching exist in the graph G?

4.1.35. A graph G contains a Hamiltonian cycle, while a graph H contains a Hamiltonian chain. Is it true that a Hamiltonian cycle exists in the graph $G \times H$?

4.1.36. Prove that if the condition $d(u) + d(v) \geqslant n$ is satisfied for any two vertices u and v of a connected n-vertex graph, the latter has a Hamiltonian cycle.

4.1.37. Prove that any graph with n vertices, containing at least $\binom{n-1}{2} + 2$ edges, has a Hamiltonian cycle.

4.1.38. Prove that a graph containing two non-adjacent vertices of degree 3 and other vertices of a degree not higher than 2 does not have a Hamiltonian cycle.

4.1.39. (1) Prove that $D(G \times H) \leqslant D(G) + D(H)$.
(2) Is it true that $D(G \times H) \leqslant \max\{D(G), D(H)\}$?

4.1.40*. Prove that each regular connected graph of degree $2d$ can be represented in the form of a union of disjoint 2-factors.

4.1.41*. Prove that a graph K_{2n} can be represented in the form of a union of a 1-factor and $n - 1$ Hamiltonian cycles.

4.1.42. Prove that a K_{2n+1} graph can be represented in the form of a union of n Hamiltonian cycles.

4.1.43. Prove that a graph K_n with labelled vertices contains $\frac{(n-1)!}{2}$ different Hamiltonian cycles.

4.1.44. Prove that the number of different perfect matchings of the graph K_{2n} with the labelled vertices is $\frac{(2n)!}{2^n n!}$.

4.2. Planarity, Connectivity, and Numerical Characteristics of Graphs

A graph is called *planar* if it can be drawn on a plane so that the arcs of the curves representing the edges intersect only at points corresponding to the vertices of the graph. Moreover, irrespective of the point of intersection, only the arcs corresponding to the edges incident exactly with the vertex corresponding to a given point converge at this point. Such a representation which maps a planar graph is called a *plane* (or *topolo-*

gical planar) graph. The *internal face* of a plane connected graph is defined as a finite region of a plane, bounded by a closed walk and containing neither vertices nor edges of the graph. The walk bounding a face is called a *boundary of the face.* The walk of a plane consisting of the points which do not belong either to the graph or to any of its internal faces is called the *external face.* For 2-connec-

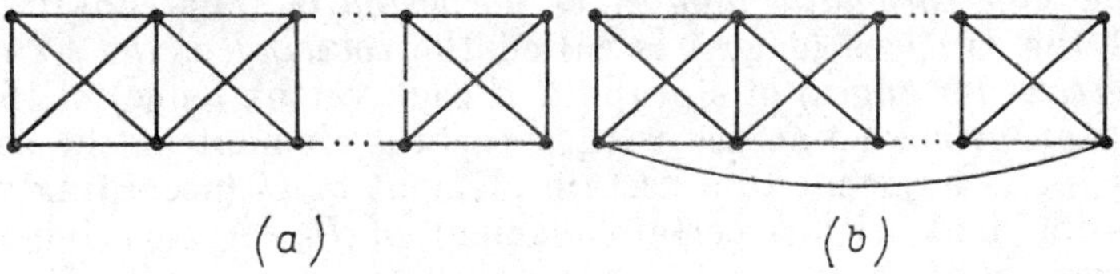

Fig. 9

ted plane graphs (multigraphs) having n vertices, m edges and r faces, the *Euler formula* $n - m + r = 2$ is valid. The following planarity criterion is also fulfilled.

Theorem (Pontryagin-Kuratowski). *A graph is planar if and only if it does not contain subgraphs homeomorphic to the graphs K_5 and $K_{3,3}$* (Fig. 8*b*, *c*).

The *thickness of a graph G* is the smallest number $t(G)$ of its planar subgraphs whose union is equal to G. Each non-trivial connected plane pseudograph G can be put in correspondence with a *dual pseudograph* G^* as follows. Within each face of the pseudograph G, we choose a vertex belonging to the graph G^*. If x is an edge of the graph G, lying on the boundary between faces g_1 and g_2, and v_1 and v_2 are vertices of the pseudograph G^* taken on these faces, the vertices v_1 and v_2 are connected by an edge in G^*. A plane pseudograph (graph) isomorphic to its dual pseudograph is referred to as *self-dual*. The cycles $Z_1, Z_2, \ldots, Z_k$ of a graph G are *linearly dependent* if for some $i_1, i_2, \ldots, i_s$ $(1 \leqslant i_1 < i_2 < \ldots < i_s \leqslant k)$ the relation $Z_{i_1} \oplus Z_{i_2} \oplus \ldots \oplus Z_{i_s} = 0$, is satisfied, where 0 is a graph without edges, and $Z \oplus Y$ is the symmetric difference between the graphs Z and Y. Otherwise, the cycles $Z_1, Z_2, \ldots, Z_k$ are referred to as *linearly independent*. The maximum number $\xi(G)$ of the cycles in the aggregate of linearly independent cycles of the graph G is termed the *cyclomatic number of the graph G.*

The colouring of the vertices (edges) of a graph is called *correct* if adjacent vertices (edges) have different colours. The smallest number $\chi(G)$ of colours for which there exists a correct colouring of the vertices of the graph G is known as the *chromatic number of the graph* G. The smallest number $\chi'(G)$ of colours for which there exists a correct colouring of the edges of the graph G is called the *edge chromatic number of the graph* G. The subset U of the vertices (edges) is called the *covering of the set of vertices* (or edges) of a graph G if each vertex (edge) of the graph either coincides with a certain element of the set U, or is adjacent to a certain element of U (accordingly, is incident with a certain element of U). A covering is *terminal* if it ceases to be a covering as a result of the omission of any element. The minimum power of a subset U of the vertices of a graph G, such that each edge of the graph is incident with at least one vertex in U, is denoted by $\alpha_0(G)$ and is called the *vertex covering number*. The minimum power of the subset Y of the edges of a graph G, such that each vertex of the graph is incident with at least one edge of Y is denoted by $\alpha_1(G)$ and is called the *edge covering number of the graph* G. The set of vertices (edges) of a graph G is called *independent* if none of its two elements are adjacent. By $\beta_0(G)$ (resp. $\beta_1(G)$) we denote the maximum power of an independent set of vertices (resp. edges) of a graph G. The number $\beta_0(G)$ (number $\beta_1(G)$) is called the *vertex* (resp. *edge*) *independence number of the graph* G. By $\alpha_{00}(G)$ we denote the minimum power of a set of vertices U such that each vertex of the graph G which is not contained in U is adjacent to at least one vertex in U.

4.2.1. Are the graphs shown in Figs 6*a* and *b* and 7*a*, *b* and *c* planar?

4.2.2. For which $n \geqslant 2$ are the graphs shown in Figs 9*a* and *b* planar?

4.2.3. Let $G_n = (V_1, V_2, X)$ be a bipartite graph,

$$V_1 = \{a_1, a_2, \ldots, a_n\},$$

$$V_2 = \{b_1, b_2, \ldots, b_n\},$$

$$X = \{x_1, x_2 \ldots, x_n, y_1, y_2, \ldots, y_n, z_1, z_2, \ldots z_n\},$$

where

$$x_i = (a_i,\ b_i), \qquad i = \overline{1,\ n};$$
$$y_i = (a_i,\ b_{i+1}), \qquad i = \overline{1,\ n-1}, \qquad y_n = (a_n,\ b_1);$$
$$z_i = (a_i,\ b_{i+2}), \qquad i = \overline{1,\ n-2},$$
$$z_{n-1} = (a_{n-1},\ b_1), \qquad z_n = (a_n,\ b_2).$$

Determine $n > 2$ for which the graph G_n is planar.

4.2.4. (1) What is the minimal number of edges that have to be removed from the cube B^4 for the obtained graph to be planar?

(2) What is the minimal number of vertices that have to be removed from B^4 for the obtained graph to be planar?

4.2.5*. Let G be a 2-connected plane graph having at least two internal faces. Prove that there exists a simple chain belonging to the boundary of the external face and such that its removal leads to a 2-connected plane graph with fewer faces.

Remark. The removal of a chain involves the removal of all its edges and internal vertices, while end vertices of the chain remain in the graph.

4.2.6. Using the result obtained by solving Problem 4.2.5, prove by induction that the number of internal faces in a plane 2-connected graph with n vertices and m edges is $m - n + 1$.

4.2.7. Prove by induction over the number of edges that the cyclomatic number $\xi(G)$ of a pseudograph with n vertices, m edges and c connected components is equal to $m - n + c$.

4.2.8. Determine the cyclomatic number $\xi(G)$ for the graphs shown in Figs 5*a*, 6*a* and 7*a* and isolate a system consisting of $\xi(G)$ linearly independent cycles.

4.2.9. Prove that any planar graph contains a vertex of degree not higher than 5.

4.2.10. Prove that any planar graph having at least four vertices contains at least four vertices of degree not higher than 5.

4.2.11. Prove that if each simple cycle in a connected planar graph with n vertices and m edges contains at least k edges, then $m \leqslant \dfrac{k(n-2)}{k-2}$.

4.2.12. A plane connected graph whose every face (including the external face) is bounded by a cycle of length three is called a *triangulation*.

Prove that any triangulation with $n \geqslant 3$ vertices has $3n - 6$ edges and $2n - 4$ faces.

4.2.13. Let $t(G)$ be the thickness of a graph G. Prove that

(1) $t(K_n) \geqslant \left[\frac{n+7}{6}\right]$;

(2) $t(K_{n,m}) \geqslant \left]\frac{n \cdot m}{2(n+m-2)}\right[$;

(3) $t(B^n) \geqslant \left]\frac{n+1}{4}\right[$.

4.2.14*. Using Euler's formula, prove that graphs $K_{3,3}$ and K_5 are not planar.

4.2.15*. Prove that a 2-connected graph with a cyclomatic number $\xi - 1$ can be obtained from a 2-connected graph with a cyclomatic number ξ $(\xi > 1)$ by removing a chain.

4.2.16. Construct a graph dual to the graph shown in Fig. 5*a*.

4.2.17. Prove that a pseudograph dual to a connected plane graph is connected and plane.

4.2.18. Prove that the cyclomatic number of a dual graph coincides with the cyclomatic number of the original graph.

4.2.19. Prove that if G has a separating vertex, G^* has a separating vertex as well.

4.2.20. Prove that a pseudograph G^* dual to a 3-connected plane graph G has no loops or multiple edges.

4.2.21. Determine the number of pairwise non-isomorphic self-dual plane 2-connected graphs with six vertices.

4.2.22. Prove that there exist no 6-connected planar graphs.

4.2.23. Prove that a graph dual to an n-vertex $(n > 3)$ triangulation is a 2-connected plane cubic graph.

4.2.24. (1) Prove that a plane cubic graph each of whose faces has at least five vertices contains not less than twelve vertices.

(2) Prove that if r_i is the number of faces of a plane cubic graph, which are bounded by i edges, then $\sum\limits_i (6 - i) r_i = 12$.

4.2.25. Determine the chromatic numbers of the graphs shown in Figs 4-8.

4.2.26. Determine the edge chromatic numbers of the graphs shown in Figs 3-7.

4.2.27. Determine the chromatic and edge chromatic number of

(1) the graph K_n;

(2) the graph $K_{n,n}$;

(3) the graph B^n.

4.2.28. What is the number of the correct vertex colourings of the cube B^n for the minimum number of colours?

4.2.29. Prove that the edge chromatic number of the Petersen graph shown in Fig. 7*a* is equal to four, but for any of its subgraphs G with eight vertices $\chi'(G) \leqslant 3$. Is the inequality $\chi'(G) \leqslant 3$ valid for an arbitrary proper subgraph of the Petersen graph obtained as a result of the removal of a vertex?

4.2.30. Prove that four colours are sufficient for the correct edge colouring of any cubic multigraph.

4.2.31. Prove that all edges of a plane cubic graph can be coloured in two colours a and b so that each vertex is incident with an edge of colour a and to two edges of colour b.

4.2.32. Prove that it is enough to have six colours for a correct colouring of the vertices of any plane graph.

4.2.33. Prove by induction over the number of vertices that the inequality $\chi(G) \leqslant 5$ is valid for a plane graph G.

4.2.34. Construct a plane graph G with the minimum possible number of vertices and such that $\chi(G) = 4$.

4.2.35. The vertices of a graph G are numbered in the order of increasing degrees. Prove that if k is the highest number for which $k \leqslant d(v_k) + 1$, then $\chi(G) \leqslant k$.

4.2.36. The operation of contraction consists in the removal of two adjacent vertices from a graph and addition of a new vertex adjacent to those of the remaining vertices which were adjacent to at least one of the removed vertices. Prove that the graph obtained as a result of contraction to a planar graph is planar.

4.2.37. Let l be the length of the longest simple chain in a graph G. Prove that $\chi(G) \leqslant l + 1$.

4.2.38. Let d be the highest of the degrees of vertices of a graph G. Prove that

(1) $\chi(G) \leqslant d+1$;
(2) $\chi'(G) \leqslant d+1$.

4.2.39. Let p be the maximum number for which there exists a subgraph of a graph G isomorphic to the graph K_p. Prove that $\chi(G) \geqslant p$.

4.2.40. Prove that for a graph G with n vertices, the following inequalities are valid:
(1) $\beta_0(G) \cdot \chi(G) \leqslant n$;
(2) $\chi(G) \cdot \chi(\overline{G}) \geqslant n$.

4.2.41. What is the smallest number n for which there exists an n-vertex non-planar graph with a non-planar complement?

4.2.42. Determine the thickness of the graph K_8.

4.2.43. Prove that

$$\alpha_0(G) + \beta_0(G) = \alpha_1(G) + \beta_1(G) = n$$

for an arbitrary connected graph G with n $(n > 1)$ vertices.

4.2.44. (1) Give an example contradicting the following statement: any vertex covering contains the minimum vertex covering.
(2) Prove that any vertex covering contains a terminal vertex covering.

4.2.45. Determine the number of terminal and minimum edge coverings for
(1) a chain of length m;
(2) a cycle of length n;
(3) Petersen's graph (Fig. 7*a*).

4.2.46. Prove that for any graph G the inequalities $\alpha_0(G) \geqslant \beta_1(G)$ and $\alpha_1(G) \geqslant \beta_1(G)$ are valid.

4.2.47. Prove that for any graph G the inequality $\alpha_{00}(G) \leqslant \alpha_0(G)$ is valid.

4.2.48. Prove or refute the following inequality: $\beta_0(G) \leqslant \alpha_{00}(G)$.

4.2.49. Let $U \subseteq V$ be a subset of vertices of a graph $G=(V, X)$ and let $\nu(U)$ be a number of vertices $v \in V \setminus U$ which are not adjacent to any vertex in U. Let

$$\overline{\nu}_k(G) = \frac{1}{\binom{|V|}{k}} \sum_{U \subseteq V,\ |U|=k} \nu(U).$$

(1) Prove that $\alpha_{00}(G) \leqslant k + \overline{\nu}_k(G)$.

(2) Let d_0 be the lowest of the degrees of vertices of a graph G. Prove that

$$\bar{\nu}_k(G) \leqslant |V| \prod_{i=0}^{k-1} \left(1 - \frac{d_0}{|V|-i}\right).$$

(3) Prove that the following inequality is valid for a regular graph G of degree d with n vertices:

$$\frac{n}{d} \leqslant \alpha_{00}(G) \leqslant 1 + \frac{n}{d}(1 + \ln d).$$

4.2.50. Prove that if d_0 is the lowest of the degrees of the vertices of a graph G, then $\alpha_0(G) \geqslant d_0$.

4.2.51*. If G is a bipartite graph and m is the number of its edges, then $m \leqslant \alpha_0(G) \cdot \beta_0(G)$. Prove that the equality is attained only for complete bipartite graphs.

4.3. Directed Graphs

A *directed pseudograph* $D = D(V, X)$ is defined by specifying a non-empty (finite) set V and tuple X of ordered pairs of elements in V. The elements of the set V are called *vertices*, and the elements of the tuple X are called *arcs* (or *directed edges*) of the directed pseudograph $D(V, X)$. The tuple X may also contain pairs of the form (v, v) referred to as *loops*, and identical pairs known as *multiple* (or *parallel*) *arcs*. Pairs (u, v) and (v, u) are assumed to be identical if and only if $u = v$. A *directed multigraph* is the term applied to a directed pseudograph containing no loops. If a directed pseudograph has neither loops nor multiple arcs, it is called a *directed graph* (or simply *digraph*). An *oriented graph* is a digraph having no symmetric pairs of directed edges, i.e. the tuple X cannot contain simultaneously an arc (u, v) and an oppositely directed arc (v, u).

Let $x = (u, v)$ be an arc of a directed pseudograph. Then the vertex u is called the *initial vertex* (*origin*), and the vertex v is the *terminal vertex* (*end*) of the arc x. In this case, the arc x is also said to *emerge from* the vertex u and *terminate at* the vertex v. If a vertex v is the origin or the end of an arc x, v and x are said to be *incident*. The *out-degree of a vertex* v (of a pseudograph D) is the number of arcs of the pseudograph D emerging from the vertex v. The out-degree of the vertex v is denoted by od(v) or $d^+(v)$. Similarly, the *in-degree of a vertex* v (denoted by id (v) or $d^-(v)$) is the number of arcs of the pseudograph terminating at the vertex v.

By replacing each ordered pair (u, v) in the tuple X of a directed pseudograph $D(V, X)$ by a disordered pair $\{u, v\}$ consisting of the same elements u and v, we obtain a *pseudograph* $G = (V, X^0)$ *associated with* $D(V, X)$.

Directed pseudographs $D_1(V_1, X_1)$ and $D_2(V_2, X_2)$ are assumed to be *isomorphic* if there exist two one-to-one correspondence φ: $V_1 \leftrightarrow V_2$ and ψ: $X_1 \leftrightarrow X_2$ such that for any arc $x = (u, v) \in X_1$ the relation $\psi(x) = (\varphi(u), \varphi(v))$ is valid. The isomorphic mapping of a directed pseudograph onto itself is referred to as the *automorphism of the pseudograph*. All automorphisms of a directed pseudograph form a group relative to the operation of multiplication (successive application) of automorphisms. This group is referred to as a *group (of automorphisms) of the directed pseudograph*.

The *operation of omission of a vertex and an arc*, *as well as the concepts of subgraph*, *spanning subgraph and generated subgraph* for directed pseudographs are defined as in the case of non-directed pseudographs.

In contrast to the definitions of the corresponding "non-directed" concepts, the definitions of *directed walk*, *closed walk*, *chain*, *cycle*, *simple chain and simple cycle* contain the requirement that the sequence (of vertices and arcs) $v_1, x_1, v_2, x_2, \ldots, x_{n-2}, v_{n-1}, x_{n-1}, v_n$ $(n \geqslant 2)$ satisfy the condition that each arc x_i $(1 \leqslant i \leqslant n-1)$ has the form (v_i, v_{i+1}), i.e. the vertex v_i is the origin of the arc x_i and the vertex v_{i+1} is its end. It is assumed that a directed (u-v)-walk is directed from its initial vertex u to the terminal vertex v. The *length of a walk* is equal to the number of arcs contained in it. The *distance* $\rho(u, v)$ *from the vertex* u *to the vertex* v is defined as the shortest length of the (u-v)-walk. A directed walk is often referred to as just a *path*, while a directed simple cycle is known as a *circuit*.

A directed simple spanning chain is called a *Hamiltonian path* (*Hamiltonian chain*). A *Hamiltonian circuit* is the spanning circuit of a directed pseudograph. If a directed pseudograph contains a Hamiltonian circuit, it is called a *Hamiltonian pseudograph*.

A vertex v of a directed pseudograph is said to be *accessible* from a vertex u if there exists a $(u\text{-}v)$-path in the pseudograph, i.e. the path emerging from the vertex u and terminating at the vertex v.

A directed pseudograph is *strongly connected* (or *strong*) if any vertex in it is accessible from its any other vertex. A directed pseudograph is called *unilaterally connected* (or *unilateral*) if for any two of its vertices at least one is accessible from the other. A directed pseudograph $D(V, X)$ is assumed to be *weakly connected* (or *weak*) if a pseudograph (V, X^0) associated with it is connected. If a directed pseudograph is not connected even weakly, it is referred to as *disconnected*. A *trivial digraph* consisting of only one vertex is assumed (by definition) to be strongly connected.

A *strong component* of a digraph D is the term applied to any of its directed subgraphs which is a strong directed graph contained in no other strongly connected directed subgraph of the digraph D. Similarly, a *unilateral component* is the maximum unilateral subgraph of the digraph D, while a *weak component* is the maximum weak subgraph. The concepts of strong, unilateral and weak components are naturally generalized to the case of a directed pseudograph.

Let $\gamma = \{S_1, S_2, \ldots, S_n\}$ be a set of all strong components of a digraph D. The *condensation* D^* of the digraph D is a digraph in which the set of vertices is γ and the arc (S_i, S_j) is contained in the digraph D^* if and only if the digraph D contains at least one arc emerging from a vertex of the component S_i and terminating at a vertex of the component S_j.

If $D = D(V, X)$ is a digraph, the digraph D' inverse to it is specified by the same set of vertices V and a set of arcs X' such that an arc (u, v) belongs to X' if and only if the arc (v, u) belongs to X.

A vertex v of a digraph D is called a *source* if any other vertex of the digraph D is accessible from it. The *sink* of a digraph D is any of its vertices v which is a source in the digraph D' inverse to digraph D.

Let D be a digraph for which a digraph associated with it is a tree. Then the digraph G is called an *arborescence* (or a *growing tree*) if it has a source. A directed pseudograph is assumed to be *complete* if any two different vertices in it are joined by at least one arc.

A complete oriented graph is known as a *tournament*.

Let D be an n-vertex digraph and let $V = \{v_1, v_2, \ldots, v_n\}$ be the set of its vertices. The *adjacency matrix* of a digraph D is an $(n \times n)$ matrix $A(D) = \| a_{ij} \|$ in which $a_{ij} = 1$ if the arc (v_i, v_j) belongs to D and $a_{ij} = 0$ in the opposite case. Moreover, we assume that the set of all arcs of the digraph D is also ordered: $X = (x_1, x_2, \ldots, x_m)$. The *incidence matrix* (or *matrix of incidence*) of a digraph D is an $(n \times m)$ matrix $B(D) = \| b_{ij} \|$ in which

$$b_{ij} = \begin{cases} 1 \text{ if the vertex } v_i \text{ is the end of the arc } x_j, \\ -1 \text{ if the vertex } v_i \text{ is the origin of the arc } x_j, \\ 0 \text{ if the vertex } v_i \text{ is not incident with the arc } x_j. \end{cases}$$

4.3.1. Disprove the following statement: if out- and in-degrees of any vertex of a digraph are positive and even, there exists for each vertex of the digraph a circuit containing it.

4.3.2. Let a digraph $D(V, X)$ be at least weakly connected, $V = \{v_1, v_2, \ldots, v_n\}$, $n \geqslant 2$ and $d^+(v_1) - d^-(v_1) = 1$, $d^+(v_2) - d^-(v_2) = -1$, $d^+(v_j) = d^-(v_j)$ for $j = 3, \ldots, n$. Prove that the digraph D also contains in this case a directed $(v_1 - v_2)$-chain including all the arcs of the digraph.

4.3.3. Prove that a digraph is strongly connected if and only if it contains a directed spanning closed walk.

4.3.4. Prove that a weak digraph is strongly connected if and only if it has a directed closed walk containing each arc of the digraph at least once.

4.3.5. Let us suppose that a digraph D can be presented in the form of a union of its certain directed closed walks $D_1, D_2, \ldots, D_k$ $(k \geqslant 1)$ which satisfy the following condition: every two walks D_j and D_{j+1} $(1 \leqslant j \leqslant k-1)$ have at least one vertex in common. Prove that the digraph D is strongly connected in this case.

4.3.6. Prove that any tournament contains a Hamiltonian path.

4.3.7. Prove that a tournament T is a strong digraph if and only if T has a spanning circuit (i.e. is a Hamiltonian tournament).

4.3.8. Let $\{v_1, v_2, \ldots, v_n\}$ be a set of vertices of a tournament. Prove that $\sum_{i=1}^{n} (d^+(v_i))^2 = \sum_{i=1}^{n} (n-1-d^+(v_i))^2$.

4.3.9. Let the out-degree of the vertex v of a tournament T be not lower than the out-degree of each other vertex of the tournament. Prove that the distance from the vertex v to any other vertex of the tournament does not exceed 2.

4.3.10*. We denote by S a set of arcs of a tournament T. The arcs of the set S are called *concordant* if the vertices of the tournament T can be renumbered so that the fact that the arc (v_i, v_j) belongs to the set S implies that $i < j$. Let $f(n)$ be the largest integer such that each n-vertex tournament ($n \geqslant 3$) contains the set S consisting of $f(n)$ concordant arcs. Prove that $f(n) \geqslant \left[\frac{n}{2}\right] \cdot \left[\frac{n+1}{2}\right]$.

4.3.11*. Prove that the number of directed cycles of length 3 in an n-vertex tournament does not exceed

$$t(n) = \begin{cases} \frac{n(n^2-1)}{24} & \text{if } n \text{ is odd,} \\ \frac{n(n^2-4)}{24} & \text{if } n \text{ is even.} \end{cases}$$

4.3.12. Prove that the group of automorphisms of any tournament has an odd order (i.e. consists of an odd number of elements).

4.3.13. Prove that the condensation D^* of an arbitrary digraph has no circuits.

4.3.14. Prove that a digraph D is unilateral if and only if its condensation D^* has a unique oriented spanning chain.

4.3.15. A digraph $D(V, X)$ is called *transitive* if the fact that arcs (u, v) and (v, w) belong to the set X implies that the set X contains the arc (u, w). Prove that the condensation of any tournament is a transitive tournament.

4.3.16. A *circuitless* digraph is a digraph containing no circuits. Prove that a circuitless digraph contains a vertex with zero out-degree.

4.3.17. Prove that a digraph is isomorphic to its condensation if and only if it is circuitless.

4.3.18. Let a digraph D be weakly connected but not

unilateral. Prove that there is no vertex in D such that its omission leads to a strong digraph.

4.3.19. Prove that a weak digraph is an arborescence if and only if only one of its vertices has zero in-degree, the in-degree of any of the remaining vertices being equal to unity.

4.3.20. Let S_n be a symmetric group of permutations operating on the set $\{1, 2, \ldots, n\}$, $n \geqslant 2$. Let us consider an arbitrary aggregate T of transpositions in the group S_n. We put in correspondence to the set T a digraph $D(V, X_T)$ in which $V = \{1, 2, \ldots, n\}$ and the arc (i, j) belongs to X_T if and only if $i < j$ and the transposition (ij) is contained in the set T. Prove that the set T forms a basis in S_n (or, in other words, the set T is an *irreducible system* of generators of the group S_n) if and only if the digraph $D(V, X_T)$ is an arborescence.

4.3.21. Prove that a complete strongly connected digraph is a Hamiltonian digraph.

4.3.22. Prove that a complete digraph contains a source.

4.3.23. Verify that any transitive tournament has a single Hamiltonian path.

4.3.24. Prove that the number of all different Hamiltonian paths in each tournament is odd.

4.3.25. Let a complete strongly connected digraph have n vertices ($n \geqslant 3$). Prove that for any k ($3 \leqslant k \leqslant n$), there exists for any vertex of the digraph a circuit of length k containing this vertex.

4.3.26. Let $D(V, X)$ be a complete strongly connected digraph for which $|V| \geqslant 4$. Prove that the digraph D contains two different vertices v_1 and v_2 satisfying the following condition: digraphs D_1 and D_2 obtained from the digraph D as a result of removal of the vertices v_1 and v_2 (respectively) are strongly connected.

4.3.27. We shall denote by A^q the q-th degree of the adjacency matrix $A(D) = \| a_{ij} \|$ of a digraph D. Prove that the (i, j)-th element $a_{ij}^{(q)}$ of the matrix A^q is equal to the number of all $(v_i$-$v_j)$-walks of length q (in the digraph D).

4.3.28. Let B be the incidence matrix of a digraph $D(V, X)$. Prove that the subset X_1 of arcs of the digraph $D(X_1 \subseteq X)$ generates a simple cycle (not necessarily oriented) if and only if the set of columns corresponding to these arcs (in matrix B) is linearly dependent, and each of its proper subsets does not possess this property.

4.3.29. Prove that the determinant of any square submatrix of the incidence matrix $B(D)$ of a digraph D is equal to 0, or +1, or −1.

4.3.30. Let B be the incidence matrix of a weakly connected n-vertex digraph D, and the matrix $\bar{B}$ be obtained from B by deletion of any (one) row. Prove that there exists a one-to-one correspondence between different[1] directed trees of the digraph D (which are treated as directed subgraphs of the digraph D) and non-degenerate $(n-1)$-order submatrices of the matrix $\bar{B}$.

4.3.31. Let a matrix $\bar{B}$ be a submatrix of the incidence matrix B of a weakly connected digraph D, described in the previous problem. By $\bar{B}'$ we shall denote a matrix transposed relative to the matrix $\bar{B}$. Prove that the number of different directed trees in the digraph D is equal to the value of the determinant of the matrix $\bar{B}\cdot\bar{B}'$.

4.4. Trees and Bipolar Networks

A pseudograph $G = (V, X)$ in which k vertices known as *poles* are isolated is termed a *k-pole network*. The pseudograph G will be called a *graph* of the corresponding *k-pole network*. A network Γ with a set of poles P and a graph $G = (V, X)$ will be denoted by $(P; V, X)$. Two k-pole networks are *isomorphic* if their graphs are isomorphic and there is one-to-one correspondence between their poles. A one-pole network whose graph is a tree is called a *rooted tree*. The only pole of such a network is called a *root*. A *plane rooted tree* is the map of a graph on a plane. This concept can be defined by induction as follows. The network shown in Fig. 10*a* is a plane rooted tree. If A and B (see Fig. 10*b*) are plane rooted trees, the figures C, D and E (Figs 10*c* and *d*) are also plane rooted trees. We shall assume that an arbitrary plane rooted tree is mapped onto a plane with a cut which is a ray emerging from the root (see Fig. 10*e*). Here we can assume that the edges incident with the root are numbered

[1] Here we assume that two trees are *different* if they are *non-isomorphic* directed trees with *labelled* (*numbered*) *vertices*.

clockwise by 1, . . ., m, where m is the degree of the root. If we remove the edge with number i from such a plane rooted tree, we shall obtain a graph with two connected components. The component which does not contain a root will be called the i-th *branch* of the original rooted tree. The vertex incident with the i-th edge (in the original tree) will be treated as the root of the i-th

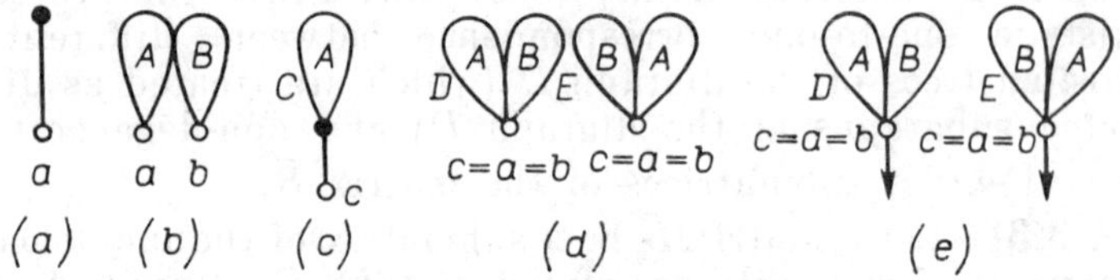

Fig. 10

branch. For an arbitrary rooted tree A, we denote by $d_0(A)$ the degree of its root. Plane rooted trees A and B are called *identical* if either $d_0(A) = d_0(B) = 0$ or $d_0(A) = d_0(B) = m > 0$ and for any $i = \overline{1, m}$ the i-th branches of the trees A and B are identical. Two trees which are not identical are referred to as *different*. Thus, trees D and E are different (see Fig. 10*d*) if the trees A and B (see Fig. 10*b*) are different. Each plane rooted tree T with m edges can be put in one-to-one correspondence with a binary vector of length $2m$, known as a *code of the tree.* A tree with a single edge corresponds to the vector 01. If vectors $\tilde{\alpha}$ and $\tilde{\beta}$ are put in correspondence with trees A and B (Fig. 10*b*) respectively, then the vector $0\tilde{\alpha}1$ is put in correspondence with the tree C (Fig. 10*c*), and vectors $\tilde{\alpha}\tilde{\beta}$ and $\tilde{\beta}\tilde{\alpha}$ are put in correspondence with the trees D and E (Fig. 10*d*).

Henceforth, unless stipulated otherwise, the term network shall mean a two-pole network. The network $\Gamma(\{a, b\}; V, X)$ will be briefly denoted by $\Gamma(a, b)$. The *subgraph* of such a network is the subgraph of the graph (V, X). A vertex of a non-trivial subgraph G of the network Γ is called *boundary vertex* if it is either a pole or incident with a certain edge of the network, which does not belong to the subgraph G. We shall call a non-trivial subgraph of a network a *shoot* if it has a unique boundary vertex. A *subnetwork* of a two-pole network is its subgraph having exactly two boundary vertices. These vertices are the poles of the subnetwork. A network is *connected* if its graph is connected. A connected network (or subnetwork) having a single edge is called *trivial*. A connected network is called *strongly connected* if each edge lies in a simple chain connecting the poles of the network. A strongly connected network is known to be *decomposable* if it has at least one non-trivial subnetwork. Otherwise, it is assumed to be *non-decomposable*. Let $\Gamma(a, b)$ be a decomposable network, $G(c, d)$ its non-trivial subnetwork, and $\Gamma_1(a, b)$ be a network obtained from $\Gamma(a, b)$ by substituting an edge (c, d) for the subnetwork $G(c, d)$. Then the network $\Gamma(a, b)$, in turn, can be obtained by substituting the network $G(c, d)$ for the edge (c, d) of the network $\Gamma_1(a, b)$. Thus, the decomposable network $\Gamma(a, b)$ can be defined by specifying the network $\Gamma_1(a, b)$, edge (c, d) of the network $\Gamma_1(a, b)$, and the network $G(c, d)$. Such a definition is referred to as the *decomposition of the network* $\Gamma(a, b)$. A network $\Gamma_1(a, b)$ is called *external* and the network $G(c, d)$ *internal network of the decomposition*. A network $\Gamma(a, b)$ is called the *superposition of the networks* $\Gamma_1(a, b)$ and $G(c, d)$. A network consisting of m parallel edges joining the poles a and b is denoted by $\Gamma_m^p(a, b)$ or simply Γ_m^p. A network whose graph is a simple chain of length m and which joins the poles a and b is denoted by $\Gamma_m^s(a, b)$ or simply Γ_m^s. A network that can be obtained from networks Γ_2^p and Γ_2^s by applying a finite number of substitutions of a network for an edge is referred to as a *series-parallel network*, or a π-network. A non-trivial non-decomposable network $\Gamma(a, b)$ differing from $\Gamma_2^p(a, b)$ and $\Gamma_2^s(a, b)$ is called an *H-network*.

A decomposable network is called *p-decomposable* (*s-decomposable*) if a certain external network of the decomposition has the form Γ_m^p (resp. Γ_m^s), $m \geqslant 2$. If a certain external network of the decomposition of a network Γ is an H-network, then Γ is referred to as *H-decomposable*. Any decomposable network is either p-, or s-, or H-decomposable. A *canonical p-decomposition* of a network is a p-decomposition for which the internal networks of the decomposition are other than networks of the type Γ_2^p and other than p-decomposable networks.

The *canonical s-decomposition* is defined in a similar way. The *canonical H-decomposition* is a *decomposition* whose external network is an H-network. Each π-network Γ with $m \geqslant 1$ edges can be put in correspondence with a plane rooted tree $T(\Gamma)$ with m pendant vertices, such that (a) each vertex of the tree $T(\Gamma)$ other than a pendant vertex is labelled by a symbol p or s; (b) the labels p and s alternate on each chain emerging from the root to a pendant vertex; and (c) vertices other than the root have a degree other than two. Pendant vertices of the tree $T(\Gamma)$ are not labelled. The tree $T(\Gamma)$ is defined by induction. If Γ has the form Γ_m^p (or Γ_m^s), then $T(\Gamma)$ is a tree

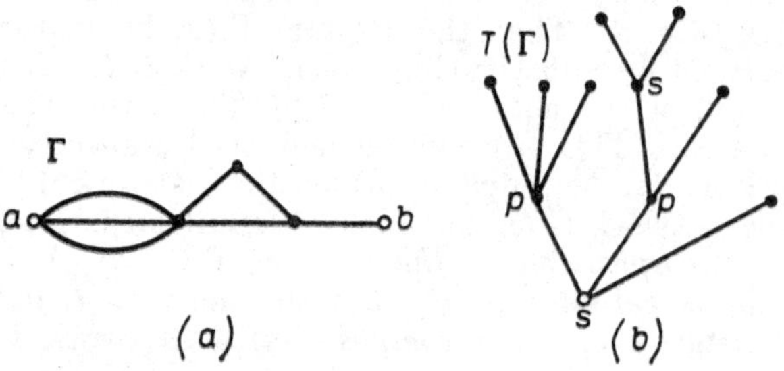

Fig. 11

whose root is labelled by the symbol p (resp. s), while the remaining m vertices are pendant, adjacent to the root and are not labelled. If a network Γ is decomposable and other than the networks of the type described above, and the external network of the decomposition has the form Γ_k^p (or Γ_k^s), while the internal networks are just $G_1, G_2, \ldots, G_k$, the tree $T(\Gamma)$ is constructed as follows. Let $T(G_1), T(G_2), \ldots, T(G_k)$ be the trees corresponding to the internal networks of the decomposition. Then for the root of $T(\Gamma)$ we take a vertex of degree k, labelled by the symbol p (resp. s). The vertices adjacent to the root are labelled by the symbol s (resp. p). The vertices $v_1, v_2, \ldots, v_k$ adjacent to the root are identified with the roots of the trees $T(G_1), T(G_2), \ldots, T(G_k)$. For example, the π-network shown in Fig. 11*a* corresponds to the tree shown in Fig. 11*b*. A tree $T(\Gamma)$ is called a *diagram of the canonical decomposition* of a π-network Γ. It should be noted that if the external network of the decomposition of a network Γ has the form $\Gamma_k^s(a, b)$, and the internal networks $G_1(a, u_1)$, $G_2(u_1, u_2)$, . . ., $G_k(u_{k-1}, b)$ are substituted for the edges (a, u_1), (u_1, u_2), . . ., (u_{k-1}, b), respectively, the vertices $v_1, v_2, \ldots, v_k$ in the tree $T(\Gamma)$, which are identified with the roots of plane rooted trees $T(G_1)$, $T(G_2)$, . . ., $T(G_k)$, follow one another from left to right in

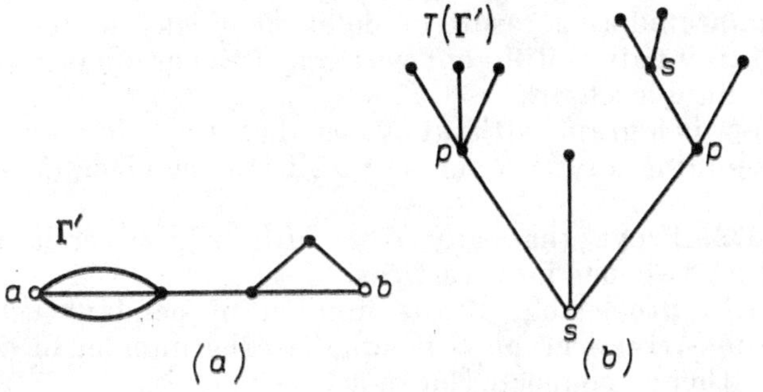

Fig. 12

the increasing order of numbers. Thus, the tree $T(\Gamma')$ shown in Fig. 12*b* is a diagram of the canonical decomposition of the network Γ' (Fig. 12*a*) but is not a diagram of the network Γ (see Fig. 11*a*).

A vertex of a network other than a pole is referred to as *internal*. A vertex v depends on a vertex u if any simple chain joining the poles and passing through v passes through u as well. Vertices v and u are *equivalent* if v depends on u and u depends on v. A vertex v is *weaker* than a vertex u, while the vertex u is *stronger* than the vertex v if v depends on u but is not equivalent to it. A vertex v is called *minimal* if it is not weaker than any other internal vertex of a network. Henceforth, we shall call a simple chain joining the poles of a network a *chain of the network*. The cases when the term "chain" is used in different sense will always be stipulated. A chain of a network will be assumed to be the *shortest* if it has the minimum possible length. The *length* of a network is equal to the length of its shortest chain. A *cut* is a set of the edges of a network whose deletion destroys all the chains. A cut is called *terminal* if none of its subsets is a cut. A cut is referred to as *minimal* if it has the minimum possible number of edges. The number of edges in the minimal cut is called the *width* of a network.

4.4.1. Let G be a graph with $n \geqslant 2$ vertices. Prove the equivalence of the following statements:

(1) G is a connected graph with $n-1$ edges;

(2) G is a connected graph which, however, becomes disconnected as a result of deletion of any of its edges;

(3) any pair of different vertices of the graph G is joined by a unique chain;

(4) G is a graph without cycles, but the addition of an edge joining any two vertices leads to the emergence of a cycle.

4.4.2. Prove that any tree with $n \geqslant 2$ vertices has at least two pendant vertices.

4.4.3. Prove that if the number of pendant vertices in a non-trivial graph G is equal to the number of edges, G is either disconnected or is a tree.

4.4.4. Let $F(G) = \{H_1, H_2, \ldots, H_n\}$ be a family of graphs in which the graph H_i is obtained from an n-vertex graph G by deleting a vertex with the number i $(i = \overline{1, n})$. The vertices in the graphs H_i are not labelled. Prove that

(1) it can be found out from the family $F(G)$ whether the graph G is a tree;

(2) if G is a tree, it can be uniquely reconstructed from $F(G)$ (accurate to an isomorphism).

4.4.5. The *intersection of* two *graphs* G and H is a graph $G \cap H$ all whose vertices and edges belong to both G and H. Prove that a non-empty intersection of two subtrees of a tree is a tree.

Let $\rho_G(v, u)$ be the distance between vertices v and u in a graph $G = (V, X)$. The vertex u_0 for which

$$\max_{v \in V} \rho_G(u_0, v) = \min_{u \in U} \max_{v \in V} \rho_G(u, v)$$

is called the *centre of the graph*, and the number $R(G) = \max_{v \in V} \rho_G(u_0, v)$ is called the *radius of the graph.*

4.4.6. (1) Determine the number of the centres in the graph

(a) G (see Fig. 6*a*);

(b) G (see Fig. 6*b*);

(c) $G = K_{n_1, n_2}$.

(2) Prove that any tree contains not more than two centres.

4.4.7. Prove that a tree has one centre if its diameter is an even number and two centres when the diameter is odd.

4.4.8. (1) Let $D(G)$ be the diameter and $R(G)$ the radius of a graph G. Prove that $R(G) \leqslant D(G) \leqslant 2R(G)$.

(2) Prove that if G is a tree, then $R(G) = \left]\frac{D(G)}{2}\right[$.

(3) Give example of a graph G for which $R(G) = D(G)$.

4.4.9*. Prove that a tree can be uniquely reconstructed (correct to an isomorphism) if pairwise distances between its pendant vertices are specified.

4.4.10. Prove that any two simple chains of maximum length in a tree with an odd diameter have at least one edge in common.

4.4.11. An *infinite tree* is the term applied to a graph with a countable set of vertices satisfying the following condition: for any two vertices u and v of the graph, there exists a unique simple (u-v)-chain with a finite length. Prove that if the degree of each vertex of an infinite tree is finite, there exists for any vertex a simple chain of infinite length, which contains this vertex.

4.4.12. Let $P = \{v_1, v_2, \ldots\}$ be an infinite simple chain. Let $G = K_2 \times P$. What is the power of the set of all spanning trees of the graph G?

Let $G = (V, X)$ be a multigraph with a set of vertices $V = \{v_1, v_2, \ldots, v_n\}$, and $M(G) = \| a_{ij} \|$ be a quadratic matrix of order n, where

$$a_{ij} = \begin{cases} d(v_i) & \text{for } i = j; \\ -1 & \text{for } i \neq j \text{ and } (v_i, v_j) \in X; \\ 0 & \text{for } i \neq j \text{ and } (v_i, v_j) \notin X. \end{cases}$$

It is well known [23] that the number of pairwise different spanning trees of the graph is equal to the minor of any element of the principal diagonal of the matrix $M(G)$.

4.4.13. Determine the number of the spanning trees of the graphs shown in Fig. 4*a*, *b* and 8*a*, *b*, if the vertices of the graphs are labelled.

4.4.14. What is the chromatic number of a tree with $n \geqslant 2$ vertices?

4.4.15. Is it true that if the diameter of a graph G is k ($k > 2$), there exists a spanning tree whose diameter is equal to k?

4.4.16. Prove that for $n \geqslant 3$, the number of pairwise non-isomorphic rooted trees with n vertices is at least double the number of pairwise non-isomorphic trees with n vertices having no roots.

4.4.17. Let a rooted tree with n ($n \geqslant 2$) pendant vertices have no vertices of power 2 other than a root. Prove

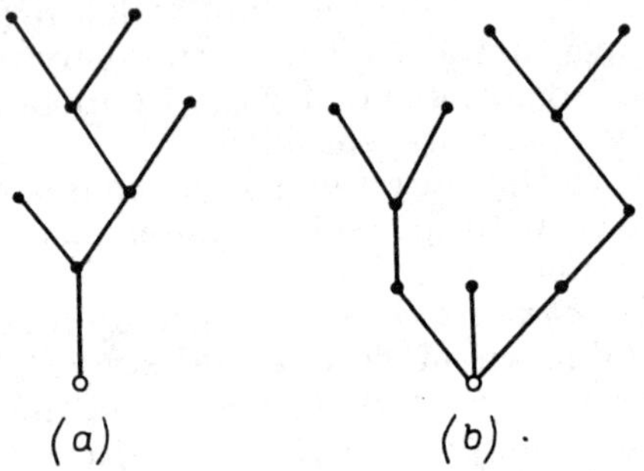

Fig. 13

that the total number of vertices of the tree does not exceed $2n - 1$.

4.4.18. Construct the codes of plane rooted trees shown in Fig. 13*a* and *b*.

4.4.19. Using a given code $\tilde{\alpha}$, construct a plane rooted tree if:

(1) $\tilde{\alpha} = (001010011011)$;

(2) $\tilde{\alpha} = (0100011001101011)$;

(3) $\tilde{\alpha} = (0001010110011011)$.

4.4.20. (1) Determine the number of pairwise non-isomorphic rooted trees with four edges.

(2) Determine the number of pairwise different plane rooted trees with four edges.

4.4.21. Divide the set of vectors $\{\tilde{\alpha}_1, \tilde{\alpha}_2, \tilde{\alpha}_3, \tilde{\alpha}_4, \tilde{\alpha}_5\}$ into equivalence classes so that the vectors of one class are the codes of pairwise isomorphic trees: $\tilde{\alpha}_1 =$

(0010101101), $\tilde{\alpha}_2 = (0100101101)$, $\tilde{\alpha}_3 = (0101001011)$, $\tilde{\alpha}_4 = (0100101011)$, $\tilde{\alpha}_5 = (0010110101)$.

4.4.22. Prove that the code $\tilde{\alpha} = (\alpha_1, \alpha_2, \ldots, \alpha_{2n})$ of a plane rooted tree with n edges possesses the following properties:

(1) $\sum_{i=1}^{2n} \alpha_i = n$;

(2) for any k $(1 \leqslant k \leqslant 2n)$, the inequality $\sum_{i=1}^{k} \alpha_i \leqslant k/2$ is valid.

4.4.23. Prove that any binary vector $\tilde{\alpha}$ of length $2n$, which satisfies conditions (1) and (2) of the previous problem, is a code of a plane rooted tree with n edges.

4.4.24. (1) Prove that the following recurrence relation is valid for the number $\psi(n)$ of vectors $\tilde{\alpha} \in B^{2n}$ that satisfy conditions (1) and (2) of Problem 4.4.22:

$$\psi(n) = \sum_{i=1}^{n} \psi(i-1)\,\psi(n-i), \text{ where } \psi(0) = \psi(1) = 1.$$

(2*) Derive an analytic expression for $\psi(n)$.

4.4.25. Let G be a connected plane graph. Starting from a certain vertex v belonging to an external face, we shall go around (circumvent) this boundary so that it always remains on the right. At a certain moment of time, we shall return to the initial vertex v. The circumvention will be continued if there remain uncircumvented edges incident with the vertex v and belonging to the boundary of the external face. Otherwise, the circumvention is terminated.

(1) Prove that G is a tree if and only if each edge is circumvented twice in the above process.

(2) Let G be a tree with a root v. In accordance with the circumvention procedures described above, the tree G can be put in correspondence with a binary vector. By circumventing edges consecutively, we shall write 0 if an edge is circumvented for the first time and 1 if it is circumvented for the second time. Prove that the vector obtained in such a way is a code of the tree G.

4.4.26. Let G be a graph with n vertices and m edges. The graph is called *balanced* if neither of its subgraphs has vertices of power higher than $2m/n$.

(1) Prove that for $n > 2$, a tree is a non-balanced graph.

(2) Give an example of a balanced graph for which $m = n + 3$.

4.4.27. Let G be a graph with a real non-negative number ascribed to each edge as a weight. The *weight of a subgraph* of the graph G is the sum of the weights of the edges of this subgraph. The *minimal spanning tree of the graph* G is its spanning tree having the minimum weight. Prove that the minimal covering tree can be obtained by using the following algorithm. At the first stage, we choose the edge with the minimum weight. Then in the next stage we choose the edge that has the minimum weight among the edges which do not form a cycle with the edges chosen earlier. The algorithm is terminated when there is no such edge.

4.4.28. Is it true that any subnetwork of a strongly connected network is strongly connected?

4.4.29. Let a connected network have a unique shoot with k vertices, which is not contained in another shoot with a larger number of vertices. Prove that it is sufficient to draw $k - 1$ additional edges to make the network strongly connected. Can we always manage with the minimum number of additional edges?

4.4.30. Is it true that for any n and m ($m \geqslant n \geqslant 3$) there exist decomposable networks with n vertices and m edges?

4.4.31. Is it true that the graph of a strongly connected network is 2-connected?

4.4.32. Is it true that if we choose two non-adjacent vertices in an arbitrary 2-connected cubic graph as poles, we shall obtain a non-decomposable network?

4.4.33. Determine the number of pairwise non-isomorphic non-decomposable networks that can be obtained by choosing two vertices in an n-dimensional cube as poles.

4.4.34. (1) Prove that if a non-decomposable network has $n > 2$ vertices and m edges, then

$$3n \leqslant 2m + 2 \leqslant n(n-1).$$

(2) Prove that for any n and m such that $m \geqslant \frac{3}{2}n - 1$, $n \geqslant 4$, there exists a non-decomposable network with n vertices and m edges.

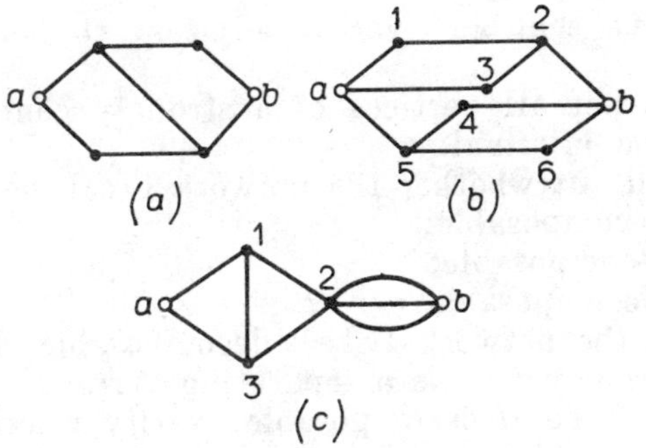

Fig. 14

4.4.35. For the networks shown in Fig. 14*a*, *b* and *c* determine

(1) the type of decomposition;

(2) external networks of the canonical decomposition.

4.4.36. Let Γ be the network shown in Fig. 15.

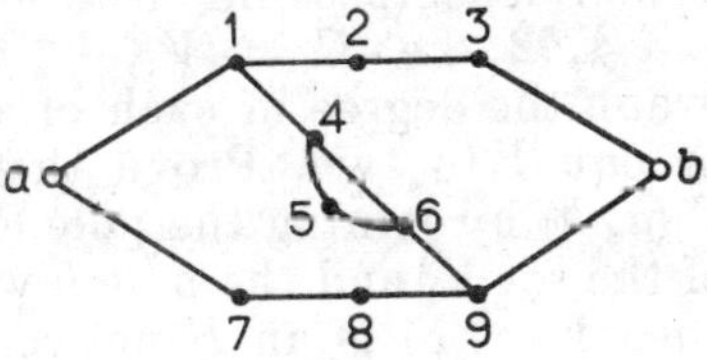

Fig. 15

(1) Indicate all minimal vertices of the network.

(2) Divide all the internal vertices of the network Γ into the classes consisting of pairwise equivalent vertices.

(3) Verify whether there exists a vertex in this network which is weaker than any other vertex.

4.4.37. Prove that for any vertex v of a strongly connected network there exists a chain containing all the vertices that are stronger than or equivalent to it.

4.4.38. Let S (Γ) be a set of all vertices v of the network Γ which are not minimal.

(1) Is it true that if Γ is an H-decompoable network without multiple edges, we obtain a non-decomposable network by joining each vertex $v \in S$ (Γ) through edges with each pole to which v is not adjacent?

(2) For obtaining a non-decomposable network is it sufficient to join each vertex v from S (Γ) with exactly one pole from those which are not adjacent to v?

4.4.39. (1) Prove that any separating vertex is minimal.

(2) Prove that any vertex adjacent to both poles is minimal.

4.4.40. Let all vertices of a strongly connected network Γ be minimal.

(1) Find out whether the network Γ can be

(a) p-decomposable;

(b) s-decomposable;

(c) H-decomposable.

(2) Let the network Γ be s-decomposable. Prove that each internal vertex is a separating vertex.

(3) Let Γ be H-decomposable. Verify whether or not any of internal networks can be

(a) an H-network;

(b) an H-decomposable network;

(c) a network other than networks Γ_m^p $(m \geqslant 1)$.

4.4.41. Let a strongly connected network Γ with eight edges be neither p- nor s-decomposable and have no subnetworks of the type Γ_m^p, Γ_m^s $(m > 1)$. Prove that Γ is a non-decomposable network.

4.4.42. Let $G = (V_1, V_2, X)$ be a connected bipartite graph the degree of each of whose vertices is higher than or equal to two. Prove that if we construct a network Γ (a, b) by joining the pole a by an edge with each vertex of the set V_1 and the pole b with each vertex of the set V_2, then Γ (a, b) is an H-network.

4.4.43. Does there exist a p-decomposable network for which any subnetwork containing at least three edges is p-decomposable?

4.4.44. Construct canonical decomposition diagrams for the networks shown in Fig. 16*a* and *b*.

4.4.45. Construct π-networks having the canonical desomposition diagrams shown in Fig. 17*a* and *b*.

4.4.46. Prove that if π-networks Γ_1 and Γ_2 are not isomorphic, they have different canonical decomposition diagrams.

4.4.47. Let A and B be two chains of a network Γ (a, b) and let a vertex u belong to chain A and not belong to chain B, while a vertex v belongs to chain B and does not belong to A. Further, let $[u, v]$ be a simple chain which joins u and v and does not intersect the chains A and B at any point other than its ends.

(1) Prove that there exist at least two chains of the network Γ whose vertices belong to the union of the chains A, B and $[u, v]$ and which contain the chain $[u, v]$. Do at least three such chains always exist?

(2) Prove that if the chains A and B have internal vertices in common but have no edges in common, there

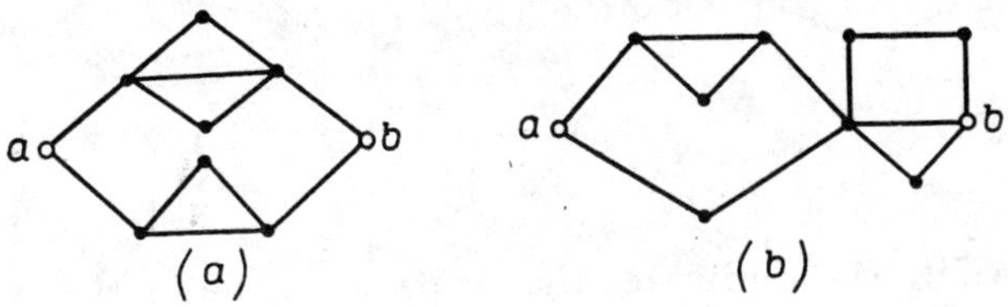

Fig. 16

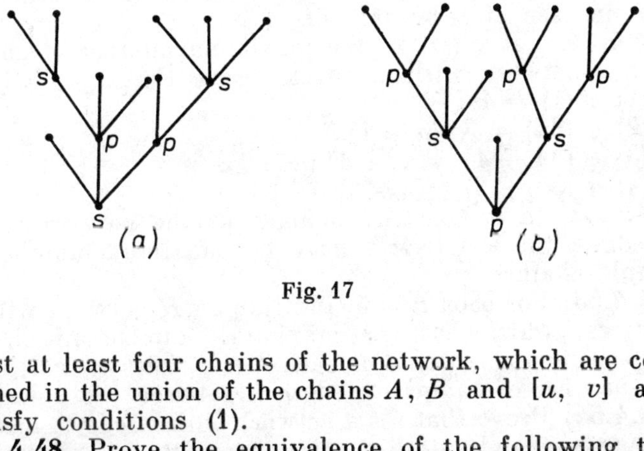

Fig. 17

exist at least four chains of the network, which are contained in the union of the chains A, B and $[u, v]$ and satisfy conditions (1).

4.4.48. Prove the equivalence of the following two definitions of a π-network:

(1) A network $\Gamma(a, b)$ is a π-network if its edges can be oriented so that all the edges in each simple chain joining the poles a and b are directed from a to b.

(2) The π-networks are those and only those networks which are obtained as a result of the following inductive process:

(a) The networks Γ_2^p and Γ_2^s (Fig. 18a) are π-networks.

(b) If the networks A and B (Fig. 18b) are π-networks, the networks shown in Fig. 18c are also π-networks.

4.4.49. Prove that among all π-networks with m ($m > 4$) edges, the s-decomposable networks have the largest number of shortest chains.

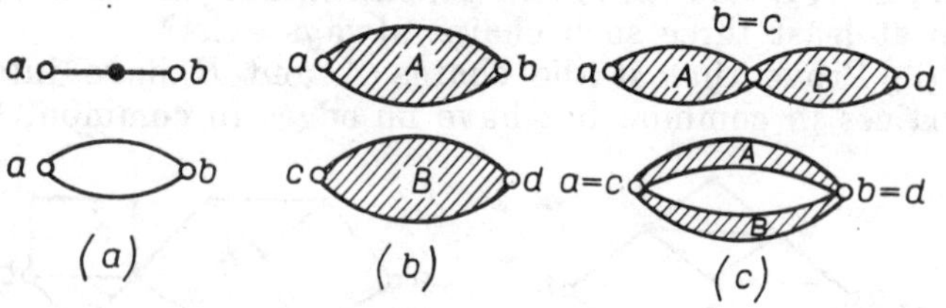

Fig. 18

4.4.50. (1) Indicate the type of π-networks which have the maximum number of simple chains joining the poles.

(2) Indicate the type of π-networks having the maximum number of terminal cuts.

4.4.51. Let $\varphi(m)$ be the maximum number of chains of a π-network with m edges. Prove that

(1) $\varphi(1) = 1$;

(2) $\varphi(3n) = 3^n$ $(n \geqslant 1)$;

(3) $\varphi(3n + 1) = 4 \times 3^n$ $(n \geqslant 1)$;

(4) $\varphi(3n + 2) = 2 \times 3^n$ $(n \geqslant 1)$.

4.4.52. Is it true that among all the networks with m edges, the π-networks have the maximum number of simple chains?

4.4.53. For each $n \geqslant 5$, indicate an H-network with n vertices, which has the maximum number of chains connecting the poles. For each k $(1 \leqslant k < n)$, calculate the number of chains of length k between the poles.

4.4.54. Prove that for a network with m edges, having a length l and width t, the inequality $m \geqslant l \times t$ is valid.

4.4.55. Prove that in a π-network, the intersection of any simple chain between poles with any terminal cut contains exactly one edge.

4.4.56*. The set of chains in a network Γ is called a *defining* set for a vertex v if $\{v\}$ is the intersection of the sets of the internal vertices of these chains. Prove or disprove the following statement: the necessary and sufficient condition for a non-p-decomposable network without multiple edges to be non-decomposable is that there must exist a defining set of chains for any internal vertex.

4.4.57. Prove the following statement: the necessary and sufficient condition for making a decomposable network Γ non-decomposable by adding edges is that Γ must not have multiple edges and must have at least four vertices.

4.4.58. Construct a decomposable network with the smallest number of edges and n vertices ($n \geqslant 4$), which cannot be made non-decomposable by successively replacing subnetworks of the type Γ_2^s and Γ_2^p by edges.

4.5. Estimates in the Theory of Graphs and Networks

A graph (digraph, pseudograph, etc.) is called *labelled* (or *numbered*) if its vertices are assigned labels (or numbers). We shall denote by $\mathcal{G}_n$ the set of all n-vertex graphs (in short n-graphs) whose vertices are labelled by numbers 1, 2, . . ., n. The subset of all graphs from $\mathcal{G}_n$, each of which has exactly m edges, will be denoted by $\mathcal{G}_{n,m}$. A graph with n vertices and m edges will be briefly called an *(n, m)-graph*. Graphs G and H in $\mathcal{G}_n$ are assumed to be *different* if there exist two vertices j and k that are adjacent in one graph but not in the other.

Let $\varphi_n(P)$ denote the number of all graphs in $\mathcal{G}_n$ having the property P. It is said that *almost all n-graphs have the property P* if $\lim\limits_{n\to\infty} \frac{\varphi_n(P)}{|\mathcal{G}_n|} = 1$. Let $m = m(n)$ be an integral non-negative function, and let $\varphi_{n,m}(P)$ be the number of all graphs in $\mathcal{G}_{n,m}$ having the property P. It is said that *almost all the $(n, m(n))$-graphs have the property P* if $\lim\limits_{n\to\infty} \frac{\varphi_{n,m}(P)}{|\mathcal{G}_{n,m}|} = 1$.

4.5.1. Prove that

$$(1)\ |\mathcal{G}_n| = 2^{\binom{n}{2}}; \quad (2)\ |\mathcal{G}_{n,m}| = \binom{\binom{n}{2}}{m}.$$

4.5.2. (1) Find the number of different tournaments with n vertices, labelled by 1, 2, . . ., n.

(2) Find the number of oriented pseudographs with n labelled vertices and m arcs.

4.5.3. (1) Show that the number of graphs in $\mathcal{G}_n$, whose k given vertices are isolated, is equal to $2^{\binom{n-k}{2}}$.

(2) Show that the number of graphs without isolated vertices in $\mathcal{G}_n$ is equal to

$$\sum_{k=0}^{n} (-1)^k \binom{n}{k} 2^{\binom{n-k}{2}}.$$

(3) Show that almost all n-graphs have no isolated vertices.

4.5.4. Let the subset $\mathcal{G} \subset \mathcal{G}_n$ consist of N pairwise different graphs. Show that the number of pairwise non-isomorphic graphs in $\mathcal{G}$ is not less than $N/n!$.

4.5.5. Let $\psi(m)$ be the number of pairwise non-isomorphic connected graphs with m edges. Show that:

(1) $\psi(m) \leqslant \sum\limits_{\frac{1}{2}(1+\sqrt{1+8m}) \leqslant n \leqslant m+1} \binom{\binom{n}{2}}{m}$;

(2) $\psi(m) \lesssim (2m)^m$ for $m \to \infty$.

4.5.6. Show that the number of pairwise non-isomorphic pseudographs that do not have isolated vertices but have m edges does not exceed $(cm)^m$, where c is a constant independent of m.

4.5.7. Show that the number of pairwise non-isomorphic k-pole networks with m edges without loops and without isolated vertices does not exceed $(2m)^k (cm)^{2m}$, where c is a constant independent of m and k.

4.5.8. Prove that the number of pairwise non-isomorphic trees with m edges does not exceed the number of pairwise different plane rooted trees with m edges.

4.5.9. (1) Prove that the number of pairwise different plane rooted trees with m edges does not exceed $\binom{2m}{m}$.

(2) Determine the asymptotic behavior of the number $q(m)$ of plane rooted trees with m edges for $m \to \infty$.

4.5.10. Using Cayley theorem, according to which the number of pairwise different trees with n labelled vertices is equal to n^{n-2}, prove that the number of pairwise non-isomorphic trees with n vertices is not less than $c_n n^{-1.5} e^n$, where $\lim_{n\to\infty} c_n = \sqrt{2\pi}$.

4.5.11. Show that the number of pairwise different trees with n labelled vertices, where the vertex with number 1 has a degree k, is equal to $\binom{n-2}{k-1}(n-1)^{n-k-1}$.

4.5.12. Find the number of graphs in $\mathcal{G}_n$ constituting forests.

4.5.13*. Show that the number of forests in $\mathcal{G}_n$, whose given vertices j and k belong to different components, is equal to $2n^{n-3}$.

4.5.14. Let $\varphi(n)$ be the number of pairwise different rooted trees with n pendant vertices, such that the degree of a root is equal to 2, and the degree of each vertex other than a root or a pendant vertex is equal to 3.

(1) Show that the number $\varphi(n)$ is equal to the number of ways in which brackets can be arranged in the expression $b_1 : b_2 : \ldots : b_n$ so that the new expression obtained in this way is meaningful.

(2) Show that $\varphi(n) = \frac{1}{n}\binom{2n-2}{n-1}$.

4.5.15. (1) Show that the number of pairwise non-isomorphic bipolar π-networks with m edges does not exceed twice the number of pairwise different plane rooted trees with m pendant vertices.

(2) Show that the number of pairwise non-isomorphic π-networks with m edges does not exceed $2\binom{4m-2}{2m-1}$.

4.5.16. Find the number of pairwise non-isomorphic networks $\Gamma(a, b)$ with n vertices and m edges, having the following properties:

(1) the network $\Gamma(a, b)$ is s-decomposable,

(2) all the vertices in the network $\Gamma(a, b)$ are minimal.

Remark. The poles a and b of the network $\Gamma(a, b)$ are not equal: the former is the entrance to the network and the latter is the exit from the network. Upon an isomorphic mapping of the network Γ to the network G, the entrance (exit) of the network Γ must correspond to the entrance (exit) of the network G.

4.5.17. Let $\Phi(n, m)$ be the number of different formulas generated by the set of connectives $\{\&, \vee\}$ and the set of variables $\{x_1, x_2, \ldots, x_n\}$ with m entries of the connectivity symbols.

(1) Show that $\Phi(n, m)$ is equal to the number of pairwise different rooted trees each of whose pendant vertices is labelled by a certain symbol in the set $\{x_1, x_2, \ldots, x_n\}$, while each non-pendant vertex is labelled by one of the symbols & or $\vee$.

(2) Show that $\Phi(n, m) = \frac{1}{m+1}\binom{2m}{m} 2^m n^{m+1}$.

Remark. Formulas are assumed to be different if they form different words in the alphabet $\{\&, \vee, (,), x_1, x_2, \ldots, x_n\}$.

4.5.18. Let $\Phi(n, m, k)$ be the number of different formulas generated by the set of connectives $\{\&, \vee, {}^{-}\}$ and the set of variables $\{x_1, x_2, \ldots, x_n\}$ with m entries of the symbols of variables and k entries of the symbol $^{-}$. Prove that

$$\Phi(n, m, k) \leqslant \frac{1}{m}\binom{2m-2}{m-1} 8^{m-1} n^m.$$

4.5.19. Show that the number of disconnected graphs in $\mathcal{G}_{n,m}$ does not exceed

$$\sum_{k=1}^{[n/2]} \binom{n}{k} \binom{\binom{n}{2} - k(n-k)}{m}.$$

4.5.20. Show that the number of graphs in $\mathcal{G}_{n,m}$ having exactly two connectivity components does not exceed

$$\sum_{k=1}^{[n/2]} \binom{n}{k} k^{k-2}(n-k)^{n-k-2} \sum_{j=k-1}^{\binom{k}{2}} \binom{\binom{k-1}{2}}{j-k+1} \binom{\binom{n-k-1}{2}}{m-j-n+k+1}.$$

4.5.21. Show that the number of k-connected graphs in $\mathcal{G}_{n,m}$ which are not $(k+1)$-connected $(k \leqslant n-2)$ does not exceed

$$\binom{n}{k} \sum_{j=\binom{k+1}{2}}^{m} \binom{\binom{k}{2} + k(n-k)}{j} \sum_{r=1}^{\left[\frac{n-k}{2}\right]} \binom{n-k}{r} \times \binom{\binom{n-k}{2} - r(n-k-r)}{m-j}.$$

Hint. For $k \leqslant n-2$, a graph that is not $(k+1)$-connected contains k vertices whose deletion results in a disconnected graph.

4.5.22. Show that the number of pairwise different connected subgraphs of a cube B^n, which are generated by subsets with k vertices does not exceed $2^n (4n)^{k-1}$. (Assume that the vertices of the cube B^n are labelled by the numbers from 1 to 2^n).

4.5.23. By majorizing the number of graphs in $\mathcal{G}_n$ having a vertex of degree $n-1$, show that almost all n-graphs have a radius larger than unity.

Let $p(G)$ be a certain numerical parameter of graph G. Let $\bar{p}(n) = 2^{-\binom{n}{2}} \sum\limits_{G \in \mathcal{G}_n} p(G)$ be the mean value of parameter p, and $\mathrm{D}p(n) = 2^{-\binom{n}{2}} \sum\limits_{G \in \mathcal{G}_n} (p(G) - \bar{p}(n))^2$ be the variance of p. In the same way, we can determine the mean value and variance of parameters of the graphs in $\mathcal{G}_{n,m}$. Let $\theta > 0$, $\delta_n(\theta)$ be the fraction of those graphs G in $\mathcal{G}_n$ for which $p(G) \geqslant \theta$, and $\Delta_n(\theta)$ be the fraction of such graphs G in $\mathcal{G}_n$, that $|p(G) - \bar{p}(n)| \geqslant \theta$. Various estimates and proofs of the properties of nearly all graphs are frequently obtained with the help of the following (Chebyshev's) inequalities:

$$\delta_n(\theta) \leqslant \frac{\bar{p}(n)}{\theta}; \qquad (1)$$

$$\Delta_n(\theta) \leqslant \frac{\mathrm{D}p(n)}{\theta^2}. \qquad (2)$$

For example, let $p(G)$ be the number of isolated vertices of the graph G. We must show that $p(G)=0$ for almost all n-graphs. Let $g_n(i)$ be the number of graphs in $\mathcal{G}_n$, in which the i-th vertex is isolated. In this case,

$$\bar{p}(n) = 2^{-\binom{n}{2}} \sum_{G \in \mathcal{G}_n} p(G) = 2^{-\binom{n}{2}} \sum_{i=1}^{n} g_n(i).$$

Obviously, $g_n(i) = 2^{\binom{n-1}{2}}$ for all $i = \overline{1, n}$. Hence $\bar{p}(n) = n \cdot 2^{-n}$. Putting $\theta = 1/2$ in (1), we find that the fraction of graphs G in $\mathcal{G}_n$, for which $p(G) \geqslant 1/2$, does not exceed $n2^{-n+1}$. But $\lim\limits_{n\to\infty} n\cdot 2^{-n+1}=0$. Hence $p(G)<1/2$ for almost all n-graphs, i.e. $p(G)=0$.

Let us now suppose that $p(G)$ is the number of edges in the graph G. We shall show that for nearly all n-graphs $p(G)=1/2\binom{n}{2}(1+\varepsilon_n)$, where $\lim\limits_{n\to\infty}\varepsilon_n=0$. We have $\bar{p}(n)=2^{-\binom{n}{2}}\sum\limits_{G\in\mathcal{G}_n} p(G)=2^{-\binom{n}{2}}\sum\limits_{(i,j)} g_n(i,j)$, where $g_n(i,j)=2^{\binom{n}{2}-1}$ is the number of graphs in which the pair (i,j) of vertices is joined through an edge. Thus, $\bar{p}(n)=1/2\binom{n}{2}$. Let us calculate the variance:

$$\mathrm{D}p(n)=2^{-\binom{n}{2}}\sum_{G\in\mathcal{G}_n}(p(G)-\bar{p}(n))^2=2^{-\binom{n}{2}}\sum_{G\in\mathcal{G}_n}p^2(G)-(\bar{p}(n))^2.$$

Let us label all pairs of the type (i,j) $1\leqslant i<j\leqslant n$ by numbers from 1 to $\binom{n}{2}$, and let $\tilde{g}_n(\nu,\mu)$ be the number of graphs G in $\mathcal{G}_n$, in which pairs with numbers ν and μ are edges. Then

$$\sum_{G\in\mathcal{G}_n}p^2(G)=\sum_{\nu=1}^{\binom{n}{2}}\sum_{\mu=1}^{\binom{n}{2}}\tilde{g}_n(\nu,\mu)=\sum_{\nu=1}^{\binom{n}{2}}\tilde{g}_n(\nu,\nu)+2\sum_{\nu<\mu}\tilde{g}_n(\nu,\mu).$$

But $\tilde{g}_n(\nu,\mu)=2^{\binom{n}{2}-2}$ if $\nu\neq\mu$. Hence

$$\mathrm{D}p(n)=\frac{1}{2}\binom{n}{2}+\frac{1}{4}\binom{n}{2}\left(\binom{n}{2}-1\right)-\left(\frac{1}{2}\binom{n}{2}\right)^2=\frac{1}{4}\binom{n}{2}.$$

Putting $\theta=\sqrt{n\bar{p}(n)}$ in (2), we find that the fraction of those graphs $G\in\mathcal{G}_n$ for which $\left|p(G)-\frac{1}{2}\binom{n}{2}\right|\geqslant\sqrt{\frac{n}{2}\binom{n}{2}}$ does not exceed $1/n$. Hence for almost all graphs $p(G)=1/2\binom{n}{2}(1+\varepsilon_n)$, where $\lim\limits_{n\to\infty}\varepsilon_n=0$.

4.5.24. Let $p(G)$ be the number of pairs of different vertices of graph G in $\mathcal{G}_n$, for which there is no chain of length smaller than 3 joining these vertices. Let $\overline{p}(n) = 2^{-\binom{n}{2}} \sum_{G \in \mathcal{G}_n} p(G)$.

(1) Show that $\overline{p}(n) = \frac{1}{2}\binom{n}{2}\left(\frac{3}{4}\right)^{n-2}$.

(2) Show that almost all n-graphs have no vertices separated by a distance larger than 2.

(3) Using the results of problems 4.5.23 and 4.5.24 (2), show that the radii and diameters of almost all n-graphs are equal to two.

4.5.25. Show that the average number of Hamiltonian cycles in graphs G in $\mathcal{G}_n$ is equal to $\frac{(n-1)!}{2^{n+1}}$.

4.5.26. Find the average number of cycles of length 3 in graphs G of $\mathcal{G}_{n,m}$.

4.5.27*. Using the Chebyshev inequality (2), show that nearly all in $(n, m(n))$-graphs, where $m(n) = [n/\ln(n \ln n)]$, the number of isolated vertices is equal to $n(1 - \varepsilon(n))$, where $\lim_{n\to\infty} \varepsilon(n) = 0$.

4.5.28*. Let k be an integer ($k \geqslant 2$). Show that if $m = m(n) = \varphi(n) \cdot n^{2-\frac{2}{k-1}}$ (where $\varphi(n) \to \infty$ for $n \to \infty$), nearly all (n, m)-graphs contain a complete subgraph with k vertices.

4.5.29. Find the average number of k-vertex independent sets in graphs G in $\mathcal{G}_n$.

4.5.30. Let k be a natural number. Calculate the average number of vertices of degree k in graphs G of $\mathcal{G}_{n,m}$.

4.5.31. Let $p(G)$ be an integral non-negative parameter and let $\overline{p}(n)$ be its average value for graphs G in $\mathcal{G}_n$. Show that if $\lim_{n\to\infty} \overline{p}(n) = 0$, then $p(G) = 0$ for almost all graphs.

4.6. Representations of Boolean Functions by Contact Schemes and Formulas

A network Γ with k poles in which each edge is labelled by a letter from the alphabet $\{x_1, x_2, \ldots, x_n, \overline{x}_1, \overline{x}_2, \ldots, \overline{x}_n\}$ is called a *k-pole contact scheme* (*circuit*) *repre-*

senting Boolean functions of variables $x_1, x_2, \ldots, x_n$ or, in short, a *⟨k, n⟩-scheme.* The ⟨2, n⟩-schemes will be called *X^n-schemes.* The network Γ is called the *contact scheme network.* A contact scheme is called *connected* (*strongly connected*, *series-parallel*, etc.), if its network has the same property. A series-parallel contact scheme is briefly denoted as *π-scheme.* The edges of a scheme labelled by the symbols of variables or their negations are called *contacts.* A contact marked by the symbol of a variable (or its negation) is called a *make* (resp. *break*) *contact.* Let Σ_1 and Σ_2 be two k-pole contact schemes whose poles are labelled by the letters $a_1, a_2, \ldots, a_k$. The schemes Σ_1 and Σ_2 are called *isomorphic* if their networks are isomorphic and if (a) the corresponding edges are labelled in the same way, and (b) the corresponding poles are labelled in the same way. Let a and b be two poles of the contact scheme Σ, and let $[a, b]$ be a chain joining a and b. Let $K_{[a, b]}$ be the conjunction of letters assigned to the edges of the chain $[a, b]$. The function $f_{ab}(\widetilde{x^n})$, defined by the formula

$$f_{ab}(\widetilde{x^n}) = \bigvee_{[a, b]} K_{[a, b]}, \tag{3}$$

in which the disjunction is taken over all simple chains in the scheme joining the poles a and b, is called the *conductivity function between poles a and b* of the scheme Σ. The scheme Σ is said to represent the function $g(\widetilde{x^n})$ if it contains the poles a and b such that $g(\widetilde{x^n}) = f_{ab}(\widetilde{x^n})$.

A contact scheme with $k + 1$ poles is called a *(1, k)-pole network* if one of its poles is isolated (this pole will be denoted by a) and the remaining poles are interchangeable (these poles will be designated by b_i ($i = \overline{1, k}$)). The function $g(\widetilde{x^n})$ is said to be *represented* by a (1, k)-pole network if there exists a pole b_i ($1 \leqslant i \leqslant k$) such that $f_{ab_i}(\widetilde{x^n}) = g(\widetilde{x^n})$. If the number of poles in a scheme is not indicated, we shall always mean two-pole contact schemes. Two contact schemes are called *equivalent* if they represent the same Boolean function. The *complexity* of a contact scheme is the number of its contacts. A contact scheme having the lowest complexity among all equivalent schemes is called *minimal*. The *complexity of a Boolean function f in a class of contact schemes* (notation $L_k(f)$) is the complexity of the minimal contact scheme representing f. The *complexity of a Boolean function f in the class of π-schemes* is the number of contacts in the minimal π-scheme realizing f (notation $L_\pi(f)$). The *complexity of a Boolean function f in a class of formulas* generated by a set of connectives $\{\vee, \&, -\}$ is the number of entries of the symbols of variables. In this class of formulas, the complexity of a function f is denoted by $L_\Phi(f)$.

A directed contourless network whose poles are divided into input and output poles is called a *scheme of functional elements*. The input poles are labelled by variable symbols. The output poles are called *outputs of the scheme*. Each vertex other than an input is labelled by a functional (logical connective) symbol. The following conditions must be satisfied in this case: (1) the in-degree of each entrance pole is equal to zero; (2) the in-degree of each vertex other than the input pole is equal to the number of arguments of a functional symbol (or connective) which labels a given vertex.

The concept of a *function f_i represented at the vertex i of a scheme* Σ is defined as follows. If the vertex i coincides with the input pole labelled by x, then $f_i = x$. Let the vertex i be labelled by a functional symbol φ of r arguments, and let $\varphi_1, \ldots, \varphi_r$ be functions represented at the vertices from which the arcs terminating at the vertex i emerge. In this case, $f_i = \varphi(\varphi_1, \ldots, \varphi_r)$. It is said that the *function f is represented by a scheme* Σ if the latter has an output where this function is represented. Functional element schemes with one exit whose entrance poles are labelled by the symbols $x_1, \ldots, x_n$, and the vertices other than the input poles are labelled by the symbols $\vee$, &, — will be called *X^n-functional schemes*. The *complexity of a functional element scheme* is the number of its vertices other than entrance poles. An X^n-functional scheme Σ representing a function f is called *minimal* if every other X^n-functional scheme representing f has a complexity not lower than that of the scheme Σ. The *complexity of a Boolean function f in a class of functional element schemes* is the complexity of a minimal X^n-functional scheme representing the function f. The complexity of a function f in this class of schemes will be denoted by $L(f)$.

4.6.1. Let $f(\widetilde{x^2}) = x_1 \oplus x_2$. Prove that $L(f(\widetilde{x^2})) = 4$.

4.6.2. Show that for a Boolean function other than a constant, the minimal contact scheme representing this function is strongly connected.

4.6.3. Find the number of Boolean functions $f(x_1, x_2)$ represented by contact schemes of complexity 3.

4.6.4. (1) Show that for each natural m there exists a minimal contact scheme of complexity m.

(2) Show that there are no minimal contact schemes of complexity 4 containing only closing contacts with labels in the set $\{x_1, x_2, x_3\}$.

4.6.5. Show that if $m > n \cdot 2^{n-1}$, none of the X^n-schemes of complexity m is minimal.

4.6.6. Show that the function f depends essentially on variable x if and only if the minimal scheme repre-

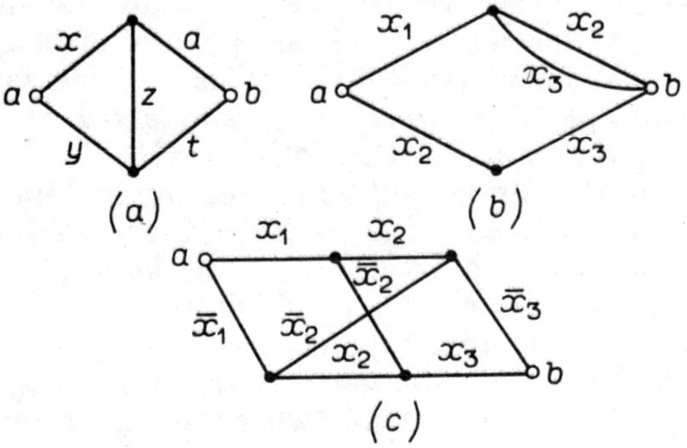

Fig. 19

senting f contains a contact labelled by the variable x or by its negation.

4.6.7. For the contact schemes shown in Figs. 19a, b, c and 20a, b, c find the conductivity function f_{ab}.

4.6.8. Construct the contact schemes representing the functions f in the following cases:

(1) $f(\widetilde{x^3}) = (x_1 \vee x_2 \vee x_3)(\bar{x}_1 \vee \bar{x}_2 \vee \bar{x}_3)$;

(2) $f(\widetilde{x^3}) = x_1x_2 \oplus x_2x_3 \oplus x_3x_1$;

(3) $f(\tilde{x}^3) = (00011111)$;

(4) $f(\tilde{x}^3) = (11010001)$;

(5) $f(\tilde{x}^3) = x_1 \oplus x_2 \oplus x_3$.

4.6.9. For each function in Problem 4.6.8, construct X^3-functional schemes.

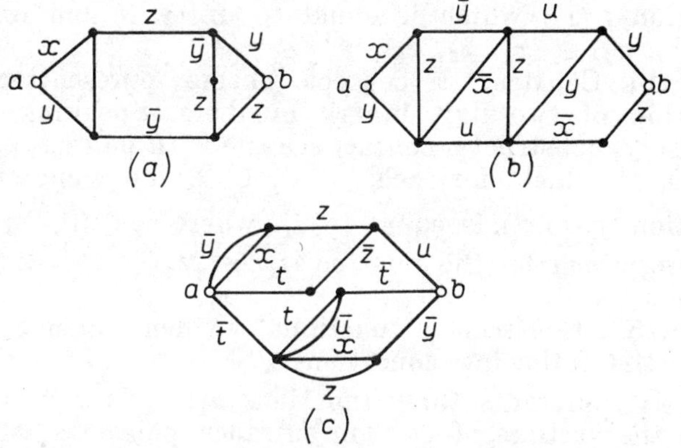

Fig. 20

4.6.10. Construct a contact scheme with a complexity not exceeding L for a function f, after simplifying its formula if:

(1) $(x_1 \vee x_2x_3)((x_1 \vee x_2)(\bar{x}_2 \vee x_4) \vee (x_3 \vee \bar{x}_4) \times (\bar{x}_1\bar{x}_2x_3 \vee x_4x_5))$, $L = 5$;

(2) $(x_1 \vee x_2)((x_1x_2 \vee \bar{x}_2\bar{x}_3)x_6 \vee (\bar{x}_2 \vee \bar{x}_3\bar{x}_5) \times (\bar{x}_2 \vee \bar{x}_3) \vee (\bar{x}_1\bar{x}_2 \vee x_1x_2 \vee x_6)x_3 \vee x_1x_2 \vee x_5x_6)$, $L = 5$;

(3) $\bar{x}_1\bar{x}_2x_3((\bar{x}_4 \vee \bar{x}_1x_5)(x_6 \vee x_1x_4x_7) \vee (x_6x_7 \vee x_2x_4x_5) \,\&\, (\bar{x}_4x_5 \vee \bar{x}_3x_6\bar{x}_7) \vee (\bar{x}_1 \vee \bar{x}_2)(\bar{x}_4 \vee x_6) \times (x_5 \vee x_7))$, $L = 6$.

4.6.11. For each function in Problem 4.6.10, construct a functional element scheme of complexity not exceeding 6 and representing this function.

4.6.12. Let $\nu(\tilde{\alpha}) = \sum_{i=1}^{n} 2^i \alpha_i$ be the number of the tuple $\tilde{\alpha} = (\alpha_1, \alpha_2, \ldots, \alpha_n)$. Construct a contact scheme with not more than 10 contacts representing the function

$$f(\tilde{\alpha}^6) = \begin{cases} 1 & \text{if } \nu(\alpha_1, \alpha_2, \alpha_3) \leqslant \nu(\alpha_4, \alpha_5, \alpha_6), \\ 0 & \text{otherwise.} \end{cases}$$

4.6.13. Construct a contact scheme representing the function $f(\tilde{x}^6)$ which is equal to unity if and only if $(x_1, x_2, x_3) \leqslant (x_4, x_5, x_6)$.

4.6.14. Construct a contact scheme representing the addition of two-digit binary numbers. Speaking more precisely, construct a contact scheme with poles a, b_0, b_1, and b_2 in which, for each $i = 0, 1, 2$, the conductivity function $f_{ab_i}(\tilde{x}^4)$ is equal to z_i, where $z_i \in \{0, 1\}$ and is determined by the relation $4z_1 + 2z_2 + z_3 = 2(x_1 + x_2) + x_2 + x_4$.

4.6.15*. Construct a functional element scheme satisfying the following conditions:

(1) it represents three functions $\bar{x}_1$, $\bar{x}_2$, $\bar{x}_3$,

(2) the vertices other than entrance poles are labelled by $\vee$, &, and —;

(3) the scheme has not more than two vertices labelled by negation symbols.

4.6.16. Show that $L_\Phi(f) \geqslant L_k(f)$ for any Boolean function f which is not a constant.

4.6.17*. Give an example of a function $f(\tilde{x}^3)$ for which $L_\pi(f) > L_k(f)$.

Hint. Consider the function represented by a scheme shown in Fig. 19*a*.

Let a 2-connected two-pole contact scheme Σ be a planar scheme (i.e. let its network $\Gamma(a, b)$ be plane) and let its poles a and b lie on one face. We draw an edge (a, b) on this face in such a way that the network Γ' obtained from Γ by adding the edge (a, b) remains planar. We choose one vertex each on the faces of the network Γ'. On the vertices chosen in this way, we construct a graph G^* dual to the graph G of the network Γ'. Each edge of the graph G^* other than (a, b) intersects a contact of the scheme Σ. We label this edge by the same letter that is used to label the contact intersected by it. Let us denote the vertices of the graph G^* located on the faces of the network Γ' divided by the edge (a, b) by a^*, b^* and call them poles. We delete the edge (a^*, b^*) from G^*. As a result, we obtain a 2-connected scheme Σ^* with poles a^* and b^*. The scheme Σ^* is called a *scheme dual to* Σ.

4.6.18*. Show that the scheme Σ^* dual to a planar 2-connected scheme Σ represents a Boolean function dual to the function represented by the scheme Σ.

Hint. Establish a one-to-one correspondence between the chains of scheme Σ and the cuts of scheme Σ^*.

4.6.19. Construct schemes dual to those shown in Figs 19*a*, *b*, and 20*a*.

4.6.20. Prove that for every Boolean function f the equality $L_\pi(f) = L_\pi(f^*)$ is satisfied.

4.6.21. Show that if $L_k(f) \leqslant 7$, then $L_k(\bar{f}) = L_k(f)$.

A contact scheme is called *simple* if the labelling of an edge by a letter x or $\bar{x}$ means that every other edge has a label other than x or $\bar{x}$.

4.6.22. Show that a strongly connected simple contact scheme represents a function that depends essentially on all variables encountered in the scheme.

4.6.23. Show that a strongly connected simple scheme is minimal.

4.6.24. Show that if a minimal scheme is supplied with a new contact labelled by a new variable in such a way that a strongly connected scheme is represented, the newly formed scheme will also be minimal.

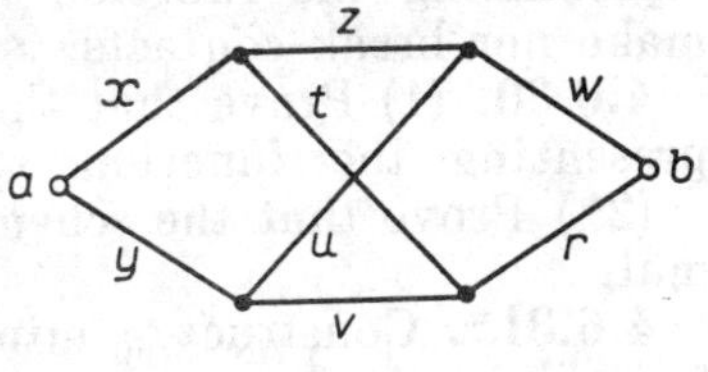

Fig. 21

4.6.25*. Let f be a function represented by the scheme shown in Fig. 21. Prove that a function dual to f cannot be represented by a simple scheme.

4.6.26*. Is the relation $L_k(f) = L_k(\bar{f})$ true for all Boolean functions f?

4.6.27. Let f_{ab} and f_{ad} be the conductivity functions of a three-pole contact scheme Σ_1, while f_{eg} and f_{eh} are the conductivity functions of a three-pole contact scheme Σ_2. Let Σ be a scheme with poles a and e, obtained from the schemes Σ_1 and Σ_2 by identifying pole b with pole g and pole d with pole e. Is it true that the scheme Σ represents the function

$$f_{ae} = (f_{ab} \,\&\, f_{eg}) \vee (f_{ab} \,\&\, f_{eg})?$$

4.6.28. (1) Show that the function f is monotonic if and only if there exists a contact scheme representing f and not containing opening contacts.

(2) Is it true that the minimal contact scheme representing a monotonic function does not contain opening contacts? The Boolean function $f(\tilde{x}^n)$ is called *monotonic in variable* x_1 if it can be presented in the form

$$f(\tilde{x}^n) = g(x_2, x_3, \ldots, x_n) \vee x_1 h(x_2, x_3, \ldots, x_n).$$

The function $f(\tilde{x}^n)$ is *quasi-monotonic in variable* x_1 if it is monotonic in x_1 or becomes monotonic in x_1 after substituting $\bar{x}_1$ for x_1. The monotonicity and quasi-monotonicity in x_i $(i \neq 1)$ are defined in an identical manner.

4.6.29. Prove that the function $f(\tilde{x}^n)$ is quasi-monotonic in x_1 if and only if there exists a contact scheme representing the function $f(\tilde{x}^n)$ and containing neither make nor break contacts.

4.6.30. (1) Prove that the minimal contact scheme representing the function $x_1 \oplus x_2$ contains 4 contacts.

(2*) Prove that the scheme shown in Fig. 20*c* is minimal.

4.6.31*. Construct a minimal contact scheme for the function f if:

(1) $f(\tilde{x}^4) = (x_1 \vee x_2 \vee x_3)\, x_4 \vee x_1 x_2 x_3$;

(2) $f(\tilde{x}^4) = x_1 x_2 x_3 \oplus x_1 x_2 x_4 \oplus x_1 x_3 x_4 \oplus x_2 x_3 x_4$;

(3) $f(\tilde{x}^4) = (0001011101111111)$.

4.6.32. Show that the number $S(n, m)$ of connected pairwise non-isomorphic X^n-contact schemes of complexity not more than m does not exceed $(cnm)^m$, where c is a constant independent of n and m.

4.6.33. Show that the number $P(n, m)$ of connected pairwise non-isomorphic π-schemes of complexity not more than m and realizing Boolean functions of variables $x_1, x_2, \ldots, x_n$ does not exceed $(cn)^m$, where c is a constant independent of n and m.

4.6.34. Show that the number $\Phi(n, m)$ of pairwise different formulas of complexity m generated by a set of connectives ($\vee$, &, —) and by a set of variables x_1, $x_2, \ldots, x_n$ does not exceed $(cn)^m$, where c is a constant independent of n and m.

The lower estimates of complexity in representing different classes of functions by schemes and formulas are frequently obtained by using "power considerations". The following statement is an example.

Let $S(n, m)$ be the number of schemes from a class K, each of which represents a Boolean function depending on the variables $x_1, x_2, \ldots, x_n$, and has a complexity higher than m. Let $\varphi(n)$ be the number of Boolean functions $f(\tilde{x}^n)$ in a set $\mathfrak{M}$. Then, if $S(n, m) < \varphi(n)$, there exists in $\mathfrak{M}$ a function $f(\tilde{x}^n)$ that cannot be represented in class K by a scheme of complexity lower than or equal to m.

4.6.35. Show that for any $\varepsilon > 0$ and a sufficiently large n, there exists a self-dual function $f(\tilde{x}^n)$ for which the following inequalities are satisfied simultaneously:

(a) $L_k(f) \geqslant \frac{2^{n-1}}{n}(1-\varepsilon)$;

(b) $L_\pi(f) \geqslant \frac{2^{n-1}}{\log_2 n}(1-\varepsilon)$.

4.6.36. Show that for any $\varepsilon > 0$ and a sufficiently large n, there exists a function $f(\tilde{x}^n)$ which is a superposition of functions $\varphi(x, y, z) = xy \vee z$ and is such that

$$L_\Phi(f) \geqslant \frac{\binom{n-1}{\left[\frac{n-1}{2}\right]}}{\log_2 n}(1-\varepsilon).$$

4.6.37. Let $L(n) = \max\limits_{f \in P_2^n} L(f)$. Show that for any self-dual function $f(\tilde{x}^n)$ the inequality $L(f(\tilde{x}^n)) \leqslant L(n-1) + 4n$ is valid.

4.6.38. A Boolean function $F(\tilde{y}^m)$ has the property U_n if any Boolean function $f(\tilde{x}^n)$ can be obtained from $F(\tilde{y}^m)$ by substitution of constants and redesignation of variables (identifications are admissible).

(1) Show that the function $y_1\bar{y}_2$ has the property U_1.

(2) Find the function with the smallest possible number of variables having the property U_2.

(3) Let $m(n)$ be the smallest possible number of variables of a function having the property U_n. Show that $\frac{2^n}{\log_2(n+2)} \leqslant m(n) \leqslant 3\cdot 2^{n-1}$.

4.6.39. Show that there exists an X^n-functional scheme having a complexity $2\cdot 2^{2^n} - n$ and representing all functions depending on variables $x_1, \ldots, x_n$.

For constructing two-pole contact schemes representing Boolean functions, the *method of cascades* is frequently used. We shall describe this method here. Let $f(x_1, x_2, \ldots, x_n)$, $n \geqslant 2$ be a Boolean function which has to be represented by a contact scheme. By $\mathfrak{A}_i$ ($i = \overline{1, n-1}$) we denote the set of all Boolean functions, each of which depends only on variables $x_{i+1}, x_{i+2}, \ldots, x_n$, and can be obtained from the function $f(\tilde{x}^n)$ by an appropriate substitution of zeros and unities for the variables $x_1, x_2, \ldots, x_i$. Each set $\mathfrak{A}_i$ is put in a one-to-one correspondence with the set V_i whose elements are points in the plane, called i-th *rank vertices*. We add two more poles, viz. the input pole a and the output pole b. Pole a is a zero rank vertex, while pole b is a vertex of rank n. The set of vertices in the scheme Σ representing the function $f(\tilde{x}^n)$ will coincide with $\{a\} \cup \{b\} \cup \bigcup_{i=1}^{n-1} V_i$. The set of contacts of the scheme may be described as follows: let v_i be an arbitrary i-th rank vertex ($n-2 \geqslant i \geqslant 0$) and let the function $\varphi(x_{i+1}, x_{i+2}, \ldots, x_n)$ in the set $\mathfrak{A}_i$ correspond to it (for $i = 0$, the function φ is identical to the function $f(\tilde{x}^n)$). The functions $\varphi(0, x_{i+2}, \ldots, x_n)$ and $\varphi(1, x_{i+2}, \ldots, x_n)$ belong to the set $\mathfrak{A}_{i+1}$ and have certain vertices v'_{i+1} and v''_{i+1} respectively corresponding to them (if the functions $\varphi(0, \overline{x_{i+2}}, \ldots, x_n)$ and $\varphi(1, x_{i+2}, \ldots, x_n)$ are identical, the vertices v'_{i+1} and v''_{i+1} coincide). The vertex v_i is joined in scheme Σ to the vertex v'_{i+1} through the contact $\overline{x}_{i+1}$ and to the vertex v''_{i+1} through x_{i+1}. Finally, the $(n-1)$-th rank vertices are joined with the n-th rank vertex (pole b) according to the following rule:

(1) if a vertex v in V_{n-1} corresponds to a function x_n, it is connected to the pole b through the contact x_n;

(2) if a vertex v corresponds to a function $\overline{x}_n$, it is joined to b through the contact $\overline{x}_n$;

(3) if a vertex v is put in correspondence with a function identically equal to unity, it is joined to b through two parallel contacts x_n and $\bar{x}_n$; and

(4) if a vertex v is put in correspondence with zero, it is not joined to pole b.

4.6.40. Using the method of cascades, construct a scheme representing the function f:

(1) $f(\tilde{x}^3) = x_1x_2 \oplus x_3$;

(2) $f(\tilde{x}^4) = x_1x_2 \vee x_2x_3 \vee x_1x_4$;

(3) $f(\tilde{x}^n) = x_1 \oplus x_2 \oplus \ldots \oplus x_n$;

(4) $f(\tilde{x}^n) = x_1x_2 \ldots x_n \vee \bar{x}_1\bar{x}_2 \ldots \bar{x}_n$;

(5) $f(\tilde{x}^n) = \bigvee_{i=1}^{n} x_1 \ldots x_{i-1}x_{i+1} \ldots x_n$.

4.6.41. (1) Show that if $f(\tilde{x}^n) \not\equiv 0$, we can delete from all sets $\mathfrak{A}_i$ the functions that are identically equal to zero when constructing by the cascade method a contact scheme representing the function f.

(2) Show that the contact scheme obtained through such a construction is strongly connected.

4.6.42. (1) Suppose that the function $f(\tilde{x}^n)$, $n \geqslant 2$, depends essentially on all its variables. Prove that the scheme representing the function f and constructed by the cascade method is strongly connected and does not contain parallel contacts of the type x_j, $\bar{x}_j$ if and only if f is a linear function.

(2) Give an example of a nonlinear function $f(\tilde{x}^n)$, $n \geqslant 2$, depending essentially on all its variables and such that the scheme representing it and constructed by the cascade method does not contain parallel contacts of the type x_j, $\bar{x}_j$.

4.6.43. Find out if the following statement is true or false: if a function $f(\tilde{x}^n)$, $n \geqslant 2$, depends essentially on all its variables and the contact scheme representing f and constructed by the modified cascade method described in Problem 4.6.41 (1) contains n contacts, then $f = x_1^{\sigma_1}x_2^{\sigma_2} \ldots x_n^{\sigma_n}$, where $\sigma_i \in \{0, 1\}$.

4.6.44. Disprove the following statement: if a function f is such that for a certain i each of the sets $\mathfrak{A}_i$ and $\mathfrak{A}_{i+1}$ used in the cascade method does not contain functions identically equal to zero and unity, but does contain at least three functions, then the scheme representing f and constructed by the method of cascades is not plane[1].

4.6.45. Give an example of a function f for which the cascade scheme realizing it is not plane.

[1] We mean the schemes without an additional edge joining the poles.

Chapter Five

Fundamentals of Coding Theory

5.1. Codes with Corrections

Let A and B be two finite alphabets and let R be a certain set of finite words in the alphabet A. A single-valued mapping φ of the set R into a set of words in the alphabet B is called a *coding of the set* R. The counterpart C of the set R for the mapping φ is referred to as a *code of the set* R. Words in C are known as *code words*. If a word w in R maps a word v in C, then v is a *code of the word* w. Words in R are called *messages*, and the alphabets A and B are called a *message alphabet* and a *coding alphabet respectively*. If a coding alphabet B consists of two letters (in this case we shall assume that $B = \{0, 1\}$), the coding φ and the corresponding code C are *binary*. A code is referred to as a *uniform*, or *block*, *code* if all code words are of the same length. A block binary code in which every code word has length n is a subset of vertices of a unit n-dimensional cube. A Boolean function $f_C(\tilde{x}^n)$ which is equal to unity in the set C and to zero outside C is called the *characteristic function of the binary block code* C. Let $\rho(\tilde{\alpha}, \tilde{\beta})$ be the ordinary Hamming distance between vertices $\tilde{\alpha}$ and $\tilde{\beta}$ in B^n, which is equal to the number of coordinates in which $\tilde{\alpha}$ and $\tilde{\beta}$ differ. The quantity $d(C) = \min \rho(\tilde{\alpha}, \tilde{\beta})$, where the minimum is taken over all pairs of different vertices belonging to the code $C \subseteq B^n$, is called the *code distance* of the code C. The code $C \subseteq B^n$ with a code distance d is abbreviated as the *$\langle n, d\rangle$-code*. The maximum possible power of an $\langle n, d\rangle$-code is denoted by $m(n, d)$, while an $\langle n, d\rangle$-code whose power is $m(n, d)$ is known as the *maximum code*. A *compact code* is an $\langle n, 2d + 1\rangle$-code which satisfies the following condition: for any vertex $\tilde{\alpha} \in B^n$, there exists a code word $\tilde{\beta}$ for which $\rho(\tilde{\alpha}, \tilde{\beta}) \leqslant d$. A binary block code is referred to as *equidistant* if the distance between any two code words is constant. A code $C \subseteq B^n$ is known to be an *equal-weight code* if any code word has the same weight, i.e. if there exists an integer k ($0 \leqslant k \leqslant n$) such that $C \subseteq B_k$. This number k is called the *weight of an equal-weight code*. The quantity max $|C|$, where the maximum is taken over all $\langle n, d\rangle$-codes of weight k is denoted by $m(n, k, d)$.

Let the words of a binary block code C be transmitted through a communication channel where a distortion of a transmitted word may occur. The transmission via such a channel can be regarded as transformations of the words being transmitted. Here we shall consider only such transformations of binary words which do not change the length of a word and consist in the replacement of certain letters by the opposite, i.e. in the replacement of 0 by 1 and 1 by 0. If a word $\tilde{\alpha}$ is transformed into a word β differing from $\tilde{\alpha}$ as a result of transmission, this means that *errors have been introduced into the channel*. If the i-th letter of a word $\tilde{\alpha}$ being transmitted differs from the i-th letter of the obtained word $\tilde{\beta}$, the *error* is said *to occur in the i-th digit*. If the obtained word differs from the word being transmitted in t digits, t *errors* are said *to have been introduced*. Obviously, the number of errors made during a transmission is equal to the Hamming distance between the transmitted and received words.

Let $C \subseteq B^n$ be a binary code. An arbitrary single-valued mapping ψ of the set B^n onto the set C is called a *decoding*. Let $\tilde{\alpha} \in C$, and let $\psi^{-1}(\tilde{\alpha})$ be a set of all vertices $\tilde{\beta}$ in B^n, such that $\psi(\tilde{\beta}) = \tilde{\alpha}$. Let $S_t^n(\tilde{\alpha})$ be a set of all words which are obtained from a code word $\tilde{\alpha}$ as a result of not more than t errors (obviously, $S_t^n(\tilde{\alpha})$ is a sphere of radius t with center at $\tilde{\alpha}$). A code C is said *to correct t errors* if there exists a decoding ψ such that $S_t^n(\tilde{\alpha}) \subseteq \psi^{-1}(\tilde{\alpha})$ for any $\tilde{\alpha} \in C$. A code C *reveals t errors* if any word which can be obtained from an arbitrary code word $\tilde{\alpha}$ as a result of not more than t distortions differs from any word in $C \setminus \{\tilde{\alpha}\}$.

5.1.1. Prove that a code $C \subseteq B^n$ corrects t errors if and only if $\rho(v, w) \geqslant 2t + 1$ for any two different code words v and w in C.

5.1.2. Is it true that a code $C \subseteq B^n$ correcting t errors reveals

(1) at least $2t + 1$ errors;

(2) at least $2t$ errors;

(3) not more than $2t$ errors?

5.1.3. Prove that it is possible to obtain from any subset $C \subseteq B^n$ a code detecting an error by deleting from C not more than half the vertices.

5.1.4. Determine the number of errors corrected and revealed by a code with a characteristic function f if:

(1) $f(\tilde{x}^n) = x_1 \oplus x_2 \oplus \ldots \oplus x_n$;

(2) $f(\tilde{x}^n) = x_1 x_2 \ldots x_n \vee \bar{x}_1 \bar{x}_2 \ldots \bar{x}_n$;

(3) $f(\tilde{x}^{3n}) = x_1 x_2 \ldots x_{3n} \vee x_1 x_2 \ldots x_{2n} \bar{x}_{2n+1} \bar{x}_{2n+2} \ldots \bar{x}_{3n} \vee x_1 x_2 \ldots x_n \bar{x}_{n+1} \bar{x}_{n+2} \ldots \bar{x}_{3n} \vee \bar{x}_1 \bar{x}_2 \ldots \bar{x}_{3n}$;

(4) $f(\tilde{x}^n) = x_1 x_2 \ldots x_{n-1} \oplus x_1 x_2 \ldots x_{n-2} x_n \oplus \ldots \oplus x_2 x_3 \ldots x_n$.

5.1.5. Let the words of a binary code C be transmitted via a transmission channel so that not more than one error can be introduced during the transmission of a code word. For each code word $\tilde{\alpha}$, construct the set of words which can be obtained as a result of the transmission of $\tilde{\alpha}$ through the channel:

(1) $C = \{01100,\ 00111,\ 11010,\ 10001\}$;

(2) $C = \{11110,\ 10100,\ 01011,\ 11001\}$.

Let p be the probability of the fact that an error is introduced in the i-th digit of an arbitrary word w in B^n during its transmission via a channel for all $i = \overline{1, n}$. Let $C \subseteq B^n$ be a code of power m, and ψ: $B^n \to C$ be a decoding. The quantity

$$Q_\psi(p, C) = \frac{1}{m} \sum_{v \in C} \sum_{w \in \psi^{-1}(v)} p^{\rho(v, w)} (1 - p)^{n - \rho(v, w)}$$

is called the *authenticity of the decoding* ψ *for the code* C.

5.1.6. Prove that for $0 < p < 1/2$, the maximum of the quantity $Q_\psi(p, C)$ over all possible decodings for a fixed C is attained provided that for each $w \in B^n$ the equality $\rho(w, \psi(w)) = \min_{v \in C} \rho(w, v)$ is satisfied.

5.1.7. (1) For the code C in Problem 5.1.5 (1), construct a decoding ψ with the maximum authenticity $Q_\psi(p, C)$, $0 < p < 1/2$ and indicate the set $\psi^{-1}(w)$ for each $w \in C$. Determine $\max_\psi Q_\psi(1/4, C)$.

(2) Determine the number of different decodings ψ with the maximum authenticity for the code C in Problem 5.1.5 (1) and $0 < p < 1/2$.

5.1.8. Determine the maximum possible power of a code $C \subseteq B^n$ possessing the following property: for any $\tilde{\alpha}$ and $\tilde{\beta}$ in C, $\rho(\tilde{\alpha}, \tilde{\beta})$ are even.

5.1.9. Determine the number of maximum $\langle n, 2\rangle$-codes.

5.1.10. Let $n = 3k$. Prove that $m(n, 2n/3) = 4$.

5.1.11. Prove that the power of a compact $\langle n, 2d+1\rangle$-code is $2^n \Big/ \sum_{i=0}^{d} \binom{n}{i}$.

5.1.12. Does there exist a compact $\langle n, 3\rangle$-code for $n = 147$?

5.1.13. Prove that for $n > 7$, there exists no compact $\langle n, 7\rangle$-codes.

5.1.14. Prove that a code $C \subseteq B^n$ is not maximum if $|C| = 3$.

5.1.15. Prove that if there exists a compact $\langle n, 3\rangle$-code in B^n, there exists a partition of the cube B^{n+1} into disjoint spheres of radius 1.

5.1.16. Let C be a compact $\langle n, 2d+1\rangle$-code. Prove that $\binom{n}{d+1}$ in this case is exactly divisible by $\binom{2d+1}{d}$.

5.1.17. Prove that there exist no equidistant $\langle n, 2d+1\rangle$-codes of power exceeding 2.

5.1.18. Prove that for an even d there exists an equidistant code of power $[2n/d]$.

5.1.19. Prove that $m(n, d)$ is a non-decreasing function of parameter n.

5.1.20. Prove that

(1) $m(n+d, d) \geqslant 2m(n, d)$;
(2) $m(2n, d) \geqslant (m(n, d))^2$;
(3) $m(n, d) \leqslant 2m(n-1, d)$.

5.1.21. Prove that

$$m(n, 2t+1) \geqslant 2^n \Big/ \sum_{i=0}^{2t} \binom{n}{i}.$$

5.1.22. Prove that

$$m(n, k, 2d) \leqslant \frac{\binom{n}{k}}{\sum_{i=0}^{d-1} \binom{k}{i}\binom{n-k}{i}}.$$

5.1.23. Prove that for $n < 2d$, the following inequality holds

$$m(n, d) \leqslant \frac{2d}{2d-n}.$$

5.1.24. Prove that

$$m(n, k, d) \leqslant \left[\frac{nd}{2k^2 - n(2k-d)}\right],$$

if $2k^2 - n(2k-d) > 0$.

5.1.25. Prove that

(1) $m(n, k, d) \leqslant \left[\frac{n}{k} m(n-1, k-1, d)\right]$;

(2) $m(n, k, d) \leqslant \left[\frac{n}{k}\left[\frac{n-1}{k-1}\left[\cdots\left[\frac{n-d}{k-d}\right]\cdots\right]\right]\right]$;

(3) $m(n, k, d) \leqslant \left[\frac{n}{n-k} m(n-1, k, d)\right]$.

5.1.26. Let $q(n, d)$ be the maximum number of vertices in B^n, the pairwise distances between which do not exceed d. Prove that

$$m(n, d+1)\, q(n, d) \leqslant 2^n.$$

5.1.27. Prove that $m(n, d) \leqslant 2^{n-d+1}$.

5.1.28. Prove that from any set $C \subseteq B^n$ such that $|C| \geqslant 2^d$, it is possible to isolate a subset D of power not less than $2^{-d+1} |C|$ which is an $\langle n, d\rangle$-code.

5.2. Linear Codes

The expression of the form

$$\lambda_1\tilde{\alpha}_1 \oplus \lambda_2\tilde{\alpha}_2 \oplus \ldots \oplus \lambda_s\tilde{\alpha}_s, \qquad (1)$$

where $\tilde{\alpha}_i \in B^n$, $\lambda_i \in \{0, 1\}$, $i = \overline{1, s}$, is called the *linear combination of the vectors* $\tilde{\alpha}_1, \tilde{\alpha}_2, \ldots, \tilde{\alpha}_s$. The linear combination (1) is called *trivial* if $\lambda_1 = \lambda_2 = \ldots = \lambda_s = 0$ and *non-trivial* otherwise. Any linear combination[1] of vectors in B^n is obviously a vector in B^n. Vectors $\tilde{\alpha}_1, \tilde{\alpha}_2, \ldots, \tilde{\alpha}_s$ in B^n are *linearly independent* if any of their non-trivial linear combinations differs from $\tilde{0} = (0, 0, \ldots, 0)$. Otherwise, the vectors $\tilde{\alpha}_1, \tilde{\alpha}_2, \ldots, \tilde{\alpha}_s$ are said to be *linearly dependent*. A subset $G \subseteq B^n$ is called a *group* if G is closed with respect to the exclusive sum, i.e. for any $\tilde{\alpha}$ and $\tilde{\beta}$ in G, the vector $\tilde{\alpha} \oplus \tilde{\beta}$ belongs to G. It follows from the closure of G with respect to the operation $\oplus$ that any linear combination of vectors in G also belongs to G (in particular, $\tilde{0} \in G$). Thus, any group G in B^n is a *linear vector space generated by the field* $F_2 = \langle\{0, 1\}, \oplus, \cdot\rangle$. The maximum number $k = k(G)$ for which there exist k linearly independent vectors in the group (linear space) G is called the *dimensionality of* G. The set of k linearly independent vectors of a linear k-dimensional space is known as the *basis* of this space. If a code $G \subseteq B^n$ forms a group, it is called a *linear*, or *group*, *code*. If a linear code in B^n has a dimensionality k, it is referred to as an (n, k)*-code*. A binary linear code correcting one error is known as a *Hamming code*.

Linear codes can be conveniently defined in terms of matrices. A matrix $H(C)$ whose rows are code words of a code $C \subseteq B^n$ is called a *code-C matrix*. A matrix $M(C)$ formed by k arbitrary linearly independent vectors which are code words of an (n, k)-code C is called a *generating matrix of the code* C. If H is an arbitrary matrix formed

---

[1] The definitions of mod 2 sum of vectors and mod 2 product of a scalar by a vector are given in the section "Boolean Vectors and a Unit n-dimensional Cube".

by zeros and unities with n columns, the set $C(H)$ of all vertices of the cube B^n, which are linear combinations of rows in the matrix H, is called a *code generated by the matrix* H. Vectors $\tilde{\alpha} = (\alpha_1, \alpha_2, \ldots, \alpha_n)$ and $\tilde{\beta} = (\beta_1, \beta_2, \ldots, \beta_n)$ are referred to as *orthogonal* if $\alpha_1\beta_1 \oplus \alpha_2\beta_2 \oplus \ldots \oplus \alpha_n\beta_n = 0$. The set $V(H)$ of all vectors in B^n which are orthogonal to each of the rows of the matrix H is known as the *null space of the matrix* H. Let C be a binary code whose each word is orthogonal to each row of a certain matrix H. If C is an (n, k)-code, and the matrix H consists of $n - k$ linearly independent rows, then H is called a *checking matrix of the code* C. The set C^* of all vectors that can be represented as a linear combination of the rows of the checking matrix of the (n, k)-code C is called a *code dual to the code* C. By $g(n, d)$ we shall denote $\max |C|$, where the the maximum is taken over all linear codes $C \subseteq B^n$ with a code distance d.

5.2.1. Let a set $C \subseteq B^n$ consist of k linearly independent vectors. Prove that any two linear combinations of vectors in the set C, having different coefficients, are different vertices of the cube B^n.

5.2.2. Prove that there exists in B^n a system of n linearly independent vectors, but there exists no system of $n + 1$ linearly independent vectors.

5.2.3. Prove that the number of vectors in B^n that can be represented by linear combinations of type (1), where $\sum_{i=1}^{s} \lambda_i \leqslant t$, does not exceed $\sum_{i=0}^{t} \binom{s}{i}$.

5.2.4. Prove that any (n, k)-code has a power 2^k.

5.2.5. Prove that in a binary linear code either each code vector or half the code vectors have even weights.

5.2.6. Prove that the number of different bases in B^n is

$$\frac{(2^n-1)(2^n-2)\ldots(2^n-2^{n-1})}{n!}.$$

5.2.7. Prove that the number of different (n, k)-codes in B^n is

$$\frac{(2^n-1)(2^n-2)\ldots(2^n-2^{k-1})}{(2^k-1)(2^k-2)\ldots(2^k-2^{k-1})}.$$

5.2.8. Determine the number of vectors in B^n which are orthogonal to a given vector $\tilde{\alpha}$ in B^n_k.

5.2.9. Prove that the set of all vectors in B^n which are orthogonal to each row of a $(k \times n)$ binary matrix H form a linear space. Is this space always $(n - k)$-dimensional?

5.2.10. Is it true that for any linear code there exists a matrix H which is

(1) a generating matrix;

(2) a checking matrix?

5.2.11. From a given matrix H, determine the power $m(C(H))$ of a code $C(H)$ generated by it, as well as the code distance $d(C(H))$:

$$(1)\ H = \begin{pmatrix} 1 & 0 & 0 & 1 & 1 \\ 1 & 1 & 0 & 0 & 1 \\ 1 & 1 & 1 & 0 & 0 \end{pmatrix}; \quad (2)\ H = \begin{pmatrix} 0 & 0 & 1 & 0 & 1 \\ 1 & 1 & 1 & 0 & 0 \\ 0 & 1 & 0 & 1 & 0 \\ 1 & 0 & 0 & 1 & 1 \end{pmatrix};$$

$$(3)\ H = \begin{pmatrix} 1 & 1 & 0 & 0 & \ldots & 0 & 0 & 0 \\ 0 & 1 & 1 & 0 & \ldots & 0 & 0 & 0 \\ 0 & 0 & 1 & 1 & \ldots & 0 & 0 & 0 \\ \cdot & \cdot & \cdot & \cdot & \cdot & \cdot & \cdot & \cdot \\ \cdot & \cdot & \cdot & \cdot & \cdot & \cdot & \cdot & \cdot \\ \cdot & \cdot & \cdot & \cdot & \cdot & \cdot & \cdot & \cdot \\ 0 & 0 & 0 & 0 & \ldots & 0 & 1 & 1 \\ 1 & 0 & 0 & 0 & \ldots & 0 & 0 & 1 \end{pmatrix}.$$

Here H is an $(n \times n)$ matrix;

(4) $H = (I_k P)$, where I_k is a unit $(k \times k)$ matrix and P is an arbitrary binary $(k \times (n - k))$ matrix, formed by different rows, each of which contains at least two unities, and $k \leqslant n - \log_2 (n + 1)$;

(5) $H = (I_5 Q)$, where I_5 is a (5×5) unit matrix and

$$Q = \begin{pmatrix} 1 & 1 & 1 & 0 & 1 & 1 & 0 & 0 & 1 & 0 \\ 0 & 1 & 1 & 1 & 1 & 0 & 0 & 1 & 0 & 1 \\ 1 & 1 & 0 & 1 & 1 & 0 & 1 & 1 & 1 & 0 \\ 0 & 1 & 1 & 0 & 0 & 1 & 1 & 1 & 1 & 1 \\ 1 & 1 & 0 & 1 & 0 & 1 & 1 & 1 & 0 & 1 \end{pmatrix}.$$

5.2.12. Let $V \subseteq B^n$ be a space consisting of linear combinations of rows in a matrix $H = (I_k P)$, where I_k is a unit $(k \times k)$ matrix and P is a $(k \times (n - k))$ matrix consisting of zeros and unities. Prove that V is a zero space of the matrix $G = (P^T I_{n-k})$, where I_{n-k} is a unit matrix of dimensionality $(n - k) \times (n - k)$ and P^T is the transpose of the matrix P.

5.2.13. Construct a checking matrix for a code generated by the matrix H from Problem 5.2.11 (1).

5.2.14. Prove that if $C \subseteq B^n$ is an (n, k)-code, the code dual to C is an $(n, n - k)$-code.

5.2.15. Let $H(C)$ be a matrix of an (n, k)-code $C \subseteq B^n$, which does not contain null columns. Prove that

(1) each column of the matrix $H(C)$ has 2^{k-1} unities and the same number of zeros;

(2) the sum of the weights of rows of the matrix $H(C)$ is $n \cdot 2^{k-1}$.

5.2.16. Prove that the code distance of a linear code $C \subseteq B^n$ is equal to the minimum weight of its nonzero vectors.

5.2.17. Prove that the code distance of an (n, k)-code does not exceed $[n2^{k-1}/(2^k - 1)]$.

5.2.18. Prove that for $n = 2d - 1$, the power of a linear $\langle n, d\rangle$-code does not exceed $2d$.

5.2.19. (1) Prove that the maximum possible power $g(n, d)$ of a linear $\langle n, d\rangle$-code satisfies the inequality $g(n, d) \leqslant 2g(n - 1, d)$.

(2) Using the result obtained by solving Problem 5.2.18, prove that $g(n, d) \leqslant d \cdot 2^{n-2d+2}$.

5.2.20. Let a code C be a zero space of a matrix H. Prove that the code distance of the code C is not smaller than d if and only if any combination of $d - 1$ or a smaller number of columns of the matrix H is linearly independent.

5.2.21. Prove that if $\sum\limits_{i=0}^{d-2} \binom{n-1}{i} < 2^k$, there exists a $(k \times n)$ matrix consisting of zeros and unities, in which any $d - 1$ columns are linearly independent, and hence there exists an $(n, n - k)$-code with a code distance equal to or longer than d.

5.2.22. Let $H_{k,n}$ be a $(k \times n)$ matrix, where $k =]\log_2 (n + 1)[$, in which the i-th column is a binary decomposition of the number i $(i = \overline{1, n})$. For example, for $n = 6$, the matrix $H_{3,6}$ has the form

$$H = \begin{pmatrix} 0 & 0 & 0 & 1 & 1 & 1 \\ 0 & 1 & 1 & 0 & 0 & 1 \\ 1 & 0 & 1 & 0 & 1 & 0 \end{pmatrix}.$$

(1) Prove that the zero space of the matrix is a Hamming code, i.e. a linear code correcting one error.

(2) Construct a code which is a zero space of the matrix $H_{3,6}$.

(3) Determine the number of errors that are corrected by the code generated by the matrix $H_{k,n}$.

5.2.23. The spectrum of a set $C \subseteq B^n$ is a vector $\tilde{s} = (s_0, s_1, \ldots, s_n)$, where s_r is the number of pairs of vertices in C that are r apart. Prove that for any (n, k)-code C, there exists an (n, k)-code C' having the same spectrum and a generating matrix of the form $(I_k P)$, where I_k is a unit $(k \times k)$ matrix and P is a $(k \times (n - k))$ matrix consisting of zeros and unities.

5.2.24. (1) Prove that $g(9, 5) = 4$.

(2) Prove that $m(9, 5) \geqslant 5$.

5.2.25. Let an (n, k)-code have a code distance d. Is it true that there exists an (n, k)-code with a code distance $d - 1$?

5.3. Alphabet Coding

Let A be an alphabet; then A^* is a set of all finite words in the alphabet A, including an empty word. The *length* (number of letters) of a word w is denoted by $\lambda(w)$. An *empty word* is denoted by Λ. A *concatenation of words* w_1 *and* w_2 obtained by writing the word w_2 just to the right of the word w_1 is denoted by w_1w_2. The word w_1 is called the *prefix*, and the word w_2 the *suffix* of the word w_1w_2. The prefix w_1 (suffix w_2) of the word w_1w_2 is called *proper* if $w_1 \neq \Lambda$ and $w_2 \neq \Lambda$ simultaneously. A word v is known as a *subword of a word* w if there exist words u_1 and u_2 such that $w = u_1vu_2$.

Let $A = \{a_1, a_2, \ldots, a_m\}$ be a message alphabet and B a coding alphabet. Let φ be a single-valued mapping of letters of the alphabet A onto B^*. The coding of words in the alphabet A when each word (message)

$a_{i_1}a_{i_2} \ldots a_{i_k}$ is juxtaposed to a word $\varphi(a_{i_1})\varphi(a_{i_2}) \ldots \varphi(a_{i_k})$ is called *alphabetic* (or *letter*) *coding*. An alphabetic coding is completely determined by the mapping φ generating it and is denoted by K_φ. The set $\{\varphi(a): a \in A\}$ is called an *alphabet code* and is denoted by $\varphi(A)$. An alphabetic coding K_φ and the corresponding code $\varphi(A)$ are assumed to be *uniquely decoded* or *divisible* if each relation of the type

$$\varphi(a_{i_1})\varphi(a_{i_2}) \ldots \varphi(a_{i_k}) = \varphi(a_{j_1})\varphi(a_{j_2}) \ldots \varphi(a_{j_l})$$

for the words in the coding alphabet B implies that $l = k$ and $j_t = i_t$ $(t = \overline{1, k})$. A code $\varphi(A)$ is called a *prefix code* if none of the words in $\varphi(A)$ is the beginning of any other word in $\varphi(A)$. A divisible alphabet code $\varphi(A)$ is assumed to be *complete* if for every word w in a coding alphabet B the following statement is valid: either w is a proper prefix of a certain word in $\varphi(A)$, or a certain word in $\varphi(A)$ is (not necessarily proper) a prefix of the word w.

One of the algorithms of detecting an alphabet code divisibility consists in the following. Let $\varphi(A) = \{w_1, w_2, \ldots, w_m\}$ be an alphabet code. Let S_1 be the set of all proper suffixes of code words and S_2 the set of all words each of which is a prefix of a code word. Let us consider an oriented multigraph G_φ whose vertices are the elements of the set $S = (S_1 \cap S_2) \cup \{\Lambda\}$. Let σ and τ be two different vertices in S. The arc joining σ and τ in the graph G_φ exists if and only if there exist such a code word w and such a sequence $P = w_{i_1}, w_{i_2}, \ldots, w_{i_k}$ of code words that w and $\sigma w_{i_1} w_{i_2} \ldots w_{i_k}\tau$ are equivalent as the words in a coding alphabet. Moreover, if $\sigma \neq \Lambda$, the sequence P may be empty. The arc joining σ with τ is assigned the word $w_{i_1} w_{i_2} \ldots w_{i_k}$. The multigraph G_φ has no loops and is referred to as the *graph of the alphabet code* φ. The following theorem is valid.

Theorem 1 (Al. A. Markov). *A code φ is divisible if and only if the graph G_φ contains no contours passing through the vertex Λ.*

Example 1. Let $\varphi(A) = \{cc, cca, bcca, aa, ab\}$. Then $S = \{\Lambda, c, cca, a, b\}$. The graph G_φ is shown in

Fig. 22*a*. There exists a contour in G_φ passing through the vertices Λ, a and b. By writing the words assigned to the vertices and arcs of this contour, we obtain a word which can be decoded in two ways:

$$ccabcca = (cca)\,(bcca) = (cc)\,(ab)\,(cca).$$

Example 2. Let $\varphi(A) = \{a, ab, acbb, bb, bbacc\}$. Then $S = \{\Lambda, b, bb\}$. The graph G_φ shown in Fig. 22*b* has no contours passing through Λ. The code is divisible.

Let us suppose that we have a message source which consecutively generates at random the letters of the

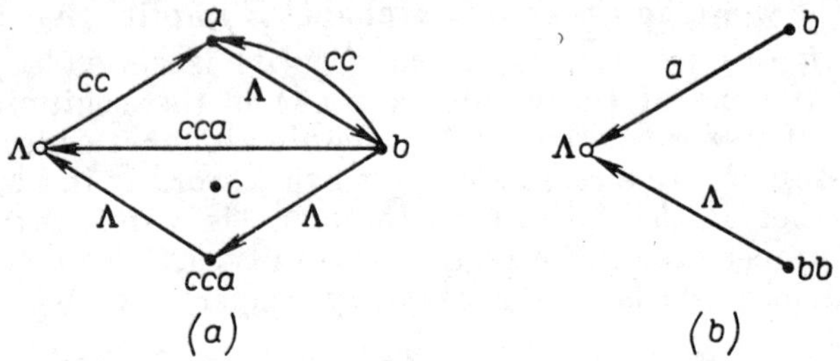

Fig. 22

alphabet $A = \{a_1, a_2, \ldots, a_m\}$. It is assumed that the emergence of letters of the alphabet A is statistically independent and obeys the probability distribution $P = \{p_1, p_2, \ldots, p_m\}$, $p_i > 0$, $\sum_{i=1}^{m} p_i = 1$. Any binary alphabet code $C = \{w_1, w_2, \ldots, w_m\}$ can be associated with a number

$$\mathcal{L}_C(P) = \sum_{i=1}^{m} p_i \lambda(w_i),$$

which is called the *cost of the code C for the distribution P*. The number $\mathcal{L}_C(P)$ is equal to the average number of letters of the coding alphabet per letter of the alphabet A. The prefix code C_0 is assumed to be *optimal for the distribution* P if $\mathcal{L}_{C_0}(P) = \inf_C \mathcal{L}_C(P)$, where the lower bound is taken over the set of prefix binary codes consisting of m words.

The Huffman method for constructing an optimal code is based on the following theorem.

Theorem 2. *If* $C = (w_1, w_2, \ldots, w_m)$ *is an optimal binary code for the distribution* $P = \{p_1, p_2, \ldots, p_m\}$ *and* $p_j = q_1 + q_2$, *where* $p_1 \geqslant p_2 \geqslant \ldots \geqslant p_{j-1} \geqslant p_j \geqslant p_{j+1} \geqslant \ldots \geqslant p_m \geqslant q_1 \geqslant q_2$, *the code* $C' = \{w_1, w_2, \ldots, w_{j-1}, w_{j+1}, \ldots, w_m, w_j0, w_j1\}$ *is optimal for the distribution*

$$P' = \{p_1, p_2, \ldots, p_{j-1}, p_{j+1}, \ldots, p_m, q_1, q_2\}.$$

The code C' is known as an *extension of the optimal code* C. The Huffman method consists in the following. Let p_{m-1} and p_m be the last two probabilities in the original list of probabilities compiled in a non-increasing order. These probabilities are excluded from the list, and their sum is inserted into the list in such a way that the probabilities in the new list are arranged in a non-increasing order. This procedure is repeated until we obtain a list of two probabilities, in which symbol 0 is assigned to one probability and symbol 1 to the other (this is the optimal code for a two-letter alphabet of messages for any probability distribution). Then, in accordance with the theorem, an optimal code is constructed for three letters for the corresponding list of probabilities, and so on until we obtain an optimal code for the original list of probabilities. The Huffman method is illustrated by the following example:

i	P_i							w_i
1	0.4	0.4	0.4	0.6	0	1	1	1
2	0.2	0.2	0.4	0.4	1	00	01	000
3	0.2	0.2	0.2			01	000	001
4	0.1	0.2					001	010
5	0.1							011

For a given probability distribution, there also exist other optimal codes such as

Table 6

i	p_i	w_i	w'_i	i	p_i	w_i	w'_i
1	0.4	1	00	4	0.1	0010	110
2	0.2	01	01	5	0.1	0011	111
3	0.2	000	10				

Fano's method of constructing codes close to optimal consists in the following. A list of probabilities compiled in a non-increasing order is divided into two (consecutive) parts so that the sums of the probabilities constituting these parts are as close as possible. Each letter of the message alphabet corresponding to a probability in the first part is assigned the symbol 0 (or 1), and the remaining letters are assigned the symbol 1 (resp. 0). This procedure is continued on these parts as long as a part contains at least two probabilities. We continue in the same manner until the entire list is divided into parts containing one probability each. The examples of codes constructed by Fano's method are given in Table 6.

5.3.1. Using a given alphabet code $\varphi(A)$, construct the graph G_φ and find out whether or not the code is divisible:

(1) $\varphi(A) = \{ab, dc, a, bcadd, ca\}$, $A = \{i, i = \overline{1, 5}\}$;

(2) $\varphi(A) = \{ddac, dd, cddab, a, cddd, b\}$, $A = \{i, i = \overline{1, 6}\}$;

(3) $\varphi(A) = \{a, ab, acbb, abb, bbacc\}$, $A = \{i, i = \overline{1, 5}\}$;

(4) $\varphi(A) = \{abc, bbc, bcb, caa, acbb, cbcb, bccabb, abcacbbb\}$, $A = \{i, i = \overline{1, 8}\}$;

(5) $\varphi(A) = \{abc, abb, bcc, ccaa, bcabbbcc, bbccaaabca, abcabbabbbcca\}$, $A = \{i, i = \overline{1, 7}\}$;

(6) $\varphi(A) = \{ab, bb, ca, cba, abb, bac, aabc, cabba\}$, $A = \{i, i = \overline{1, 8}\}$.

5.3.2. Let the numbers 1, 2, 4, 17 and 98 be encoded by their binary divisions of the minimum possible length.

For example, the code of unity is 1, the code of two is 10, and the code of four is 100. Is this coding divisible?

5.3.3. Using a given indivisible code $\varphi(A) = \{aa, ab, cc, cca, bcca\}$ and the word w in the coding alphabet $B = \{a, b, c\}$, find out whether the word w is a code of a message. If this is the case, find out whether the word w is a code of exactly one message:

(1) $w = ccabccabccabcc$;
(2) $w = bccaccabccabccacabcca$;
(3) $w = abbccaccabccaabab$.

5.3.4. Let $\varphi(A)$ be an alphabet code and G_φ be a graph of this code. Let a graph G'_φ be obtained from G_φ by deleting all the vertices which are code words. Will Theorem 1 remain valid if G'_φ is substituted for G_φ?

5.3.5. Let $\varphi(A)$ be an alphabet code and G_φ the graph of this code. Let a graph G'_φ be obtained from G_φ by deleting all the arcs which are labelled by Λ and which terminate at the vertex Λ. Will Theorem 1 remain in force if we substitute G'_φ for G_φ?

5.3.6. For a given divisible code $\varphi(A)$, construct a prefix code with the same set of lengths of the code words:

(1) $\varphi(A) = \{01, 10, 100, 111, 011\}$;
(2) $\varphi(A) = \{1, 10, 100, 0100\}$;
(3) $\varphi(A) = \{10, 101, 111, 1011\}$.

5.3.7. Let $\varphi(A)$ be an alphabet code of power m, in which the sum of the lengths of the code words is equal to N, and the maximum length of the code words is l. Using Theorem 1, prove that the code $\varphi(A)$ is divisible if and only if each word in the coding alphabet having a length not larger than $(l-1)(N-m+1)+1$ is either a code of exactly one word composed of the letters of the alphabet of messages A, or is not a code of any word in A^*.

5.3.8. Let k be the minimum and l the maximum length of the code words of an alphabet code $\varphi(A)$ and let N be the sum of the lengths of the code words. Prove that in order to establish the divisibility of the code $\varphi(A)$, it is sufficient to verify the uniqueness of decoding of all codes of words whose length in the alphabet of messages does not exceed Nl/k.

5.3.9. We shall assume that two words w and v in the alphabet $\{0, 1\}$ are *equivalent* if there exists a word u

which can be obtained both from w and from v by using the following finite number of operations:

(a) deleting subwords of the form 10 and 1001;

(b) inserting words of the form 10 and 1001 into the spaces between letters.[1]

Are the following pairs of words w and v equivalent:

(1) $w = 1010101$, $v = 0101010$;

(2) $w = 10010010$, $v = 010010010$;

(3) $w = 11011010$, $v = 01101101$?

5.3.10. Let M be a set consisting of m non-empty words in an alphabet A consisting of k letters. Prove that

(1) there exists a word in M, whose length is not less than $\log_k (1 + m(k - 1))$;

(2) for any $\varepsilon > 0$, the fraction of the words in M, whose length is smaller than $(1 - \varepsilon) \log_k (1 + m(k - 1))$, does not exceed $\left(\frac{4}{3}\right)^{1-\varepsilon} m^{-\varepsilon}$ for $m \geqslant 2$, $k \geqslant 2$.

5.3.11. Let each word in an alphabet binary code C of power $2^n + 1$ have a length not exceeding n.

(1) Prove that the code C is not a prefix code.

(2) Can the code C be divisible?

5.3.12. For given probability distribution of occurrence of letters, construct optimal codes by Huffman's method:

(1) $P = (0.34; 0.18; 0.17; 0.16; 0.15)$;

(2) $P = (0.6; 0.1; 0.09; 0.08; 0.07; 0.06)$;

(3) $P = (0.4; 0.4; 0.1; 0.03; 0.03; 0.02; 0.02)$;

(4) $P = (0.3; 0.2; 0.2; 0.1; 0.1; 0.05; 0.05)$.

5.3.13. (1) Construct the codes for the probability distributions of the previous problem by using Fano's method.

(2) Give an example of a probability distribution for which a code constructed by Fano's method is not optimal.

5.3.14. Let $C = \{w_1, w_2, \ldots, w_m\}$ be an optimal binary code corresponding to the distribution $P = (p_1, p_2, \ldots, p_m)$, $p_1 \geqslant p_2 \geqslant \ldots \geqslant p_m$. Prove that:

(1) $\lambda(w_i) \leqslant \lambda(w_j)$ if $p_i > p_j$;

(2) the code C is complete;

(3) there exist two code words of length $\lambda(w_m)$ which have identical prefixes of length $\lambda(w_m) - 1$.

5.3.15. Prove that if m is not a power of two, there exist two words of different lengths in the optimal binary

[1] It is also allowed to write the words 10 and 1001 just to the right and to the left of the word being transformed.

code for any probability distribution $P = (p_1, p_2, \ldots, p_m)$.

5.3.16. Using the solutions of Problems 5.3.14 and 5.3.15 and the definition of an optimal code, explain why the following codes are not optimal for given probability distributions:

(1)

p_i	w_i
0.6	0
0.2	10
0.1	11
0.1	01

(2)

p_i	w_i
0.6	1
0.2	01
0.15	001
0.05	0001

(3)

p_i	w_i
0.2	000
0.2	001
0.2	010
0.2	011
0.2	100

5.3.17. Determine the smallest m and a probability distribution $P = (p_1, p_2, \ldots, p_m)$ for which there exist optimal codes differing in tuples of the lengths of the code words.

5.3.18. (1) Prove that the maximum length of a code word in an optimal code of power m does not exceed $m - 1$.

(2) Prove that for any integer m ($m \geqslant 2$) there exists a probability distribution $P = (p_1, p_2, \ldots, p_m)$ satisfying the following condition: there exists an optimal code corresponding to the distribution P such that the maximum length of the code word in it is equal to $m - 1$.

5.3.19. Prove that there exist not more than $\binom{2^m - 1}{m}$ complete prefix binary codes of power m.

5.3.20. A code is referred to as *almost uniform* if the lengths of its code words differ by not more than unity. Prove that for any natural m a nearly uniform code is optimal for the distribution $P = (1/m, 1/m, \ldots, 1/m)$. Is the converse true?

5.3.21. (1) Prove (by induction in m) that the sum of the lengths of the code words in an optimal code with m messages does not exceed $\frac{1}{2}(m + 2)(m - 1)$, $m \geqslant 2$.

(2) Prove that for any integer $m \geqslant 2$, the estimate in the previous problem is accessible.

5.3.22. Let $g_1, g_2, \ldots, g_m$ be arbitrary pairwise disjoint faces of the cube B^n with dimensionalities $r_1, r_2, \ldots, r_m$ respectively.

(1) Using the obvious inequality $\sum_{i=1}^{m} 2^{r_i} \leqslant 2^n$, prove that the lengths $\lambda(w_1), \lambda(w_2), \ldots, \lambda(w_m)$ of the words of an arbitrary binary prefix code $C = \{w_1, w_2, \ldots, w_m\}$ satisfy the condition $\sum_{i=1}^{m} 2^{-\lambda(w_i)} \leqslant 1$.

(2) Prove that a complete prefix code satisfies the equality $\sum_{i=1}^{m} 2^{-\lambda(w_i)} = 1$.

(3) Prove that if $\sum_{i=1}^{m} 2^{-\lambda_i} \leqslant 1$, where λ_i are natural numbers, $i = \overline{1, m}$, there exists a prefix code with words of lengths $\lambda_1, \lambda_2, \ldots, \lambda_m$.

5.3.23. (1) Is it true that the number of code words of maximum length in an optimal binary code is even?

(2) Is it true that this number is a power of two?

5.3.24. Let $C = \{w_1, w_2, \ldots, w_m\}$ be a complete prefix binary code and let $\lambda(w_i) \geqslant \lambda(w_{i+1})$ for all $i = \overline{1, m-1}$. Prove that for $m \geqslant 2$ and all $i = \overline{1, m-1}$ the inequality $\lambda(w_{i+1}) - \lambda(w_i) \leqslant \log_2 m - 1$ is valid.

5.3.25*. Let L_m be the smallest integer for which there exists a binary prefix code of power m, with the sum of the lengths of the code words equal to L_m. Prove that $L_m \geqslant m [\log_2 m]$ for $m \geqslant 2$.

5.3.26. Prove that the lengths of the code words $\lambda(w_1), \lambda(w_2), \ldots, \lambda(w_m)$ of an optimal binary code $C = \{w_1, w_2, \ldots, w_m\}$ satisfy the condition $\sum_{i=1}^{m} 2^{-\lambda(w_i)} = 1$.

5.3.27. Let $P = (p_1, p_2, \ldots, p_m)$ be a probability distribution ($p_i > 0$, $i = \overline{1, m}$), and $\mathcal{L}(P) = \inf_C \mathcal{L}_C(P)$, where the lower bound is taken over all binary prefix codes of power m ($m \geqslant 3$).

(1) Prove that $\mathcal{L}(P) > 1$.

(2) Prove that for any $\varepsilon > 0$ and any integer $m \geqslant 1$, there exists a probability distribution $P = (p_1, p_2, \ldots, p_m)$ such that $\mathcal{L}(P) < 1 + \varepsilon$.

Shannon's method for constructing almost optimal codes consists in the following.

Let $P = (p_1, p_2, \ldots, p_m)$, $p_i > 0$, $\sum_{i=1}^{m} p_i = 1$, $p_i \geqslant p_{i+1}$, $i = \overline{1, m-1}$, be the probability distribution for the occurrence of the letters in the alphabet of messages $A = \{a_1, a_2, \ldots, a_m\}$. Then a letter a_i is assigned a code word of length $l_i = \left] \log \frac{1}{p_i} \right[$, composed of the first (after the decimal point) l_i digits of the decomposition of a number $q_{i-1} = \sum_{j=1}^{i-1} p_j$ into an infinite binary fraction (rounded down).

For example, if $P = (0.4, 0.3, 0.3)$, the code of the letter a_1 is 00, the code of a_2 is 01, and the code of a_3 is 10.

5.3.28. Using Shannon's method, construct the codes for the probability distributions in Problems 5.3.12 (1) and (2).

5.3.29. Indicate the smallest m for which there exists a probability distribution $P = (p_1, p_2, \ldots, p_m)$ such that the code constructed by Shannon's method for the given probability distribution is not optimal.

5.3.30. Prove that the code constructed by Shannon's method is a prefix code.

5.3.31. Let $\mathcal{L}^*(m) = \sup_P \inf_C \mathcal{L}_C(P)$, where the upper bound is taken over all distributions $P = (p_1, p_2, \ldots, p_m)$ such that $p_i > 0$, $i = \overline{1, m}$, $\sum_{i=1}^{m} p_i = 1$.

(1) Prove that $\mathcal{L}^*(m) \geqslant [\log_2 m]$.

(2) Prove that $\mathcal{L}^*(m) \leqslant [\log_2 m] + 1$.

Chapter Six

Finite Automatons

6.1. Determinate and Boundedly Determinate Functions

Let A be a non-empty finite alphabet. The elements of the alphabet are called *letters* (or *symbols*). A *word in an alphabet* A is a sequence formed by the letters of this alphabet. The *length of a word* w (the number of letters in the word) is denoted by $\lambda(w)$. The set of all words $\tilde{x}^s = x(1)\, x(2) \ldots x(s)$ of length s ($s \geqslant 1$) in the alphabet A will be denoted by A^s. A word of length 0 (*empty word*) is denoted by the symbol Λ. By A^* we denote the set $\{\Lambda\} \cup \bigcup\limits_{s\geqslant 1} A^s$, while A^ω denotes the set of all words $\tilde{x}^\omega = x(1)\, x(2) \ldots$, where $x(t) \in A$, $t = 1, 2, \ldots$. The words in the set A^ω are called *infinite words in the alphabet* A.

The word w, obtained by writing a word w_2 to the right of a finite (or empty) word w_1 is called a *concatenation of words* w_1 and w_2 and is denoted by $w_1 w_2$. The word w_1 is called the *beginning* (*prefix*), and w_2 the *end* (*suffix*) of the word w.

Let A and B be finite non-empty alphabets. The mapping $\varphi\colon A^\omega \to B^\omega$ is called a *determinate function* or a *determinate operator* (in abbreviated form, a *d-function* or a *d-operator*) if it satisfies the following condition:

for any $s \geqslant 1$, the s-th symbol $y(s)$ in a word $\tilde{y}^\omega = \varphi(\tilde{x}^\omega)$ is a single-valued function of the first s symbols $x(1), x(2), \ldots, x(s)$ of the word $\tilde{x}^\omega$.

If the words $\tilde{x}_1^\omega$ and $\tilde{x}_2^\omega$ have identical prefixes of length s ($s \geqslant 1$), the words $\tilde{y}_1^\omega = \varphi(\tilde{x}_1^\omega)$ and $\tilde{y}_2^\omega = \varphi(\tilde{x}_2^\omega)$ have also identical prefixes of length s.

The set of all d-functions of the type $\varphi\colon A^\omega \to B^\omega$ will be denoted by $\Phi_{A, B}$.

If $A = A_1 \times A_2 \times \ldots \times A_n$ and $B = B_1 \times B_2 \times \ldots \times B_m$, then the mapping $\varphi: A^\omega \to B^\omega$ induces m functions each of which depends on n variables, the variable X_j running through the set A_j^ω $(j = 1, 2, \ldots, n)$. These functions are defined as follows. Let $\tilde{x}^\omega = x(1)\, x(2) \ldots x(t) \ldots$ be a word in A^ω and $\tilde{y}^\omega = \varphi(\tilde{x}^\omega) = y(1)\, y(2) \ldots y(t) \ldots$. In this case, $x(t) = (x_1(t),\ x_2(t), \ldots,\ x_n(t))$, $x_j(t) \in A_j$, $\tilde{x}_j^\omega = x_j(1)\, x_j(2) \ldots x_j(t) \ldots$, $\tilde{x}^\omega = (\tilde{x}_1^\omega, \tilde{x}_2^\omega, \ldots, \tilde{x}_n^\omega)$, $y(t) = (y_1(t),\ y_2(t), \ldots,\ y_m(t))$, $y_i(t) \in B_i$, $\tilde{y}_i^\omega = y_i(1)\, y_i(2) \ldots y_i(t) \ldots$, $\tilde{y}^\omega = (\tilde{y}_1^\omega, \tilde{y}_2^\omega, \ldots, \tilde{y}_m^\omega)$, $\varphi_i(\tilde{x}_1^\omega, \tilde{x}_2^\omega, \ldots, \tilde{x}_n^\omega) = \tilde{y}_i^\omega$.

By proceeding in the reverse order, we can construct the mapping φ from the functions φ_i.

The concept of a *d-function of n arguments* arises naturally upon a consideration of d-functions of the type $\varphi: (A_1 \times A_2 \times \ldots \times A_n)^\omega \to B^\omega$.

The variable X_1 of a function $\varphi(X_1, X_2, \ldots, X_n)$: $(A_1 \times A_2 \times \ldots \times A_n)^\omega \to B^\omega$ is called *essential* if there are two tuples $(\tilde{x}_{11}^\omega, \tilde{x}_{21}^\omega, \ldots, \tilde{x}_{n1}^\omega)$ and $(\tilde{x}_{12}^\omega, \tilde{x}_{21}^\omega, \ldots, \tilde{x}_{n1}^\omega)$ of values of the variables $X_1, X_2, \ldots, X_n$ which differ only in their first components and such that $\varphi(\tilde{x}_{11}^\omega, \tilde{x}_{21}^\omega, \ldots, \tilde{x}_{n1}^\omega) \neq \varphi(\tilde{x}_{12}^\omega, \tilde{x}_{21}^\omega, \ldots, \tilde{x}_{n1}^\omega)$. If the variable X_1 is not essential, it is called *fictitious*. All other essential and fictitious variables X_i on which a given function depends are defined in an identical manner.

The function $\varphi(X_1, X_2, \ldots, X_n)$ is said *to depend essentially* (*fictitiously*) on a variable X_i $(1 \leqslant i \leqslant n)$ if this variable is an essential (resp. fictitious) variable of the function φ.

If A is a set of all vectors of length n with elements in E_k, and B is a set of all vectors of length m with elements in E_l, we shall use the notation $\Phi_{k,l}^{n,m}$ instead of $\Phi_{A,B}$. For $n = m = 1$, the superscripts in $\Phi_{k,l}^{1,1}$ will be omitted. If $k = l$, we shall write only one subscript: $\Phi_k^{n,m}$.

Sometimes it is convenient to assume that a d-function φ in $\Phi_{A,B}$ is *represented* by a *discrete device* (*automaton*) $\mathfrak{A}_\varphi$ operating at discrete instants of time $t = 1, 2, \ldots$. At each instant t of time, a signal $x(t)$ is supplied to the input of this device, and a signal $y(t)$ appears at the output (Fig. 23). The words $\tilde{x}^\omega$ are called *input words* and $\tilde{y}^\omega$ the *output words.* The alphabets A and B are called *input and output alphabets of the automaton* $\mathfrak{A}_\varphi$. If $A = A_1 \times A_2 \times \ldots \times A_n$ and $B = B_1 \times B_2 \times \ldots \times B_m$, it can be assumed that the automaton $\mathfrak{A}_\varphi$ represents m determinate functions, each of which depends on n variables.

Fig. 23

Any two functions φ_1 and φ_2 in $\Phi_{A,B}$ are called *distinguishable* if there exists an input word $\tilde{x}_0^\omega$ which is processed into different output words by them, i.e. $\varphi_1(\tilde{x}_0^\omega) \neq \varphi_2(\tilde{x}_0^\omega)$. If, however, the equality $\varphi_1(\tilde{x}^\omega) = \varphi_2(\tilde{x}^\omega)$ is satisfied for any input word $\tilde{x}^\omega$, then φ_1 and φ_2 are called *equivalent* or *indistinguishable d-functions.*

Let $\varphi \in \Phi_{A,B}$ and $\psi \in \Phi_{A,B}$. If there exists a word $\tilde{x}_0^s \in A^*$ such that[1] $\varphi(\tilde{x}_0^s\tilde{x}^\omega) = \varphi(\tilde{x}_0^s)\,\psi(\tilde{x}^\omega)$ for any word $\tilde{x}^\omega \in A^\omega$, the operator ψ is called a *residual operator of the operator* φ *generated by the word* $\tilde{x}_0^s$, and is denoted by $\varphi_{\tilde{x}_0^s}$. The set $Q(\varphi, \tilde{x}_0^s)$ of all *residual operators* of the operator φ, equivalent to the operator $\varphi_{\tilde{x}_0^s}$, forms an *equivalence class* called the *state of operator* φ *containing a residual operator* $\varphi_{\tilde{x}_0^s}$. The state containing the operator φ is called the *initial state*. If $\psi \in Q(\varphi, \tilde{x}_0^s)$, the *operator* ψ is said *to be represented by the state* $Q(\varphi, \tilde{x}_0^s)$ of the operator φ. The operator φ is called a *boundedly determinate operator* (abbreviated as a *b.d.-operator* or a *b.d.-function*) if it has a finite number of pairwise different states. The

---

[1] Here, $\varphi(\tilde{x}_0^s)$ denotes a prefix of length s to the output word $\varphi(\tilde{x}_0^s\tilde{x}^\omega)$.

number of different states of a b.d.-function is known as its *weight*. If the pairwise different states of an operator φ form an infinite set, we shall assume that the weight of the operator φ is equal to ∞. By $\hat{\Phi}_{A,B}$ we shall denote the set of all functions in $\Phi_{A,B}$ that are b.d.-functions.

It is convenient to analyze d-functions by considering their graphic representation in the form of *infinite infor-*

Fig. 24

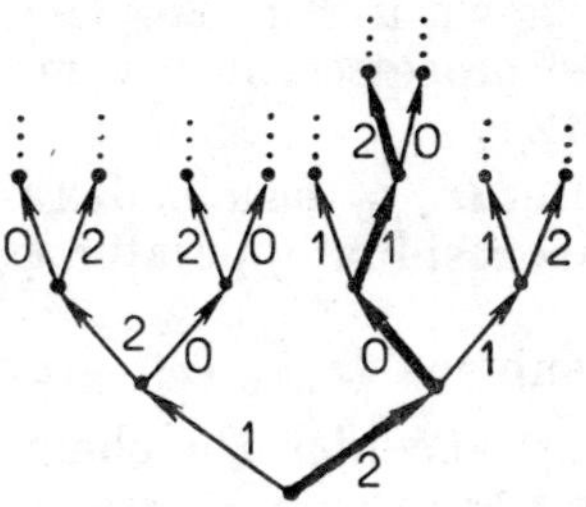

Fig. 25

mation trees. Let A be an alphabet of n letters. By D_A we shall denote an infinite oriented rooted tree satisfying the following conditions:

(a) the out-degree of each vertex, including a root, is equal to n;

(b) the in-degree of a root is equal to zero, while every other vertex has an in-degree 1;

(c) each arc of the tree D_A is assigned a letter of the alphabet A, and different arcs emanating from the same vertex of the tree (in particular from a root) are assigned different letters. A root of the tree is assumed to be a *zero rank vertex*. If a vertex v is the end of an arc emanating from an i-th rank vertex ($i \geqslant 0$), it is called a *vertex of* $(i + 1)$*-th rank*. An *arc of* j*-th level* ($j \geqslant 1$) is an arc emanating from the $(j - 1)$*-th rank vertex*. Each infinite oriented chain in the tree D_A corresponds to a word in A^ω. Figure 24 shows a *fragment of the tree* D_A, $A = \{0, 1\}$, consisting of the first three levels of this tree (here and below, we assume that the left and right arcs emanating from a vertex are assigned the symbols 0 and 1 respectively). Thick lines are used to isolate the chain corresponding to the word 101.

A *loaded tree* $D_{A,B}$ is obtained from the tree D_A by assigning a letter from the alphabet B to each arc. Each oriented infinite chain in the tree $D_{A,B}$ has a corresponding word in B^ω, formed by the letters assigned to the arcs of this chain. Hence it can be assumed that the loaded tree $D_{A,B}$ defines (represents) the mapping $\varphi: A^\omega \to B^\omega$, which is a d-function. The fragment of a loaded tree $D_{A,B}$, where $A = \{0, 1\}$ and $B = \{0, 1, 2\}$, is shown in Fig. 25. The d-function corresponding to this tree processes, for example, the word 1010 into the word 2012.

Let $D_{A,B}$ be a loaded tree representing the d-function φ. The residual operator $\varphi_{\tilde{x}_0^s}$ $(s \geqslant 0)$ of the operator φ has a subtree $D_{A,B}(\tilde{x}_0^s)$ growing from an s-th rank vertex $v(\tilde{x}_0^s)$ at which the chain starting from the root and containing exactly s arcs terminates. In the i-th level, this chain has a corresponding arc labelled by the letter $x_0(i) \in A$.

If the residual operators $\varphi_{\tilde{x}_1^{s_1}}$ and $\varphi_{\tilde{x}_2^{s_2}}$ are equivalent, the vertices $v(\tilde{x}_1^{s_1})$ and $v(\tilde{x}_2^{s_2})$ corresponding to them and the subtrees growing from these vertices are also called *equivalent*. The weight of a tree representing a d-function is equal to the weight of this function, and hence to the maximum number of pairwise non-equivalent vertices (or subtrees) of the given tree.

6.1.1. Find out if the mapping $\varphi: \{0, 1\}^\omega \to \{0, 1\}^\omega$ is a d-function:

(1) $\varphi(x(1)x(2)\ldots) = x(2)x(3)\ldots$, i. e. $y(t) = x(t+1)$, $t \geqslant 1$;

(2) $\varphi(\tilde{x}^\omega) = 10100100010\ldots\underbrace{010\ldots010}_{n \text{ times}}$ for any input word $\tilde{x}^\omega$;

(3) $\varphi(x(1)x(2)x(3)\ldots) = x(1)x(2)x(1)x(2)x(3)\ldots$, i.e. $y(1) = x(1)$, $y(2) = x(2)$ and $y(t) = x(t-2)$ for $t \geqslant 3$;

(4) $\varphi(x(1)x(2)\ldots) = 0x(1+x(2))x(1+x(3))\ldots$, i.e. $y(1) = 0$, $y(t) = x(1+x(t))$ for $t \geqslant 2$.

6.1.2. Is the function φ: $\{0, 1\}^\omega \to \{0, 1\}^\omega$ described by the following description a d-function?

(1) The word $\tilde{x}^\omega = x(1)\,x(2)\ldots$ is processed into a word $\tilde{0}^\omega = 000\ldots 0\ldots$, if there exists a t such that $x(t) = 0$; otherwise, $\varphi(\tilde{x}^\omega) = \tilde{1}^\omega = 111\ldots 1\ldots$.

(2) The s-th letter $y(s)$ in a word $\tilde{y}^\omega = \varphi(\tilde{x}^\omega)$ is equal to zero if the inequality $3x(t) \leqslant x(t+1) + x(t+2)$ is satisfied for a certain $t \leqslant s$; otherwise, $y(s) = 1$.

(3) The s-th letter $y(s)$ in a word $\tilde{y}^\omega = \varphi(\tilde{x}^\omega)$ is equal to 0 if there exists a $t > s$ such that $x(t) \leqslant x(s)$; otherwise, $y(s) = 1$.

(4) $y(1) = 0$, and for $s \geqslant 2$ the prefix $y(1)\,y(2)\ldots y(s)$ in the word $\tilde{y}^\omega = \varphi(\tilde{x}^\omega)$ contains one more zero than in the word $x(2)\,x(3)\ldots x(s)$.

Each word $\tilde{x}^\omega = x(1)\,x(2)\ldots x(t)\ldots$ from $\{0, 1\}^\omega$ corresponds to a number $\nu(\tilde{x}^\omega)$ in the segment $[0, 1]$. The binary expansion of this number is $0, x(1)\,x(2)\ldots x(t)\ldots$. If $a \in [0, 1)$, its binary expansion[2] $0, a_1a_2\ldots a_t\ldots$ generates the word $\langle a\rangle = x(1)\,x(2)\ldots x(t)\ldots$, where $x(t) = a_t$ $(t \geqslant 1)$.

6.1.3. Find out if the function φ: $\{0, 1\}^\omega \to \{0, 1\}^\omega$ is a determinate function:

(1) $\varphi(\tilde{x}^\omega) = \langle 1/3\rangle$;

(2) $\varphi(\tilde{x}^\omega) = \langle 7/15\rangle$;

(3) $\varphi(\tilde{x}^\omega) = \langle 2/7\rangle$;

(4) $\varphi(\tilde{x}^\omega) = \langle \sqrt{2}/2\rangle$;

(5) $\varphi(\tilde{x}^\omega) = \langle 1/\sqrt{3}\rangle$.

6.1.4*. Find out if the function $\varphi(X_1, X_2, \ldots, X_n)$: $\underbrace{\{0, 1\}^\omega \times \{0, 1\}^\omega \times \ldots \times \{0, 1\}^\omega}_{n \text{ times}} \to \{0, 1\}^\omega$ is determinate:

(1) $\varphi(\tilde{x}_1^\omega, \tilde{x}_2^\omega) = \begin{cases} \tilde{x}_2^\omega & \text{if } \nu(\tilde{x}_1^\omega) \geqslant \nu(\tilde{x}_2^\omega), \\ \tilde{x}_1^\omega & \text{otherwise;} \end{cases}$

[2] If $a = p/2^n$ $(n \geqslant 1)$, we consider a binary expansion containing an infinitely large number of zeros $(p < 2^n)$. If $a = 1$, then, by definition, $\langle a\rangle = 1^\omega$.

(2) $\varphi(\tilde{x}_1^\omega, \tilde{x}_2^\omega) = \begin{cases} \tilde{x}_1^\omega & \text{if } \nu(\tilde{x}_1^\omega)\,\nu(\tilde{x}_2^\omega) \leqslant 1/2, \\ \tilde{x}_2^\omega & \text{otherwise;} \end{cases}$

(3) $\varphi(\tilde{x}_1^\omega, \tilde{x}_2^\omega) = \begin{cases} \tilde{1}^\omega & \text{if } \nu(\tilde{x}_1^\omega) \leqslant \nu(\tilde{x}_2^\omega), \\ \tilde{0}^\omega & \text{otherwise;} \end{cases}$

(4) $\varphi(\tilde{x}_1^\omega, \tilde{x}_2^\omega, \tilde{x}_3^\omega) = \begin{cases} \tilde{1}^\omega & \text{if } \nu(\tilde{x}_1^\omega) + \nu(\tilde{x}_2^\omega) \leqslant \nu(\tilde{x}_3^\omega), \\ \tilde{x}_3^\omega & \text{otherwise.} \end{cases}$

6.1.5. The partial functions $\varphi\colon \{0, 1\}^\omega \to \{0, 1\}^\omega$ given below are not defined only on the word $\tilde{0}^\omega = 00\ldots0\ldots$. Which of these functions can be redefined to determinate functions and which ones cannot?

(1) $\varphi(\tilde{x}^\omega) = \begin{cases} \tilde{x}^\omega & \text{if the number of zeros in each prefix of the word } \tilde{x}^\omega \text{ is not less than the number of unities,} \\ \tilde{1}^\omega & \text{otherwise;} \end{cases}$

(2) $\varphi(\tilde{x}^\omega) = y(1)\,y(2)\ldots y(t)\ldots$, where

$$y(t) = \begin{cases} 0, & \text{if } \exists s\,((s \leqslant t)\,\&\,(x(s) = 1)), \\ 1 & \text{otherwise;} \end{cases}$$

(3) $\varphi(\tilde{x}^\omega) = y(1)\,y(2)\ldots y(t)\ldots$, where

$$y(t) = \begin{cases} 1 & \text{if for a certain } s \leqslant t \text{ in the prefix } x(1)\,x(2)\ldots x(s), \text{ the number of zeros is larger than the number of unities,} \\ x(t) & \text{otherwise;} \end{cases}$$

(4) $\varphi(\tilde{x}^\omega) = y(1)\,y(2)\ldots y(t)\ldots$, where

$$y(t) = \begin{cases} 1 & \text{if } \nu(x(1)\,x(2)\ldots x(t)\,00\ldots0\ldots) \leqslant 1/2, \\ 0 & \text{otherwise.} \end{cases}$$

6.1.6*. (1) Disprove the following statement: if the function $\varphi(X_1, X_2)\colon \{0, 1\}^\omega \times \{0, 1\}^\omega \to \{0, 1\}^\omega$ depends essentially on the variable X_1 and if, for each (fixed) word $\tilde{x}_2^\omega \in \{0, 1\}^\omega$, $\varphi(X_1, \tilde{x}_2^\omega)$ is a d-function, the function $\varphi(X_1, X_2)$ is also a d-function.

(2) Let the function $\varphi(X_1, X_2)\colon \{0, 1\}^\omega \times \{0, 1\}^\omega \to$

$\{0, 1\}^\omega$ satisfy the following condition: for any words $\tilde{x}_1^\omega$ and $\tilde{x}_2^\omega$ in $\{0, 1\}^\omega$, the functions $\varphi(X_1, \tilde{x}_2^\omega)$ and $\varphi(\tilde{x}_1^\omega, X_2)$ are determinate. Will the function $\varphi(X_1, X_2)$ be determinate as well?

6.1.7. Construct a fragment of a loaded tree containing s first levels for the function $\tilde{y}^\omega = \varphi(\tilde{x}^\omega)$ belonging to the set Φ_2:

(1) $y(1) = 1$ and $y(t) = x(t-1)$ for $t \geqslant 2$, $s = 3$;

(2) $y(1) = 0$ and $y(t) = x(t) \oplus y(t-1)$ for $t \geqslant 2$, $s = 4$;

(3) $\varphi(\tilde{x}^\omega) = \left\langle \frac{2}{3} \right\rangle$, $s = 3$;

(4) $\varphi(\tilde{x}^\omega) = \left\langle \frac{1}{4} \right\rangle$, $s = 4$.

6.1.8. For a given function $\varphi(\tilde{x}^\omega) \in \Phi_2$, represent in a loaded tree a chain corresponding to the prefix $\tilde{x}^s$ of the input word $\tilde{x}^\omega$ and write the prefix $\tilde{y}^s$ of the output word $\tilde{y}^\omega$:

(1) $\varphi(\tilde{x}^\omega) = 10100100010\ldots$ $\left(\text{i.e. } y(t) = 1 \text{ only for } t = \binom{i}{2},\ i = 2, 3, \ldots\right)$, $\tilde{x}^7 = 0101001$;

(2) $\varphi(\tilde{x}^\omega) = \left\langle \frac{2}{9} \right\rangle$, (a) $\tilde{x}^7 = 1111101$, (b) $\tilde{x}^7 = 1010110$;

(3) $y(t) = \begin{cases} 1 & \text{if } x(1) + x(2) + \ldots + x(t) > t/2, \\ 0 & \text{otherwise,} \end{cases}$

(a) $\tilde{x}^{10} = 0101010110$, (b) $\tilde{x}^{10} = 1100101110$.

6.1.9. The loaded tree corresponding to a certain function $\varphi(\tilde{x}^\omega) \in \Phi_2$ has the following form: the left and right arcs emanating from a root are labelled by 0 and 1 respectively; if v is an i-th rank vertex ($i \geqslant 1$) and the i-th level arc terminating at the vertex v is labelled by $\sigma \in \{0, 1\}$, the left arc emanating from v is labelled by the same symbol σ, while the right arc is labelled by $\bar{\sigma}$ (negation of σ).

(1) Is it true that the s-th letter of the output word

$\tilde{y}^{\omega} = \varphi(\tilde{x}^{\omega})$ for this function can be determined from the relation

(a) $y(1) = x(1)$, $y(s) = x(s-1) \oplus x(s)$ for $s \geqslant 2$;

(b) $y(1) = x(1)$, $y(2) = x(1) \oplus x(2)$, $y(s) = x(s-2) \oplus x(s)$ for $s \geqslant 3$;

(c) $y(s) = x(1) \oplus x(2) \oplus \ldots \oplus x(s)$?

(2) Find the weight of the function φ.

6.1.10. Are the residual operators $\varphi_{\tilde{x}_0^s}$ and $\varphi_{\tilde{x}_1^r}$ of the d-function $\varphi \in \Phi_2$ equivalent?

(1) $\varphi(\tilde{x}^{\omega}) = 10100100010 \ldots \left(\text{i.e. } y(t) = 1 \text{ only for } t = \binom{i}{2},\ i = 2, 3, \ldots\right)$, $\tilde{x}_0^3 = 101$, $\tilde{x}_1^3 = 010$;

(2) $\varphi(\tilde{x}^{\omega}) = \left\langle \frac{8}{15} \right\rangle$, $\tilde{x}_0^2 = 10$, $\tilde{x}_1^5 = 00101$;

(3) $\varphi(\tilde{x}^{\omega}) = \left\langle \frac{2}{5} \right\rangle$, $\tilde{x}_0^1 = 1$, $\tilde{x}_1^3 = 001$;

(4) $\varphi(\tilde{x}^{\omega}) = y(1)\, y(2) \ldots$ and

$$y(t) = \begin{cases} 0 \text{ if } x(1) + x(2) + \ldots + x(t) < t/2, \\ 1 \text{ otherwise}, \end{cases}$$

$\tilde{x}_0^2 = 10$, $\tilde{x}_1^6 = 010110$.

6.1.11. Find out if φ_1 is a residual operator of the function $\varphi \in \Phi_2$:

(1) φ: $\begin{cases} y(1) = 0, \\ y(t) = x(t) \oplus \bar{y}(t-1), \quad t \geqslant 2, \end{cases}$

φ_1: $\begin{cases} y(1) = 1, \\ y(t) = \bar{y}(t-1), \quad t \geqslant 2; \end{cases}$

(2) φ: $\begin{cases} y(1) = 1, \\ y(t) = \bar{y}(t-1), \quad t \geqslant 2, \end{cases}$

φ_1: $\begin{cases} y(1) = 0, \\ y(t) = \bar{y}(t-1), \quad t \geqslant 2; \end{cases}$

(3) φ: $\begin{cases} y(1) = 0, \\ y(t) = x(t)\,\bar{x}(t-1) \vee \bar{x}(t)\, x(t-1), \quad t \geqslant 2, \end{cases}$

φ_1: $\begin{cases} y(1) = 0, \\ y(t) = x(t) \oplus y(t-1), \quad t \geqslant 2; \end{cases}$

(4) $\varphi(\tilde{x}^\omega) = \left\langle \frac{13}{15} \right\rangle$, $\varphi_1(\tilde{x}^\omega) = \left\langle \frac{7}{15} \right\rangle$.

6.1.12. Find out if the function $\varphi \in \Phi_2^{n, m}$ is a b.d.-function, and determine its weight

(1) $\varphi(\tilde{x}^\omega)$: $\begin{cases} y(1) = y(2) = 1, \\ y(t) = x(t-2), \quad t \geqslant 3; \end{cases}$

(2) $\varphi(\tilde{x}^\omega)$: $\begin{cases} y(2t) = x(t+1), \quad t \geqslant 1, \\ y(2t-1) = \bar{x}(t), \quad t \geqslant 1; \end{cases}$

(3) $\varphi(\tilde{x}^\omega)$: $\begin{cases} y(1) = 1, \\ y(2t-1) = x(2t-1) \oplus y(2t-3), \quad t \geqslant 2, \\ y(2t) = x(2t-1), \quad t \geqslant 1; \end{cases}$

(4) $\varphi(\tilde{x}^\omega)$: $\begin{cases} y_1(1) = x_1(1) \oplus x_2(1), \\ y_1(t) = x_1(t) \oplus x_2(t) \oplus y_2(t-1), \quad t \geqslant 2, \\ y_2(1) = x_1(1)\, x_2(1), \\ y_2(t) = x_1(t)\, x_2(t) \oplus x_1(t)\, y_2(t-1) \\ \quad \oplus\, x_2(t)\, y_2(t-1), \quad t \geqslant 2; \end{cases}$

(5) $\varphi(\tilde{x}^\omega)$: $\begin{cases} y_1(1) = 1, \\ y_1(t) = \bar{y}_2(t-1), \quad t \geqslant 2, \\ y_2(1) = 0, \\ y_2(t) = y_1(t-1), \quad t \geqslant 2. \end{cases}$

6.1.13*. Let $D_{A,B}$ be a loaded tree representing a b.d.-function φ: $A^\omega \to B^\omega$. Each vertex of the tree $D_{A,B}$ is assigned a number equal to the weight of its subtree. We obtain a new tree $\bar{D}_{A,B}$.

(1) For any $r \geqslant 1$, give an example of a b.d.-function φ, such that each vertex of the tree $\bar{D}_{A,B}$ corresponding to the function φ is labelled by the number r.

(2) For any $r \geqslant 2$, give an example of a b.d.-function such that for $j = 0, 1, \ldots, r-1$, each j-th rank vertex of the tree $\bar{D}_{A,B}$ corresponding to the function φ is labelled by the number $r - j$.

6.1.14. (1) Prove that the tree $\bar{D}_{A,B}$ constructed according to the conditions of Problem 6.1.13 has the following property: the sequence of numbers $\nu(v_1)$, $\nu(v_2)$, ... assigned to the vertices of the oriented chain v_1, (v_1, v_2), v_2, (v_2, v_3), v_3, ... (finite or infinite) is monotonically non-increasing, and that this sequence is stabilized for an infinite chain.

(2) Show that any residual operator of the function $\varphi \in \Phi_{A,B}$ has a weight not exceeding the weight of the function φ.

6.1.15. The tree representing the function $\varphi_0(\tilde{x}^\omega) \in \Phi_2$ is of the form: the symbol 1 is assigned only to the arcs belonging to the oriented chain emanating from the root and corresponding to the input word 10100100010 . . . $\left(\text{here } x(t) = 1 \text{ only for } t = \binom{i}{2},\ i = 2, 3, \ldots\right)$; the remaining arcs are labelled by the symbol 0. Prove that the function φ_0 has an infinite weight and is therefore not boundedly determinate.

6.1.16. For each $r \geqslant 2$, obtain an example of a b.d.-function of weight r satisfying the following condition: in the loaded tree representing this function, the symbol 1 is assigned only to some (not necessarily all) arcs of a *single* infinite oriented chain Z emanating from the root. The remaining arcs (not belonging to the chain Z) are assigned the symbol 0.

The word $\tilde{x}^\omega \in A^\omega$ is called *quasi-periodic* if there exist integers n_0 and T, such that $n_0 \geqslant 1$, $T \geqslant 1$, and $x(n + T) = x(n)$ for $n \geqslant n_0$. In this case, the prefix $x(1)\, x(2) \ldots x(n_0 - 1)$ of the word $\tilde{x}^\omega$ is called a *pre-period*, the number $n_0 - 1$ the *length of this pre-period*, the word $x(n_0)\, x(n_0 + 1) \ldots x(n_0 + T - 1)$ the *period of the word* $\tilde{x}^\omega$, and the number T the *length of this period*. Such a quasi-periodic word can be conveniently presented in the form

$$x(1)\, x(2) \ldots x(n_0 - 1)\, [x(n_0)\, x(n_0 + 1) \ldots x(n_0 + T - 1)]^\omega.$$

6.1.17. (1) Prove that if the function $\tilde{\varphi} \in \hat{\Phi}_{A,B}$, then any quasi-periodic word in A^ω is transformed by the function φ into a quasi-periodic word in B^ω.

(2) Confining yourself to the set Φ_2, show that the statement converse to the one formulated in part (1) is not true.

Hint. See Problem 6.1.15.

6.1.18*. Disprove the following statement: if a d-function $\varphi(X_1, X_2)$: $\{0, 1\}^\omega \times \{0, 1\}^\omega \to \{0, 1\}^\omega$ depends essentially on the variable X_1 and if for every (fixed) word $\tilde{x}_2^\omega \in \{0, 1\}^\omega$, $\varphi(X_1, \tilde{x}_2^\omega)$ is a b.d.-function, the function $\varphi(X_1, X_2)$ is also boundedly determinate.

6.1.19. Suppose that all the vertices of a loaded tree are divided in the normal way into equivalence classes. Prove that any equivalence class has a vertex v satisfying the condition: all the vertices in an oriented chain emanating from the root of a tree and terminating at the vertex v are not pairwise equivalent.

6.1.20. (1) Prove that for each vertex of a loaded tree having a weight r, there exists an equivalent vertex having a rank not higher than $r - 1$.

(2) Can we replace $r - 1$ in the previous problem by $r - 2$ if $r \geqslant 2$?

Let $f_1(x_1, x_2, \ldots, x_n)$, $f_2(x_1, x_2, \ldots, x_n)$, ..., $f_m(x_1, x_2, \ldots, x_n)$ be a function of the type $\underbrace{E_k \times E_k \times \ldots \times E_k}_{n \text{ times}} \to E_l$, where k and l are not lower than 2. The operator $\varphi_{f_1, f_2, \ldots, f_m}$ in $\Phi_{k,l}^{n,m}$ is called an *operator generated by the functions* $f_1, f_2, \ldots, f_m$ if the following equalities hold for all $t \geqslant 1$:

$$y_i(t) = f_i(x_1(t), x_2(t), \ldots, x_n(t)),$$
$$i = 1, 2, \ldots, m.$$

6.1.21. Show that an operator in $\Phi_{k,l}^{n,1}$ is generated (by a function of the type $\underbrace{E_k \times E_k \times \ldots \times E_k}_{n \text{ times}} \to E_l$) if and only if its weight is equal to 1.

6.1.22. Find out if the operator $\varphi \in \Phi_2^{1,2}$ defined by the following relations is a generated operator:

$$\begin{cases} y_1(1) = 0, \\ y_1(t) = x(t-1) \oplus y_2(t-1), \quad t \geqslant 2, \\ y_2(t) = x(t). \end{cases}$$

6.1.23. A partial operator φ: $\{0, 1\}^\omega \to \{0, 1\}^\omega$ is defined only at the input words $\tilde{0}^\omega$ and $\tilde{1}^\omega$. Redefine it in such a way that the resulting operator has the minimum possible weight.

(1) $\varphi(\tilde{0}^\omega) = 0101\,[100]^\omega$, $\varphi(\tilde{1}^\omega) = 11\,[10]^\omega$;

(2) $\varphi(\widetilde{0^\omega}) = 01\,[011]^\omega$, $\varphi(\widetilde{1^\omega}) =$ $1010010001000010 0\ldots$
$\left(\text{i.e. } y(t) = 1 \text{ only for } t = \binom{i}{2},\quad i = 2, 3, \ldots\right)$

6.1.24. (1) Prove that if $|A| > 1$ and $|B| > 1$, the power of the set $\Phi_{A,B}$ is equal to the power $\mathfrak{c}$ (of the continuum).

(2) Find the power of a set $\Phi_{A,B}$ for the case when either $|A| = 1$ or $|B| = 1$.

6.1.25. The words $\widetilde{x_1^\omega}$ and $\widetilde{x_2^\omega}$ in A^ω are called s-equivalents (notation: $\widetilde{x_1^\omega} \underset{s}{\sim} \widetilde{x_2^\omega}$) if their prefixes of length s ($s \geqslant 1$) are equal. The relation $\underset{s}{\sim}$ is the equivalence relation and divides the set A^ω into classes $K_j(s)$ of s-equivalent words.

(1) Find the power of each class $K_j(s)$ (s is a fixed number).

(2) How many different classes $K_j(s)$ exist for a fixed s?

(3) How many different classes $K_{ji}(s + l)$ exist in each class $K_j(s)$ (here $l \geqslant 1$)?

(4) Prove that the mapping $\varphi: A^\omega \to B^\omega$ is a d-function if and only if for all $s \geqslant 1$ and for any words $\widetilde{x_1^\omega}$ and $\widetilde{x_2^\omega}$ belonging to the set A^ω, the relation $\widetilde{x_1^\omega} \underset{s}{\sim} \widetilde{x_2^\omega}$ leads to the relation $\varphi(\widetilde{x_1^\omega}) \underset{s}{\sim} \varphi(\widetilde{x_2^\omega})$.

6.1.26. (1) Show that the set $\hat{\Phi}_{A,B}$ is countable infinite for $|B| \geqslant 2$.

(2) Find the power of the set $\hat{\Phi}_{A,B}$ for $|B| = 1$.

6.1.27. Let $D_{A,B}$ be a loaded tree representing a b.d.-function $\varphi: A^\omega \to B^\omega$ of weight r. Let us change the output symbol of some arc in the j-th level of the tree $D_{A,B}$ (we assume that $|B| \geqslant 2$). This leads to a tree $D'_{A,B}$, representing a certain (new) b.d.-function φ' of weight r'.

(1) Show that $1 \leqslant r' \leqslant r + j$.

(2) Give an example of a b.d.-function φ at which the upper estimate $r' = r + j$ is attained.

6.2. Representation of Determinate Functions by Moore Diagrams, Canonical Equations, Tables and Schemes. Operations Involving Determinate Functions

Let $Q = \{Q_0, Q_1, \ldots, Q_{w-1}\}$ be a set of all states of the function φ in $\hat{\Phi}_{A,B}$. We associate a digraph Γ_φ with the function φ:

(1) the set of vertices of the digraph Γ_φ is the set $E_w = \{0, 1, \ldots, w-1\}$; the vertex j is assumed to correspond to the state Q_j;

(2) if $\varphi^{(i)}$ and $\varphi^{(j)}$ are residual operators of the function φ represented by the states Q_i and Q_j respectively ($\varphi^{(j)}$ is also a residual operator of $\varphi^{(i)}$ and corresponds to the prefix $\tilde{x}^1 = a$ and if $\varphi^{(i)}(a\tilde{x}^\omega) = b\varphi^{(j)}(\tilde{x}^\omega)$), then the digraph Γ_φ contains the arc (i, j) and is assigned the expression $a\ (b)$;

(3) the arc (i, j) exists in Γ_φ only when the conditions in (2) are satisfied.

The vertex in Γ_φ, corresponding to the initial state of the function φ, is often marked by an asterisk. Suppose that m arcs, which are assigned the expressions $a_1\ (b_1)$, $a_2\ (b_2), \ldots, a_m\ (b_m)$, pass from vertex i to vertex j (here, it is essential that $a_p \neq a_q$ for $p \neq q$, although some or all of the symbols b_s may be identical); we shall then join i with j through a single arc (i, j) which is assigned all the expressions $a_p\ (b_p)$, $p = \overline{1, m}$. The digraph Γ_φ is called the *Moore diagram of function* φ. Two functions can be associated with this diagram:

(a) F: $A \times Q \to B$ *is output function*,

(b) G: $A \times Q \to Q$ *is transition function*.

The functions F and G in Γ_φ are defined as follows: from the pair (a, j), we determine the vertex j and an arc emanating from j which is assigned the input symbol a (suppose that this arc has the form (j, r)); the value of the function F on the pair (a, j) is an output symbol which is assigned to the arc (j, r) and which appears after the symbol a in the parentheses. The value of the function G on the pair (a, j) coincides with r, i.e. it is equal to the "number" of the state which "is the end" of the arc (j, r).

The system of equations

$$\begin{cases} y(t) = F(x(t), \quad q(t-1)), \\ q(t) = G(x(t), \quad q(t-1)), \\ q(0) = q_0, \end{cases} \tag{1}$$

where $x(t) \in A$, $y(t) \in B$, $q(t) \in Q$ $(t = 1, 2, \ldots)$ and $q_0 \in Q$ is called the *canonical equations for the operator* φ *with the initial condition* q_0.

With the help of Moore's diagram and canonical equations, we can also specify determinate operators that are not boundedly determinate. The set of vertices of the digraph Γ_φ corresponding to such a d-operator coincides with the natural scale $N = \{0, 1, 2, \ldots\}$.

If φ is a boundedly determinate operator, the functions $F(x(t), q(t-1))$ and $G(x(t), q(t-1))$ (see Eqs. (1)) and the arguments on which these functions depend assume a finite number of values. Hence it is possible to compile a *canonical table* of the b.d.-operator φ.

Table 7

$x(t)$	$q(t-1)$	$y(t)$	$q(t)$
.	.	.	.
.	.	.	.
.	.	.	.
a	j	$F(a, j)$	$G(a, j)$
.	.	.	.
.	.	.	.
.	.	.	.

Instead of the canonical equations (1), it is convenient to consider canonical equations in which the input and transition functions are functions of the k-valued logic P_k $(k \geqslant 2)$. In order to obtain a corresponding representation of the operator φ the alphabets A, B and Q are coded by vectors whose coordinates belong to the set E_k $(k \geqslant 2)$. If φ is a b.d.-operator and $n =]\log_k |A|[$, $m =]\log_k |B|[$, $r =]\log_k |Q|[$, then it is sufficient to take vectors (whose coordinates belong to E_k) of length n, m, and r to code letters from the alphabets A, B and Q respectively.[3] In this case, the system (1) is transformed as follows:

$$\begin{cases} y_1(t) = F_1(x_1(t), \ldots, x_n(t), q_1(t-1), \ldots, q_r(t-1)), \\ \ldots\ldots\ldots\ldots\ldots\ldots\ldots\ldots\ldots\ldots \\ y_m(t) = F_m(x_1(t), \ldots, x_n(t), q_1(t-1), \ldots, q_r(t-1)), \\ q_1(t) = G_1(x_1(t), \ldots, x_n(t), q_1(t-1), \ldots, q_r(t-1)), \\ \ldots\ldots\ldots\ldots\ldots\ldots\ldots\ldots\ldots\ldots \\ q_r(t) = G_r(x_1(t), \ldots, x_n(t), q_1(t-1), \ldots, q_r(t-1)), \\ q_1(0) = q_{01}, \ldots, q_r(0) = q_{0r}. \end{cases} \tag{2}$$

Henceforth, we shall also use the "vector form" of systems similar to (2). In such a notation, the system (2) has the form

$$\begin{cases} \boldsymbol{y}^{(m)}(t) = \boldsymbol{F}^{(m)}(\boldsymbol{x}^{(n)}(t), \boldsymbol{q}^{(r)}(t-1)), \\ \boldsymbol{q}^{(r)}(t) = \boldsymbol{G}^{(r)}(\boldsymbol{x}^{(n)}(t), \boldsymbol{q}^{(r)}(t-1)), \\ \boldsymbol{q}^{(r)}(0) = \boldsymbol{q}_0^{(r)}. \end{cases} \tag{3}$$

The functions F_i and G_l in (2) are, in general, *partial* functions, i.e. they are not defined everywhere. Usually, they are redefined in such a way that the right-hand sides of Eqs. (2) acquire the simplest possible form.

The canonical table describing the b.d.-operator φ corresponding to (2) has $m + n + 2r$ columns and k^{n+r} rows.

The equations in system (2) are called *canonical equations in scalar form*, while the equations in (3) are *canonical equations in vector form*.

It can be assumed that system (2) defines an operator in the set $\Phi_k^{n,m}$.

Let a d-operator φ be defined by system (2), and let each of the functions F_i and G_l be a function of the k-valued logic P_k $(k \geqslant 2)$, which are defined everywhere. We shall consider the operator φ as an element of the set $\Phi_k^{n,m}$. The scheme Σ_φ of the operator φ is defined as follows: Σ_φ is a network whose poles are assigned the symbols of input and output variables, while some of the vertices other than the poles are assigned symbols of some

[3] If the d-operator φ is not a b.d.-operator, $r = \infty$.

d-operators $\left(\text{in the set } \bigcup\limits_{s=1}^{\infty} \bigcup\limits_{s'=1}^{\infty} \Phi_k^{s,s'}\right)$. The scheme of φ in $\Phi_k^{n,m}$ will be presented in the form of a rectangle (Fig. 26) with *n input and m output channels.* The input channels are represented by arrows emanating from the input poles, while the output channels are represented by arrows terminating at the output poles. The poles are

Fig. 26

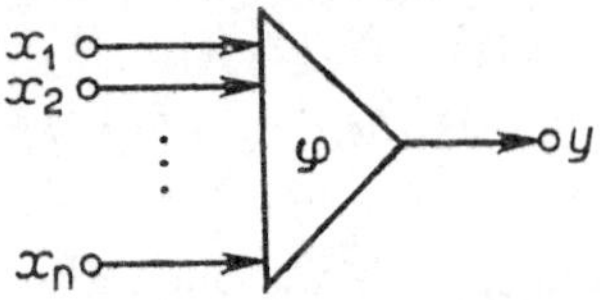

Fig. 27

represented in the form of circles. If $m = 1$, the scheme Σ_φ of the operator φ can sometimes be represented in the form of a triangle (Fig. 27) with n input poles and one output pole.

We assume that an input symbol $x_i(t) \in E_k$ is "*supplied*" to the i-th input x_i at each instant of time $t = 1, 2, \ldots$, while the value $y_i(t) = F_j(x_1(t), \ldots, x_n(t), q_1(t-1), \ldots, q_r(t-1))$ is *delivered* (*represented*) at the j-th output y_j. The *output* y_j is said to *have a delayed dependence on the input* x_i if the function $F_j(\boldsymbol{x}^{(n)}(t), \boldsymbol{q}^{(r)}(t-1))$ does not depend essentially on the variable $x_i(t)$.

The concept of delayed dependence can be introduced in a different way. Let us consider, for example, a d-function of the type $\varphi(X_1, X_2, \ldots, X_n)$: $A_1 \times A_2 \times \ldots \times A_n \to B$ and define its delayed dependence on the variable X_1. The function φ *has a delayed dependence on* X_1 if for any input words $\tilde{x}_1^{\omega}, \tilde{x}_2^{\omega}, \ldots, \tilde{x}_n^{\omega}$ $(\tilde{x}_j^{\omega} \in A_j^{\omega}, j = \overline{1, n})$ the s-th letter of the $\tilde{y}^{\omega} = \varphi(\tilde{x}_1^{\omega}, \tilde{x}_2^{\omega}, \ldots, \tilde{x}_k^{\omega})$ is defined by the first s symbols of the words $\tilde{x}_2^{\omega}, \ldots, \tilde{x}_n^{\omega}$ and $s - 1$ first symbols of the word $\tilde{x}_1^{\omega}$.

Suppose that a d-function φ is defined by Eqs. (2) and that Σ_φ is the scheme of this function. We define three operations involving φ and Σ_φ.

(1) *Operation* O_1 involves the *identification of two* or *more input variables* in the function φ and the *identification of the input poles* corresponding to these variables *in the scheme* Σ_φ. The identified poles are considered as a single

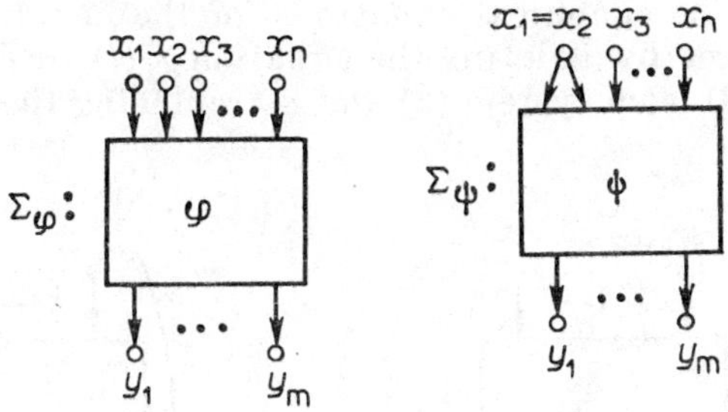

Fig. 28

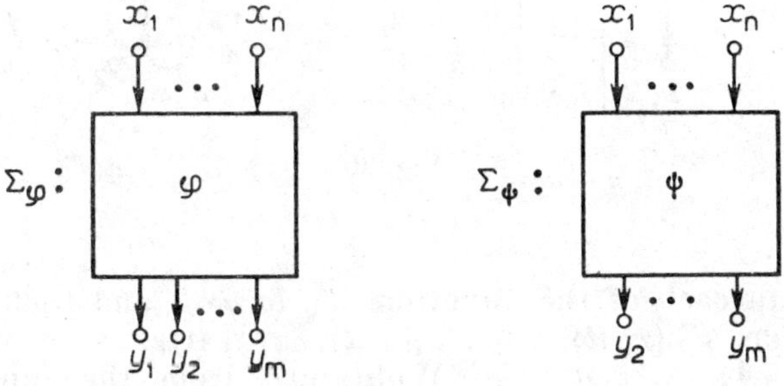

Fig. 29

pole in the new scheme. Figure 28 shows a scheme Σ_ψ obtained from Σ_φ by identifying poles x_1 and x_2.

(2) *Operation* O_2 involves the *deletion of an output variable* y_j in the function φ (which is equivalent to the deletion of the equation $y_j(t) = F_j(\boldsymbol{x}^{(n)}(t), \boldsymbol{q}^{(r)}(t-1))$ from (2)) and the *deletion of the output channel and the pole* corresponding to the output variable y_j *from the scheme* Σ_φ (see Fig. 29, which shows the scheme Σ_ψ obtained from Σ_φ after the deletion of the output channel and the pole y_1).

Remark. If $m = 1$, the deletion of the variable y_1 (the only output variable) leads to an *automaton without an output*.

(3) *Operation* O_3 involves the *introduction of a feedback* of one input and one output variables. Let us take the variables x_i and y_j as the input and output variables. The operation O_3 can be applied (to the function φ and scheme Σ_φ) only if the output y_j has delayed dependence on the input x_i. The canonical equations for the new function ψ are obtained by deleting the equation $y_j(t) = F_j(\boldsymbol{x}^{(n)}(t), \boldsymbol{q}^{(r)}(t-1))$ from system (2) and by replacing the variable

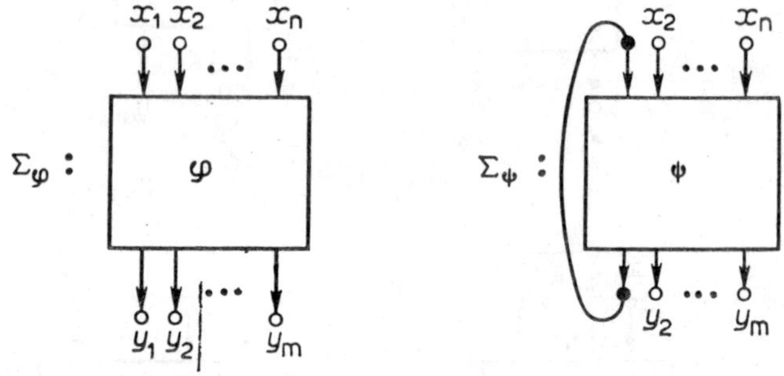

Fig. 30

$x_i(t)$ in each of the functions F_q $(q \neq j)$ and G_l by the function $F'_j(x_1(t), \ldots, x_{i-1}(t), x_{i+1}(t), \ldots, x_n(t), q_1(t-1), \ldots, q_r(t-1))$ obtained from the function $F_j(\boldsymbol{x}^{(n)}(t), \boldsymbol{q}^{(r)}(t-1))$ by omitting the fictitious variable $x_i(t)$. The initial conditions remain the same as before. The scheme Σ_ψ is obtained from the scheme Σ_φ by identifying the output y_j with the input x_i. In this case, the identified poles are referred to as the internal vertex of the scheme Σ_ψ. Figure 30 shows the scheme Σ_ψ obtained from Σ_φ by introducing the feedback in variables x_1 and y_1.

Remark 1. If $n = 1$, the introduction of a feedback in variable x_1 (and any output variable) leads to an *automaton without input*.

Remark 2. While applying the operations described above, it is convenient to indicate in parentheses (following the notation of these operations) the channels (poles and variables) to which these operations are applied.

For example, we can use the notation $O_1(x_1, x_3)$, $O_2(y_5)$ and $O_3(x_{10}, y_2)$.

Let us define two more operations involving d-functions.

(4) *Operation* O_4 is the *union of two (or more) functions.* Let $\varphi_1 \in \Phi_k^{n_1, m_1}$ and $\varphi_2 \in \Phi_k^{n_2, m_2}$. We shall assume that these functions have input variables $x'_1, x'_2, \ldots, x'_{n_1}$ and $x''_1, x''_2, \ldots, x''_{n_2}$ respectively. For the output variables, we take $y'_1, y'_2, \ldots, y'_{m_1}$ and $y''_1, y''_2, \ldots, y''_{m_2}$ respectively. All these variables are assumed to be pairwise different. Let Σ_{φ_1} and Σ_{φ_2} be the schemes of the functions φ_1 and φ_2 respectively. In this case, the scheme Σ_ψ of the function ψ, which is equal to the union of the functions φ_1 and φ_2, will appear as shown in Fig. 31.

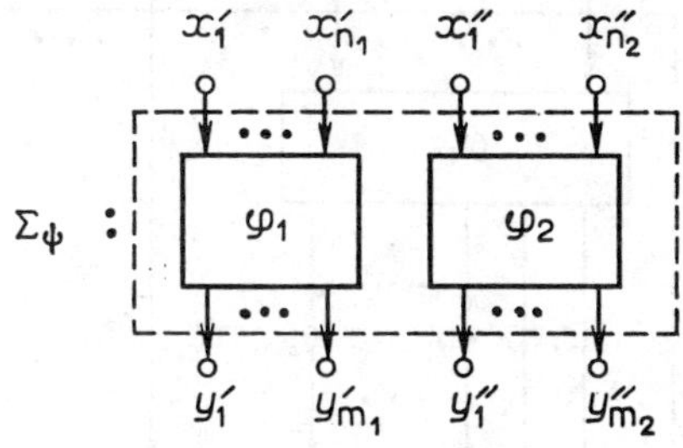

Fig. 31

The output (input) poles of the scheme Σ_ψ are all output (input) poles of the initial schemes Σ_{φ_1} and Σ_{φ_2}. The system of canonical equations (and initial conditions) of the function ψ is obtained simply by a union of the corresponding systems of functions φ_1 and φ_2 (naturally, it is assumed that the sets $Q^{(1)}$ and $Q^{(2)}$ of all the states of the functions φ_1 and φ_2 are disjoint).

(5) *Operation* S is the *superposition operation.* Let $\varphi_1 \in \Phi_{A,B}$ and $\varphi_2 \in \Phi_{B,C}$. The *superposition* $\varphi_2(\varphi_1)$ *of operators* φ_1 and φ_2 is an operator $\psi \in \Phi_{A,C}$, such that $\psi(\tilde{x}^\omega) = \varphi_2(\varphi_1(\tilde{x}^\omega))$ for any input word $\tilde{x}^\omega$ in A^ω. Let the operators $\varphi_i \in \Phi_k^{n_i, m_i}$ $(i = 1, 2)$, and let the schemes Σ_{φ_1} and Σ_{φ_2} correspond to them (we assume that the input and output variables and the states of the functions φ_1

and φ_2 are the same as in the preceding paragraph). In this case, we can consider a "different" superposition for these operators: for example, we identify the input pole x''_1 of the scheme Σ_{φ_2} with the output pole y'_{l+1} of the scheme Σ_{φ_1}, the pole x''_2 with the pole y'_{l+2}, and so on. Finally, we identify the pole x''_{m_1-l} with the pole y'_{m_1}. As a result, we get a scheme Σ_ψ (Fig. 32) for which: (a) all

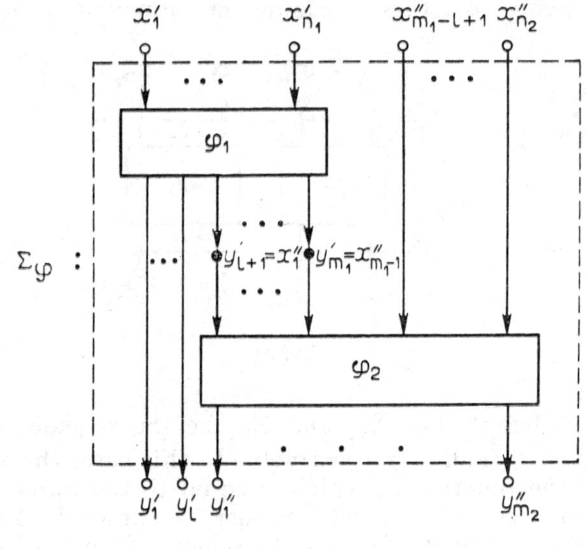

Fig. 32

the input poles will be the input poles of the scheme Σ_{φ_1} and the input poles of the scheme Σ_{φ_2} not participating in the above identification procedure; (b) the output poles are all output poles of the scheme Σ_{φ_2} and those output poles of the scheme Σ_{φ_1} which were not identified with any of the input poles of the scheme Σ_{φ_2}. The identified poles are called the internal vertices of the scheme Σ_ψ. The scheme Σ_ψ is called the *superposition of schemes* Σ_{φ_2} and Σ_{φ_1} (in variables $x''_1 - y'_{l+1}$, $x''_2 - y'_{l+2}$, . . ., x''_{m_1-l} —

y'_{m_1}). If the operator φ_1 is defined by the system of equations

$$
\begin{cases}
y'_1(t) = F'_1(x'_1(t), \ldots, x'_{n_1}(t), q'_1(t-1), \\
\qquad\qquad \ldots, q'_{r_1}(t-1)), \\
\ldots\ldots\ldots\ldots\ldots\ldots\ldots\ldots \\
y'_{m_1}(t) = F'_{m_1}(x'_1(t), \ldots, x'_{n_1}(t), q'_1(t-1), \\
\qquad\qquad \ldots, q'_{r_1}(t-1)), \\
q'_1(t) = G'_1(x'_1(t), \ldots, x'_{n_1}(t), q'_1(t-1), \\
\qquad\qquad \ldots, q'_{r_1}(t-1)), \\
\ldots\ldots\ldots\ldots\ldots\ldots\ldots\ldots \\
q'_{r_1}(t) = G'_{r_1}(x'_1(t), \ldots, x'_{n_1}(t), q'_1(t-1), \\
\qquad\qquad \ldots, q'_{r_1}(t-1)), \\
q'_1(0) = q'_{01}, \ldots, q'_{r_1}(0) = q'_{0r_1},
\end{cases}
\tag{4'}
$$

and the operator φ_2 is defined by the system of equations

$$
\begin{cases}
y''_1(t) = F''_1(x''_1(t), \ldots, x''_{n_2}(t), q''_1(t-1), \\
\qquad\qquad \ldots, q''_{r_2}(t-1)), \\
\ldots\ldots\ldots\ldots\ldots\ldots\ldots\ldots \\
y''_{m_2}(t) = F''_{m_2}(x''_1(t), \ldots, x''_{n_2}(t), q''_1(t-1), \\
\qquad\qquad \ldots, q''_{r_2}(t-1)), \\
q''_1(t) = G''_1(x''_1(t), \ldots, x''_{n_2}(t), q''_1(t-1), \\
\qquad\qquad \ldots, q''_{r_2}(t-1)), \\
\ldots\ldots\ldots\ldots\ldots\ldots\ldots\ldots \\
q''_{r_2}(t) = G''_{r_2}(x''_1(t), \ldots, x''_{n_2}(t), q''_1(t-1), \\
\qquad\qquad \ldots, q''_{r_2}(t-1)), \\
q''_1(0) = q''_{01}, \ldots, q''_{r_2}(0) = q''_{0r_2},
\end{cases}
\tag{4''}
$$

then the operator ψ represented by the scheme Σ_ψ has the following system corresponding to it:

$$\left\{\begin{array}{l}
y_1'(t) = F_1'(x_1'(t), \ldots, x_{n_1}'(t), q_1'(t-1) \\
\qquad\qquad \ldots, q_{r_1}'(t-1)), \\
\ldots\ldots\ldots\ldots\ldots\ldots\ldots\ldots \\
y_l'(t) = F_l'(x_1'(t), \ldots, x_{n_1}'(t), \quad q_1'(t-1), \\
\qquad\qquad \ldots, q_{r_1}'(t-1)), \\
y_1''(t) = F_1''(F_{l+1}', \ldots, F_{m_1}', x_{m_1-l+1}''(t), \\
\quad \ldots, x_{n_2}''(t), \quad q_1(t-1), \ldots, q_{r_2}(t-1)), \\
\ldots\ldots\ldots\ldots\ldots\ldots\ldots\ldots \\
y_{m_2}''(t) = F_{m_2}''(F_{l+1}', \ldots, F_{m_1}', x_{m_1-l+1}''(t), \\
\quad \ldots, x_{n_2}''(t), q_1(t-1), \ldots, q_{r_2}(t-1)), \\
q_1'(t) = G_1'(x_1'(t), \ldots, x_{n_1}'(t), q_1'(t-1), \\
\qquad\qquad \ldots, q_{r_1}'(t-1)), \\
\ldots\ldots\ldots\ldots\ldots\ldots\ldots\ldots \\
q_{r_1}'(t) = G_{r_1}'(x_1'(t), \ldots, x_{n_1}'(t), q_1'(t-1), \\
\qquad\qquad \ldots, q_{r_1}'(t-1)), \\
q_1(t) = G_1'(F_{l+1}', \ldots, F_{m_1}', x_{m_1-l+1}''(t), \\
\quad \ldots, x_{n_2}''(t), q_1(t-1), \ldots, q_{r_2}(t-1)), \\
\ldots\ldots\ldots\ldots\ldots\ldots\ldots\ldots \\
q_{r_2}(t) = G_{r_2}''(F_{l+1}', \ldots, F_{m_1}', x_{m_1-l+1}''(t), \\
\quad \ldots, x_{n_2}''(t), q_1(t-1), \ldots, q_{r_2}(t-1)), \\
q_1'(0) = q_{01}', \ldots, q_{r_1}'(0) = q_{0r_1}', q_1(0) = q_{01}'', \\
\qquad\qquad \ldots, q_{r_2}(0) = q_{0r_2}'',
\end{array}\right. \tag{5}$$

where

$$F_j' = F_j'(x_1'(t), \ldots, x_{n_1}'(t), q_1'(t-1), \ldots, q_{r_1}'(t-1)),$$

$$j = l+1, \ldots, m_1.$$

The *element of unit delay* (in the set $\hat{\Phi}_k$) is a b.d.-operator φ_d defined by the system of equations

$$\begin{cases} y(t) = q(t-1), \\ q(t) = x(t), \\ q(0) = 0. \end{cases} \tag{6}$$

Example. The operator φ in $\hat{\Phi}_2$ is defined with the help of Moore's diagram presented in Fig. 33. The canonical

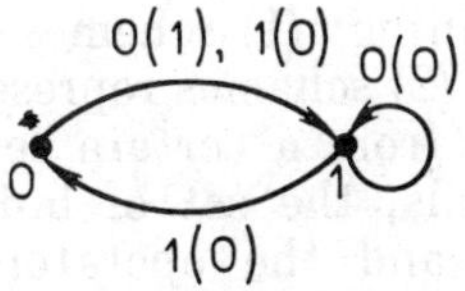

Fig. 33

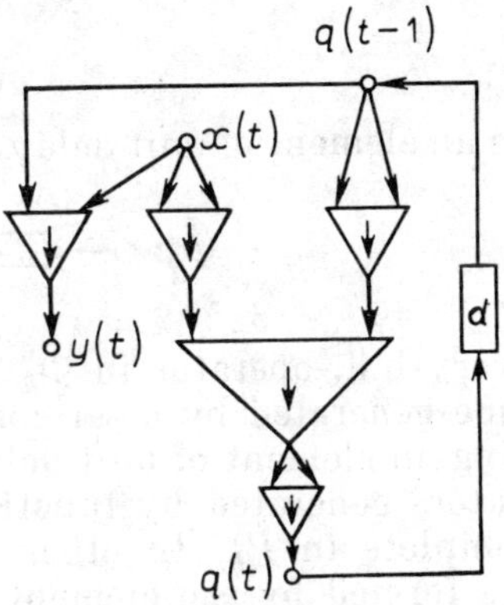

Fig. 34

table (see Table 8), canonical equations, and the initial condition for this operator have the following form:

Table 8

$x(t)$	$q(t-1)$	$y(t)$	$q(t)$
0	0	1	1
0	1	0	1
1	0	0	1
1	1	0	0

$$\begin{cases} y(t) = \overline{x}(t) \cdot \overline{q}(t-1), \\ q(t) = \overline{x}(t) \vee \overline{q}(t-1), \\ q(0) = 0, \end{cases} \tag{7}$$

Figure 34 shows a scheme representing this operator. This diagram has been constructed by using an element of unit delay and an operator in $\hat{\Phi}_2^{2,1}$, generated by a symbol of joint negation $x \downarrow y$. Here, the operator generated by the joint negation symbol is represented through the scheme

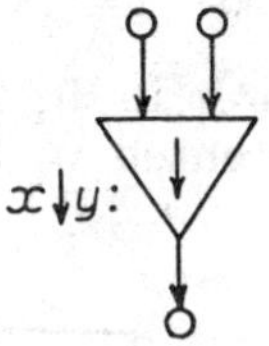

while an element of unit delay is represented by the scheme

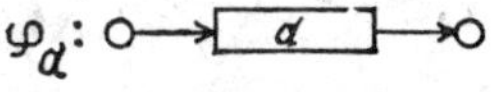

Every b.d.-operator in $\hat{\Phi}_k^{n,m}$ may be represented by a scheme generated by a set containing: (1) schemes representing an element of unit delay; (2) schemes representing operators generated by functions from a certain set that is complete in P_k. In other words, the set of b.d.-operators formed by the element φ_d and the operators generated by functions from a certain set complete in P_k form a set that is complete in $\bigcup\limits_{n,m} \hat{\Phi}_k^{m,n}$ with respect to the operations O_1, O_2, O_3, O_4, and S.

Remark. Henceforth, while constructing a scheme of any b.d.-operator φ generated by a certain given set M of b.d.-operators, we shall adopt the following procedure: the scheme Σ_φ of the operator φ is constructed by using only single-output schemes representing operators in the set M (in this case, unless stipulated otherwise, we can only carry out the operations O_1, O_2, O_3, O_4, and S).

6.2.1. Construct the Moore diagram, the canonical table, and the canonical equations for the function $\varphi \in \hat{\Phi}_2$.

(1) $\varphi(\tilde{x}^\omega)$: $\begin{cases} y(2t-1) = x(2t-1), & t \geqslant 1, \\ y(2t) = x(2t) \oplus y(2t-1), & t \geqslant 1; \end{cases}$

(2) $\varphi(\tilde{x}^\omega)$: $\begin{cases} y(1) = 1, \\ y(t) = x(t-1) \oplus x(t), & t \geqslant 2; \end{cases}$

(3) $\varphi(\widetilde{x^\omega})$: $\begin{cases} y(3t-2)=\bar{x}(3t-2), & t\geqslant 1, \\ y(3t-1)=x(3t-2), & t\geqslant 1, \\ y(3t)=x(3t)\,y(3t-1), & t\geqslant 1; \end{cases}$

(4) $\varphi(\widetilde{x^\omega})=\langle 2/3\rangle$;

(5) $\varphi(\widetilde{x^\omega})=\langle 5/8\rangle$.

6.2.2. Redefine the partial operator φ: $\{0, 1\}^\omega \to \{0, 1\}^\omega$ in such a way as to obtain a b.d.-operator. Construct the Moore diagram, the canonical table and the canonical equations for the new operator if:

(1) $\varphi([011010]^\omega) = [0111]^\omega$, $\varphi(0\,[1]^\omega) = 0\,[1]^\omega$;

(2) $\varphi(\widetilde{0^\omega}) = \widetilde{1}^\omega$, $\varphi(1\,[0]^\omega) = [10]^\omega$;

(3) $\varphi(1\,[10]^\omega) = [01]^\omega$, $\varphi([001]^\omega) = 1\,[10]^\omega$;

(4) φ: $\begin{cases} y(3t-1) = x(3t-1), & t\geqslant 1, \\ y(3t) = \bar{x}(3t-1), & t\geqslant 1. \end{cases}$

6.2.3. Find the weight of the b.d.-operator φ in $\hat{\Phi}_2$, defined by the canonical equations:

(1) φ: $\begin{cases} y(t)=\bar{x}(t)\,\bar{q}_1(t-1)\vee x(t)\,q_2(t-1), \\ q_1(t)=x(t)\,q_1(t-1)\vee\bar{x}(t)\,q_2(t-1), \\ q_2(t)=\bar{x}(t)\,q_2(t-1)\vee\bar{x}(t)\,\bar{q}_1(t-1)\,\bar{q}_2(t-1), \\ q_1(0)=q_2(0)=0; \end{cases}$

(2) φ: $\begin{cases} y(t)=x(t)\oplus q_1(t-1)\oplus q_2(t-1), \\ q_1(t)=x(t)\sim q_1(t-1), \\ q_2(t)=(x(t)\to\bar{q}_1(t-1))\to\bar{x}(t)\,\bar{q}_1(t-1), \\ q_1(0)=q_2(0)=1; \end{cases}$

(3) φ is defined by the cannonical equations of the preceding problem, but with different initial conditions: $q_1(0) = 0$, $q_2(0) = 1$;

(4) φ: $\begin{cases} y(t)=(x(t)\to q_2(t-1))\to q_1(t-1), \\ q_1(t)=q_1(t-1)\to(x(t)\to\bar{q}_1(t-1)), \\ q_2(t)=q_2(t-1)\to x(t), \\ q_1(0)=q_2(0)=0. \end{cases}$

6.2.4. Let $(\hat{\Phi}_k^{n,m};\ X_1, X_2, \ldots, X_n; Y_1, Y_2, \ldots, Y_m)$ denote the set of all functions in $\hat{\Phi}_k^{n,m}$ with input variables $X_1, X_2, \ldots, X_n$ and output variables $Y_1, Y_2, \ldots, Y_m$. Show that the number of functions with a weight w

in the set $(\hat{\Phi}_k^{n,m}; X_1, X_2, \ldots, X_n, Y_1, Y_2, \ldots, Y_m)$ does not exceed $(wk^m)^{wk^n}$.

6.2.5. Let the operator φ in $\hat{\Phi}_2^{n,m}$ be defined by (2) and let the functions $G_1, G_2, \ldots, G_r$ be connected through

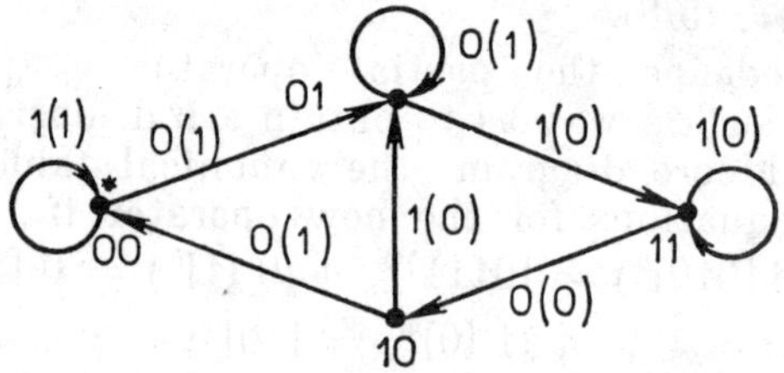

Fig. 35

the relation $G_1 \oplus G_2 \oplus \ldots \oplus G_r \equiv 0$. Show that the weight of the operator φ does not exceed 2^{r-1}.

6.2.6. Represent the operator $\varphi \in \hat{\Phi}_2^{n,m}$ through a scheme generated by a set formed by unit delay elements and the operator generated by Sheffer's stroke:

$$(1)\quad \varphi: \begin{cases} y(t) = q(t-1), \\ q(t) = x(t), \\ q(0) = 1; \end{cases}$$

$$(2)\quad \varphi: \begin{cases} y(t) = (x(t) \vee q_1(t-1)) \to q_2(t-1), \\ q_1(t) = x(t) \to q_2(t-1), \\ q_2(t) = \bar{q}_1(t-1) \vee \bar{q}_2(t-1), \\ q_1(0) = 0, \quad q_2(0) = 1; \end{cases}$$

$$(3)\quad \tilde{\varphi}(\tilde{x}^\omega) = \begin{cases} 10\tilde{1}^\omega & \text{if } \tilde{x}^\omega = 00x(3)\,x(4)\,\ldots, \\ \tilde{1}^\omega & \text{otherwise}; \end{cases}$$

(4) the operator φ is defined by Moore's diagram shown in Fig. 35;

$$(5)\quad \varphi: \begin{cases} y_1(t) = x(t) \to \bar{q}_1(t-1), \\ y_2(t) = q_1(t-1)\,q_2(t-1), \\ q_1(t) = \bar{x}(t), \\ q_2(t) = (q_1(t-1) \vee q_2(t-1)) \cdot \bar{x}(t), \\ q_1(0) = q_2(0) = 0. \end{cases}$$

6.2.7. Using the scheme of the operator $\varphi \in \hat{\Phi}_2^{n,m}$, construct the canonical equations, the canonical table and Moore's diagram.

(1) see Fig. 36*a*; (2) see Fig. 36*b*; (3) see Fig. 36*c*.

6.2.8. For the superposition $\psi = \varphi_1(\varphi_2)$ of operators φ_2 and φ_1 in $\hat{\Phi}_2$, construct the canonical equations,

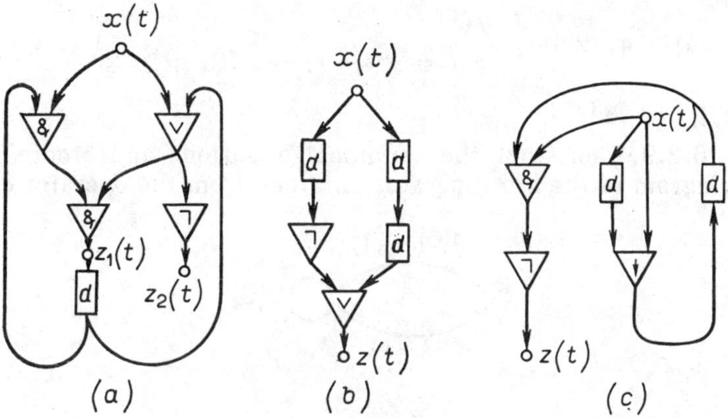

Fig. 36

Moore's diagram, and the scheme Σ_ψ. The scheme Σ_ψ must be generated by the set consisting of a unit delay element and operators generated by the implication $x \to y$ and the negation $\bar{x}$. Consider

(1) $$\varphi_1: \begin{cases} y_1(t) = x_1(t)\, q_1(t-1), \\ q_1(t) = q_1(t-1) \to x_1(t), \\ q_1(0) = 0, \end{cases}$$

$$\varphi_2: \begin{cases} y_2(t) = x_2(t) \oplus q_2(t-1), \\ q_2(t) = x_2(t) \vee q_2(t-1), \\ q_2(0) = 1; \end{cases}$$

(2) $$\varphi_1(\widetilde{x^\omega}) = \langle 1/3 \rangle,$$

the operator φ_2 is defined by Moore's diagram shown in Fig. 37;

(3) $\varphi_1(\tilde{x}^\omega)$: $\begin{cases} y(1)=0, \\ y(t)=y(1-1)\oplus x(t-1), \quad t\geqslant 2, \end{cases}$

$\varphi_2(\tilde{x}^\omega)$: $\begin{cases} y(1)=1, \\ y(t)=x(t)\to x(t-1), \quad t\geqslant 2; \end{cases}$

(4) $\varphi_1(\tilde{x}^\omega)$: $\begin{cases} y(1)=1, \\ y(t)=y(t-1)\to x(t), \quad t\geqslant 2, \end{cases}$

$\varphi_2(\tilde{x}^\omega)=\langle 1/7\rangle$.

6.2.9. Construct the canonical equations and Moore's diagram of the b.d.-operator obtained from the operator φ

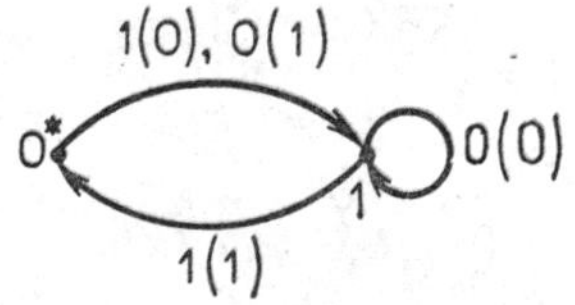

Fig. 37

by introducing the feedback in variables x_2 and y_1 if

(1) φ: $\begin{cases} y_1(t)=q(t-1)\to x_1(t)\,\overline{x}_3(t), \\ y_2(t)=x_2(t)\vee(x_1(t)\to q(t-1)), \\ q(t)=\overline{x}_2(t)\oplus q(t-1), \\ q(0)=0; \end{cases}$

(2) φ: $\begin{cases} y_1(t)=\overline{q}(t-1)\to x_1(t)\,x_3(t), \\ y_2(t)=(x_1(t)\downarrow q(t-1))\vee x_2(t), \\ q(t)=x_2(t)\oplus q(t-1), \\ q(0)=0. \end{cases}$

6.2.10. Find the weight of the b.d.-operator obtained from the b.d.-operator φ by introducing the feedback in variables x_3 and y_2 if:

(1) φ: $\begin{cases} y_1(t)=x_1(t)\to(x_3(t)\to q(t-1)), \\ y_2(t)=x_2(t)\to x_1(t), \\ q(t)=x_2(t)\to(x_1(t)\,\overline{x}_3(t)\to(x_1(t)\to q(t-1))), \\ q(0)=0; \end{cases}$

$$(2)\ \varphi: \begin{cases} y_1(t) = x_1(t)\, x_3(t) \to q(t-1), \\ y_2(t) = x_2(t)\, q(t-1), \\ q(t) = \bar{x}_2(t) \vee x_3(t) \vee q(t-1), \\ q(0) = 0. \end{cases}$$

6.2.11. Suppose that the operator ψ is obtained from the operator φ through the operation O_1 (identification of input variables).

(1) Give an example of a pair of operators φ and ψ for which the following inequality holds: the weight of the operator φ is more than the weight of the operator ψ.

(2) Can the weight of the operator φ be less than the weight of the operator ψ?

6.2.12. Let φ be a b.d.-operator of weight r and let ψ be an operator of weight r', obtained from φ by introducing the feedback in a certain pair of variables. Is it always true that

(1) $r \geqslant r'$; (2) $r = r'$; (3) $r \leqslant r'$?

6.2.13. Suppose that the b.d.-operators φ_1 and φ_2 have weights r_1 and r_2 respectively. What will be the

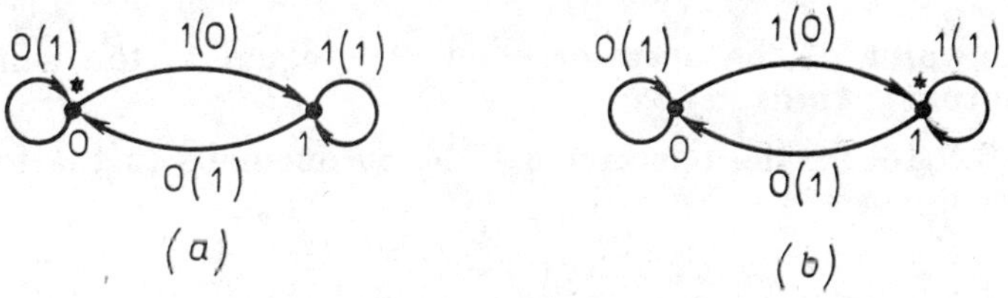

Fig. 38

weight of the operator ψ obtained from φ_1 and φ_2 with the help of the operator O_4 (union)?

6.2.14. The b.d.-operators φ_1 and φ_2 have weights equal to r_1 and r_2 respectively. Will the weight of the superposition $\varphi_1(\varphi_2)$ be

(a) larger than

(1) r_1, (2) r_2, (3) $r_1 + r_2$, (4) $r_1 \cdot r_2$?

(b) smaller than

(5) r_1, (6) r_2?

6.2.15. Find the weight of the superposition $\varphi_1(\varphi_2)$ if

(1) the operators φ_1 and φ_2 are defined by Moore's diagrams shown in Fig. 38*a* and *b*;

(2) the operators φ_1 and φ_2 are defined by Moore's diagrams shown in Fig. 39*a* and *b*;

$$(3)\quad \varphi_i:\begin{cases} y_i(t)=x_i(t)\rightarrow q_i(t-1),\\ q_i(t)=q_i(t-1)\rightarrow x_i(t), \quad i=1,\ 2,\\ q_i(0)=0.\end{cases}$$

The operator φ in $\Phi_{A,B}$ is called *autonomous* (*constant*, or *without input operator*) if for any input word $\widetilde{x^\omega}\in A^\omega$

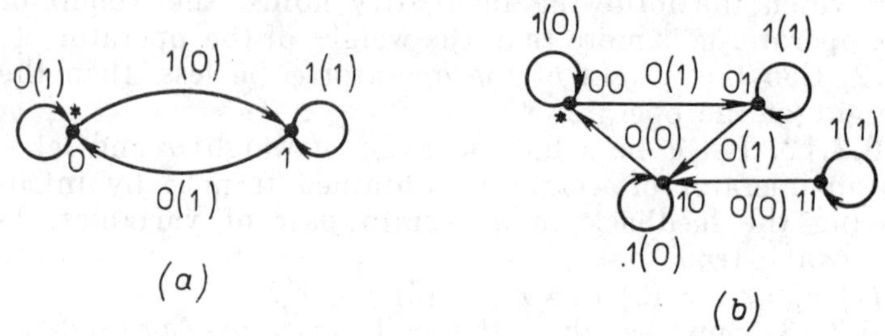

Fig. 39

the value of the operator φ on $\widetilde{x^\omega}$ is equal to the same (output) word in B^ω.

6.2.16. Is the operator $\varphi\in\hat{\Phi}_2$ autonomous in the following cases?

$$(1)\quad \varphi(\widetilde{x^\omega}):\begin{cases} y(1)=y(2)=1,\\ y(t)=y(t-1)\oplus y(t-2), \quad t\geqslant 3;\end{cases}$$

$$(2)\quad \varphi:\begin{cases} y(t)=x(t)\rightarrow q(t-1),\\ q(t)=x(t)\vee q(t-1),\\ q(0)=1;\end{cases}$$

$$(3)\quad \varphi:\begin{cases} y(t)=x(t)\rightarrow q(t-1),\\ q(t)=x(t)\vee q(t-1),\\ q(0)=0;\end{cases}$$

$$(4)\quad \varphi(\widetilde{x^\omega}):\begin{cases} y(1)=0,\\ y(2t-1)=\overline{y}(2t-2), \quad t\geqslant 2,\\ y(2t)=\left|\cos\frac{\pi}{2}t\right|, \quad t\geqslant 1.\end{cases}$$

6.2.17. Let φ be an autonomous operator of weight r $(r < \infty)$.

(1) Show that the output word of the operator φ is quasiperiodic.

(2) Prove that the sum of lengths of the period and pre-period of the output word of the operator φ does not exceed r.

6.2.18. (1) Construct a scheme Σ_φ using the set

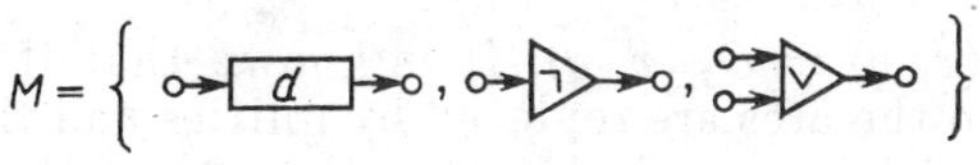

such that it contains exactly four elements in M and represents a certain autonomous operator φ in $\hat{\Phi}_2$.

(2) Is it possible to construct a scheme Σ_φ with one input and one output (using the set M) representing a certain autonomous operator φ in $\hat{\Phi}_2$ and containing less than four elements?

Operation O_5, called the *branching operation* (for a certain output of a b.d.-function) can be defined as follows. Let φ be an operator in $\hat{\Phi}_k^{n,m}$ and let Σ_φ be the scheme representing it. Applying the operation O_5 to the output y_j of the function φ, we obtain an operator ψ, represented by a scheme Σ_ψ which acquires, instead of one channel (and pole) y_j, several "equally operative" channels $y_j^{(1)}, \ldots, y_j^{(s)}$ $(s \geqslant 2)$, each of which represents the same function as y_j does in the scheme Σ_φ.

6.2.19. (1) What changes must be made in the system of canonical equations and initial conditions describing a function φ in order to obtain a corresponding system for the function ψ resulting from the application of the operation O_5 to the output y_j of the function φ?

(2) Is it possible to construct by using operations O_1-O_5, S and the set M (see Problem 6.2.18) a scheme Σ_φ which contains not more than three elements and which represents the autonomous operator φ in $\hat{\Phi}_2$?

6.2.20. Suppose that the operator ψ is obtained from $\varphi \in \hat{\Phi}_k^{n,m}$ by applying the operation O_5 to a certain output of the operator φ. Show that the operator ψ can be constructed with the help of the set $\{\varphi\}$ (i.e. by using the operator φ a few times) through the operations O_1, O_2, and O_4.

6.2.21. The operators φ_1 and φ_2 in $\hat{\Phi}_2$ have a weight 2 each, and are defined by the following canonical equations:

$$\varphi_i: \begin{cases} y_i(t) = F_i(x_i(t),\ q_i(t-1)), \\ q_i(t) = G_i(x_i(t),\ q_i(t-1)), \quad i = 1,\ 2, \\ q_i(0) = 0, \end{cases}$$

where $G_1(x, q) \equiv G_2(x, q)$. It is known that if the zeros assigned to the arcs are replaced by unities and the unities are replaced by zeros, in Moore's diagram of the operator φ_1, we obtain Moore's diagram of the operator φ_2. Prove that if the output values of the operators φ_1 and φ_2 coincide on a certain prefix of length 2 (i.e., if $\varphi_1(\sigma_1\sigma_2) = \varphi_2(\sigma_1\sigma_2)$ for some values of σ_1 and σ_2), these operators are identically equal.

6.2.22. Count the number of b.d.-functions in $\hat{\Phi}_2$ which satisfy the following conditions: (a) a function depends on its input variable X; (b) the weight of a function is equal to 3; (c) in the Moore diagram describing a function, the in-degree of each vertex is the same and equal to the out-degree.

6.2.23. For each $r \geqslant 2$, give an example of an operator φ in $\hat{\Phi}_2$ such that the weight of the superposition $\varphi(\varphi)$ is equal to r. Is it possible to do the same for non-autonomous operators?

6.2.24. The weight of a function $\varphi \in \hat{\Phi}_2$ is equal to r and the weight of the superposition $\varphi(\varphi)$ is equal to $2r$. Will the superposition $\varphi(\varphi(\varphi))$ have a weight equal to $3r$?

6.3. Closed Classes and Completeness in the Sets of Determinate and Boundedly Determinate Functions

Let M be a set of d-functions (or b.d.-functions), and let $\mathfrak{O}$ be a *set of operations* the results of whose application also remain in the set of all d-functions (or b.d.-functions). In other words, if $\sigma \in \mathfrak{O}$ the application of σ to arbitrary (or admissible) d- (or b.d.) functions will again result in d- (or b.d.) functions. The *closure* $[M]_{\mathcal{O}}$ *of the set* M *relative to the set of operations* $\mathcal{O}$ is the set of all d- (or b.d.) functions which may be obtained from the functions of set M with the help of operations in $\mathcal{O}$, and these operations can be applied any finite number of times. Usually, it is assumed that $[M]_{\mathcal{O}} \supseteq M$. Henceforth, we shall always assume that this inclusion is satisfied (if this is not the case, a special mention will be made). The operation of obtaining the set $[M]_{\mathcal{O}}$ from M is called the *closure operation*. The set M is called *(functionally) closed class relative to the set of operations* $\mathcal{O}$ if $[M]_{\mathcal{O}} = M$. Let M be a class of d- (or b.d.-) functions closed relative to the set of operations $\mathcal{O}$. The subset $\mathcal{P}$ in M is called a *(functionally) complete set in* M *relative to the set of operations* $\mathcal{O}$ if $[\mathcal{P}]_{\mathcal{O}} = M$. The set $\mathcal{P}$ of d- (or b.d.-) functions is called an *irreducible set relative to the operations* $\mathcal{O}$ if for every proper subset $\mathcal{P}'$ in $\mathcal{P}$ the (strict) inclusion $[\mathcal{P}']_{\mathcal{O}} \subset [\mathcal{P}]_{\mathcal{O}}$ is observed. Any complete and irreducible set in M is called a *basis* of the closed class M relative to the set of operations $\mathcal{O}$. The set M', contained in the closed class M is called a *precomplete class in* M if it is not a complete set in M, but if the equality $[M' \cup \{\varphi\}]_{\mathcal{O}} = M$ is satisfied for every function $\varphi \in M \setminus M'$.

Remark 1. Henceforth, we shall omit the expression "relative to the set of operations $\mathcal{O}$" if the set $\mathcal{O}$ relative to which the closure, complete sets, bases, etc. are considered is known.

Remark 2. Usually, we shall use the set $\{O_1, O_2, O_3, O_4, S\}$ formed by the operations described in Sec. 6.2, for the set of operations $\mathcal{O}$.

It was mentioned in Sec. 6.2 that the set of b.d.-functions containing a unit delay element φ_d and the operators generated by the functions of a certain set that is complete in P_k ($k \geqslant 2$) forms a complete set in the set $\bigcup\limits_{n=0}^{\infty} \bigcup\limits_{m=0}^{\infty} \hat{\Phi}_k^{n,m}$ relative to the operations O_1, O_2, O_3, O_4, and S (i.e., in a set of all *k-valued b.d.-functions*).

The set $\bigcup\limits_{n=0}^{\infty} \bigcup\limits_{m=0}^{\infty} \hat{\Phi}_k^{n,m}$ ($k \geqslant 2$) is known to contain bases relative to the set $\{O_1, O_2, O_3, O_4, S\}$, which consist of a single function. An example of such a basis is the set $\{\varphi_0(X_1, X_2, X_3, \varphi_d(X_4))\}$, where φ_d is a unit delay element from $\hat{\Phi}_k$, and φ_0 is an operator in $\hat{\Phi}_k^{4,1}$ generated by the function $\max(x_1 \cdot x_4 + x_2(1 - x_4), x_3) + 1$ (mod k sum, difference and product).

Everywhere in this section, we shall use the symbol $\hat{\Phi}_{(k)}$ ($\Phi_{(k)}$) to denote the set of all k-valued b.d.-(resp. d-) functions including functions without inputs, or without outputs, or without both, i.e.

$$\hat{\Phi}_{(k)} = \bigcup_{n=0}^{\infty} \bigcup_{m=0}^{\infty} \hat{\Phi}_k^{n,m} \left(\text{and } \Phi_{(k)} = \bigcup_{n=0}^{\infty} \bigcup_{m=0}^{\infty} \Phi_k^{n,m}\right).$$

6.3.1. Is the set M a closed class relative to the set of operations Θ?

(1) M consists of all k-valued b.d.-functions without outputs, and of all functions φ belonging to $\bigcup_{n=0}^{\infty} \bigcup_{m=1}^{\infty} \hat{\Phi}_k^{n,m}$ and "preserving $\tilde{0}^{\omega}$", i.e. $\varphi(\underbrace{\tilde{0}^{\omega}, \ldots, \tilde{0}^{\omega}}_{n \text{ times}}) = (\underbrace{\tilde{0}^{\omega}, \ldots, \tilde{0}^{\omega}}_{m \text{ times}})$, $\Theta = \{O_3, O_5, S\}$.

(2) M consists of all k-valued b.d.-functions belonging to the set $\bigcup_{m=0}^{\infty} \hat{\Phi}_k^{1,m}$ and having a weight that is multiple of k; $\Theta = \{S\}$.

(3) M consists of all k-valued d-functions belonging to the set $\bigcup_{n=1}^{\infty} \bigcup_{m=1}^{\infty} \Phi_k^{n,m}$, and having either an even or infinite weight; $\Theta = \{O_1, O_2, O_4, O_5\}$.

(4) $M = \bigcup_{n=0}^{1} \bigcup_{m=0}^{\infty} (\Phi_k^{n,m} \setminus \hat{\Phi}_k^{n,m})$; $\Theta = \{O_3, S\}$.

6.3.2. (1) Let φ_d be a unit delay element in the set $\hat{\Phi}_2$ and $\Theta = \{O_3, O_5, S\}$. Prove that the class $[\varphi_d]_{\Theta}$ contains only functions that are identically equal to zero (i.e. operators whose each output yields the word $\tilde{0}^{\omega}$), functions without inputs and outputs which are always in the same initial state, and functions having an even weight.

(2) Give an example of a function in $\hat{\Phi}_2$ having an

even weight and not belonging to the class $[\varphi_d]_{\mathcal{O}}$ described in Problem 6.3.2 (1).

6.3.3. Let $\varphi_0(X_1, X_2)$ be an operator in $\hat{\Phi}_{(2)}$ generated by the Sheffer stroke $x_1 \mid x_2$, and let φ_d be a unit delay element in $\hat{\Phi}_2$. Prove that the operator $\varphi(\widetilde{x^\omega}) = \langle 1/3 \rangle$ in $\hat{\Phi}_2$ does not belong to the class $[\varphi_0, \varphi_d]_{\mathcal{O}}$ if $\mathcal{O} = \{S\}$.

6.3.4. A periodic word $\widetilde{x^\omega}$ of period 3 is supplied at the input of a b.d.-operator $\varphi \in \hat{\Phi}_2$. Find the maximum period of the output word (the majorization is carried out over all input periodic words of period 3):

$$(1)\quad \varphi: \begin{cases} y(t) = (x(t) \oplus q_1(t-1))\ q_2(t-1), \\ q_1(t) = x(t)\ \bar{q}_1(t-1)\ \bar{q}_2(t-1) \vee \bar{x}(t)\ q_2(t-1), \\ q_2(t) = x(t)\ q_1(t-1)\ \bar{q}_2(t-1) \vee \bar{x}(t)\ q_2(t-1), \\ q_1(0) = q_2(0) = 0; \end{cases}$$

$$(2)\quad \varphi: \begin{cases} y(t) = x(t)\ q_1(t-1), \\ q_1(t) = \bar{x}(t)\ \bar{q}_1(t-1) \vee q_2(t-1), \\ q_2(t) = q_1(t-1), \\ q_1(0) = q_2(0) = 0. \end{cases}$$

6.3.5. Let φ be a b.d.-function of weight r, belonging to the set $\hat{\Phi}_{A,B}$.

(1) Prove that if each input word $\widetilde{x^\omega}$ is quasi-periodic with a period T, the output word $\varphi(\widetilde{x^\omega})$ is also quasi-periodic and its period does not exceed the number $r \cdot T$.

(2) Find an upper estimate for the length of the pre-period of the output word $\varphi(\widetilde{x^\omega})$ if the pre-period of the word $\widetilde{x^\omega}$ is known to have a length p.

6.3.6. Let the functions φ_0, φ_d and φ be the same as in the Problem 6.3.3. Does there exist an $s \geqslant 1$ satisfying the condition $\varphi(\varphi_d^s(X)) \in [\varphi_0, \varphi_d]_{\mathcal{O}}$, where $\mathcal{O} = \{S\}$? (Here, $\varphi_d^s(X) = \varphi_d(\varphi_d(\ldots \varphi_d(X) \ldots))$ is the superposition of s functions φ_d.)

If $f(\widetilde{x^n}) \in P_k$, then $\varphi_{f(\widetilde{x^n})}(X_1, \ldots, X_n)$ will denote a b.d.-operator (in $\hat{\Phi}_k^{n,1}$) generated by the function $f(\widetilde{x^n})$.

6.3.7. Are the following sets of b.d.-functions complete in $\hat{\Phi}_{(k)}$ relative to the set of operations $\{O_1, O_2, O_3, O_4, S\}$?

(1) $\{\varphi_{\max(x_1, x_2)}(X_1, X_2), \varphi_d(X)\}$, $k \geqslant 2$;

(2) $\{\varphi_{x_1 \cdot x_2}(X_1, X_2), \varphi_{x_1 \dot{-} x_2}(X_1, X_2), \varphi_{x_1 - x_2}(\varphi_d(X), X_2), \varphi_{\equiv 1}(X), \varphi_{\equiv k-2}(X)\}$, $k \geqslant 3$ (here, $\varphi_{\equiv j}(X)$ is a b.d.-operator generated by a function in P_k, identically equal to j);

(3*) $\{\varphi_{x+1}(X), \varphi_d(\varphi_d(X)), \varphi_{x \dot{-} y}(\varphi_d(X), (X_2))\}$.

6.3.8. Isolate at least one basis in the set $\mathcal{P}$ which is complete in $\hat{\Phi}_{(k)}$ relative to the set of operations $\{O_1, O_2, O_3, O_4, S\}$

(1) $\mathcal{P} = \{\varphi_d(X), \varphi_{\equiv 0}(X), \varphi_{\equiv 1}(X), \varphi_{x_1 \cdot x_2}(X_1, X_2), \varphi_{x_1 \oplus x_2 \oplus x_3}(X_1, X_2, X_3)\}$, $k = 2$;

(2) $\mathcal{P} = \{\varphi_d(X), \varphi_{\equiv 0}(X), \varphi_{\equiv 1}(X), \varphi_{j_0(x)}(X), \varphi_{x_1 + x_2}(X_1, X_2)\}$, $k \geqslant 3$;

(3) $\mathcal{P} = \{\varphi_{\equiv 0}(X), \varphi_{x_1 \dot{-} x_2}(\varphi_d(X), X_2), \varphi_{x+1}(X), \varphi_{\min(x_1, x_2)}(X_1, X_2)\}$, $k \geqslant 3$.

6.3.9*. Does $\hat{\Phi}_{(2)}$ contain a basis relative to the set of operations $\{O_1, O_2, O_3, O_4, S\}$ with five functions?

6.3.10*. Prove that any closed class in $\hat{\Phi}_{(k)}$ except the entire set $\hat{\Phi}_{(k)}$ can be extended to a precomplete class.[4]

6.3.11*. Using the Problems 6.3.8 (1) and 6.3.10, show that $\hat{\Phi}_{(2)}$ contains at least four precomplete classes.

6.3.12. Let $\varphi_d \in \hat{\Phi}_2$. Enumerate all the classes that are precomplete in $[\varphi_d]_{\{O_3, O_5, S\}}$ relative to the set of operations $\{O_5, S\}$.

6.3.13. Let $\varphi_x = \varphi_x(X)$ be a b.d.-operator in $\Phi_{(k)}$, generated by the function x. Prove that whatever the type of the closed class[4] M in $\Phi_{(k)}$, the equality $[M \cup \{\varphi_x\}] = M \cup \{\varphi_x\}$ is always satisfied (here, as usual, we take together with the function all functions that are equal to it or congruent to it).

6.3.14. Can there exist a function belonging to each precomplete class in $\hat{\Phi}_{(k)}$?

---

[4] The closure and precompletion are taken relative to the set of operations $\{O_1, O_2, O_3, O_4, S\}$.

6.3.15. Can the set $\hat{\Phi}_{(k)}$ be presented in the form of a union $\bigcup_{i=1}^{s} M_i$ ($s \geqslant 2$) of pairwise disjoint closed[5] classes in $\hat{\Phi}_{(k)}$?

6.3.16*. Prove that the power of the set of all closed classes in $\hat{\Phi}_{(k)}$ is continual.

6.3.17. Determine the power of the set of all closed[5] classes in $\hat{\Phi}_{(k)}$, having finite complete sets.

6.3.18*. Prove that the power of the set of all closed[5] classes in $\Phi_{(k)}$ is equal to $2^{\mathfrak{c}}$ (the power of a hypercontinuum).

6.3.19*. Find the power of the set of all closed classes[5] in $\Phi_{(k)}$ having finite bases.

6.3.20. Disprove the following statements:

(1) the set $\hat{\Phi}_{(k)}$ forms a precomplete class in $\Phi_{(k)}$;

(2) the set $\Phi_{(k)}$ contains a function forming together with the set $\hat{\Phi}_{(k)}$ a complete set in $\Phi_{(k)}$.

6.3.21. Let M_0 be a set consisting of all k-valued d-functions that do not have outputs, and of all functions φ belonging to $\bigcup_{n=0}^{\infty} \bigcup_{m=1}^{\infty} \Phi_k^{n,m}$ and assuming the value 0 for $t = 1$, i.e. $\varphi(\tilde{x}_1^{\omega}, \ldots, \tilde{x}_n^{\omega}) = (\tilde{y}_1^{\omega}, \ldots, \tilde{y}_m^{\omega})$ and $y_j(1) = 0$ for $j = \overline{1, m}$.

(1) Is M_0 a precomplete class in $\Phi_{(k)}$?

(2) Is the intersection $M_0 \cap \hat{\Phi}_{(k)}$ a precomplete class in $\hat{\Phi}_{(k)}$?

6.3.22. (1) Is the set $\bigcup_{n=0}^{1} \bigcup_{m=0}^{\infty} \hat{\Phi}_k^{n,m}$ a precomplete class in $\hat{\Phi}_{(k)}$?

(2) Is the set $\bigcup_{n=0}^{1} \bigcup_{m=0}^{\infty} \Phi_k^{n,m}$ a precomplete class in $\Phi_{(k)}$?

---

[5] See footnote on p. 210.

Chapter Seven

Fundamentals of the Algorithm Theory

7.1. Turing's Machines and Operations with Them. Functions Computable on Turing's Machines

A *Turing's machine* is an (abstract) device consisting of a *tape*, a *controlling device* and a *scanning head.*

The tape is divided into *squares*. Any square at each (discrete) instant of time contains exactly one *symbol* in an *external alphabet* $A = \{a_0, a_1, \ldots, a_{n-1}\}$, $n \geqslant 2$. A certain symbol of the alphabet A is called a *dummy symbol*, and a square containing a dummy symbol at a given moment is known as a *blank square* (at this moment). A dummy symbol is usually represented by 0 (zero). The tape is assumed to be infinite in both directions.[1]

At any instant of time the controlling device is in a certain *state* q_j, which belongs to a set $Q = \{q_0, q_1, \ldots, q_{r-1}\}$, $r \geqslant 1$. The set Q is referred to as an *internal alphabet* (or a *set of internal states*). Sometimes, disjoint subsets Q_1 and Q_0 of *initial* and *final states* respectively are isolated from the set Q.

Remark. Henceforth, unless stated otherwise, we shall assume that $|Q| \geqslant 2$, and take only one state q_1 for the initial state. As a rule, the state q_0 will be the final state.

The head moves along the tape so that at each instant it scans exactly one square of the tape. The head can *scan* a symbol in the square and *record* (*print*) on it a new symbol of the external alphabet instead. The symbol "printed" in the square can, in particular, coincide with that being scanned (at that instant).

During its operation, the controlling device changes or does not its (internal) state depending on its state and on the symbol scanned by the head, orders the head to print

---

[1] This should be understood as follows: at each instant the tape is finite (contains a finite number of squares), but the "length" of the tape (the number of squares in it) can be increased (if required).

in the square scanned a symbol of the external alphabet, and either to remain at the same square, or to move by one square to the left or right.

The operation of the controlling device can be characterized by three functions:

$$G: \ Q \times A \to Q,$$
$$F: \ Q \times A \to A,$$
$$D: \ Q \times A \to \{S, \ L, \ R\}.$$

The function G is called a *transition function*, F an *output function*, and D is a *function of displacement* (*of the head*). The symbols S, L and R denote respectively the absence of head displacement, a left movement and a right movement by a square.

The functions G, F and D can be specified by a list of quintuples

$$q_i a_j G(q_i, \ a_j) \ F(q_i, \ a_j) \ D(q_i, \ a_j) \tag{1}$$

which is abbreviated as $q_i a_j q_{ij} a_{ij} d_{ij}$. These quintuples are referred to as *commands*. The functions G, F and D are, in general, *partial* functions (specified not everywhere). This means that a quintuple of the type (1) is defined not for each pair $(q_1, \ a_j)$. The list of all quintuples which determine the operation of a Turing's machine is known as a *program* of this machine. The program is often given in tabular form (see Table 9).

Table 9

	q_0	. . .	q_i	. . .	q_{r-1}
			.		
			.		
a_0			.		
. . .					
a_j		. . .	$q_{ij}a_{ij}d_{ij}$	. . .	
. . .					
a_{n-1}			.		
			.		
			.		

If there is no quintuple of the type (1) in the program for a pair (q_i, a_j), a dash appears in the table at the intersection of the row a_j and the column q_i.

The operation of Turing's machine can be conveniently described in the "language of configurations".

Let at instant t the extreme left non-blank square C_1 of the tape contain a symbol a_{j_1}, while the extreme right non-blank square C_s $(s \geqslant 2)$ contains a symbol a_{j_s} (there are $s-2$ squares between C_1 and C_s). In this case, the *word* $P = a_{j_1}a_{j_2} \ldots a_{j_p} \ldots a_{j_s}$, where a_{j_p} is the symbol contained at instant t in the square C_p $(1 \leqslant p \leqslant s)$, is said *to be recorded* on the tape *at instant t*. For $s = 1$, i.e. when the tape contains only one non-dummy symbol, $P = a_{j_1}$. Let us suppose that at this instant of time the controlling device is in a state q_i and the head scans a symbol a_{j_l} of the word P $(l \geqslant 2)$. Then the word

$$a_{j_1} \ldots a_{j_{l-1}} q_i a_{j_l} \ldots a_{j_s} \qquad (2)$$

is called the *configuration of the machine* (*at a given instant t*). For $l = 1$, the configuration has the form $q_i a_{j_1} \ldots a_{j_s}$. If at instant t the head scans a blank square located to the left (or right) of the word P, and there are $v \geqslant 0$ blank squares between this square and the first (last) square of the word P, the *configuration of the machine at the moment t* is presented by the word

$$q_i \underbrace{\Lambda \ldots \Lambda}_{v+1 \text{ times}} a_{j_1} \ldots a_{j_s} \qquad (3)$$

(resp. the word $a_{j_1} \ldots a_{j_s} \underbrace{\Lambda \ldots \Lambda}_{v \text{ times}} q_i \Lambda$), where Λ denotes the dummy symbol of the alphabet A. If at the moment t the tape is blank, i.e. a blank word consisting of only dummy symbols of the external alphabet is recorded on it, the word $q_i\Lambda$ will be the *configuration of the machine at the instant t*.

Let the configuration of the machine at a moment t have the form (2), and suppose that the program of the machine contains the command

$$q_i a_{j_l} q_{ij_l} a_{ij_l} d_{ij_l}.$$

Then for $d_{ij_l} = L$, the configuration of the machine at the next instant will be expressed by the word

(a) $q_{ij_1} \Lambda a_{ij_1} a_{j_2} \ldots a_{j_s}$ if $l = 1$;

(b) $q_{ij_2} a_{j_1} a_{ij_2} a_{j_3} \ldots a_{j_s}$ if $l = 2$;

(c) $a_{j_1} \ldots a_{j_{l-2}} q_{ij_l} a_{j_{l-1}} a_{ij_l} a_{j_{l+1}} \ldots a_{j_s}$ if $l > 2$.

The cases when $d_{ij_l} = R$ or $d_{ij_l} = S$, or when the machine configuration corresponds to the head outside the word P (as in the word (3)), or when the word P is dummy are described in a similar way.

If a program of the machine contains no quintuple of the form (1) for a pair (q_i, a_{jl}), or if a "new" state q_{ij_l} is the *terminal* state, the machine *stops*, and the "resulting" configuration is referred to as the *terminal* configuration. The configuration corresponding to the beginning of operation of the machine is referred to as the *initial* configuration.

Let K be the configuration of the machine at a certain instant and let K' be its configuration at the next instant. Then the configuration K' is called a configuration *directly derived from* K (and denoted by $K \models K'$). If K_1 is the initial configuration, the sequence $K_1, K_2, \ldots, K_m$, where $K_i \models K_{i+1}$ for $1 \leqslant i \leqslant m - 1$, is called the *Turing computation*. In this case, K_m is said to be *derivable from* K_1, and the following notation is used: $K_1 \vdash K_m$. If, in addition, K_m is a terminal configuration, it is said to be *terminally derivable from* K_1 and the notation $K_1 \mid\!\!\dot{-}\, K_m$ is used.

Suppose that a Turing machine T starts operating at a certain (initial) instant of time. The word recorded at this moment on the tape is called the *initial* word. To put the machine T in operation, it is necessary to place the scanning head against a square of the tape and indicate the state of the machine T at the initial moment.

If P_1 is an initial word, the machine T which starts to operate at the "word" P_1 will either stop after a certain number of steps, or will not stop at all. In the former case, the machine T is said to be *applicable to the word* P_1, and the *result of application of the machine* T *to the word* P_1 is a word P corresponding to the terminal configuration (notation $P = T(P_1)$). In the latter case the machine T is said to be *inapplicable to the word* P_1.

Henceforth, unless otherwise stipulated, we shall assume that (1) the initial word is non-dummy, (2) at the initial moment the head is against the extreme left non-blank square on the tape, and (3) the machine starts to operate when it is in the state q_1.

The *zone of operation* of the machine T (*on the word* P_1) is the set of squares which are scanned by the head at least once during the operation.

We shall often use the notation $[P]^m$ for words of the form $PP \ldots P$ (m times). If $P = a$ is a word of length 1, we shall write a^m instead of $aa \ldots a$ (m times) and $[a]^m$.

We shall denote by W an arbitrary finite word in the external alphabet of a Turing machine (in particular, a *dummy* word, i.e. that consisting of dummy symbols of the external alphabet).

While describing the operation of a Turing machine in terms of "configuration language", we shall use an expression similar to

$$q_1 1^x 0 1^y O W \models 1^y 0 q_0 W,.$$

$x \geqslant 1$ and $y \geqslant 1$. The expression should be understood as follows: the machine "erases" the word 1^x and stops at the first symbol of the word W. If, however, W is a dummy word, the machine stops at the second 0 (zero) following the word 1^y.

Turing's machines T_1 and T_2 are called *equivalent* (in alphabet A) if for any input word P (in alphabet A) the relation $T_1(P) \simeq T_2(P)$ holds. This expression has the following meaning: $T_1(P)$ and $T_2(P)$ are either both specified or both not specified[2], and if they are specified, then $T_1(P) = T_2(P)$. The symbol $\simeq$ is called the *symbol of conditional equality*.

Let machines T_1 and T_2 have programs Π_1 and Π_2 respectively. We shall assume that the internal alphabets of these machines do not intersect and that q_1' is a final state of the machine T_1 while q_2' is an initial state of the machine T_2. We shall replace the state q_1' by the state q_2' throughout the program Π_1 and combine the obtained program with Π_2. The new program Π determines the machine T called the *composition of the machines* T_1 and T_2 (*over the pair of states* (q_1', q_2')) and denoted by

[2] $T(P)$ is defined (not defined) if the machine T is applicable (inapplicable) to the word P.

$T_1 \bigcirc T_2$ or T_1T_2 (in detail, $T = T\ (T_1,\ T_2,\ (q_1',\ q_2')))$. The external alphabet of the composition T_1T_2 is the union of the external alphabets of T_1 and T_2.

Let q' be a final state of a machine T, and q'' be a state of this machine, which is other than final. We shall replace the symbol q' throughout the program Π of the machine T by q''. We shall obtain a program Π' determining the machine $T'\ (q',\ q'')$. The machine T' is called the *iteration of the machine* T (*over the pair of states* $(q',\ q'')$).

Let Turing's machines T_1, T_2 and T_3 be defined by programs Π_1, Π_2 and Π_3 respectively. We assume that the internal alphabets of these machines are pairwise non-intersecting. Let q_1' and q_1'' be two different final states of the machine T_1. We shall replace the state q_1' throughout the program Π_1 by an initial state q_2' of the machine T_2, and the state q_1'' by an initial state q_3' of T_3. Then we combine the new program with Π_2 and Π_3. We obtain a program Π defining the Turing machine $T = T\ (T_1,\ (q_1',\ q_2'),\ T_2,\ (q_1'',\ q_3'),\ T_3)$. This machine is known as a *branching of the machines* T_2 *and* T_3, *controlled by the machine* T_1.

In defining complicated Turing machines, the *operator notation of algorithms* is often used in the form of a line consisting of symbols of machines, transition symbols $\left(\text{of the form } \left|\frac{q'}{k}\right. \text{ and } \left.\frac{q''}{k}\right|\right)$, and also symbols α and ω used to denote the beginning and end of algorithm operation. In operator notation (of an algorithm), the expression $T_i \left|\frac{q_{i0}}{k}\right. T_j \ldots T_m \left.\frac{q_{n1}}{k}\right| T_n$ denotes the branching of machines T_j and T_n, which is controlled by the machine T_i. Here the final state q_{i0} of the machine T_i is replaced by the initial state q_{n1} of the machine T_n, while any other final state of the machine T_i is replaced by the (same) initial state of T_j. If the machine T_i has a single final state, the symbols $|\underline{q_{i0}}$ and $\underline{q_{n1}}|$ denote unconditional transition. The symbols q_{i0} and q_{n1} are omitted whenever a confusion is unlikely.

Example. The operator scheme

$$\left.\frac{}{2}\right| \left.\frac{}{1}\right| T_1 \alpha T_2 \left|\frac{q_{20}}{1}\right. T_3 \left|\frac{q_{30}}{2}\right. T_4 \omega$$

describes the following "computational process". At first the machine T_2 starts to operate. If it stops in the state

q_{20}, the machine T_1 begins to operate. When it stops, T_2 starts to operate again. If, however, T_2 stops in a final state other than q_{20}, the "work is continued" by the machine T_3. If this machine comes to the final state q_{30}, then T_1 starts to operate. If, however, T_3 terminates at a final state other than q_{30}, the "work is continued" by T_4. If the machine T_4 stops in a certain state, the computational process is terminated.

Let $\tilde{\alpha} = (\alpha_1, \alpha_2, \ldots, \alpha_n)$, $n \geqslant 1$, be an arbitrary tuple of non-negative integers. The word $1^{\alpha_1+1}\,01^{\alpha_2+1}\,0\ldots 01^{\alpha_n+1}$ is called the *basic machine code* (or just a *code*) *of the tuple* $\tilde{\alpha}$ (in the alphabet $\{0, 1\}$) and is denoted by $k(\tilde{\alpha})$. In particular, the word $1^{\alpha+1}$ is the basic machine code of the number α.

Henceforth, we shall mainly consider *partial numerical functions*. The function $f(x_1, x_2, \ldots, x_n)$, $n \geqslant 1$, is a *partial numerical function* if the variables x_i assume the values in the natural scale $N = \{0, 1, 2, \ldots, m, \ldots\}$, and if $f(\tilde{\alpha}) \in N$ when the function f is defined on the tuple $\tilde{\alpha} = (\alpha_1, \alpha_2, \ldots, \alpha_n)$.

A numerical partial function is said to be *computable* (*in the Turing sense*) if there exists a Turning machine T_f having the following properties:

(a) if $f(\tilde{\alpha})$ is specified, then $T_f(k(\tilde{\alpha})) = k(f(\tilde{\alpha}))$;

(b) if $f(\tilde{\alpha})$ is not specified, then either $T_f(k(\tilde{\alpha}))$ is not a code of any number in N, or the machine T_f cannot be applied to the word $k(\tilde{\alpha})$.

Remark. Henceforth, we shall assume that at the initial moment the head scans the extreme left unity of the word $k(\tilde{\alpha})$. It is known that this restriction does not make the class of computable functions narrow.

If a function f is computable in the Turing sense by using a machine T_f, the *machine* T_f will be said *to compute the function* f.

7.1.1. Find out whether a Turing machine T specified by a program Π is applicable to a word P. If it is, determine the result of application of the machine T to the word P. It is assumed that q_1 is the initial and q_0 the final states, and at the initial moment the head scans the extreme left unity on the tape.

(1) $$\Pi: \begin{cases} q_1 0 q_2 0R \\ q_1 1 q_1 1R \\ q_2 0 q_3 0R \\ q_2 1 q_1 1L & \text{(a) } P = 1^3 0^2 1^2, \\ q_3 0 q_0 0S & \text{(b) } P = 1^3 0 1^3, \\ q_3 1 q_2 1R & \text{(c) } P = 10\,[01]^2\,1. \end{cases}$$

(2) $$\Pi: \begin{cases} q_1 0 q_2 1R \\ q_1 1 q_3 0R \\ q_2 0 q_3 1L & \text{(a) } P = 1^4 01, \\ q_2 1 q_2 1S & \text{(b) } P = 1^3 01^2, \\ q_3 1 q_1 1R & \text{(c) } P = 1^6. \end{cases}$$

(3) $$\Pi: \begin{cases} q_1 0 q_1 1R \\ q_1 1 q_2 0R \\ q_2 0 q_1 1R & \text{(a) } P = 101^2, \\ q_2 1 q_3 1L & \text{(b) } P = 1^2 0^2 1, \\ q_3 0 q_1 1L & \text{(c) } P = [10]^2 1. \end{cases}$$

7.1.2. In the alphabet $\{0, 1\}$, construct a Turing machine having the following property (for dummy symbol we take 0):

(1) the machine is applicable to any non-dummy word in the alphabet $\{0, 1\}$;

(2) the machine is inapplicable to any word in the alphabet $\{0, 1\}$, and the zone of operation in each word is infinite;

(3) the machine is inapplicable to any word in the alphabet $\{0, 1\}$ and the zone of operation is limited by the same number of squares, which does not depend on the chosen word;

(4) the machine is applicable to words of the form 1^{3n} $(n \geqslant 1)$ and inapplicable to any word of the form $1^{3n+\alpha}$, where $\alpha = 1, 2$ and $n \geqslant 1$;

(5) the machine is applicable to words of the form $1^\alpha 0 1^\alpha$, where $\alpha \geqslant 1$, and inapplicable to $1^\alpha 0 1^\beta$ if $\alpha \neq \beta$ $(\alpha \geqslant 1$ and $\beta \geqslant 1)$.

7.1.3. Using a given Turing machine T and an initial configuration K_1, determine the terminal configuration (q_0 is the final state):

(1) T:

	q_1	q_2
0	q_01S	q_10R
1	q_20R	q_21L

(a) $K_1 = 1^2q_11^301$, (b) $K_1 = 1q_11^4$.

(2) T:

	q_1	q_2	q_3
0	q_00S	q_01L	q_11L
1	q_21R	q_30R	q_10R

(a) $K_1 = 1q_11^5$, (b) $K_1 = q_11^301$, (c) $K_1 = 10q_11^4$.

7.1.4. Construct a Turing machine in the alphabet $\{0, 1\}$, which transforms the configuration K_1 into K_0:

(1) $K_1 = q_11^n$, $K_0 = q_01^n01^n$ $(n \geqslant 1)$,
(2) $K_1 = q_10^n1^n$, $K_0 = q_0\,[01]^n$ $(n \geqslant 1)$,
(3) $K_1 = 1^nq_10$, $K_0 = q_01^{2n}$ $(n \geqslant 1)$,
(4) $K_1 = 1^nq_101^m$, $K_0 = 1^m\,q_001^n$ $(m \geqslant 1, n \geqslant 1)$.

7.1.5. (1) Prove that for any Turing machine there exists an equivalent machine whose program does not contain the symbol S.

(2) Prove that for any Turing machine it is possible to construct an equivalent machine whose program does not contain final states.

7.1.6. The program of a Turing machine T has the form (the dummy symbol is 0 and the initial state is q_1).

	q_1	q_2	q_3	q_4
0	q_21R	q_31R	q_12L	q_23R
1	q_41L	q_31R	q_21R	q_31R
2	q_12L	q_20R	q_32R	—
3	—	q_23R	q_30R	q_43L

(1) Prove that starting the operation with a blank tape, the machine T will construct a word of the form $P_n = 11010^2 10^3 \ldots 10^n 1$ (n is an arbitrary positive integer) in $t(n)$ steps, and at any instant $t \geqslant t(n)$ (for $n \geqslant 3$) the head of the machine will be to the right of the word P_{n-2}.

(2) Construct a machine possessing the same property and having the set $\{0, 1\}$ as an external alphabet.

(3) Prove that if the program of a Turing machine does not contain the symbol L (naturally, it may contain symbols S and R), the machine, starting the operation with a blank tape, cannot construct a prefix of the word $11010^2 10^3 1 \ldots 0^n 10^{n+1} 1 \ldots$, however long.

7.1.7. Prove that for any Turing machine T (in an alphabet A) there exists a countable set of machines $T_1, T_2, \ldots, T_m, \ldots$ equivalent to it (in the alphabet A) and differing from one another in their programs.

7.1.8. Let $A = \{a_0, a_1, \ldots, a_{n-1}\}$ be a certain alphabet containing at least two letters. We shall encode its characters as follows: the code of a_i is the word $10^{i+1}1$ in the alphabet $\{0, 1\}$. Accordingly to this coding, the code of an arbitrary word $P = a_{i_1} a_{i_2} \ldots a_{i_s}$ ($s \geqslant 1$) in the alphabet A will have the form $10^{i_1+1}110^{i_2+1}1 \ldots 10^{i_s+1}1$. Prove that irrespective of the type of a Turing machine T with the external alphabet A there exists a Turing machine T_0 with the external alphabet $\{0, 1\}$, which satisfies the following condition: for any word P (in the alphabet A) the machine T_0 is applicable to the code of this word if and only if the machine T is applicable to the word P. Moreover, if $T(P)$ is specified, the code of the word $T(P)$ coincides with the word which is a result of application of the machine T_0 to the code of the word P.

7.1.9. Construct the composition $T_1 T_2$ of Turing machines T_1 and T_2 (for the pair of states (q_{10}, q_{21})) and determine the result of application of the composition $T_1 T_2$ to the word P (q_{20} is the final state of the machine T_2).

(1) T_1:

	q_{11}	q_{12}
0	$q_{12}0R$	$q_{10}1L$
1	$q_{12}1R$	$q_{11}0R$

T_2:

	q_{21}	q_{22}
0	$q_{22}1R$	$q_{21}1R$
1	$q_{21}0L$	$q_{20}1S$

(a) $P = 1^3 0^2 1^2$, (b) $P = 1^4 01$.

(2) T_1:

	q_{11}	q_{12}	q_{13}
0	$q_{10}0L$	$q_{13}0R$	$q_{11}0R$
1	$q_{12}1R$	$q_{13}1R$	$q_{11}0R$

T_2:

	q_{21}	q_{22}
0	$q_{22}1L$	$q_{20}0R$
1	$q_{22}1L$	$q_{21}0L$

(a) $P = 1^4 0^2 1^3 0 1^2$, (b) $P = 1^2 0 1 0 1^3$.

(3) T_1:

	q_{11}	q_{12}	q_{13}
0	$q_{12}0R$	$q_{13}0R$	$q_{10}1L$
1	$q_{11}1R$	$q_{11}1R$	—

T_2:

	q_{21}	q_{22}	q_{23}
0	$q_{22}0L$	$q_{23}0L$	$q_{20}0R$
1	$q_{21}1L$	$q_{22}1L$	$q_{23}1L$

(a) $P = 1^2 0 1^3 0 1^2$, (b) $P = 1^2 0 1^2 0^2 1^2$.

7.1.10. Determine the result of application of iteration of a machine T (for a pair of states (q_0, q_i)) to the word P (q_0 and q_0' are the final states):

(1) $i = 1$, T:

	q_1	q_2	q_3	q_4	q_5
0	$q_0 0S$	$q_4 0S$	$q_5 0S$	$q_4 1R$	$q_0' 1L$
1	$q_2 0R$	$q_3 0R$	$q_1 0R$	—	—

(a) $P = 1^{3k}$, (b) $P = 1^{3k+1}$, (c) $P = {}^{3k+2}$, $k \geqslant 1$.

(2) $i = 1$, T:

	q_1	q_2	q_3	q_4	q_5	q_6
0	$q_0' 0R$	$q_0' 0R$	$q_4 0R$	$q_5 1L$	$q_6 0L$	$q_0 0R$
1	$q_2 0R$	$q_3 0R$	$q_3 1R$	$q_4 1R$	$q_5 1L$	$q_6 1L$

(a) $P = 1^{2x}$, (b) $P = 1^{2x+1}$, $x \geqslant 1$.

(3) $i=3$, T:

	q_1	q_2	q_3	q_4	q_5	q_6
0	$q_2 0L$	—	$q_4 0R$	$q_5 1L$	$q_6 0L$	$q_0' 0R$
1	$q_1 2R$	$q_2 1R$	—	$q_4 1R$	$q_5 1L$	$q_6 1L$
2	—	$q_3 1R$	—	—	—	$q_0 1R$

$P=1^x 0 1^y (x\geqslant 1,\ y\geqslant 1)$.

7.1.11. Detemine the result of application of the machine $T=T(T_1, (q_{10}', q_{21}), T_2, (q_{10}'', q_{31}), T_3)$ to the word P (q_{20} and q_{30} are the final states of the machines T_2 and T_3 respectively):

(1) T_1:

	q_{11}	q_{12}
0	$q_{12} 0R$	$q_{10}' 0R$
1	$q_{12} 1R$	$q_{10}'' 1L$

T_2:

	q_{21}
0	$q_{20} 1S$
1	$q_{21} 0R$

T_3:

	q_{31}	q_{32}
0	$q_{32} 1L$	$q_{30} 1L$
1	$q_{31} 1L$	—

(a) $P=101^3$, (b) $P=1^3 01$.

(2) T_1:

	q_{11}	q_{12}	q_{13}
0	$q_{12} 0R$	$q_{10}' 0L$	$q_{10}'' 0R$
1	$q_{11} 1R$	$q_{13} 1R$	$q_{13} 1R$

T_2:

	q_{21}	q_{22}
0	$q_{22} 0L$	$q_{20} 0R$
1	$q_{21} 1L$	$q_{22} 1L$

T_3:

	q_{31}	q_{32}
0	$q_{32} 0R$	$q_{30} 1S$
1	$q_{31} 1R$	$q_{31} 1R$

(a) $P=1^x 0^2 1 \quad (x\geqslant 1)$,
(b) $P=1^x 0101^y 01^z \quad (x\geqslant 1,\ y\geqslant 1,\ z\geqslant 1)$.

7.1.12. Using machines T_1, T_2, T_3, T_4, and T_5 construct the operator scheme of the algorithm $\mathfrak{A}$ (here q_{10}, q'_{10}, q_{20}, q_{30}, q_{40}, and q_{50} are the final states of the corresponding machines):

T_1:

	q_{11}	q_{12}
0	$q_{12}0R$	$q_{10}0R$
1	$q'_{10}0R$	$q_{11}1S$

T_2:

	q_{21}	q_{22}	q_{23}
0	$q_{22}0R$	$q_{23}0R$	$q_{20}0S$
1	$q_{21}1R$	$q_{22}1R$	$q_{23}1R$

T_3:

	q_{31}	q_{32}
0	$q_{32}1R$	$q_{30}1S$
1	$q_{31}1R$	—

T_4:

	q_{41}	q_{42}	q_{43}
0	$q_{42}0L$	$q_{43}0L$	$q_{40}0R$
1	$q_{41}1L$	$q_{42}1L$	$q_{43}1L$

T_5:

	q_{51}
0	$q_{50}1S$
1	$q_{51}1R$

(1) $\mathfrak{A}$: $q_1 1^x \models q_0 1^{2x}$ $(x \geqslant 1)$,

(2) $\mathfrak{A}$: $q_1 1^{x+1} \models q_0 1^{3x}$ $(x \geqslant 0)$,

(3) $\mathfrak{A}$: $W0q_1 1^{x+1} \models W0q_0 1^{2x+1}$ $(x \geqslant 0)$.

Remark. If $x = 0$, then 1^x is assumed to be a dummy word.

7.1.13. Using the operator scheme of the algorithm $\mathfrak{A}$ and the description of the machines appearing in the scheme of the algorithm, construct the program of a machine and determine the result of application of the machine specified by this scheme to the word P.

(1) $\mathfrak{A} = \alpha T_1 \underset{1}{\underline{\quad}}\,|\, T_2 T_3 \,|\, \underset{1}{\underline{q_{30}}}\, \omega,$

T_1:

	q_{11}
0	$q_{10}0L$
1	$q_{11}2R$
2	—

T_2:

	q_{21}	q_{22}	q_{23}
0	$q_{23}0R$	$q_{23}0R$	$q_{20}1L$
1	$q_{21}1R$	—	$q_{23}1R$
2	$q_{22}1R$	—	—

T_3:

	q_{31}	q_{32}
0	$q_{32}0L$	$q'_{30}0R$
1	$q_{31}1L$	$q_{32}1L$
2	—	$q_{30}1R$

$$P = 1^x 0 1^y \quad (x \geqslant 1,\ y \geqslant 1)$$

(q_{11}, q_{21} and q_{31} are the initial and q_{10}, q_{20}, q_{30}, and q'_{30} the final states of the machines).

(2) $\mathfrak{A} = \alpha \underset{2}{\underline{\quad}}\,|\, T_1 T_2 \,|\, \underset{1}{\underline{q_{20}}}\, T_3 \,|\, \underset{2}{\underline{q_{30}}}\, \underset{1}{\underline{\quad}}\,|\, \omega,$

T_1:

	q_{11}	q_{12}	q_{13}	q_{14}
0	—	$q_{13}0R$	$q_{14}0L$	—
1	$q_{12}0R$	$q_{12}1R$	$q_{13}1R$	$q_{10}0L$

T_2:

	q_{21}	q_{22}	q_{23}
0	$q_{22}0L$	$q_{20}1S$	$q_{20}1S$
1	$q'_{20}1S$	$q_{23}1L$	$q_{23}1L$

T_3:

	q_{31}	q_{32}	q_{33}	q_{34}
0	$q_{32}0L$	$q_{33}0R$	$q'_{30}1S$	$q_{30}1R$
1	$q_{31}1L$	$q_{34}1L$	—	$q_{34}1L$

$$P = 1^x 0 1^y \quad (x \geqslant 1,\ y \geqslant 1)$$

(q_{11}, q_{21} and q_{31} are the initial and q_{10}, q_{20}, q'_{20}, q_{30}, and q'_{30} the final states of the machines).

7.1.14. Construct a Turing machine computing the function f:

(1) $f(x) = \operatorname{sgn} x = \begin{cases} 0 & \text{if } x = 0 \\ 1 & \text{if } x \geqslant 1; \end{cases}$

(2) $f(x) = \left[\frac{1}{x}\right] = \begin{cases} 1 & \text{if } x = 1, \\ 0 & \text{if } x \geqslant 2, \\ \text{is not defined} & \text{if } x = 0; \end{cases}$

(3) $f(x) = \left[\frac{x}{2}\right] = m$ if $x = 2m$ or $x = 2m + 1$, $m \geqslant 0$;

(4) $f(x, y) = x \dot{-} y = \begin{cases} 0 & \text{if } x \leqslant y, \\ x - y & \text{if } x > y; \end{cases}$

(5) $f(x, y) = x - y$;

(6) $f(x, y) = \frac{4 - x}{y^2}$.

Remark. Here (and below), some well known "elementary" functions are often used for "analytic" representation of numerical functions. An "analytically" presented function is assumed to be *specified* only on integral tuples of values of variables (belonging to the natural scale N) for which *all the "elementary functions"* appearing in a given analytic representation of the function under consideration *are specified and assume integral non-negative values.* For example, the function $x^2/(3 - y/2)$ is specified only when $y/2$ is a non-negative integer, $3 - y/2$ is a positive integer and $x^2/(3 - y/2)$ is a non-negative integer.

A Turing machine T is said to *correctly compute a function* $f(\tilde{x}^n)$ if:

(1) $f(\tilde{\alpha}^n)$ is specified, $T(k(\tilde{\alpha}^n)) = k(f(\tilde{\alpha}^n))$, and the head of the machine in the terminal configuration scans the left unity of the code $k(f(\tilde{\alpha}^n))$;

(2) $f(\tilde{\alpha}^n)$ is not specified, and the machine T starting to operate from the left unity of the code $k(\tilde{\alpha}^n)$ does not stop.

7.1.15. Prove that for any computable function, there exists a Turing machine that correctly computes this function.

7.1.16. Construct a Turing machine which correctly computes the function f:

(1) $f(x)=x \dot{-} 1$; (4) $f(x, y)=x+y$;

(2) $f(x)=\overline{\operatorname{sgn}} x=1-\operatorname{sgn} x$; (5) $f(x, y)=\frac{x}{2-y}$.

(3) $f(x)=\frac{1}{x-2}$;

7.1.17. Using the program of the Turing machine T, write the analytic expression for the functions $f(x)$ and $f(x, y)$ computable by the machine T. (For the initial and final states everywhere, we take q_1 and q_0 respectively.)

(1) T:

	q_1	q_2
0	$q_2 1L$	$q_0 0R$
1	$q_1 1R$	$q_2 1L$

(2) T:

	q_1	q_2	q_3	q_4	q_5
0	$q_2 0R$	$q_1 0L$	$q_4 0L$	$q_4 0L$	$q_0 0R$
1	$q_1 1R$	$q_3 0R$	$q_3 0R$	$q_5 1L$	$q_5 1L$

(3) T:

	q_1	q_2	q_3	q_4	q_5	q_6	q_7	q_8	q_9
0	$q_2 0R$	$q_3 0R$	$q_0 1S$	$q_5 0R$	$q_3 0L$	$q_7 0L$	—	$q_9 0L$	$q_1 0R$
1	$q_2 0R$	$q_4 1R$	$q_3 1L$	$q_4 1R$	$q_6 1R$	$q_6 1R$	$q_8 0L$	$q_8 1L$	$q_9 1L$

7.1.18. Which functions of one variable can be computed by Turing's machines (in the alphabet $\{0, 1\}$) whose programs contain only one command?

7.1.19. Is it true that two different computable functions $f_1(\tilde{x}^m)$ and $f_2(\tilde{x}^n)$ can be computed on the same Turing machine if and only if $m \neq n$?

7.1.20. Let M be a countable set of computable functions and let $T(M)$ be the minimum possible set of Turing's machines, such that for any function f in M there exists a machine in the set $T(M)$ that computes the function f.

(1) Prove that if for a certain $n \geqslant 1$ there exists in the set M an infinite subset consisting of functions of n

variables, then there exists in $T(M)$ a machine with an indefinitely large number of states (i.e. for any $l_0 \geqslant 1$, there exists in $T(M)$ a machine with a number of states exceeding l_0).

(2) What are the necessary and sufficient conditions for the set $T(M)$ to be finite?

7.1.21. Construct an operator scheme of a Turing machine computing a function f. For elementary operators, use the machines T_i ($i = 4, 5, 6, 7$, and 8). In Problems (1) and (2), first construct an operator scheme only through the machines T_1, T_2, T_3 and then represent each of them by an operator scheme using $T_4, T_5, \ldots, T_8$. The initial states in the machines under consideration are $q_{11}, q_{21}, \ldots, q_{81}$ and the final states are $q_{10}, q'_{10}, q_{20}, q'_{20}, q_{30}, q_{40}, \ldots, q_{80}, q'_{80}$.

T_4:

	q_{41}
0	$q_{40}0R$
1	$q_{40}1R$

T_5:

	q_{51}
0	$q_{50}0L$
1	$q_{50}1L$

T_6:

	q_{61}
0	$q_{60}0S$
1	$q_{61}1R$

T_7:

	q_{71}
0	$q_{70}0S$
1	$q_{71}1L$

T_8:

	q_{81}
0	$q_{80}1S$
1	$q'_{80}0S$

(1) $f(x, y) = x + y \pmod 2$,

T_1:

	q_{11}	q_{12}
0	$q_{12}1L$	$q_{10}0R$
1	$q_{11}1R$	$q_{12}1L$

T_2:

	q_{21}	q_{22}	q_{23}
0	$q_{23}1L$	$q_{23}0L$	$q_{20}1S$
1	$q_{22}0L$	$q_{21}0L$	—

T_3:

	q_{31}
0	$q_{30}0L$
1	$q_{31}1R$

(2) $f(x) = 2^x$,

$$T_1: \begin{cases} q_{11}1 \models q_{10}1^2, \\ q_{11}1^{x+1} \models 101^x0q'_{10}1^2 \quad \text{for} \quad x > 0; \end{cases}$$

$$T_2: \begin{cases} 1^x010^t q_{21}1^y \models q_{20}1^y, \\ 1^x0^{z+1}10^t q_{21}1^y \models 1^{x+1}01^z0^t q'_{20}1^y \quad \text{for} \quad z > 0. \end{cases}$$

Here $x > 0$, $y > 0$, $t > 0$;

$T_3: W0q_{31}1^{y+1} \models W0q_{30}1^{2y+1}$, $y \geqslant 0$;

(3) $f(x) = 3x$;

(4) $f(x, y) = x \cdot y$;

(5) $f(x, y) = x - y$.

The *distance between two squares* C and C' on a tape is equal to the number of squares between C and C' plus one. In particular, adjacent squares are separated from each other by unity. Let l be a positive integer. The subset of all squares of the tape, such that every two of them are separated by a distance multiple of l, is called a *lattice with spacing* l. Thus, the tape can be regarded as a union of l lattices with spacing l. Let $R_{(l)}$ be a lattice with spacing l. Two squares of this lattice will be called *adjacent* if the distance between them (relative to the entire tape) is l. The *word* $P = a_1a_2 \ldots a_m$ is said to be *recorded on the lattice* $R_{(l)}$ if

(1) the symbol a_1 is recorded in a square C_1 of this lattice;

(2) the symbol a_2 is recorded in the square C_2 which is adjacent to C_1 in the lattice $R_{(l)}$ and is located to the right of C_1, and so on;

(m) the symbol a_m is recorded in the square C_m separated from the square C_1 by a distance $(m - 1)l$ and located to the right[3] of C_1.

A *Turing machine* T_1 is said to *simulate a Turing machine* T *on a lattice* $R_{(l)}$ (with spacing l) if irrespective of the word P (in an alphabet A), the following condition is satisfied: let the word P be recorded on the lattice $R_{(l)}$ and let at the initial instant the head of the machine T_1 scan the extreme left character of the word P; the machine T_1 stops if and only if the machine T is applicable to the word P. If $T(P)$ is defined, the word $T(P)$ will be

---

[3] We assume that only dummy symbols of the external alphabet are outside the squares $C_1, C_2, \ldots, C_m$ on the lattice $R_{(l)}$.

recorded on the lattice $R_{(l)}$ after the machine T_1 terminates its operation.

7.1.22. Simulate the operation of a machine T computing the function f on a lattice with spacing l:

(1) $f(x)=\left[\frac{x}{2}\right]$, $l=4$;

(2) $f(x, y)=\frac{\operatorname{sgn} x}{y}$, $l=3$;

(3) $f(x, y)=x+y$, $l=3$.

7.1.23. Prove that for any Turing machine T and any integer $l\geqslant 2$, there exists a machine T' simulating the machine T on a lattice with spacing l.

A word in the alphabet $\{0, 1\}$ having the form $1^{l(\alpha_1+1)}0^l1^{l(\alpha_2+1)}0^l \ldots 0^l1^{l(\alpha_n+1)}$ $(l\geqslant 2)$ will be called an *l-multiple code of the tuple* $\tilde{\alpha}^n=(\alpha_1, \alpha_2, \ldots, \alpha_n)$.

7.1.24. (1) Prove that a machine transforming the basic code of a tuple $\tilde{\alpha}^n$ to an l-multiple code of this tuple $(l\geqslant 2)$ can be defined by the following operator scheme:

$$\alpha T_1 \underset{1}{\underline{\quad}}\big| T_2 \big|\underset{1}{\underline{\,q'_{20}\,}}\,\omega,$$

where:

(a) T_1 has the initial state q_{11} and the final state q_{10}, and

$$q_{11}1^{\alpha_1+1}01^{\alpha_2+1}0 \ldots 01^{\alpha_n+1}$$
$$\models q_{10}1^{\alpha_1+1}0^21^{\alpha_2+1}01^{\alpha_3+1}0 \ldots 01^{\alpha_n+1}0^{l+1}1^l;$$

(b) T_2 has the initial state q_{21} and two final states q_{20} and q'_{20} and

$$q_{21}1^{x+1}0^21^{\alpha_{i+1}+1}01^{\alpha_{i+2}+1}0 \ldots 01^{\alpha_n+1}0^{l+1}1^{l(\alpha_1+1)}0^l1^{l(\alpha_2+1)}0^l$$
$$\ldots 0^l1^{l(\alpha_i+1-x)} \models q'_{20}1^x0^21^{\alpha_{i+1}+1}01^{\alpha_{i+2}+1}0$$
$$\ldots 01^{\alpha_n+1}0^{l+1}1^{l(\alpha_1+1)}0^{l+1}1^{l(\alpha_2+1)}0^l \ldots 0^l1^{l(\alpha_i+2-x)}$$

for $x>0$,

$$q_{21}10^21^{\alpha_{i+1}+1}01^{\alpha_{i+2}+1}0 \ldots 01^{\alpha_n+1}0^{l+1}1^{l(\alpha_1+1)}0^l$$
$$\ldots 0^l1^{l(\alpha_i+1)} \models q'_{20}1^{\alpha_{i+1}+1}0^21^{\alpha_{i+2}+1}01^{\alpha_{i+3}}0$$
$$\ldots 01^{\alpha_n+1}0^{l+1}1^{l(\alpha_1+1)}0^l \ldots 0^l1^{l(\alpha_i+1)}0^l1^l,$$

$$q_{21}10^{l+1}1^{l(\alpha_1+1)}0^l1^{l(\alpha_2+1)}0^l \ldots 0^l1^{l(\alpha_n+1)}$$
$$\vdash q_{20}1^{l(\alpha_1+1)}0^l1^{l(\alpha_2+1)}0^l \ldots 0^l1^{l(\alpha_n+1)}.$$

(2) Construct the programs of the machines $T_3, T_4, \ldots, T_{11}$ using the following description:

the machine T_3, which starts to operate from the last unity of the array of unities, "shifts" the array by one square to the left (without changing the "remaining content" of the tape[4]); the head stops against the first unity of the "shifted" array;

for a given $l \geqslant 1$, the head of the machine T_4, which starts to operate from an arbitrary square containing unity, moves to the right until it passes through an array of $l + 1$ zeros; the head stops in the first square behind this array and prints 1 on it;

for a given $l \geqslant 1$, the head of the machine T_5, which starts to operate from any square and moves to the right, prints l unities in succession and stops at the last unity;

the machine T_6 starts to operate at the extreme left non-blank square; for a given $l \geqslant 1$, the first left array of $l + 1$ zeros is "detected", and the head stops at the last of these zeros ("the content of the initial piece of the tape" remains unchanged);

the machine T_7 starts to operate from the extreme left non-blank square, and detects the unity to the left of the first array consisting of three zeros "bordered" by unities, the head stops at the unity detected ("the content of the initial piece of the tape" remains unchanged);

the machine T_8 prints 0 in the initial square, and the head stops after shifting to the left by a square;

the head of the machine T_9 is shifted to the right by two squares of the "initial" square, and the machine stops at the state q_{90} if the new square contains 0 and at the state q'_{90} if the "new" square contains 1 (the content of the tape remains unchanged);

the head of the machine T_{10} is shifted to the left by one square (after that the machine stops, and no changes occur in the tape);

---

[4] In other words, we assume that not a single "new" unity has appeared and the changes in the initial piece of the tape occurred *only* in the indicated array.

the head of T_{11} starts to move to the right of an "initial" square, "detects" the first (in this displacement) unity and, having made another step, stops at the square located to the right of the "detected" unity[5] (the content of the tape remains unchanged).

(3) Taking $T_3, T_4, \ldots, T_{11}$ for initial machines, construct the operator scheme for the machines T_1 and T_2 and for the machine transforming the basic code of the tuple into an l-multiple code.

7.1.25. Construct the operator scheme of a Turing machine transforming the l-multiple code of the tuple $\tilde{\alpha}^n$ into the basic code of this tuple. For initial machines corresponding to elementary operators, use the machines T_4, $T_5, \ldots, T_8$ of Problem 7.1.21. and the following three machines:

the machine T_1, such that for $l \geqslant 1$, its head, moving to the right of a blank square, detects the first (in this displacement) array containing at least l unities; then the head erases the first l unities in this array and stops at the square containing the last unity which has been erased ("the remaining" content of the tape remains unchanged);

the machine T_2 whose head is shifted from the "initial position" to the left by l squares (l is specified); the machine stops at the l-th square, and the content of the tape remains unchanged in this case;

the machine T_3 operates in the same way as machine T_2, but the head is shifted to the right.

The *lattice code of the tuple* $\tilde{\alpha}^n = (\alpha_1, \alpha_2, \ldots, \alpha_n)$ is the word in the alphabet $\{0, 1\}$ recorded on n lattices with spacing n so that the first lattice contains the word 1^{α_1+1}, the second contains the word 1^{α_2+1}, etc., and the n-th lattice contains the word 1^{α_n+1}. The beginnings of the words on the lattices must be *agreed with each other*, i.e. the extreme left unity on the first lattice immediately precedes (on the tape) the extreme left unity in the second lattice, and this unity immediately precedes the extreme left unity of the third lattice, etc.

7.1.26. (1) Construct the operator scheme of a Turing machine transforming the basic code of the tuple $\tilde{\alpha}^n$ into

---

[5] If unity is recorded in the "initial" square, the head stops at the next square to the right.

a lattice code of this tuple. For the initial machines corresponding to elementary operators, use the machines $T_4, T_5, \ldots, T_8$ in Problem 7.1.21. and the following two machines:

the machine T_1, whose head is shifted from the "initial position" to the right by n squares, stops at the n-th square; the content of the tape remains unchanged;

the machine T_2 operates in a similar way, but its head is shifted to the left.

(2) Using the same machines as in the previous problem, construct the operator scheme of a Turing machine that transforms the lattice code of a tuple $\tilde{\alpha}^n$ into the basic code of this tuple.

7.2. Classes of Computable and Recursive Functions

The functions considered in this section are partial numerical functions.

The function $F(x_1, \ldots, x_n) = f(g_1(x_1, \ldots, x_n), \ldots, g_m(x_1, \ldots, x_n))$ is called the *superposition of functions f and* $g_1, \ldots, g_m$ and is denoted by $S(f^{(m)}; g_1^{(n)}, \ldots, g_m^{(n)})$. Here the function F is defined on the tuple $\tilde{\alpha}^n$ and $F(\tilde{\alpha}^n) = f(g_1(\tilde{\alpha}^n), \ldots, g_m(\tilde{\alpha}^n))$ if and only if each function g_i $(1 \leqslant i \leqslant m)$ is defined on the tuple $\tilde{\alpha}^n$ and, besides, the function f is defined on the tuple $(g_1(\tilde{\alpha}^n), \ldots, g_m(\tilde{\alpha}^n))$.

Let $g(x_1, \ldots, x_{n-1})$ and $h(x_1, \ldots, x_{n-1}, x_n, x_{n+1})$ be two functions and $n \geqslant 2$. We shall define the third function $f(x_1, \ldots, x_{n-1}, x_n)$ with the help of the following scheme:

$$\begin{cases} f(x_1, \ldots, x_{n-1}, 0) = g(x_1, \ldots, x_{n-1}), \\ f(x_1, \ldots, x_{n-1}, y+1) \\ \quad = h(x_1, \ldots, x_{n-1}, y, f(x_1, \ldots, x_{n-1}, y)), \quad y \geqslant 0. \end{cases} \tag{1}$$

This scheme is known as a *primitive recursive scheme* for the function $f(\tilde{x}^n)$ *in variable* x_n (and x_{n+1}) and gives a *primitive recursive description of the function* $f(\tilde{x}^n)$ in terms of functions g and h. The *function* f is also said to *be obtained from g and h by using the primitive recursion operation in variable* x_n (and x_{n+1}). In this case, the following notation is used: $f = R(g, h)$ (the variables in which the recursion is carried out are indicated separately).

In the primitive recursive description of the function $f(x)$ depending on a single variable, the primitive recursion scheme has the form

$$\begin{cases} f(0) = a, \\ f(y+1) = h(y, f(y)), \quad y \geqslant 0, \end{cases} \tag{2}$$

where a is a constant (a number from the natural scale $N = \{0, 1, 2, \ldots\}$).

Let $f(x_1, \ldots, x_{n-1}, x_n)$, $n \geqslant 1$ be a certain function. We shall define the function $g(x_1, \ldots, x_{n-1}, x_n)$ as follows: let $\tilde{\alpha} = (\alpha_1, \ldots, \alpha_{n-1}, \alpha_n)$ be an arbitrary tuple of non-negative integers. We shall consider the equation

$$f(\alpha_1, \ldots, \alpha_{n-1}, y) = \alpha_n. \tag{3}$$

(a) If equation (3) has a solution $y_0 \in N$ and for all $y \in N$ such that $0 \leqslant y < y_0$ the function $f(\alpha_1, \ldots, \alpha_{n-1}, y)$ is defined, and its values differ from α_n, we assume that $g(\tilde{\alpha}) = y_0$.

(b) If Eq. (3) does not have a solution in non-negative integers, we assume that $g(\tilde{\alpha})$ is non-defined.

(c) If y_0 is the smallest non-negative integral solution of Eq. (3) and for a certain $y_1 \in N$ and $y_1 < y_0$, the value of $f(\alpha_1, \ldots, \alpha_{n-1}, y_1)$ is non-defined, we assume that $g(\tilde{\alpha})$ is non-defined.

The function $g(\tilde{x}^n)$ constructed in this way from the function $f(\tilde{x}^n)$ is said *to be obtained from the function* $f(x_1, \ldots, x_{n-1}, x_n)$ *by using the operation of minimization with respect to the variable* x_n (or just the *minimization in* x_n). In this case, the following notation is used: $g = Mf$, or $g(\tilde{x}^n) = M_{x_n}(f(\tilde{x}^n))$, or $g(\tilde{x}^n) = \mu_y(f(x_1, \ldots, x_{n-1}, y) = x_n)$, or $g(\tilde{x}^n) = \mu_{x_n}(f(\tilde{x}^n))$.

Remark. The operations of primitive recursion and minimization can be applied with respect to any variables in the functions f, g and h (however, the variables should always be indicated).

The following functions will be henceforth referred to as *simplest*:

(a) $s(x) = x + 1$ is the *successor function*,

(b) $o(x) \equiv 0$ is the *nullary function*, and

(c) $I_m^n(x_1, \ldots, x_n) = x_m (1 \leqslant m \leqslant n, n = 1, 2, \ldots)$ is the *selection function*, or the *function of selecting arguments*.

The *class* $K_{\text{pr.r}}$ *of all primitive recursive functions* is the set of all functions which can be obtained from the simplest ones by operations of superposition and primitive recursion.

The *class* $K_{\text{par.r}}$ *of all partial recursive functions* is the set of all functions that can be obtained from the simplest ones by operations of superposition, primitive recursion and minimization.

Remark. In the definition of the $K_{\text{pr.r}}$ and $K_{\text{par.r}}$ classes it is assumed that while constructing each concrete function, the corresponding operations are applied a finite number of times including zero (some or all of the operations may not be applied at all).

The *class* $K_{\text{g.r}}$ *of all general recursive functions* is a set of all partial recursive functions defined everywhere.

It can easily be shown that $K_{\text{par.r}} \supset K_{\text{g.r}}$ (the inclusion is strict!). The following strict inclusion is also valid: $K_{\text{g.r}} \supset K_{\text{pr.r}}$.

We shall denote by K_{c} the *class of all partial numerical functions that can be computed on Turing's machines*.

The following statement is valid: the classes $K_{\text{par.r}}$ and K_{c} coincide.

Theorem (R. Robinson). *All primitive recursive functions of one variable, and only these functions, can be obtained from the functions* $x + 1$ *and* $\operatorname{sgn}(x \dot{-} [\sqrt{x}]^2)$ *by applying a finite number of times the following three operations*:

(a) *absolute difference* $f(x) = |f_1(x) - f_2(x)|$;

(b) *composition* $f(x) = f_1(f_2(x))$;

(c) *iteration* $\begin{cases} f(0) = 0, \\ f(x+1) = f_1(f(x)). \end{cases}$

7.2.1. Apply the primitive recursion to the functions $g(x_1)$ and $h(x_1, x_2, x_3)$ with respect to variables x_2 (and x_3). Write the function $f(x_1, x_2) = R(g, h)$ in the "analytic" form:

(1) $g(x_1) = x_1$, $h(x_1, x_2, x_3) = x_1 + x_3$;
(2) $g(x_1) = x_1$, $h(x_1, x_2, x_3) = x_1 + x_2$;
(3) $g(x_1) = 2^{x_1}$, $h(x_1, x_2, x_3) = x_3^{x_1}$ (we put $0^0 = 1$);
(4) $g(x_1) = 1$, $h(x_1, x_2, x_3) =$
$x_3 (1 + \text{sgn} \mid x_1 + 2 - 2x_3 \mid)$;
(5) $g(x_1) = x_1$, $h(x_1, x_2, x_3) = (x_3 + 1) \text{ sgn} \left(1 + \frac{x_3}{3}\right)$.

7.2.2. Prove that the functions $f(\widetilde{x^n})$ are primitive recursive:

(1) $f(x_1, x_2) = x_1 \dot{-} x_2^2$;
(2) $f(x_1) = 3^{x_1}$;
(3) $f(x_1, x_2, x_3) = x_1^2 x_2 \oplus x_3$ (mod 2 sum).

7.2.3. Prove the validity of the relation $f = R(g, h)$ if:

(1) $f(x_1, x_2) = \text{rest}(x_1, x_2) = \begin{cases} x_1 \text{ if } x_2 = 0, \\ \text{the remainder of division} \\ \text{of } x_1 \text{ by } x_2 \text{ if } x_2 > 0; \end{cases}$

$g(x_2) = 0$, $h(x_1, x_2, x_3) = (x_3 + 1) \text{ sgn} \mid x_2 - (x_3 + 1) \mid$, the recursion is carried out in variables x_1 (and x_3).

(2) $f(x_1, x_3) = \left[\frac{x_1}{x_2}\right] = \begin{cases} x_1 \text{ if } x_2 = 0, \\ \text{the quotient of division of} \\ x_1 \text{ by } x_2 \text{ if } x_2 > 0; \end{cases}$

$g(x_2) = 0$, $h(x_1, x_2, x_3)$
$= x_3 + \overline{\text{sgn}} \mid x_1 + 1 - (x_3 + 1) x_2 \mid + \overline{\text{sgn}}\, x_2$,
the recursion is carried out in variables x_1 (and x_3).

(3) $f(x_1) = x_1 \dot{-} [\sqrt{x_1}]^2$,
$g = 0$, $h(x_1, x_2) = (x_2 + 1) \text{ sgn} (4x_1 \dot{-} (x_2^2 + 4x_2))$.

(4) $f(x_1) = [\sqrt{x_1}]$,
$g = 0$, $h(x_1, x_2) = x_2 + \overline{\text{sgn}} ((x_2 + 1)^2 \dot{-} (x_1 + 1))$.

7.2.4. Prove that if the functions $g(y)$, $\varphi_1(x)$, $\varphi_2(x)$, and $\varphi_3(x)$ are primitive recursive, the function

$$f(x, y) = \begin{cases} \varphi_1(x) & \text{if } g(y) \leqslant a, \\ \varphi_2(x) & \text{if } a < g(y) \leqslant b, \\ \varphi_3(x) & \text{if } g(y) > b, \end{cases}$$

where $0 \leqslant a \leqslant b$, is also primitive recursive.[6]

[6] The conditions imposed on the function $g(y)$ must be understood as follows: we consider all such values of y for which the function $g(y)$ satisfies the indicated relation.

7.2.5. Let $g_1(y)$, $g_2(x)$ and $g_3(x, y)$ be primitive recursive functions. Prove that in this case the function $f(x, y)$ defined by the scheme

$$\begin{cases} f(0, y) = g_1(y), \\ f(x+1, 0) = g_2(x), \\ f(x+1, y+1) = g_3(x, y) \end{cases}$$

is also primitive recursive (here $x \geqslant 0$ and $y \geqslant 0$).

7.2.6. Let the functions $g(x_1, \ldots, x_{n-1}, x_n)$, $h_1(x_1, \ldots, x_{n-1}, x_n)$ and $h_2(x_1, \ldots, x_{n-1}, x_n)$, $n \geqslant 1$, be primitive recursive. Prove that the following functions are also primitive recursive in this case:

(1) $f(x_1, \ldots, x_{n-1}, x_n) = \sum_{i=0}^{x_n} g(x_1, \ldots, x_{n-1}, i)$;

(2) $f(x_1, \ldots, x_{n-1}, x_n) = \prod_{i=0}^{x_n} g(x_1, \ldots, x_{n-1}, i)$;

(3) $$f(x_1, \ldots, x_{n-1}, y, z) = \begin{cases} \sum_{i=y}^{z} g(x_1, \ldots, x_{n-1}, i) & \text{for } y \leqslant z, \\ 0 & \text{for } y > z; \end{cases}$$

(4) $$f(x_1, \ldots, x_{n-1}, y, z) = \begin{cases} \prod_{i=y}^{z} g(x_1, \ldots, x_{n-1}, i) & \text{for } y \leqslant z, \\ 1 & \text{for } y > z; \end{cases}$$

(5) $$f(x_1, \ldots, x_{n-1}, x_n) = \sum_{i=h_1(x_1, \ldots, x_{n-1}, x_n)}^{h_2(x_1, \ldots, x_{n-1}, x_n)} g(x_1, \ldots, x_{n-1}, i)$$

(here, as usual, the sum is considered to be zero if the upper limit of summation is smaller than the lower one);

(6) $$f(x_1, \ldots, x_{n-1}, x_n) = \prod_{i=h_1(x_1, \ldots, x_{n-1}, x_n)}^{h_2(x_1, \ldots, x_{n-1}, x_n)} g(x_1, \ldots, x_{n-1}, i)$$

(in the case when the upper limit of the product is smaller than the lower limit, the product is assumed to be unity).

7.2.7. Apply the minimization operation to the function f in x_i. Represent the resultant function in an "analytic" form:

(1) $f(x_1) = 3, \quad i = 1;$

(2) $f(x_1) = \left[\frac{x_1}{2}\right], \quad i = 1;$

(3) $f(x_1, x_2) = I_1^2(x_1, x_2), \quad i = 2;$

(4) $f(x_1, x_2) = x_1 \dot{-} x_2, \quad i = 1, 2;$

(5) $f(x_1, x_2) = x_1 - \frac{1}{x_2}, \quad i = 1, 2;$

(6) $f(x_1, x_2) = 2^{x_1}(2x_2 + 1), \quad i = 1, 2.$

7.2.8. Applying the minimization operation to an appropriate primitive recursive function, prove that the function f is partially recursive:

(1) $f(x_1) = 2 - x_1;$ (3) $f(x_1, x_2) = x_1 - 2x_2;$

(2) $f(x_1) = \frac{x_1}{2};$ (4) $f(x_1, x_2) = \frac{x_1}{1 - x_1 x_2}.$

7.2.9. Is the following statement correct: if at least one of partial recursive functions g and h is not defined everywhere, then $f = R(g, h) \notin K_{\text{g.r}}$?

7.2.10. (1) Can a function that is not defined anywhere be obtained by a single application of minimization of a function which is defined everywhere?

(2) Give an example of a primitive recursive function which leads to a function that is not defined anywhere by applying the minimization operation twice.

7.2.11. Prove the computability of the following functions:

(1) $f(x, y, z) = \left[\frac{2}{x+1}\right](x - \overline{\text{sgn}}(2^x \dot{-} y)) \dot{-} (x+1)^z;$

(2) $f(x, y, z) = \left(\frac{yz}{x-1} + 2^{[x/2]}\right)(y^2 \dot{-} xz);$

(3) $f(x, y, z) = 4^{x^2 \dot{-} y^2} - (x^2 + 1)^{z \dot{-} 1};$

(4) $f(x, y, z) = \frac{x^2 - y^2}{z+1} 2^{(x^3+y)\overline{\text{sgn}}(x \dot{-} yz)}.$

7.2.12. What are the powers of the classes $K_{\text{pr.r}}$, $K_{\text{g.r}}$, $K_{\text{par.r}}$ and K_{c}?

The function $\varphi(x, y)$, defined by the scheme

$$\begin{cases} \varphi(0, y) = y + 1, \\ \varphi(x+1, 0) = \varphi(x, 1), \\ \varphi(x+1, y+1) = \varphi(x, \varphi(x+1, y)), \end{cases}$$

where $x \geqslant 0$ and $y \geqslant 0$ is usually called *Ackermann's function*.

7.2.13. Prove that the Ackermann function satisfies the following conditions:

(a) $\varphi(x, y) > y$ for any x and y,

(b) $\varphi(x, y)$ is strictly monotonic in both variables,

(c) $\varphi(x + 1, y) \geqslant \varphi(x, y + 1)$ for any x and y.

7.2.14. Using the solution of Problem 7.2.13. and Robinson's theorem (stating that the set $\{x + 1, \overline{\text{sgn}}\,(x \dot{-} [\sqrt{x}]^2)\}$ is complete relative to the composition, iteration and absolute difference operations in the class of all primitive recursive functions of one variable) prove that irrespective of the form of the primitive recursive function $f(y)$ of one variable, there exists an x for which $f(y) < \varphi(x, y)$ for any value of the variable y.

7.2.15. Prove that the Ackermann function is general recursive but is not primitive recursive.

7.2.16*. Let us denote by $K^{(1)}_{\text{pr.r}}$ and $K^{(1)}_{\text{g.r}}$ the sets of all unary primitive recursive and all general recursive functions of one variable respectively. Prove that the sets $K^{(1)}_{\text{pr.r}} \cup \{x + y\}$ and $K^{(1)}_{\text{g.r}} \cup \{x + y\}$ are complete relative to superposition in the classes $K_{\text{pr.r}}$ and $K_{\text{g.r}}$ respectively.

7.2.17. Let a Turing machine T compute the function $f_1(x) \in K_{\text{g.r}} \setminus K_{\text{pr.r}}$. Is it always true that the function $f_2(x, y)$ computable by this machine does not belong to $K_{\text{pr.r}}$?

7.2.18. (1) Let Turing machines T_1 and T_2 compute primitive recursive functions $f_1(x)$ and $f_2(x)$ respectively. Is it always true that the composition T_1T_2 also computes a primitive recursive function $f(x)$? What will happen if the machines T_1 and T_2 correctly compute the functions f_1 and f_2?

(2) Let a machine T compute a primitive recursive function $f(x)$. Is it always true that if an iteration of the machine T computes a function $g(x)$ defined everywhere, this function is necessarily primitive recursive?

7.2.19. Are the following relations valid?

(1) $\mu_x (x \dot{-} 1) = (\mu_x (x \dot{-} 2)) \dot{-} 1$.

(2) $\mu_{x_2} (x_1 + (x_2 \dot{-} x_1)) = \mu_{x_1} ((x_1 \dot{-} x_2) + x_2)$.

(3) $\mu_x \left(x \dot{-} \left[\frac{x}{2}\right]\right) \in K_{\text{pr.r}}$.

(4) $\mu_x (x \dot{-} [\sqrt{x}]^2) \in K_{\text{pr.r}}$.

(5) $\mu_x ([\sqrt[3]{x^2}]) \in K_{\text{pr.r}}$.

7.2.20*. Let the functions $f_1(x)$ and $f_2(x)$ belong to the set $K_{\text{g.r}} \setminus K_{\text{pr.r}}$. Can the following statements be correct?

(1) $f_1 (f_2 (x)) \in K_{\text{pr.r}}$, but $f_2 (f_1 (x)) \notin K_{\text{pr.r}}$.

(2) $f_1 (x^2) \in K_{\text{pr.r}}$, but $[\sqrt{f_1 (x^2)}] \notin K_{\text{pr.r}}$.

(3) $f_1 (x) + f_2 (x) \in K_{\text{pr.r}}$, but $f_1 (x) + 2f_2 (x) \notin K_{\text{pr.r}}$.

7.2.21*. Let $f(x) \in K_{\text{g.r}} \setminus K_{\text{pr.r}}$. Are the following relations always satisfied?

(1) $f(2x) \notin K_{\text{pr.r}}$;

(2) $f(x + y) \in K_{\text{pr.r}}$;

(3) $f(x \cdot y) \notin K_{\text{p.r}}$;

(4) $1 \dot{-} f(x) \notin K_{\text{pr.r}}$;

(5) $f(x \dot{-} y) \in K_{\text{pr.r}}$.

7.2.22. (1) Both variables of a function $f(x, y)$ in $K_{\text{g.r}}$ are essential. We assume that $\mu_x f(x, y)$ and $\mu_y f(x, y)$ are the functions defined everywhere. Can at least one of these functions essentially depend only on one variable?

(2) The function $f(x, y) \in K_{\text{g.r}}$ has one fictitious variable. Can both variables of the function $\mu_x f(x, y)$ be essential if we assume in addition that it is a general recursive function?

7.2.23. Formulate the necessary and sufficient condition that a function $\mu_x f(x)$ is not defined anywhere.

7.2.24. (1) What is the necessary and sufficient condition for the relation $\mu_x f(x) \in K_{\text{g.r}}$ to be fulfilled?

(2) Can at least one of the functions $\mu_x f(x, y)$ or $\mu_y f(x, y)$ be general recursive if $f(x, y) \in K_{\text{par.r}} \setminus K_{\text{g.r}}$?

7.2.25*. Let $f(x) \in K_{\text{g.r}} \setminus K_{\text{pr.r}}$. Can the relation $\mu_x f(x) \in K_{\text{pr.r}}$ be true?

7.2.26. It is known that $f(x) \in K_{\text{g.r}}$ and that for any $x \geqslant 0$, $f(2x + 1) = f(x)$ and $f(2x) = f(x + 1)$. It is true that $f(x) \in K_{\text{pr.r}}$?

7.2.27. Find out whether the set $K_{\text{g.r}} \setminus K_{\text{pr.r}}$ is complete relative to superposition in the class $K_{\text{g.r}}$?

7.2.28. Does a set M form a complete set relative to the set of operations Θ in the class $K_{par.r}$?

(1) $M = K_{par.r} \setminus K_{pr.r}$, $\Theta = \{R, \mu\}$;

(2) $M = K_{par.r} \setminus K_{g.r}$, $\Theta = \{S\}$;

(3) $M = K_{g.r} \setminus K_{pr.r}$, $\Theta = \{S, \mu\}$;

(4) $M = K_{pr.r}$, $\Theta = \{\mu\}$.

2.7.29. Let a general recursive function $f(x)$ be such that $f(N) = \{f(x): x \in N\}$ is an infinite proper subset of the set N. Will the following equality hold

$$[(K^{(1)}_{par.r} \setminus K^{(1)}_{g.r}) \cup \{f(x)\}] = K^{(1)}_{par.r},$$

where $K^{(1)}_{par.r}$ and $K^{(1)}_{g.r}$ are the sets of all unary partially recursive and all general recursive functions of one variable respectively and the closure is taken relative to superposition and minimization?

7.3. Computability and Complexity of Computations

A partial function $F(x_0, x_1, \ldots, x_n)$ is called *universal* for a family $\mathcal{G}$ of functions of n variables if the following two conditions are satisfied:

(a) for any i ($i = 0, 1, \ldots$) the function $F(i, x_1, \ldots, x_n)$ of n variables, belongs to $\mathcal{G}$;

(b) for any function $f(x_1, \ldots, x_n)$ in $\mathcal{G}$, there exists a number i such that for all values of variables $x_1, \ldots, x_n$,

$$F(i, x_1, \ldots, x_n) = f(x_1, \ldots, x_n).$$

The number i is called the *number of the function* $f(x_1, \ldots, x_n)$, and the numbering of functions of the family $\mathcal{G}$, obtained in this way, is called a *numeration corresponding to the universal function* $F(x_0, x_1, \ldots, x_n)$. Conversely, if a numeration of the family $\mathcal{G}$ is specified, i.e. if a mapping φ: $i \to f_i$ of the natural scale on $\mathcal{G}$ is defined, the function $F(x_0, x_1, \ldots, x_n)$ defined by the formula

$$F(x_0, x_1, \ldots, x_n) = f_{x_0}(x_1, \ldots, x_n)$$

is universal for $\mathcal{G}$. Each primitive (partial) recursive function $f(x_1, \ldots, x_n)$ can be associated with a term reflecting the way of representation of the function $f(x_1, \ldots,$

x_n) in terms of the functions $I_m^n(x_1, \ldots, x_n) = x_m$, $s(x) = x + 1$ and $o(x) \equiv 0$ using superposition, primitive recursion (and minimization). By numbering all the terms we can obtain the numeration for all primitive (partial) recursive functions. A numeration of this type is called a *Gödel numbering*[7]. Henceforth, we shall assume that a certain Gödel numbering is fixed. A partial recursive function of n variables, having a number x in this numbering, will be denoted by $\varphi_x^{(n)}$. The superscript will be omitted if there are no other stipulations concerning the number of variables of the function $\varphi_x^{(n)}$.

The following statements are valid.

Theorem 1 (on a function which is universal for the set of all primitive recursive functions of n variables). *The class of all primitive recursive functions of n variables has a general recursive universal function.*

This universal function (corresponding to a chosen Gödel numbering) will be denoted by $D(x_0, x_1, \ldots, x_n)$.

Theorem 2 (on a universal function). *There exists a partial recursive function $U(x_0, x_1, \ldots, x_n)$ which is universal for the set of all partial recursive functions of n variables.*

The concept of universal function is frequently used in proofs involving "diagonalization". As an example, we can consider the following proof of the existence of a general recursive function which is not primitive recursive. Let $D(x_0, x_1)$ be a universal function in the class of primitive recursive functions of one variable. If follows from part (a) of the definition of the universal function that $D(x_0, x_1)$ is defined everywhere. Let us consider $g(x) = D(x, x) + 1$. The function $g(x)$ is not primitive recursive. Indeed, if this were so, the equality $D(j, x) = g(x)$ would be satisfied for a certain j and for all x. However, for $x = j$, this equality is transformed into an inconsistent relation

$$D(j, j) = g(j) = D(j, j) + 1.$$

The following (s-m-n)-theorem, put forth by Kleene, plays an important role.

---

[7] For a formal definition of the Gödel numbering, see, for example, "Theory of Recursive Functions and Effective Computability" by Rogers H., McGraw-Hill, New York, 1967.

Theorem 3. *For any* $m, n \geqslant 1$, *there exists a primitive recursive function* $s^{(m+1)}$ *such that for all* $x, y_1, \ldots, y_m$, *the following relation holds*:

$$\varphi_x^{(m+n)}(y_1, \ldots, y_m, z_1, \ldots, z_n) = \varphi_{s(x, y_1, \ldots, y_m)}^{(n)}(z_1, \ldots, z_n).$$

Let T be a Turing machine and K be a certain configuration. The *time complexity* $t_T(K)$ *of a computation process* is defined as the number of steps taken by the machine T during a transition from K to the final configuration if T can be applied to K. The function $t_T(K)$ is not defined if T is inapplicable to K. The *computation process zone* is the minimum part of the tape containing all squares in at least one configuration encountered in the computation process. The *storage complexity* $s_T(K)$ is defined as the length of the computation process zone with the initial configuration K if T is applicable to this configuration. The function $s_T(K)$ is not defined if T is inapplicable to K. By $\omega_T(K)$ we denote the number of changes in the direction of the head displacement, and by $r_T(K)$ the maximum number of passages of the head across the boundary between two adjacent squares of the zone during computation (the maximum is taken over all pairs of adjacent squares). The functions $\omega_T(K)$ and $r_T(K)$ are not defined if T is inapplicable to K. If for K we take the initial configuration for the word P, the complexity functions are denoted by $t_T(P)$, $s_T(P)$, $\omega_T(P)$ and $r_T(P)$ respectively. If q_T is a complexity function, then $q_T(n) = \max\limits_{P:\ \lambda(P) \leqslant n} q_T(P)$.

7.3.1. Prove that a function that is universal for primitive recursive functions of one variable

(a) assumes all the values in $N = \{0, 1, 2, \ldots\}$;

(b) assumes each value in N an infinite number of times.

7.3.2. Prove that each primitive recursive function of one variable in a Gödel numbering has a countable set of numbers.

7.3.3. Prove that there is no partial recursive universal function for a family of all general recursive functions of n variables.

7.3.4. Prove that there exist partial numerical functions that are not partial recursive. Give the proof based on a comparison of powers of two sets: that of partial recursive functions and that of all partial numerical functions. Give also the other proof based on "diagonalization".

7.3.5. Give an example of a partial numerical function which assumes exactly one value and which is not partial recursive.

7.3.6. Prove that the function f is not a partial recursive function:

(1) $f(x, y) = \begin{cases} 1 & \text{if the value of } \varphi_x(y) \text{ is defined,} \\ 0 & \text{otherwise;} \end{cases}$

(2) $f(x) = \begin{cases} 1 & \text{if } \varphi_x(x) \text{ is defined,} \\ 0 & \text{otherwise;} \end{cases}$

(3) $f(x, y) = \begin{cases} \varphi_x(y) & \text{if } \varphi_x(y) \text{ is defined,} \\ 0 & \text{otherwise;} \end{cases}$

(4) $f(x, y, z) = \begin{cases} 1 & \text{if } \varphi_x(y) = z, \\ 0 & \text{otherwise;} \end{cases}$

(5) $f(x, y, z) = \begin{cases} 1 & \text{if there exists } y \text{ for which } \varphi_x(y) = z, \\ 0 & \text{otherwise.} \end{cases}$

7.3.7. Is the function f partial recursive in the following cases:

(1) $f(x) = \begin{cases} 1 & \text{if } \varphi_x(x) = 1, \\ 0 & \text{otherwise;} \end{cases}$

(2) $f(x) = \begin{cases} \text{is not defined} & \text{if } \varphi_x(x) \text{ is defined,} \\ 1 & \text{otherwise;} \end{cases}$

(3) $f(x) = \begin{cases} 1 & \text{if the unity belongs to the set of values of function } \varphi_x, \\ 0 & \text{otherwise;} \end{cases}$

(4) $f(x) = \begin{cases} 1 & \text{if } \varphi_x \text{ is primitive recursive,} \\ 0 & \text{otherwise;} \end{cases}$

(5) $f(x)$

$= \begin{cases} 1 & \text{if the decimal decomposition of the number } \pi \text{ contains an infinite number of zeros,} \\ 0 & \text{otherwise.} \end{cases}$

7.3.8. Let $u(x_0, x_1, \ldots, x_n)$ be a partial recursive function that is universal for a non-empty subset M of general recursive functions, such that $K_{g.r} \setminus M$ is infinite. Using the "diagonalization" process, indicate a countable set of general recursive functions not belonging to M.

7.3.9. Show that if a function $f(x)$ is partial recursive, any function differing from $f(x)$ on a finite set of values of the argument is partial recursive.

7.3.10. Let $U(x, y)$ be a function that is universal for the set of all partial recursive functions of a single variable. Prove that the function $f(x) = U(x, x) + 1$ does not have recursive extensions (in other words, a function defined everywhere, coinciding with $f(x)$ wherever $f(x)$ is defined and arbitrary otherwise, is not a partial recursive function).

7.3.11. Let a partial recursive function $f(x)$ be such, that the function $h(x)$, defined by the condition

$$h(x) = \begin{cases} 0 & \text{at points where } f(x) \text{ is defined,} \\ 1 & \text{at points where } f(x) \text{ is not defined,} \end{cases}$$

is recursive. Prove that the function $f(x)$ has a recursive extension.

7.3.12. Give an example of a binary sequence $\alpha_0, \alpha_1, \ldots, \alpha_n, \ldots$ generated by an appropriate autonomous deterministic operator and satisfying the following condition: the function $f(x)$, that is defined by the equality $f(n) = \alpha_n$ for all $n \geqslant 0$, is not general recursive.

7.3.13. (1) Prove that if the sequence $\alpha_0, \alpha_1, \ldots, \alpha_n, \ldots$ is the output for a certain autonomous b.d.-operator, the function $f(x)$, defined by the relation $f(n) = \alpha_n$ for all $n \geqslant 0$, is general recursive.

(2) Is such a function always primitive recursive?

7.3.14. Give an example of an infinite binary sequence $\tilde{\alpha} = \alpha_0, \alpha_1, \ldots, \alpha_n, \ldots$ satisfying the following conditions:

(1) there is no autonomous b.d.-operator for which $\tilde{\alpha}$ is an output sequence;

(2) there exists a Turing machine which starts operating on a blank tape and constructs the sequence $\tilde{\alpha}$; moreover, for each n there exists an instant of time $t_0 = t_0(n)$, such that for $t \geqslant t_0$, the head of the machine does not scan the squares of the tape to the left of the square, where the symbol α_n is recorded.

7.3.15. Give an example of transformation of finite words which may be carried out by a suitable Turing machine, but cannot be carried out by any deterministic operator.

7.3.16. For a given function f, construct a Turing machine T which can correctly compute f with an upper estimate for its time complexity, and majorize the remaining complexity functions. The input data are given in a unary form. The external alphabet is given by $A = \{0, 1\}$.

(1) $f(x, y) = x + y, \quad t_T(n) \leqslant cn;$
(2) $f(x) = 2x, \quad t_T(n) \leqslant cn^2;$
(3) $f(x) = |x - y|, \quad t_T(n) \leqslant cn^2;$
(4) $f(x, y) = \left[\frac{x}{y}\right], \quad t_T(n) \leqslant cn^2;$
(5) $f(x) =]\log_2 x[, \quad t_T(n) \leqslant cn^2.$

7.3.17. Construct a Turing machine T transforming the unary notation of a number into a binary notation with given restrictions on the complexity function: $t_T(n) \leqslant c_1 n^2$, $s_T(n) \sim n$, $\omega_T(n) \leqslant c_2 n$, $r_T(n) \leqslant c_3 n$, where c_1, c_2 and c_3 are constants.

7.3.18. Construct a Turing machine T which transforms a binary notation into a unary notation, and such that $t_T(n) \leqslant c_1 n 2^n$, $s_T(n) \sim c_2 n + 2^n$, $\omega_T(n) \leqslant c_3 n 2^n$, $r_T(n) \leqslant c_4 n 2^n$, where c_1, c_2, c_3 and c_4 are constants.

7.3.19. Construct a Turing machine T with an input alphabet A of m characters, transforming a word P into $P*P$, where the symbol $*$ does not appear in P, and such that $t_T(n) \leqslant c_1 n^2$, $s_T(n) \leqslant c_2 n$, $\omega_T(n) \leqslant c_3 n$, $r_T(n) \leqslant c_4 n$.

7.3.20. (1) Construct a machine T, recognizing the linearity[8] of an arbitrary Boolean function $f(\tilde{x}^n)$. The function $f(\tilde{x}^n)$ is defined by a binary vector $\tilde{\alpha}_f$ of length $N = 2^n$. The input alphabet of the machine is $A = \{0, 1, \Lambda\}$. The functions $t_T(N)$ and $s_T(N)$ must satisfy the inequalities $t_T(N) \leqslant c_1 N^2$, $s_T(N) \leqslant c_2 N$.

(2) Construct a machine T, recognizing the self-duality of an arbitrary Boolean function $f(\tilde{x}^n)$, and such that $t_T(N) \leqslant c_1 N^2$, $s_T(N) \leqslant c_2 N$.

[8] A machine recognizing a property of an input word has two final states, q_0' and q_0''. The machine stops at the state q_0' if the property is observed, and at the state q_0'' if it is not.

(3) Construct a machine T, recognizing the monotonicity of an arbitrary Boolean function $f(\tilde{x}^n)$, and such that $t_T(N) \leqslant c_1 N^2$, $s_T(N) \leqslant c_2 N$.

7.3.21. Show that for any Turing machine T and for any word P

(1) $s_T(P) \leqslant t_T(P) + |P|$;

(2) there exists a constant c, such that $t_T(P) \leqslant c^{s_T(P)}$.

Chapter Eight

Elements of Combinatorial Analysis

8.1. Permutations and Combinations. Properties of Binomial Coefficients

The tuple of elements $a_{i_1}, \ldots, a_{i_r}$ in the set $U = \{a_1, \ldots, a_n\}$ is called an *r-multisubset of n-set* U, or an (n, r)*-multiset*. If the sequential order of elements in a multiset is specified, we call it an *ordered multiset*. Two ordered multisets, differing only in the sequential order of their elements, are assumed to be *different*. If the sequential order of elements is not significant, the multiset is called a *disordered multiset*. Elements may, or may not, repeat themselves in a multiset. An ordered (n, r)-multiset in which the elements may be repeated is called a *permutation of n elements with repetitions taken r at a time*, or an (n, r)*-permutation with repetitions*. If an ordered (n, r)-multiset has pairwise different elements, we call it an (n, r)*-permutation without repetitions*, or simply an (n, r)*-permutation*. The number of (n, r)-permutations will be denoted by the symbol $P(n, r)$, while the number of (n, r)-permutations with repetitions will be denoted by $\hat{P}(n, r)$. A disordered (n, r)-multiset, in which the elements can be repeated, is called a *combination of n elements taken r at a time*, or simply an (n, r)*-combination with repetitions*. If the elements of a disordered multiset are pairwise different, it is called a *combination (without repetition) of n elements taken r at a time*, or an (n, r)*-combination*. Each such combination is a subset of power r in the set U. The number of combinations of n elements taken r at a time will be denoted by $C(n, r)$, while the number of combinations (with repetitions) of n elements taken r at a time will be denoted by $\hat{C}(n, r)$.

Example 1. Let $U = \{a, b, c\}$, $r = 2$. In this case, we obtain:

nine permutations with repetitions, viz. aa, ab, ac, ba, bb, bc, ca, cb, cc, six permutations without repetitions, viz., *ab*, *ac*, *ba*, *bc*, *ca*, *cb*,

six combinations with repetitions, viz., *aa*, *ab*, *ac*, *bb*, *bc*, *cc*, and

three combinations without repetitions, viz., *ab*, *ac*, *bc*.

The product $n(n-1)\ldots(n-r+1)$, where n is real and r is a positive integer, will be denoted by $(n)_r$. By definition, we put $(n)_0 = 1$. If n is a natural number, $(n)_n$ is denoted by the symbol $n!$ and is called *n-factorial.* For $n = 0$, we assume that $0! = 1$. For any real n and a non-negative integer r, the quantity $(n)_r/r!$ is called a *binomial coefficient*[1] and is denoted by the symbol $\binom{n}{r}$. Let $r_1, r_2, \ldots, r_k$ be non-negative integers and let $r_1 + r_2 + \ldots + r_k = n$. The quantity $\frac{n!}{r_1!\,r_2!\ldots r_k!}$ is called a *polynomial coefficient*, and is denoted by $\binom{n}{r_1, r_2, \ldots, r_k}$.

The following two rules are taken into consideration while counting the number of different combinations.

Multiplication rule. If an object A can be chosen in m different ways, and if after each such choice an object B can, in turn, be chosen in n different ways, the choice "A and B" in this order can be carried out in $m \times n$ ways.

Summation rule. If an object A can be chosen in m different ways and an object B in n other ways, and if the choice "A and B" is not possible, the choice "A or B" can be made in $m + n$ ways.

Example 2. Two dice (with six faces each) are cast. In how many different ways can they be cast so that each of them shows either an even number, or each shows an odd number.

Solution. Let A be the number of ways in which an even number is shown by each die, and let B be the number of ways for an odd cast. Then, according to the sum rule, the required number is $A + B$. Let C be the number of ways of an even cast by the first die and D the number of ways for an even cast by the second die. Obviously,

[1] The notations C_n^r, nC_r, and (n, r) are also encountered in the literature.

$C = D = 3$. According to the multiplication rule, $A = C \times D = 9$. Similarly, $B = 9$. Hence the total possible number of ways is equal to 18.

Example 3. Show that the number of (n, r)-permutations without repetitions is equal to n^r.

Solution. Induction on r. For $r = 1$, the number of ways in which one object can be chosen from n is equal to n^1. Suppose that the equality $P(n, r) = (n)_r$ is satisfied for a certain $r \geqslant 1$. We shall prove that an analogous equality also holds for $r + 1$. Each set of $r + 1$ objects can be obtained by first choosing r objects comprising the (n, r)-permutation, followed by the addition of the $(r + 1)$-th object to it. If r objects are chosen, there are $n - r$ ways in which the $(r + 1)$-th object can be chosen. According to the multiplication rule, we obtain

$$P(n, r + 1) = P(n, r) \times (n - r).$$

Using the induction hypothesis, we can write

$$P(n, r + 1) = (n)_r \times (n - r) = (n)_{r+1}.$$

A large number of combinatorial problems can be reduced to a computation of the number of binary vectors.

Example 4. In how many ways can the number n be presented as a sum of k non-negative components (representations differing only in the order of the components are assumed to be different)?

Solution. Each decomposition of the number n into k non-negative integral components is assigned a vector of length $n + k - 1$ with n unities and $k - 1$ zeros, such that the number of unities just before the first zero is equal to the first component, the number of unities between the first and second zeros is equal to the second component, and so on. This is a one-to-one correspondence. It should be noted that each binary vector with $n + k - 1$ coordinates and n unities can, in turn, be associated with an n-element subset A of the set $U = \{a_1, a_2, \ldots, a_{n+k-1}\}$ as follows: the i-th coordinate of the vector is equal to 1 if and only if $a_i \in A$, $i = 1, 2, \ldots, n + k - 1$. But the number of such subsets is $C(n, r)$.

8.1.1. In how many ways can three tickets be distributed among 20 students if:

(1) The tickets are for different theatres, and each student can get not more than one ticket?

(2) The tickets are for different theatres and for different shows, and any student can obtain any number of tickets (not exceeding three)?

(3) The tickets are for the same show, and each student can get not more than one ticket?

8.1.2. In how many ways can nine persons be lined up

(1) in single file?

(2) in threes, if they are formed in each rank according to height, and no two persons have the same height?

8.1.3. Prove that

(1) $\hat{P}(n, r) = n^r$,

(2) $C(n, r) = \binom{n}{r}$,

(3) $\hat{C}(n, r) = \binom{n+r-1}{r-1}$.

8.1.4. Find the number of vectors $\tilde{\alpha} = (\alpha_1, \ldots, \alpha_n)$ whose coordinates satisfy the following conditions:

(1) $\alpha_i \in \{0, 1, \ldots, k-1\}$, $i = 1, \ldots, n$;

(2) $\alpha_i \in \{0, 1, \ldots, k_{i-1}\}$, $i = 1, \ldots, n$;

(3) $\alpha_i \in \{0, 1\}, i = 1, \ldots, n$, and $\alpha_1 + \ldots + \alpha_n = r$.

8.1.5. (1) Find the number of matrices having n rows and m columns with their elements in the set $\{0, 1\}$.

(2) Find the same, provided that the rows of the matrix are pairwise different.

8.1.6. Given n objects of one type and m objects of another. Find the number of multisets containing r objects of one type and s objects of the other type.

8.1.7. Words are formed by using n letters, of which a and b appear α and β times respectively, while the remaining letters are pairwise different. How many different r-lettered words can be formed, so that the letters a and b appear h and k times respectively?

8.1.8. A deck containing $4n$ ($n \geqslant 5$) cards has n cards each of four different suits, numbered $1, 2, \ldots, n$. In how many different ways can five cards be chosen, so that they contain

(1) five consecutive cards of the same suit?

(2) four of the five cards with the same value?

(3) three cards with one value and two cards with some other value?

(4) five cards of the same suit?

(5) five successively numbered cards?

(6) three of the five cards with the same value?

(7) not more than two cards of the same suit?

8.1.9. Solve the following problems by using the summation and multiplication rules:

(1) In how many ways can two chips be chosen from among 28 in a game of domino, such that they can be put next to each other (i.e., the same number is encountered on both chips)?

(2) Three dice are cast. In how many ways can they show the same number or pairwise different numbers?

(3) It is customary to give more than one christian name to the newly born baby in England. In how many different ways can a baby be named if he is to be given three names from a total of 300?

8.1.10. (1) In how many ways can the number n be presented as a sum of k natural components (representations differing only in the order of components are assumed to be different)?

(2) In how many ways can the number 7^n be presented in the form of a product of three cofactors (representations differing only in the order of cofactors are assumed to be different)?

(3) The same, provided that representations differing only in the order of cofactors are assumed to be identical, and $n \neq 3s$?

8.1.11. In how many ways can n zeros and k unities be arranged so that each two unities are separated by not less than m zeros?

(2) Find the number of non-negative integers not exceeding 10^n whose digits are arranged in non-diminishing order.

(3) A rectangular city is divided by streets into squares. There are n such squares from north to south, and k squares from east to west. Find the number of shortest walks from the north-eastern end of the city to the south-western end.

8.1.12. Let n be the product $p_1^{\alpha_1} \times \ldots \times p_r^{\alpha_r}$ of powers of pairwise different prime numbers. Find

(1) the number of all natural divisors of the number n_j;

(2) the number of all divisors that cannot be divided by a square of any integer other than 1; and

(3) the sum of divisors of the number n.

8.1.13. Prove the following properties of binomial coefficients:

(1) $\binom{n}{k}=\binom{n}{n-k}$;

(2) $\binom{n}{k}\binom{k}{r}=\binom{n-r}{k-r}\binom{n}{r}$;

(3) $\binom{n}{k}=\binom{n-1}{k}+\binom{n-1}{k-1}$;

(4) $\binom{n}{k-r}\Big/\binom{n}{k}=\dfrac{(k)_r}{(n-k+r)_r}$;

(5) $\binom{n}{k}=\sum\limits_{r=0}^{n}\binom{n-r-1}{k-r}$;

(6) $\binom{n-r}{k-r}\Big/\binom{n}{k}=\dfrac{(k)_r}{(n)_r}$;

(7) $\binom{n+1}{k}\Big/\binom{n}{k}=\dfrac{n+1}{n-k+1}$;

(8) $\sum\limits_{r=k}^{n}\binom{r}{k}=\binom{n+1}{k+1}$.

8.1.14. Prove that

(1) $\binom{n}{k}$ increases with n for a fixed k;

(2) $\binom{n-r}{k-r}$ decreases with r for fixed n and k;

(3) if n is fixed, then $\binom{n}{k}$ increases with k for $k <]n/2[-1$ and decreases for $k \geqslant]n/2[$;

(4) $\max\limits_{0\leqslant k\leqslant n}\binom{n}{k}=\binom{n}{[n/2]}$;

(5) the minimum value of the sum $\binom{n_1}{k}+\binom{n_2}{k}+\ldots+\binom{n_s}{k}$ provided that $\sum\limits_{i=1}^{s} n_i=n$ is equal to $(s-r)\times\binom{q}{r}+r\binom{q+1}{k}$, where $q=[n/s]$, $r=n-s[n/s]$;

(6) the maximum value of the sum $\binom{n}{k_1}+\binom{n}{k_2}+\ldots+\binom{n}{k_s}$ provided that $0\leqslant k_1<\ldots<k_s\leqslant n$, $1\leqslant s\leqslant n+1$ is equal to $\sum\limits_{\frac{n-s}{2}\leqslant j<\frac{n+s}{2}}\binom{n}{j}$;

(7) for a prime p and for any $p > k \geqslant 1$, the number $\binom{p}{k}$ is a multiple of p;

(8) $\prod\limits_{n<p_i\leqslant 2n} p_i \leqslant \binom{2n}{n}$, where the product is taken over all prime numbers p_i $(n < p_i \leqslant 2n)$.

8.1.15. (1) Let m be a non-negative integer, and let $n = n\,(m)$ be the minimum integer, such that $m < n!$. Prove that there is one and only one way in which the number m can be associated with a vector $\tilde{\alpha}\,(m) = (\alpha_1, \alpha_2, \ldots, \alpha_{n-1})$, such that $m = \alpha_1 \times 1! + \alpha_2 \times 2! + \ldots + \alpha_{n-1} \times (n-1)!$, $0 \leqslant \alpha_i < i$, $i = 1, \ldots, n-1$. Let $\mu\,(\tilde{\alpha})$ be a number, such that $\tilde{\alpha} = \tilde{\alpha}\,(\mu\,(\tilde{\alpha}))$.

(2) Find the vector $\tilde{\alpha}\,(m)$ for $m = 4, 15, 37$.

(3) Find $\mu\,(\tilde{\alpha})$ from the following values of $\tilde{\alpha}$:

(a) $\tilde{\alpha} = (0, 2, 0, 4)$; (b) $\tilde{\alpha} = (0, 2, 1)$; (c) $\tilde{\alpha} = (1, 2, 3, 2)$.

(4) The permutation of a set $Z_n = \{1, 2, \ldots, n\}$ is defined as an arbitrary mapping of Z_n onto itself. Any permutation π can be put in a one-to-one correspondence with the vector $\vec{\pi} = (\pi\,(1), \ldots, \pi\,(n))$, in which the $\pi\,(i)$ coordinate indicates the position of the element i. Each permutation π is put in correspondence with a number $\nu\,(\pi)$, $0 \leqslant \nu\,(\pi) < n!$, called the number of the permutation. For this purpose, we first construct a vector $\tilde{\alpha}_\pi = (\alpha_1, \alpha_2, \ldots, \alpha_{n-1})$. We put $\alpha_{n-1} = \pi\,(1) - 1$. If $\alpha_{n-1}, \ldots, \alpha_{n-j}$ are already defined and $s\,(j) = |\{i < j \mid \pi\,(i) < \pi\,(j)\}|$, we put $\alpha_{n-j-1} = \pi\,(j+1) - s\,(j+1) - 1$. The number $\nu\,(\pi)$ is defined as $\mu\,(\tilde{\alpha}_\pi)$. For example, if $\vec{\pi} = (3, 4, 2, 1)$, then $\tilde{\alpha}_\pi = (0, 2, 3)$, $\nu\,(\pi) = 0 \times 1! + 2 \times 2! + 3 \times 3! = 22$. From the permutation π defined by the vector $\vec{\pi}$, find the number $\nu\,(\pi)$ in the following cases:

(a) $\vec{\pi} = (2, 3, 1, 4)$, (b) $\vec{\pi} = (3, 5, 2, 1, 4)$, (c) $\vec{\pi} = (1, 4, 3, 5, 2)$.

(5) Find the algorithm for constructing the permutation π from its number $\nu\,(\pi)$.

(6) From the number m, find the permutation π on the set Z_n, $n! > m \geqslant (n-1)!$, such that $\nu(\pi) = m$:

(a) $m = 7$; (b) $m = 18$; and (c) $m = 28$.

8.1.16. (1) Let k and n be natural numbers. Prove that any integer m $\left[0 \leqslant m < \binom{n}{k}\right]$ can be uniquely associated with an integral vector $\tilde{\beta}(m) = (\beta_1, \beta_2, \ldots, \beta_k)$, satisfying the condition $n > \beta_1 > \beta_2 > \ldots > \beta_k \geqslant 0$, $m = \binom{\beta_1}{k} + \binom{\beta_2}{k-1} + \ldots + \binom{\beta_k}{1}$. In this case, the number m is called the number of the tuple $\tilde{\beta}$ (denoted by $m = \mu(\beta)$).

(2) For given m, n, and k, construct the vector $\tilde{\beta}(m)$:

(a) $m = 19$, $n = 7$, $k = 4$;

(b) $m = 25$, $n = 7$, $k = 3$;

(c) $m = 32$, $n = 8$, $k = 4$.

(3) From a given vector $\tilde{\beta} = (\beta_1, \ldots, \beta_k)$, find a number m satisfying the conditions of Problem 8.1.16 (1):

(a) $\tilde{\beta} = (6, 3, 0)$; (b) $\tilde{\beta} = (5, 4, 3, 1)$; (c) $\tilde{\beta} = (6, 4, 3, 2, 1)$.

(4) Let B_k^n be the set of all vectors of length n having k unities and $n-k$ zeros. Using Problem 8.1.16 (1), enumerate all tuples in B_k^n from 1 to $\binom{n}{k}$, i.e. construct a one-to-one mapping ν of the set B_k^n onto the set $\left\{1, 2, \ldots, \binom{n}{k}\right\}$.

8.1.17. Using the relation

$$\binom{n+1}{k} = \binom{n}{k} + \binom{n}{k-1}$$

prove the following identity by induction on n:

$$(1+t)^n = \sum_{k=0}^{n} \binom{n}{k} t^n. \tag{1}$$

8.1.18. Let n and m be positive integers. Using the identity (1) or some other method, prove the following equalities:

(1) $\sum_{k=0}^{n} \binom{n}{k} = 2^n;$

(2) $\sum_{k=0}^{n} (-1)^k \binom{n}{k} = 0;$

(3) $\sum_{k=1}^{n} k \binom{n}{k} = n2^{n-1};$

(4) $\sum_{k=2}^{n} k(k-1) \binom{n}{k} = n(n-1)\, 2^{n-2};$

(5) $\sum_{k=0}^{n} (2k+1) \binom{n}{k} = (n+1)\, 2^n;$

(6) $\sum_{k=0}^{n} \frac{1}{k+1} \binom{n}{k} = \frac{1}{n+1} (2^{n+1} - 1);$

(7) $\sum_{k=0}^{n} (-1)^k \frac{1}{k+1} \binom{n}{k} = \frac{1}{n+1};$

(8) $\sum_{k=1}^{n} \frac{(1)^{k-1}}{k} \binom{n}{k} = 1 + \frac{1}{2} + \ldots + \frac{1}{n};$

(9) $\sum_{r=0}^{k} \binom{m}{r} \binom{n}{k-r} = \binom{n+m}{k};$

(10) $\sum_{k=0}^{n} \binom{n}{k}^2 = \binom{2n}{n};$

(11) $\sum_{k=0}^{n} \frac{(2n)!}{(k!)^2\, ((n-k)!)^2} = \binom{2n}{n}^2;$

(12) $\sum_{k=0}^{n} \sum_{r=0}^{n-k} \binom{n}{k} \binom{n-k}{r} = 3^n;$

(13) $\sum_{r=k}^{n} (-1)^{k-r} \binom{n}{r} = \sum_{r=0}^{n-k} (-1)^{n-1-r} \binom{n}{r};$

(14) $\sum_{r=0}^{n-k}\binom{n}{k+r}\binom{m}{r}=\binom{m+n}{n-k}$;

(15) $\sum_{k=n}^{m}(-1)^{k-n}\binom{k}{n}\binom{m}{k}=\begin{cases}0 & \text{for} \quad m\neq n,\\ 1 & \text{for} \quad m=n.\end{cases}$

8.1.19. Prove the following identities:

(1) $\sum_k\binom{n}{2k}=\sum_k\binom{n}{2k+1}=2^{n-1}$;

(2) $4\sum_k\binom{n}{4k}=2^n+2^{\frac{n}{2}+1}\cos\frac{\pi n}{4}$;

(3) if $0\leqslant r<m$, then

$$m\sum_k\binom{n}{mk+r}=\sum_{\nu=0}^{m-1}e^{-\frac{2\pi ir\nu}{m}}\left(1+e^{\frac{2\pi i\nu}{m}}\right)^n, \text{ where } i^2=-1;$$

(4) $\sum_k\binom{n}{4k+r}=\frac{1}{4}\left(2^n+2^{\frac{n}{2}+1}\cos\frac{\pi}{4}(n-2r)\right)$,

$$0\leqslant r\leqslant 3;$$

(5) if $0\leqslant r<m$, $m\geqslant 1$, then

$$\frac{2^n}{m}\left(1-(m-1)\cos^n\frac{\pi}{m}\right)\leqslant\sum_k\binom{n}{mk+r}\leqslant\frac{2^n}{m}\left(1+(m-1)\cos^n\frac{\pi}{m}\right).$$

8.1.20. Prove that

(1) $m\sum_k\alpha^{mk+r}\binom{n}{mk+r}=\sum_{\nu=0}^{m-1}e^{-\frac{2\pi r\nu i}{m}}\left(1+\alpha e^{\frac{2\pi i\nu}{m}}\right)^n$. (*)

Using the identity (*), calculate

(2) $\sum_k 3^k\binom{n}{2k}$;

(3) $\sum_k(-1)^k 2^k\binom{n}{4k+1}$;

(4) $\sum_k(-1)^k 3^k\binom{n}{2k+1}\binom{2k+1}{r}$.

8.1.21. Find the number of rational terms in the expansion

(1) $(\sqrt{2}+\sqrt[6]{3})^{20}$; (2) $(\sqrt{3}+\sqrt[4]{5})^{50}$;

(3) $(\sqrt[3]{6}+\sqrt[4]{2})^{100}$; (4) $(\sqrt[3]{12}+\sqrt[6]{3})^{30}$.

8.1.22. Find the coefficient of t^k in the expansions

(1) $(1+2t-3t^2)^8$, $k=9$;
(2) $(1-t+2t^2)^{10}$, $k=7$;
(3) $(2+t-2t^3)^{20}$, $k=10$;
(4) $(2+t^4+t^7)^{15}$, $k=17$.

8.1.23. Show that the following identities are valid for integral $m \geqslant 0$ and $n \geqslant 0$:

(1) $$\sum_{k=0}^{n} \frac{(n)_k}{(m)_k} = \frac{m+1}{m-n+1}, \quad m \geqslant n;$$

(2) $$\sum_{k=0}^{n} \binom{m+k-1}{k} = \sum_{k=0}^{m} \binom{n+k-1}{k}.$$

8.1.24. Let a and b be real, and let k, m, n and r be non-negative integers. Prove that

(1) $$\binom{a}{k}+\binom{a}{k-1}=\binom{a+1}{k};$$

(2) $$(1+t)^a=\sum_{k=0}^{\infty}\binom{a}{k}t^k, \quad |t|<1;$$

(3) $$\binom{-a}{k}=(-1)^k\binom{a+k-1}{k}, \quad a>0;$$

(4) $$\sum_{k=0}^{n}\binom{a-k}{r}=\binom{a+1}{r+1}-\binom{a-n}{r+1};$$

(5) $$\sum_{k=0}^{n}\binom{a}{k}\binom{b}{n-k}=\binom{a+b}{n}$$ (addition theorem);

(6) $$\sum_{k=0}^{n}(-1)^{n-k}\binom{a}{k}=\binom{a-1}{n};$$

(7) $$\sum_{k=0}^{n}\binom{a+n-k-1}{n-k}\binom{b+k-1}{k}=\binom{a+b+n-1}{n};$$

(8) $\sum\limits_{0\leqslant k,\, r\leqslant n} \binom{a}{k}\binom{b}{r}\binom{c}{n-k-r}=\binom{a+b+c}{n}$;

(9) $\sum\limits_{k=0}^{\infty} (-1)^k \binom{2k}{k}\left(\frac{1}{8}\right)^k=\sqrt{\frac{2}{3}}$;

(10) $\sum\limits_{k=0}^{\infty} \binom{1/2}{2k+1}\left(\frac{1}{2}\right)^k=\sqrt{5}-\sqrt{3}$;

(11) $m\sum\limits_{k}\binom{a}{mk+r} b^{mk+r}$

$$=\sum_{\nu=0}^{m-1} e^{-\frac{2\pi i r\nu}{m}}\left(1+be^{\frac{2\pi i\nu}{m}}\right)^a, \quad |b|<1.$$

8.1.25. (1) Find the number of all words of length mn in an n-lettered alphabet, in which each letter is encountered m times.

(2) In how many ways can a set of n elements be decomposed into s subsets the first of which contains k_1 elements, the second k_2 elements, and so on?

(3) With the help of combinatorial analysis, prove that the following equality holds for any non-negative integers $k_1, k_2, \ldots, k_s, n$, such that $k_1 + k_2 + \ldots + k_s = n$:

$$\binom{n}{k_1}\binom{n-k_1}{k_2}\cdots\binom{n-k_1-k_2-\ldots-k_{s-1}}{k_s}=\frac{n!}{k_1!\,k_2!\,\ldots\,k_s!}.$$

(4) Prove the following identity by induction on s:

$$(t_1+t_2+\ldots+t_s)^n=\sum_{\substack{k_1,\,\ldots,\,k_s\\ k_1+\ldots+k_s=n}}\frac{n!}{k_1!\,k_2!\,\ldots\,k_s!}\,t_1^{k_1}t_2^{k_2}\,\ldots\,t_s^{k_s}.$$

8.2 Inclusion and Exclusion Formulas

Let us consider N objects and n properties $A_1, \ldots, A_n$. Each object may, or may not, possess any of these properties. We denote by $N_{i_1,\ldots,i_k}$ the number of objects having the properties $A_{i_1}, \ldots, A_{i_k}$ (and perhaps some other properties as well). In this case, the number $\hat{N}_0$

of objects not possessing any of the properties $A_1, \ldots, A_n$ is defined by the equality

$$\hat{N}_0 = S_0 - S_1 + S_2 - \ldots + (-1)^n S_n, \qquad (2)$$

where $S_0 = N$, and

$$S_k = \sum_{1 \leqslant i_1 < \ldots < i_k \leqslant n} N_{i_1, \ldots, i_k}, \quad k = 1, \ldots, n.$$

Formula (2) is called the inclusion and exclusion formula.

Example 1. Suppose that a deck contains n cards, numbered from 1 to n. In how many ways can the cards be arranged in the deck so that card number i does not occupy the i-th position for any value of i $(1 \leqslant i \leqslant n)$?

Solution. We have n properties α_i of the form: "the i-th card occupies the i-th position in the deck". The number of possible arrangements of the cards in the deck is equal to $n!$. The number of arrangements $N_{i_1, \ldots, i_k}$ for which a card with number i_ν occupies the position $i\nu$ $(\nu = \overline{1, k})$, equal to $(n - k)!$. This gives

$$S_0 = n!, \quad S_k = \sum_{1 \leqslant i_1 < \ldots < i_k \leqslant n} N_{i_1, \ldots, i_k}$$

$$= \binom{n}{k} (n-k)! = \frac{n!}{k!}.$$

Using formula (2) we find that the number $\hat{N}_0$ of the arrangements for which none of the properties α_i is satisfied, is equal to

$$\sum_{k=0}^{n} (-1)^k S_k = n! \sum_{k=0}^{n} (-1)^k \frac{1}{k!}.$$

If the number of properties is not very large, such problems can be conveniently solved with the help of Venn's circles.

Example 2. In a group of 25 students, 20 have successfully passed their examinations. Out of the 12 students engaged in sports, 10 have passed their exams. How many of the unsuccessful students do not take part in sports activities?

Solution. We represent the group of students who have successfully passed their examinations by a circle marked A (see Fig. 40), and the group of students participating in sports activities by a circle marked B. The intersection of the circles corresponds to the group of successful students who participate in sports, while the union of the circles corresponds to the group of students who have either passed their examinations or take part in sports

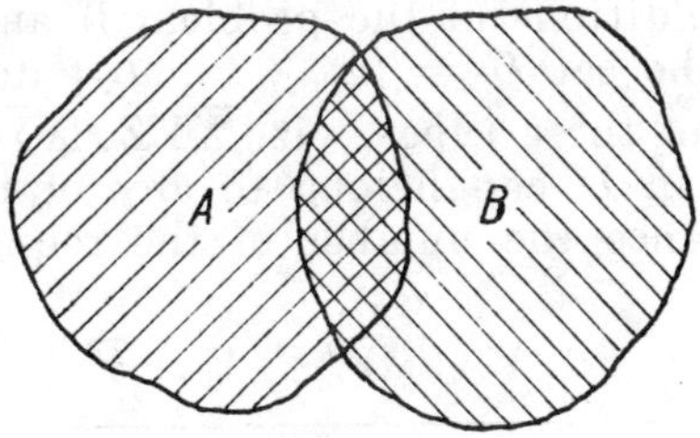

Fig. 40

activities. The number of such students is equal to $20 + 12 - 10 = 22$. The number of unsuccessful students who do not take part in sports, is equal to $25-22=3$.

Venn's diagrams are used for solving problems involving the counting of number of elements in the set $U = \{a_1, \ldots, a_N\}$ having given properties. Venn's diagram for n properties has the form of a rectangle divided into 2^n squares. Each square corresponds to one type of elements. The type of an element is determined by whether or not the i-th element possesses the i-th property for each i, $1 \leqslant i \leqslant n$. Hence it is convenient to encode the type of an element and the cell corresponding to it by a binary vector $(\alpha_1, \ldots, \alpha_n)$, in which $\alpha_i = 1$ if the i-th property takes place for a given type of elements, and $\alpha_i = 0$ otherwise $(i = 1, \ldots, n)$.

Example 3. Let X, Y and Z be the subsets of the set $U = (a_1, \ldots, a_N)$, satisfying the conditions $X \subseteq (Y \cap Z) \cup U \setminus Y$, $Z \subseteq (X \cap Y) \cup \overline{X}$, $Y \subseteq (X \cap Z) \cup \overline{Z}$, $\overline{A} = U \setminus A$, $A \in \{X, Y, Z\}$. Find the number of such triads.

Solution. For each $a \in U$, three properties are defined: $a \in X$, $a \in Y$, $a \in Z$. Each element belongs to one of the eight types depending on its affiliation to the sets X, Y, and Z. Apparently, the inclusion $A \subseteq B$ is equivalent to

$A \cap \bar{B} = \varnothing$. Hence the condition $X \subseteq (Y \cap Z) \cup \bar{Y}$ is equivalent to $X \cap \overline{(Y \cap Z) \cap \bar{Y}} = X \cap ((\bar{Y} \cup \bar{Z}) \cap Y) = X \cap (\bar{Z} \cap Y) = \varnothing$. The remaining two conditions are equivalent to the equalities $Z \cap (\bar{Y} \cap X) = \varnothing$ and $Y \cap (\bar{X} \cap Z) = \varnothing$. These conditions are presented in the diagram below as empty squares. Hence X, Y, and Z satisfy the conditions of the problem if and only if the elements of the set $U = \{a_1, \ldots, a_N\}$ do not contain the elements of three types, viz. $\bar{X}YZ$, $X\bar{Y}Z$, and $XY\bar{Z}$. Any element in U can belong to any of the remaining five types. Hence the number of the required triads is equal to 5^n.

	Y	$\bar{Y}$	Y	$\bar{Y}$
X		$\varnothing$	$\varnothing$	
$\bar{X}$	$\varnothing$			
	Z	Z	$\bar{Z}$	$\bar{Z}$

8.2.1*. (1) Prove formula (2) by induction.

(2) Let $\hat{N}_m$ be the number of objects possessing exactly m properties from among n. Prove that

$$\hat{N}_m = \sum_{k=0}^{n-m} (-1)^k \binom{m+k}{m} S_{m+k}. \tag{3}$$

(3) Let $\check{N}_m$ be the number of objects possessing not less than m properties from among n. Prove that

$$\check{N}_m = \sum_{k=0}^{n-m} (-1)^k \binom{m-1+k}{m-1} S_{m+k}. \tag{4}$$

(4) Prove that

$$S_k = \sum_{m=k}^{n} \binom{m}{k} \hat{N}_m, \tag{5}$$

$$S_k = \sum_{m=k}^{n} \binom{m-1}{k-1} \check{N}_m. \tag{6}$$

(5) Prove that

$$S_m-(m+1)\,S_{m+1}\leqslant \hat{N}_m\leqslant S_m, \qquad (7)$$

$$S_m-mS_{m+1}\leqslant \check{N}_m\leqslant S_m. \qquad (8)$$

8.2.2. Four persons deposit their hats in the cloakroom. Assuming that the hats are returned at random, find the probability that exactly k persons will get their hats back. Consider all the values of k $(0\leqslant k\leqslant 4)$.

8.2.3. Let $E\,(r,\,n,\,m)$ be the number of ways in which r different objects can be arranged in n boxes of which m are empty. Let $F\,(r,\,n,\,m)$ be the number of arrangements for which at least m boxes are empty. Show that

(1) $E\,(r,\,n,\,0)=\sum\limits_{k=0}^{n}(-1)^k\binom{n}{k}(n-k)^r$;

(2) $E\,(r,\,n,\,m)=\binom{n}{m}\sum\limits_{k=0}^{n-m}(-1)^k\binom{n-m}{k}(n-m-k)^r$;

(3) $F\,(r,\,n,\,m)$

$$=\binom{n}{m}\sum_{k=0}^{n-m}(-1)^k\binom{n-m}{k}(n-m-k)^r\,\frac{m}{m+k}.$$

8.2.4. A survey into the reading habits of students revealed that 60% read magazine A, 50% read magazine B, 50% read magazine C, 30% read magazines A and B, 20% read magazines B and C, 40% read magazines A and C, while 10% read all three magazines. What percentage of students

(1) do not read any magazine,
(2) read exactly two magazines,
(3) read at least two magazines?

8.2.5. A university department has thirteen staff members, each of whom knows at least one foreign language. Ten know English, seven German, and six French. Five know both English and German, four know English and French, and three know French and German.

(1) How many know all three languages?
(2) How many know exactly two languages?
(3) How many staff members know only English?

8.2.6. (1) Show that the number of positive integers that are divisible by n and that are not larger than x is equal to $\lfloor x/n\rfloor$.

(2) Find the number of positive integers not exceeding 1000 that are not divisible by 3, 5, and 7.

(3) Find the number of positive integers not exceeding 1000 that are not divisible by 6, 10, and 15.

(4) Show that if $n = 30m$, the number of positive integers not exceeding n and indivisible by 6, 10, and 15 is equal to $22m$.

(5) Let $p_1, \ldots, p_r$ be all prime numbers not exceeding $\sqrt{n}$. Show that the number of prime numbers p such that $\sqrt{n} < p \leqslant n$, is equal to $n - 1 + \sum_{k=1}^{r} (-1)^k S_k$, where the sum

$$S_k = \sum \left[\frac{n}{p_1^{\alpha_1} \ldots p_r^{\alpha_r}} \right]$$

is taken over all possible $\binom{r}{k}$ tuples of powers $\alpha_1, \ldots, \alpha_r$, of which exactly k powers are equal to 1, while the remaining powers are equal to zero.

(6) Find the number of prime numbers not exceeding 100.

8.2.7. Let U be a set of n $(n \geqslant 3)$ elements.

(1) Find the number of pairs (X, Y) of subsets of the set U satisfying the condition $X \cap Y = \varnothing$.

(2) Find the number of pairs (X, Y), such that $X \subseteq U$, $Y \subseteq U$, $|(X \setminus Y) \cup (Y \setminus X)| = 1$.

(3) Find the number of triads (X, Y, Z) for which $X \subseteq U$, $Y \subseteq U$, $Z \subseteq U$, $X \cup (Y \cap \overline{Z}) = \overline{X} \cup \overline{Y}$.

(4) Find the number of pairs (X, Y) of subsets of the set U for which $X \cap Y = \varnothing$, $|X| \geqslant 2$, $|Y| \geqslant 3$.

(5) Find the number of pairs (X, Y) for which $X \subseteq U$, $Y \subseteq U$, $|(X \setminus Y) \cup (Y \setminus X)| = 1$, $|X| \geqslant 2$, $|Y| \geqslant 2$.

(6) Find the number of triads (X, Y, Z), such that $X \subseteq U$, $Y \subseteq U$, $Z \subseteq U$, $X \cup (Y \cap \overline{Z}) = \overline{X} \cup \overline{Y}$, $|X| \geqslant 1$, $|Y| \geqslant 1$, $|Z| \leqslant 1$.

8.2.8. Butler's problem. n pairs of warring knights $(n \geqslant 2)$ are invited to dinner. Show that there are $\sum_{k=0}^{n} (-1)^k \binom{n}{k} 2^k (2n - k - 1)!$ ways of seating them by a round table, so that no two enemies sit side by side.

8.2.9. Married couples problem. In how many different ways can n married couples be seated around a round table so that men and women occupy alternate positions, and none of the couples is seated side by side?

8.3. Recurrent Sequences, Generating Functions, and Recurrence Relations

The sequence $a_0, a_1, \ldots, a_n, \ldots$ is called *recurrent* if, for a certain k and all n, a relation of the following type is satisfied:

$$a_{n+k} + p_1 a_{n+k-1} + \ldots + p_k a_n = 0, \tag{9}$$

where the coefficients p_i, $i = \overline{1, k}$ are independent of n. The polynomial

$$P(x) = x^k + p_1 x^{k-1} + \ldots + p^k \tag{10}$$

is a *characteristic* polynomial for recurrent sequences $\{a_n\}$.

8.3.1. (1) Prove that a recurrent sequence can be completely defined by specifying its first k terms and the relation (9).

(2) Let λ be the root of a characteristic polynomial. Prove that the sequence $\{\lambda^n\}$ satisfies the relation (9).

(3) Prove that if $\lambda_1, \ldots, \lambda_k$ are prime (not multiple) roots of the characteristic polynomial (10), the general solution of the recurrence relation (9) has the form

$$a_n = C_1\lambda_1^n + \ldots + C_k\lambda_k^n.$$

(4) Let λ_i be a root of multiplicity r_i $(l = \overline{1, s})$ of the characteristic polynomial (10). Prove that in this case, the general solution of the recurrence relation (9) has the form

$$a_n = \sum_{i=1}^{s} (C_{i_1} + C_{i_2} n + \ldots + C_{i_{r_i}} n^{r_i - 1}) \lambda_i^n,$$

where C_{ij} $(i = \overline{1, s},\ j = \overline{1, r_i})$ are arbitrary constants.

8.3.2. Find the general solutions of the following recurrence relations:

(1) $a_{n+2} - 4a_{n+1} + 3a_n = 0$;
(2) $a_{n+2} + 3a_n = 0$;
(3) $a_{n+2} - a_{n+1} - a_n = 0$;
(4) $a_{n+2} + 2a_{n+1} + a_n = 0$;

(5) $a_{n+3} + 10a_{n+2} + 32a_{n+1} + 32a_n = 0$;
(6) $a_{n+3} + 3a_{n+2} + 3a_{n+1} + a_n = 0$.

8.3.3. Find a_n from the following recurrence relations and initial conditions:
(1) $a_{n+2} - 4a_{n+1} + 3a_n = 0$,
$a_1 = 10,\ a_2 = 16$;
(2) $a_{n+3} - 3a_{n+2} + a_{n+1} - 3a_n = 0$,
$a_1 = 3,\ a_2 = 7,\ a_3 = 27$;
(3) $a_{n+3} - 3a_{n+1} + 2a_n = 0$,
$a_1 = a,\ a_2 = b,\ a_3 = c$;
(4) $a_{n+2} - 2\cos\alpha\, a_{n+1} + a_n = 0$,
$a_1 = \cos\alpha, \quad a_2 = \cos 2\alpha$;
(5) $a_{n+2} - a_n = 0$,
$a_1 = 2, \quad a_2 = 0$;
(6) $a_{n+2} - 6a_{n+1} + 9a_n = 0$,
$a_1 = 6,\ a_2 = 27$.

8.3.4. Show that
(1) if $x = 1$ is not a root of the polynomial $x^2 + px + q$, the partial solution of the recurrence relation

$$a_{n+2} + pa_{n+1} + qa_n = \alpha n + \beta, \tag{11}$$

where α, β, p, q are given numbers, is the sequence $a_n^* = an + b$; find a and b;
(2) if $x = 1$ is a simple root of the polynomial $x^2 + px + q$, the partial solution can be obtained in the form $a_n^* = n\,(an + b)$; find the values of a and b;
(3) if $x = 1$ is a multiple root of the polynomial $x^2 + px + q$, the partial solution may be obtained in the form $a_n^* = n^2\,(a_n + b)$; find a and h;
(4) in each of the above cases, find the general solution of the relation (11).

8.3.5. Solve the following recurrence relations:
(1) $a_{n+1} - a_n = n, \quad a_1 = 1$;
(2) $a_{n+2} + 2a_{n+1} - 8a_n = 27 \times 5^n$,
$a_1 = -9, \quad a_2 = 45$;
(3) $a_{n+2} - 2a_{n+1} + 2a_n = 2^n$,
$a_0 = 1, \quad a_1 = 2$;
(4) $a_{n+2} + a_{n+1} - 2a_n = n$,
$a_0 = 1, \quad a_1 = -2$;
(5) $a_{n+2} - 4a_{n+1} + 4a_n = 2^n$,
$a_0 = 1, \quad a_1 = 2$;
(6) $a_{n+2} + a_{n+1} - 6a_n = 5 \times 2^{n+1}$,
$a_0 = 2, \quad a_1 = -1$.

9.3.6. (1) Let $\{a_n\}$ and $\{b_n\}$ be two sequences whose terms are connected through the relations

$$a_{n+1} = p_1 a_n + q_1 b_n,$$
$$b_{n+1} = p_2 a_n + q_2 b_n,$$
$$\Delta = p_1 q_2 - p_2 q_1 \neq 0,$$

where p_1, q_1, p_2, q_2 are given numbers. Find the expressions for a_n and b_n, assuming a_1 and b_1 to be given.

(2) Find the solution of the following system of recurrence relations:

$$a_{n+1} = 3a_n + b_n, \quad b_{n+1} = -a_n + b_n,$$
$$a_1 = 14, \quad b_1 = -6.$$

(3) Find the general solution for the following system of recurrence relations:

$$a_{n+1} = b + 5, \quad b_{n+1} = -a_n + 3.$$

8.3.7. Fibonacci's sequence $\{F_n\}$ is defined by the recurrence relation $F_{n+2} = F_{n+1} + F_n$ and by the initial conditions $F_1 = F_2 = 1$. Prove that

(1) $F_{n+m} = F_{n-1}F_m + F_n F_{m+1}$ for any natural numbers n and m;

(2) the number F_n is divisible by F_m for any m and $n = km$;

(3) two adjacent numbers F_n and F_{n+1} are mutually prime;

(4) any natural number N ($N > 1$) may be uniquely presented as a sum of Fibonacci's numbers, such that each number enters the sum not more than once and no two adjacent numbers enter together;

(5) $F_n = \frac{1}{\sqrt{5}}\left[\left(\frac{1+\sqrt{5}}{2}\right)^n - \left(\frac{1-\sqrt{5}}{2}\right)^n\right]$;

(6) $F_1 + F_3 + \ldots + F_{2n+1} = F_{2n+2}$;

(7) $1 + F_2 + F_4 + \ldots + F_{2n} = F_{2n+1}$;

(8) $F_{n+1}^3 + F_n^3 - F_{n-1}^3 = F_{3n}$.

Each sequence $a_0, a_1, \ldots, a_n, \ldots$ can be associated with a series $A(t) = a_0 + a_1 t + \ldots + a_n t^n + \ldots$ called a *generating function for the sequence* $\{a_n\}$. If the series $A(t)$ converges to a function $f(t)$, the latter is also a *generating function for* $\{a_n\}$. The *exponential generating function for* $\{a_n\}$ is the series $E(t) = a_0 +$

$a_1(t) + \ldots + \frac{a_n t^n}{n!} + \ldots$. The operations of addition, multiplication, and multiplication by a constant can be defined for generating functions by treating them as formal series. Let $A(t)$ and $B(t)$ be the generating functions for the sequences $\{a_n\}$ and $\{b_n\}$ respectively, and let α and β be constants. In this case,

$$\alpha A(t) + \beta B(t) = \alpha a_0 + \beta b_0 + (\alpha a_1 + \beta b_1)t + \ldots + (\alpha a_n + \beta b_n)t^n + \ldots,$$

$$A(t) B(t) = a_0 b_0 + (a_0 b_1 + a_1 b_0)t + \ldots + (a_0 b_n + a_1 b_{n-1} + \ldots + a_n b_0)t^n + \ldots.$$

If $E_a(t)$ and $E_b(t)$ are exponential generating functions for the sequences $\{a_n\}$ and $\{b_n\}$ respectively, the operations of addition and multiplication by a constant are defined in the same way as for normal generating functions, while their product is defined as

$$E_a(t) E_b(t) = c_0 + c_1 t + \ldots + \frac{c_n t^n}{n!} + \ldots,$$

where $c_n = a_0 b_n + \binom{n}{1} a_1 b_{n-1} + \ldots + \binom{n}{k} a_k b_{n-k} + \ldots + a_n b_0$.

8.3.8. Find the generating function $f(t)$ for the sequence $\{a_n\}$ if

(1) $a_n = 1$ for all $n \geqslant 0$;
(2) $a_n = 1$ for $0 \leqslant n \leqslant N$ and $a_n = 0$ for $n > N$;
(3) $a_n = \alpha^n$;
(4) $a_n = \alpha^n/n!$;
(5) $a_n = (-1)^n$;
(6) $a_n = n$;
(7) $a_n = n(n-1)$;
(8) $a_n = \binom{m}{n}$, m is a natural number;
(9) $a_n = \binom{\alpha}{n}$, α is a real number;
(10) $a_n = n^2$;
(11) $a_n = \sin \alpha n$;
(12) $a_n = \cos \alpha n$.

8.3.9. Find the exponential generating functions $E(t)$ for the sequence $\{a_n\}$ if

(1) $a_n = 1$;
(2) $a_n = \alpha^n$;

(3) $a_n = n$;
(4) $a_n = n(n-1)$;
(5) $a_n = (m)_n$;
(6) $a_n = n^2$.

8.3.10. Using the identities connecting generating functions, derive the following identities for binomial coefficients:

(1) $(1+t)^n(1+t)^m = (1+t)^{n+m}$;

$$\sum_{s=0}^{k} \binom{n}{s}\binom{m}{k-s} = \binom{n+m}{k};$$

(2) $(1-t)^{-1-n}(1-t)^{-1-m} = (1-t)^{-2-n-m}$;

$$\sum_{s=0}^{k} \binom{n+s}{n}\binom{m+k-s}{m} = \binom{m+n+k+1}{k};$$

(3) $(1+t)^n(1+t)^{-m} = (1+t)^{n-m}$;

$$\sum_{s=0}^{k} (-1)^{k-s}\binom{n}{s}\binom{m+k-s-1}{k-s} = \binom{n-m}{k};$$

(4) $(1-t)^{-1-n}(1+t)^{-1-n} = (1-t^2)^{-1-n}$;

$$\sum_{s=0}^{2k} (-1)^s\binom{n+s}{n}\binom{n+2k-s}{n} = \binom{n+k}{k};$$

(5) $(1+t)^n(1-t^2)^{-n} = (1-t)^{-n}$;

$$\sum_{s=0}^{\lfloor k/2 \rfloor} \binom{n}{k-2s}\binom{n+s-1}{s} = \binom{n+k-1}{k};$$

(6) $(1+t)^n(1-t)^n = (1-t^2)^n$;

$$\sum_{s=0}^{k} (-1)^s\binom{n}{k-s}\binom{n}{s} = \begin{cases} (-1)^{k/2}\binom{n}{k/2}, & k \text{ is even}, \\ 0, & k \text{ is odd}. \end{cases}$$

8.3.11. Find the general term a_n in the sequence for which the function $A(t)$ is a generating function.

(1) $A(t) = (q+pt)^m$;
(2) $A(t) = (1-t)^{-1}$;
(3) $A(t) = \sqrt{1-t}$;
(4) $A(t) = t^m(1-t)^m$;

(5) $A(t) = (t + t^2 + \ldots + t^r)^m$;

(6) $A(t) = \left(1 + \frac{t^2}{2}\right)^{-m}$;

(7) $A(t) = (1 + 2t)^{-1/2}\left(1 - \frac{t}{2}\right)^{-m}$;

(8) $A(t) = t^2(1 - t)(1 + 2t)^{-m}$;

(9) $A(t) = \ln(1 + t)$;

(10) $A(t) = \operatorname{arc\,tan} t$;

(11) $A(t) = \operatorname{arc\,sin} t$;

(12) $A(t) = e^{-2t^2}$;

(13) $A(t) = \int_0^t e^{-x^2}\,dx$;

(14) $A(t) = \left(\frac{-t}{1+t}\right)^m$.

8.3.12. Derive the following identities[2]:

(1) $\sum_s (-1)^{n-s}\binom{n}{s}\binom{m+s}{m+1} = \binom{m}{n-1}$;

(2) $\sum_s (-1)^s\binom{m}{s}\binom{m}{2n-s} = (-1)^n\binom{m}{2n}$;

(3) $\sum_s (-1)^s\binom{m}{m-k+s}\binom{n+s}{n} = \binom{m-n-1}{k}$;

(4) $2\sum_s \binom{n}{2s}\binom{n}{2m-2s} = \binom{2n}{2m} + (-1)^m\binom{n}{m}$;

(5) $2\sum_s \binom{n+2s}{n}\binom{n+2m-2s+1}{n+1} = \binom{2n+2m+2}{2n+1}$;

(6) $\sum_s (-1)^{n-s}4^s\binom{n+s+1}{2s} = n+1$.

8.3.13. Let $A(t)$ and $E(t)$ be a generating and an exponential generating functions of the sequence $\{a_n\}$ respectively.

(1) Using the equality $n! = \int_0^\infty e^{-x}x^n\,dx$, show that

$$A(t) = \int_0^\infty e^{-x}E(xt)\,dx. \qquad (*)$$

---

[2] Summation is carried out over all s for which these expressions are meaningful.

(2) Verify that formula (*) is valid for generating functions $A(t) = (1 - t)^{-1}$ and $E(t) = e^t$ of the sequence $a_0 = a_1 = a_2 = \ldots = 1$.

(3) The same, for a sequence with a general term

$$a_n = \begin{cases} 0, & n < j, \\ (n)_j, & n \geqslant j. \end{cases}$$

8.3.14. (1) Let $(a)_n = a(a-1)\ldots(a-n+1)$. With the help of exponential generating functions, prove that

$$(a+b)_n = \sum_{k=0}^{n} \binom{n}{k} (a)_{n-k} (b)_k.$$

Hint. Use the identity $(1+t)^{a+b} = (1+t)^a (1+t)^b$.

(2) Let $(a)_{n,h} = a(a+h)\ldots(a-h(n-1))$. Prove that

$$(a+b)_{n,h} = \sum_{k=0}^{n} \binom{n}{k} (a)_{n-k,h} (b)_{k,h}.$$

8.3.15. Let $\{a_n\}$ and $\{b_n\}$ be sequences, and let $A(t)$ and $B(t)$ be their respective generating functions. Prove that if

(1) $a_n = b_n - b_{n-1}$, then $A(t) = B(t)(1-t)$;

(2) $a_n = b_{n+1} - b_n$, then $A(t) = B(t)\dfrac{1-t}{t} - b_0 t^{-1}$;

(3) $a_n = b_{n+1} + b_{n+2} + \ldots$, then $A(t) = \dfrac{B(1) - B(t)}{1-t}$;

(4) $a_n = nb_n$, then $A(t) = t\dfrac{d}{dt}B(t)$;

(5) $a_n = n^2 b_n$, then $A(t) = t\dfrac{d}{dt}\left(t\dfrac{d}{dt}B(t)\right)$;

(6) we define the operation S^k $(k \geqslant 0)$ on the sequence $\{b_n\}$ with the help of the relation

$$S^k(b_n) = b_n + \binom{k}{1} b_{n-1} + \ldots + \binom{k+j-1}{j} b_{n-j} + \ldots + \binom{k+n-1}{n} b_0$$

and we put $a_n = S^k(b_n)$, then $A(t) = (1-t)^{-k} B(t)$;

(7) $a_n = b_{2n}$, then $A(t) = \frac{1}{2}(B(t^{1/2}) + B(-t^{1/2}))$;

(8) $a_n = b_0 + b_1 + \ldots + b_{n-1}$, $a_0 = 0$, then $A(t) = B(t)\, t\, (1-t)^{-1}$.

8.3.16. Let $A(t)$ and $B(t)$ be the generating functions of the sequences $\{a_n\}$ and $\{b_n\}$ respectively, and let $A(t)\, B(t) = 1$. Find $\{b_n\}$ and $B(t)$ from the given sequence $\{a_n\}$ if

(1) $a_n = \binom{m}{n}$; (2) $a_n = a^n$; (3) $a_n = n + 1$;

(4) $a_0 = a_2 = 1$, $a_n = 0$ for $n \neq 0, 2$;

(5) $a_n = (-1)^n$; (6) $a_n = (-1)^n \binom{2n}{n} 4^{-n}$.

8.3.17. Suppose that the sequence $\{a_n\}$ satisfies the recurrence relation $a_{n+2} + pa_{n+1} + qa_n = 0$.

(1) Prove that $A(t) = \dfrac{a_0 + (a_1 + pa_0)\, t}{1 + pt + qt^2}$.

(2) Let $1 + pt + qt^2 = (1 - \lambda_1 t)(1 - \lambda_2 t)$, $\lambda_1 \neq \lambda_2$. Prove that

$$a_n = (a_1 + pa_0)\frac{\lambda_1^n - \lambda_2^n}{\lambda_1 - \lambda_2} + a_0 \frac{\lambda_1^{n+1} - \lambda_2^{n+1}}{\lambda_1 - \lambda_2}.$$

(3) Write an expression for a_n for the case when

$$1 + pt + qt^2 = (1 - \lambda t)^2.$$

8.3.18. Let

$$a_n = \sum_{j=0}^{n} \binom{n+j}{2j}, \quad b_n = \sum_{j=0}^{n-1} \binom{n+j}{2j+1}, \quad n = 0, 1, 2, \ldots,$$

and let $A(t)$ and $B(t)$ be the corresponding generating functions.

(1) Prove that a_n and b_n are connected through relations of the type

$$a_{n+1} = a_n + b_{n+1},$$
$$b_{n+1} = a_n + b_n, \quad a_0 = 1, \quad b_0 = 0.$$

(2) Prove that $A(t)$ and $B(t)$ satisfy the system of equations

$$A(t) - 1 = tA(t) + B(t),$$
$$B(t) = tA(t) + tB(t).$$

(3) Find $A(t)$ and $B(t)$.

(4) Show that

$$\lim_{n\to\infty}\left(\frac{2}{3+\sqrt{5}}\right)^n a_n=\frac{1+\sqrt{5}}{2\sqrt{5}},\quad \lim_{n\to\infty}\left(\frac{2}{3+\sqrt{5}}\right)^n b_n=\frac{1}{\sqrt{5}}.$$

8.3.19. Suppose that the terms of the sequence $\{a_n\}$ satisfy the relation

$$a_n=a_0a_{n-1}+a_1a_{n-2}+\ldots+a_{n-1}a_0,\quad a_0=1.$$

(1) Prove that the generating function $A(t)=\sum\limits_{n=0}^{\infty}a_nt^n$ satisfies the equality $tA^2(t)=A(t)-a_0$ or, if the initial conditions are taken into account, $A(t)=\dfrac{1-\sqrt{1-4t}}{2t}$.

(2) Expanding $A(t)$ into a power series in t, show that $a_n=\dfrac{1}{n+1}\dbinom{2n}{n}$.

(3) Find the sequence $\{a_n\}$ whose terms satisfy the relations

$$a_0a_{n-1}+a_1a_{n-2}+\ldots+a_{n-1}a_0=2^na_n,$$
$$a_0=a_1=1.$$

8.3.20. Derive a recurrence relation for the sequence $\{a_n\}$ and solve this relation if

(1) a_n is the number of ways in which a convex $(n+2)$-gon can be divided into triangles by diagonals that do not intersect within this polygon.

(2) a_n is the number of ways in which parentheses can be arranged in the expression $b_1:b_2:\ldots:b_{n+1}$, so that the resulting expressions are meaningful.

8.3.21. Using the method of mathematical induction, find the sequence $\{a_n\}$ from the following recurrence relations and initial conditions:

(1) $a_{n+1}=(n+1)a_n,\quad a_0=1;$

(2) $na_{n+1}+a_n=0,\quad a_1=1;$

(3) $(n+2)(n+1)a_{n+2}-n^2a_n,\quad a_0=0,\quad a_1=1;$

(4) $(n+2)^2a_{n+2}+a_n=0,\quad a_0=1,\quad a_1=0;$

(5) $n^2a_{n+2}+(n+2)^2a_n=0,\quad a_1=1,\quad a_2=0;$

(6) $a_{n+1}^2-a_na_{n+2}=(-1)^{n-1},\quad a_0=1,\quad a_1=1.$

8.3.22. Let $A_n(t)=\sum\limits_{k=0}^{\infty}a(n,k)t^k$ be an arbitrary function for a sequence satisfying the relation

$$a(n,k)=a(n,k-1)+a(n-1,k)$$

with initial conditions $a(n, 0) = 1$, $a(0, k) = 0$ for $k > 0$. Show that

(1) $(1-t) A_n(t) = A_{n-1}(t)$;
(2) $A_n(t) = (1-t)^{-n}$;
(3) $a(n, k) = \binom{n+k-1}{k}$.

8.3.23. (1) In how many ways can a 10-kopeck coin be changed into 1, 2, 3 and 5-kopeck coins?

(2) The same, but each coin must have a duplicate.

(3) The same problem under the condition that the change is to be made from four 1-kopeck coins, three 2-kopeck coins, two 3-kopeck coins, and one 5-kopeck coin.

8.3.24. Find the generating function $A(t)$ for the sequence $\{a_n\}$, where

(1) a_n is the number of solutions in non-negative integers for the equation

$$2x + 3y + 5z = n.$$

(2) a_n is the number of solutions of the same equation given that x, y, z assume values in the set $\{0, 1\}$.

(3) a_n is the number of integral solutions of the same equation given that $0 \leqslant x \leqslant p$, $0 \leqslant y \leqslant r$, $0 \leqslant z \leqslant s$.

8.3.25*. Find the value of a_n using a given generating function $A(t)$ for the sequence $\{a_n\}$:

$$(1)\ A(t) = \prod_{k=0}^{\infty} (1 - q^k t), \quad |q| < 1.$$

Hint. Show that $A(t) = (1 - qt) A(qt)$, and compare the coefficients of t^n on the left- and right-hand sides of this equation.

$$(2)\ A(t) = \prod_{k=1}^{\infty} (1 + qt^{2k}).$$

Hint. Prove that $a_n = q^{b_n}$, where b_n is the number of unities in the binary expansion of the number n.

8.3.26. Let

$$S(n, k, l) = \sum_{\nu=0}^{n} (-1)^{n-\nu} \binom{n}{\nu} (\nu + l)^k.$$

Prove that

(1) $S(n+1, k, l) = S(n, k, l+1) - S(n, k, l)$;
(2) $S(n, k, l+1) = S(n, k, l) - S(n+1, k, l)$;

(3) $S(n, k+1, l) = (n+l)\, S(n, k, l) + nS(n-1, k, l)$;

(4) $S(n, k, l) = 0$ for $n > k$;

(5) $S(n, n, l) = n!$;

(6) $S(n, k, l) > 0$ for $n \leqslant k$;

(7) $S(n, k, l)$ is an increasing function of parameters k and l for $n \leqslant k$;

(8) $S(n, n+1, l) = \sum\limits_{k=0}^{n} (k+l)\, k!$;

(9) $S(1, k, 0) = 1$ for $1 \leqslant k$.

8.3.27. Let $S(n, k) = S(n, k, 0) = \sum\limits_{\nu} (-1)^{n-\nu} \binom{n}{\nu} \nu^k$ and $\sigma_n(t) = \sum\limits_{k=0}^{\infty} S(n, k)\, t^k$. Prove that for $|t| < 1$

(1) $\sigma_n(t) = \dfrac{n!\, t^n}{(1-t)(1-2t)\ldots(1-nt)}$;

(2) $\sigma_n(t) = t \sum\limits_{k=1}^{n} (-1)^{n-k}\, k \binom{n}{k} (1-kt)^{-1}$.

8.4. Polya's Theory

The *permutation on a set* $Z_n = \{1, 2, \ldots, n\}$ is the mapping of Z_n onto itself[3]. The permutation $\pi = \begin{pmatrix} 1, & 2, & \ldots, & n \\ i_1, & i_2, & \ldots, & i_n \end{pmatrix}$ will be frequently specified by the line $(i_1, i_2, \ldots, i_n)$. A permutation is called *cyclic* (or a *cycle*) if a certain number j_1 is substituted for j_2, j_2 for j_3, and so on, j_{k-1} is substituted for j_k, and j_k for j_1, while all the other numbers remain unchanged. Such a cycle is denoted by $(j_1, j_2, \ldots, j_k)$. The number k is called the *length of the cycle*. Any permutation can be presented as a product of cycles. For example, $\begin{pmatrix} 1, & 2, & 3, & 4, & 5 \\ 2, & 1, & 4, & 5, & 3 \end{pmatrix} = (1, 2)(3, 4, 5)$. A cycle of length 2 is called a *transposition*. The permutations on the set Z_n form a group relative to the multiplication operation. The multiplication operation of the permutations π_1 and π_2 involves their successive application. For example,

[3] Any set of n elements can be taken as Z_n.

if $\pi_1 = \begin{pmatrix} 1, & 2, & 3, & 4 \\ 2, & 4, & 3, & 1 \end{pmatrix}$, $\pi_2 = \begin{pmatrix} 1, & 2, & 3, & 4 \\ 1, & 4, & 3, & 2 \end{pmatrix}$, then $\pi_1\pi_2 = \begin{pmatrix} 1, & 2, & 3, & 4 \\ 4, & 2, & 3, & 1 \end{pmatrix}$. It can be easily verified that the multiplication operation is associative: $\pi(\sigma\tau) = (\pi\sigma)\tau$. The unit element of a group is the identity permutation $\begin{pmatrix} 1, & 2, & \ldots, & n \\ 1, & 2, & \ldots, & n \end{pmatrix}$. The permutation inverse to $\begin{pmatrix} 1, & 2, & \ldots, & n \\ i_1, & i_2, & \ldots, & i_n \end{pmatrix}$ is given by $\begin{pmatrix} i_1, & i_2, & \ldots, & i_n \\ 1, & 2, & \ldots, & n \end{pmatrix}$. The permutation group on the set Z_n is called a *symmetry group of n-th degree* and is denoted by S_n. The order of a symmetry group of n-th degree (the number of its elements) is equal to $n!$.

If the permutation on a set N^n is represented as a product of b_1 cycles of length 1, b_2 cycles of length 2, and so on, and b_n cycles of length n, the *permutation* is said *to be of the type* $(b_1, b_2, \ldots, b_n)$. For example, the permutation $\begin{pmatrix} 1, & 2, & 3, & 4 \\ 3, & 2, & 4, & 1 \end{pmatrix}$ is of the type (1, 0, 1, 0). If G is a subgroup of the group S_n, the polynomial

$$P_G = P_G(t_1 \ldots t_n) = |G|^{-1} \sum_{\pi \in G} t_1^{b_1} \ldots t_n^{b_n},$$

where $(b_1, \ldots, b_n)$ is the type of permutation π, is called the *cyclic index of the group G*. Let G be a group of permutations on Z_n. The elements a and b in Z_n are called *G-equivalent* (notation $a \sim b$), if there exists a permutation $\pi \in G$, such that $\pi a = b$ (or, which is the same, $\pi b = a$). The classes of G-equivalence are called *transitive sets or orbits*.

Bernside's Lemma. *The number of orbits* $\nu(G)$ *in the set* Z_n, *defined by the group* G, *is given by the equality*

$$\nu(G) = |G|^{-1} \sum_{\pi \in G} b_1(\pi).$$

Let M and N be finite sets, and let G and H be the permutation groups of M and N. The power group H^G consists of all possible pairs $(\pi; \sigma)$, where $\pi \in G$, $\sigma \in H$, and acts on the set N^M of all functions $f\colon M \to N$. Moreover, by definition, $(\pi; \sigma) f(x) = \sigma f(\pi(x))$ for all $x \in M$ and $f \in N^M$. Suppose that a weight function w: $N \to \{0, 1, \ldots\}$ is defined on the set N, and that q_n is the number of elements having a weight n in N. The generating function $Q(t) = \sum_{n=0}^{\infty} q_n t^n$ is called a *figure counting series*. The weight of the function f in N^M is determined from the equality $w(t) = \sum_{x \in M} w(f(x))$. The functions f_1 and f_2 in N^M are called *equivalent* (notation $f_1 \sim f_2$) if there exists an element $\pi \in G$, such that $f_1(\pi x) = f_2(x)$ for all $x \in M$. If f_1 and f_2 are equivalent, they have the same weight. Hence we can determine the *weight* $w(F)$ *of the equivalence class* F as the weight of any element f in F. Let φ_k be the number of equivalent classes of weight k in N^M. The generating function $\Phi(t) = \sum_{k=0}^{\infty} \varphi_k t^k$ is called a *function counting series.*

Polya's Theorem.

$$\Phi(t) = P_G(Q(t), Q(t^2), \ldots, Q(t^n)),$$

where $n = |G|$, $P_G(t_1, t_2, \ldots, t_n)$ is the cyclic index of group G, and $Q(t^k)$ is substituted into P_G in the place of variable t_k, $k = 1, 2, \ldots, n$.

8.4.1. Find the type of permutation π:

(1) $\pi = (2, 3, 4, 1)$; (3) $\pi = (3, 4, 5, 6, 1, 2)$;
(2) $\pi = (4, 2, 3, 1)$; (4) $\pi = (8, 2, 1, 7, 4, 6, 3, 5)$.

8.4.2. Represent the permutations in Problem 8.4.1 in the form of a product of transpositions.

8.4.3. Find the cyclic index of the group G, where

(1) G is a group of permutations of the vertices of a square, which are obtained by rotating the square in a plane;

(2) G is a group of permutations of the vertices of a square, which are obtained by rotating the square in space;

(3) G is a group of permutations of the vertices of a tetrahedron, which are obtained by rotating it;

(4) G is a group of permutations of the edges of a tetrahedron, which are obtained by rotating it;

(5) G is a group of permutations of the faces of a tetrahedron, which are obtained by rotating it;

(6) $G = S_3$;

(7) $G = A_4$, where A_4 is an alternating group of power 4, i.e. a subgroup of group S_4, formed by permutations representable as products of an even number of transpositions;

(8) G is a group of permutations of the faces of a cube, which are obtained as a result of rotation; and

(9) G is a group of permutations of the vertices of an octahedron.

8.4.4. (1) Prove that the group G, defined in Problem 8.4.3(1), specifies one orbit on the set N_4.

(2) Find the number of orbits in the set N_4, which are defined by the group G and formed by the permutations $\pi_1 = (1, 2, 3, 4)$, $\pi_2 = (1, 2, 4, 3)$, $\pi_3 = (2, 1, 3, 4)$ and $\pi_4 = (2, 1, 4, 3)$.

8.4.5. Prove Bernside's lemma.

8.4.6. Let $\vec{b} = (b_1, b_2, \ldots, b_n)$ be a vector corresponding to the decomposition of the number n, i.e. $b_1 + 2b_2 + \ldots + nb_n = n$, where b_k are non-negative integers ($1 \leqslant k \leqslant n$). We denote by $H(\vec{b})$ the set of all permutations of the symmetry group S_n, whose type coincides with $\vec{b}$, and let $h(\vec{b}) = |H(\vec{b})|$. Prove that

(1) $h(\vec{b}) = n!\left(\prod_{k=1}^{n} k^{b_k} b_k!\right)^{-1}$;

(2) $P_{S_n}(t_1, t_2, \ldots, t_n) = \frac{1}{n!} \sum_{\vec{b}} h(\vec{b}) \prod_{k=1}^{n} t_k^{b_k}$;

(3) the cyclic index P_{S_n} is equal to the coefficient of x^n in the expansion of the function

$$\exp(t_1 x + t_2 x^2/2 + \ldots + t_k x^k/k + \ldots)$$

into a power series in x.

8.4.7. Let A_n be an alternating group of power n, i.e. a subgroup of the group S_n, consisting of all its permutations, which can be represented as a product of an even number of transpositions. Prove that

$$P_{\bar{A}_n}(t_1, \ldots, t_n) = P_{S_n}(t_1, \ldots, t_n) + P_{S_n}(t_1, -t_2, \ldots, (-1)^n t_n).$$

8.4.8. Let G be a group of permutations of the set X, and H a group of permutations of the set Y, $X \cap Y = \varnothing$. Any pair of permutations $\pi \in G$, $\sigma \in H$ is put in correspondence with the permutation $\pi \times \sigma$ of the set $X \cup Y$, defined as $z \to \pi z$ for $z \in X$, $z \to \sigma z$ for $z \in Y$. Prove that

(1) the permutations $\pi \times \sigma$ form a group of the order $|G|\,|H|$. This group is called the *direct product of the groups G and H*, and is denoted by $G \times H$;

(2) if the permutations $\pi \in G$ and $\sigma \in H$, are of the type $(b_1, \ldots, b_n)$ and $(c_1, \ldots, c_n)$, then $\pi \times \sigma$ is of the type $(b_1 + c_1, \ldots, b_n + c_n)$;

(3) $P_{G\times H} = P_G \times P_H$.

8.4.9. (1) Find the number of necklaces that can be made from beads of two colours, if each necklace contains seven beads. If one necklace can be obtained from the other by turning it, the two are assumed to be identical (mirror reflections are not allowed).

(2) Using Polya's theorem, find the number of necklaces from beads of k colours, if each necklace consists of n beads. The necklaces are considered to be identical if they can be obtained from one another by rotation (without reflection).

8.4.10. (1) Find the number of different colourings of a tetrahedron vertices in two colours. Two colourings are assumed to be different if the colours of the vertices cannot be made to coincide through a rotation of the tetrahedron.

(2) Find the number of different colourings of the vertices of a tetrahedron using three colours.

(3) Find the number of different colourings of the faces of a cube, such that three faces are painted in red, two in blue, and one in white.

8.4.11. Let G be a group of permutations of the set Z_n, and E be a unit group acting on the set N and transforming each element $x \in N$ into itself. Find the number of orbits defined by the power group E^G on the set N^{Z_n}.

8.4.12. (1) Let T_k be the number of pairwise non-isomorphic rooted trees, and let $T(x) = \sum_{k=1}^{\infty} T_k x^k$ be the generating function for the sequence $\{T_k\}$. Prove that

$$T(x) = x + x \sum_{n=1}^{\infty} P_{S_n}(T(x), \ldots, T(x^n)).$$

(2) Prove the validity of the following recurrence relation:

$$T_{n+1}=\frac{1}{n}\sum_{k=1}^{n}\sum_{r\mid k} rT_rT_{n-k+1}, \quad T_1=1.$$

8.4.13. Let g_n be the number of pairwise non-isomorphic graphs with n vertices, and let l_n be the number of pairwise non-isomorphic connected graphs with n vertices. Let

$$g(t)=\sum_{n=1}^{\infty} g_n t^n, \quad l(t)=\sum_{n=1}^{\infty} l_n t^n$$

be the corresponding generating functions. Prove that

$$g(t)=\sum_{n=1}^{\infty} G_{s_n} \quad (l(t), \ldots, l(t^n)).$$

8.4.14. Let $p(n, k)$ be the number of permutations of the group S_n, which consist of k cycles, and let $p_n(t)=\sum_{n=0}^{\infty} p(n, k)\, t^k$ be the corresponding generating function. Prove that

(1) $p_n(t)=\prod_{i=0}^{n-1}(t+i)$;

(2) $p(n, k)=p(n-1, k-1)+(n-1)\,p(n-1, k)$.

8.5. Asymptotic Expressions and Inequalities

The following notation will be used while estimating the growth of functions. The notation $\varphi(x)=O(\psi(x))$ for $x \in X$ means that there exists a constant C such that $|\varphi(x)| \leqslant C|\psi(x)|$ for $x \in X$. If $\varphi(x)=O(\psi(x))$ and $\psi(x)=O(\varphi(x))$ for $x \in X$, we can write $\varphi(x) \asymp \psi(x)$ for $x \in X$. The notation $\varphi(x)=O(\psi(x))$ for $x \to a$ means that $\lim_{x\to a}\frac{\varphi(x)}{\psi(x)}=0$. The functions $\varphi(x)$ and $\psi(x)$ are said to be *asymptotically equal* (notation $\varphi(x) \sim \psi(x)$) for $x \to a$, if $\varphi(x)=\psi(x)+O(\psi(x))$ for $x \to a$. For different types of estimates it is convenient to use Stirling's formula

$$n! \sim \sqrt{2\pi n}\, n^n e^{-n}. \tag{12}$$

For more accurate estimates, the following inequalities are used:

$$\sqrt{2\pi n}\, n^n \exp\left(-n+\frac{1}{12n}-\frac{1}{360n^3}\right)<n!$$
$$<\sqrt{2\pi n}\, n^n \exp\left(-n+\frac{1}{12n}\right). \qquad (13)$$

8.5.1. Prove the following inequalities:[4]

(1) $n^{n/2}<n!<\left(\frac{n+1}{2}\right)^n$ for $n>2$;

(2) $(2n)!<[n(n+1)]^n$;

(3) $\left(1+\frac{1}{n}\right)^n<3$;

(4) $\left(\frac{n}{3}\right)^n<n!$;

(5) $(n!)^2<\left(\frac{n(n+1)}{2}\right)^n, \quad n>1$;

(6) $1\times 2^2\times 3^3\times\ldots\times n^n\leqslant\left(\frac{2n+1}{3}\right)^{\frac{n(n+1)}{2}}$;

(7) $\frac{(2n-1)!!}{(2n)!!}<\frac{1}{\sqrt{3n+1}}$; (10) $(1+\alpha)^n\geqslant 1+\alpha n,\ \alpha\geqslant -1$;

(8) $(2n-1)!!<n^n,\ n>1$; $(n+2/n+1)^{n+1}>(n+1/n)^n$.

(9) $n!>e^{-n}n^n$.

8.5.2. Prove the following inequalities:

(1) $\left(2\frac{n-k}{k+1}\right)^k<\binom{n}{k}<\left(\frac{3n}{k}\right)^k, \quad n>k>1$;

(2) $\left(\frac{n}{k}\right)^k<\binom{n}{k}<\frac{n^n}{k^k(n-k)^{n-k}}, \quad n>k>0$;

(3) $\frac{4^n}{2\sqrt{n}}<\binom{2n}{n}\leqslant\frac{4^n}{\sqrt{3n+1}}, \quad n>1$.

8.5.3. Using Stirling's formula, show that for $n\to\infty$, the following asymptotic equalities are valid:

(1) $(2n-1)!!\sim\sqrt{2}\,(2n)^n e^{-n}$;

(2) $\binom{2n}{n}\sim\frac{1}{\sqrt{\pi n}}4^n$;

(3) $\frac{n!}{\left(\left[\frac{n}{3}\right]!\right)^2\left(n-2\left[\frac{n}{3}\right]\right)!}\sim\frac{3\sqrt{3}}{2\pi}\frac{3^n}{n}$;

(4) $\frac{(m+1)(m+2)\ldots(m+n)}{(k+1)(k+2)\ldots(k+m)}\sim\frac{k!}{m!}n^{m-k}$ for integral

[4] The symbol $(2n-1)!!$ is used to denote the number $1\times 3\times 5\times\ldots\times(2n-1)$, while $(2n)!!=2\times 4\times 6\times\ldots\times(2n)$.

non-negative k and m;

(5) $\frac{(2n)!!}{(2n-1)!!} \sim \sqrt{\frac{\pi}{2} n}$.

8.5.4. Prove that for $n \to \infty$, the following asymptotic equalities are valid:

(1) $\sum_{k=1}^{n} \frac{1}{k+1} \binom{n}{k} \sim \frac{2^{n+1}}{n}$;

(2) $\sum_{\nu} \binom{n}{r+k\nu} \sim \frac{1}{k} 2^n, \quad 0 \leqslant r < k$;

(3) $\sum_{\nu} \binom{n}{r+k\nu} \alpha^{r+k\nu} \sim \frac{1}{k} (1+\alpha)^n, \quad 0 \leqslant r < k, \quad \alpha > 0$;

(4) $\sum_{k=1}^{n} \frac{(-1)^{k-1}}{k} \binom{n}{k} \sim \ln n$;

(5) $\sum_{k=1}^{n} k \binom{n}{k}^2 \sim \frac{1}{2} \sqrt{\frac{n}{\pi}}\, 4^n$.

8.5.5. Let $b_0, b_1, \ldots, b_n$ be such numbers that $0 < a^k \leqslant b_k \leqslant c^k < 1$. Are the following inequalities valid?

(1) $(1+a)^n \leqslant \sum_{k=0}^{n} \binom{n}{k} b_k \leqslant (1+c)^n$,

(2) $(1-c)^n \leqslant \sum_{k=0}^{n} (-1)^k \binom{n}{k} b_k \leqslant (1-a)^n$.

8.5.6. (1) Prove the Chebyshev inequality in the following form. Let $A = \{a_1, a_2, \ldots, a_n\}$ be a set of numbers, and $\bar{a} = \frac{1}{n} \sum_{i=1}^{n} a_i$, $Da = \frac{1}{n} \sum_{i=1}^{n} (a_i - \bar{a})^2$. Then the fraction δ_t of those a_i for which $|a_i - \bar{a}| \geqslant t$, does not exceed $\frac{Da}{t^2}$.

(2) Using Chebyshev's inequality, prove that

$$\sum_{0 \leqslant k \leqslant \frac{n}{2} - t\sqrt{n}} \binom{n}{k} + \sum_{\frac{n}{2} + t\sqrt{n} \leqslant k \leqslant n} \binom{n}{k} \leqslant \frac{2^n}{(2t)^2}.$$

(3) Prove that

$$\sum_{k=1}^{n} \frac{1}{k}\binom{n}{k} \sim \frac{2^{n+1}}{n}.$$

8.5.7. Using Stirling's formula, show that

(1) if $k \to \infty$ and $k - \frac{n}{2} = O(n^{3/4})$ for $n \to \infty$, then $\binom{n}{k} \sim \frac{2^{n+1}}{\sqrt{2\pi n}} e^{-(2k-n)^2/2n}$;

(2*) if $a > 0$, $k \to \infty$ and $k - \frac{an}{a+1} = O(n^{2/3})$ for $n \to \infty$, then $\binom{n}{k} a^k = \frac{(1+a)^{n+1}}{\sqrt{2\pi n a}} e^{-\frac{(k(a+1)-an)^2}{2an}} \left(1 + O\left(\frac{1}{n} + \frac{(k(a+1)-an)^3}{n^2}\right)\right)$;

(3*) if $a > 0$, $k < m$, $k \to \infty$ and $k,\ m = \frac{an}{a+1} + O(n^{2/3})$ as $n \to \infty$, then

$$\sum_{\nu=k}^{m} \binom{n}{\nu} a^{\nu} \sim \frac{(a+1)^n \sqrt{na}}{\sqrt{2\pi}} \times \left(\frac{e^{-\frac{((a+1)m-an)^2}{na}}}{m(a+1)-na} - \frac{e^{-\frac{((a+1)k-an)^2}{na}}}{k(a+1)-na} \right);$$

(4*) if $a > 0$, $k \to \infty$ and $k - \frac{an}{a+1} = O(n^{2/3})$ as $n \to \infty$, then

$$\sum_{\nu > \frac{na}{a+1} + t\frac{\sqrt{na}}{a+1}} \binom{n}{\nu} a^{\nu} \sim \frac{1}{\sqrt{2\pi}\, t} e^{-\frac{t^2}{2}} (1+a)^n.$$

8.5.8. Let $0 < \lambda < 1$, λ_n be an integer, $\mu = 1 - \lambda$, and let $G(n, \lambda) = \frac{\lambda^{-\lambda n} \mu^{-\mu n}}{\sqrt{2\pi\lambda\mu n}}$. Using formulas (12) and (13), show that

(1) $\binom{n}{\lambda n} \sim G(n, \lambda)$ as $n \to \infty$;

(2) $G(n, \lambda) \exp\left(-\frac{1}{12n}\left(\frac{1}{\lambda} + \frac{1}{\mu}\right)\right) < \binom{n}{\lambda n} < G(n, \lambda)$;

(3) $\frac{\sqrt{\pi}}{2} G(n, \lambda) \leqslant \binom{n}{\lambda n}$;

(4) $\binom{n}{\lambda n} < \sum_{k=\lambda n}^{n} \binom{n}{k} \frac{\lambda}{2\lambda - 1}\binom{n}{\lambda n}$ for $\lambda > \frac{1}{2}$;

(5) $\sum_{k=\lambda n}^{n} \binom{n}{k} < \lambda^{-\lambda n}\mu^{-\mu n}$ for $\lambda > \frac{1}{2}$, $n \geqslant 3$.

8.5.9. Let k and n be natural numbers ($k < n$). Show that

(1) $(n)_k = n^k \exp\left(-\sum_{\nu=1}^{\infty} \frac{1}{\nu n^\nu} \sum_{i=1}^{k-1} i^\nu\right)$;

(2) if for $n \to \infty$ $k = 0(\sqrt{n})$, then $(n)_k \sim n^k$;

(3) if $k = o(n)$ as $n \to \infty$, then for any $m > 1$

$$(n)_k = n^k \exp\left(-\sum_{\nu=1}^{m-1} \frac{k^{\nu+1}}{\nu(\nu+1)n^\nu} + O\left(\frac{k^{m+1}}{n^m}\right)\right);$$

(4) for $n \to \infty$ and $k = o(n^{3/4})$

$$(n)_k = n^k \exp\left(-\frac{k^2}{2n} - \frac{k^3}{6n^2} + o(1)\right);$$

(5) if $\frac{r}{n+m} \to t$, $\frac{n}{n+m} \to p$, $\frac{m}{n+m} \to q$ and $h \times (k - rp) \to x$ for $m + n \to \infty$, then

$$\binom{n}{k}\binom{m}{r-k}\Big/\binom{m+n}{r} = h\varphi(x),$$

where $h = ((n+m)\,pqt\,(1-t))^{-1/2}$, $\varphi(x) = (2\pi)^{-1/2} e^{-x^2/2}$.

8.5.10. (1) Let $k = k(n)$ and $s = s(n)$ be such that for $n \to \infty$ $s = o(\sqrt{k})$. Show that $\binom{n-s}{k-s}\Big/\binom{n}{k} \sim \left(\frac{k}{n}\right)^s$.

(2) Show that if $s = o\left(k^{\frac{r}{r+1}}\right)$, then

$$\binom{n-s}{k-s}\Big/\binom{n}{k} \sim \left(\frac{k}{n}\right)^s \exp\left(-\sum_{\nu=1}^{r} \frac{s^{\nu+1}}{\nu(\nu+1)}\left(\frac{1}{k^\nu} - \frac{1}{n^\nu}\right)\right).$$

8.5.11. Let $s = s(n)$ and $k = k(n)$ be non-negative integral functions of a natural argument. Show that

(1) if $s + k = o(n)$ for $n \to \infty$, then

$$\binom{n-s}{k}\Big/\binom{n}{k} = \exp\left(-\frac{sk}{n} - \frac{s^2k + sk^2}{2n^2} - s\sum_{\nu=3}^{\infty} \frac{1}{\nu}\left(\frac{k}{n}\right)^\nu \right.$$

$$\left. - k\sum_{\nu=3}^{\infty} \frac{1}{\nu(\nu-1)}\left(\frac{s}{n}\right)^\nu + o(1)\right);$$

(2) if $s^2+k^2=o(n)$ for $n\to\infty$, then $\binom{n-s}{k}\Big/\binom{n}{k}\sim 1$;

(3) $\exp\left(-\frac{sk}{n}\left(1+\frac{k}{n-k}+\frac{sn}{2(n-k)(n-k-s)}\right)\right)<$

$$\frac{\binom{n-s}{k}}{\binom{n}{k}}<\exp\left(-\frac{sk}{n}\right),\quad n>k+s.$$

8.5.12. Let $f(x)$ be a continuous, monotonically increasing function on the segment $[n, m]$. Show that

$$f(n)\leqslant\sum_{k=n}^{m}f(k)-\int_{n}^{m}f(x)\,dx\leqslant f(m).$$

8.5.13. Using the previous problem, show that for $m\to\infty$, the following relations are valid:

(1) $\sum_{k=1}^{m}\ln k\sim m\ln m-m+O(\ln m)$;

(2) $\sum_{k=1}^{m}k^n=\frac{1}{n+1}m^{n+1}+O(m^n),\quad n>1$;

(3) $\sum_{k=2}^{m}\frac{1}{k\ln k\times\ln\ln k}=\ln\ln\ln m+c+O\left(\frac{1}{m\ln m\ln\ln m}\right)$,

c is a constant;

(4) $\sum_{k=1}^{m}\frac{\log k}{k}=\frac{1}{2}\log^2 m+c+O\left(\frac{\log m}{m}\right)$, c is a constant;

(5) $\sum_{k=1}^{m}\frac{1}{k\ln^2 k}=c+\frac{1}{\ln m}+O\left(\frac{1}{m\ln^2 m}\right)$, c is a constant;

(6) $\sum_{k=n}^{m}\frac{1}{k^{\nu}}=\frac{1}{\nu-1}\left(\frac{1}{n^{\nu-1}}-\frac{1}{m^{\nu-1}}\right)+O\left(\frac{1}{n^{\nu}}-\frac{1}{m^{\nu}}\right)$.

8.5.14. The sequence $\{p_n\}$ is defined by the recurrence relation $p_n=p_{n-1}-\alpha p_{n-1}^{\beta}$, $p_0=1$, $0<\alpha<1$, $\beta>1$. Show that

(1) $0<p_n<1$, $n\geqslant 1$;

(2) p_n decreases monotonically with increasing n;

(3*) $p_n\sim(\alpha(\beta-1)n)^{\frac{1}{1-\beta}}$, $n\to\infty$.

Hint. Use the inequalities

$$n=\sum_{k=1}^{n}\frac{p_k-p_{k-1}}{-\alpha p_{k-1}^{\beta}}<\int_{1}^{p_n}\frac{dx}{-\alpha x^{\beta}}<\sum_{k=1}^{n+1}\frac{p_k-p_{k-1}}{-\alpha p_k^{\beta}}=n+1.$$

8.5.15. (1) Show that the solution of the equation $xe^x = t$ has the form

$$x=\ln t-\ln\ln t+\frac{\ln\ln t}{\ln t}+O\left(\left(\frac{\ln\ln t}{\ln t}\right)^2\right),\quad t\to\infty.$$

(2) Prove that the solution of the equation $e^x + \ln x = t$ for $t\to\infty$ has the form

$$x=\ln t-\frac{\ln\ln t}{t}+O\left(\left(\frac{\ln\ln t}{t}\right)^2\right).$$

8.5.16. Let $f(t)>0$ and $e^{tf(t)}=f(t)+t+O(1)$ for $0<t<\infty$. Show that $f(t)=\frac{\ln t}{t}+O(t^{-2})$ for $t\to\infty$.

8.5.17. Suppose that the generating function $A(t)$ of the sequence $\{a_n\}$ has the form $A(t)=\frac{Q(t)}{P(t)}$, where $Q(t)$ and $P(t)$ are polynomials with real coefficients. Let λ_1 be the smallest (in magnitude) root of the polynomial $P(t)$. Prove that:

(1) if λ_1 is a simple (not multiple) real root, then for $n\to\infty$

$$a_n\sim-\frac{Q(\lambda_1)}{P'(\lambda_1)}\lambda_1^{-n-1},\quad\text{where } P'(\lambda_1)=\frac{d}{dt}P(t)\,|_{t=\lambda_1};$$

(2) if λ_1 is a real root of multiplicity r, then for $n\to\infty$

$$a_n\sim\frac{(-1)^r r!\,Q(\lambda_1)}{P^{(r)}(\lambda_1)}\binom{r+n-1}{r-1}\left(\frac{1}{\lambda_1}\right)^{n+r},$$

where $P^r(\lambda_1)$ is a derivative of $P(t)$ of the order r at the point $t=\lambda_1$.

8.5.18. Let $A(t)$ be the generating function of the sequence $\{a_n\}$. Find the asymptotic behaviour of a_n for $n\to\infty$:

(1) $A(t)=\frac{1+t}{3t^2-4t+1}$;

(2) $A(t)=\frac{1}{6t^2+5t-6}$;

(3) $A(t)=\frac{12t^3+10t^2}{6t^2+5t-6}$;

(4) $A(t)=\frac{t}{4t^2+1}$;

(5) $A(t)=\frac{2t^3}{6t^4-17t^3+35t^2-22t+4}$;

(6) $A(t)=\frac{1-3t}{(8t^3-1)(t^2+1)}$;

(7) $A(t)=\frac{1}{(t^2+2t-2)^2}$;

(8) $A(t)=\frac{t^3-1}{(2t^2+1)(t^2+1.4t+0.49)}$.

8.5.19. Find the asymptotic behaviour of a_n for $n\to\infty$ from the following recurrence relations and the corresponding initial conditions:

(1) $a_{n+2}+3a_{n+1}+2a_n=0,\quad a_0=1,\ a_1=2$;

(2) $a_n=qa_{n-1}+p(1-a_{n-1}),\quad a_0=1,\quad p+q=1$, $p,\ q>0$;

(3) $a_{n+2}+2a_{n+1}+4a_n=0,\quad a_0=0,\quad a_1=2$;

(4) $a_{n+3}-9a_{n+2}+26a_{n+1}-24a_n=0,\quad a_0=a_1=1,\quad a_2=-3$;

(5) $a_{n+4}-4a_{n+2}+4a_n=2^n,\quad a_0=1,\quad a_1=0,\quad a_2=2,\quad a_3=0$.

8.5.20*. Find the limit of the sequence $\{a_n\}$, given by the recurrence relations

(1) $a_{n+1}=(a_n+b/a_n)/2,\quad b>0,\quad a_0>0$.

(2) $a_{n+1}=(2a_n+b/a_n^2)/3,\quad b>0,\quad a_0>0$.

(3) $a_{n+1}=(b-a_n^2),\quad 0<b<1,\quad a_0=b/2$.

8.5.21. Suppose that a_n satisfies the relations

$$a_n\geqslant 2^{-n-1}+(n-1)4^{-n-1}, \tag{14}$$

$$a_{n+2}\leqslant 2^{-n-3}+(n+1)4^{-n-3}+a_n\left(\left(\frac{3}{4}\right)^{n+2}+(n+2)2^{n+2}+4a_n\right), \tag{15}$$

$$a_1=0,\quad a_2=1/16.$$

Show that

(1) $a_n\leqslant 1/8$;

(2) $a_n\leqslant 9(3/4)^n$;

(3) $a_n=2^{-n-1}(1+O((3/4)^n))$.

8.5.22. Suppose that the sequence $\{a_n\}$ satisfies the condition $a_{n+m}\leqslant a_n+a_m$, $a_1>0$. Prove that $a_n<a_1n$ for $n\geqslant 1$.

8.5.23. Let k and n be integers. Calculate $k=k(n)$ to the nearest integer, for which the function $f(n,k)$

assumes the maximum value:

(1) $f(n, k) = \binom{n}{k} 2^{-2^k}$;

(2) $f(n, k) = \binom{n}{k} 2^{n-k-2^k} (1-2^{-2^k})^{n-k}$.

8.5.24. Find the minimum and maximum values of the expression

$$f(n, r, k) = \binom{n}{r}\binom{n-r}{k-r} 2^{-r+2^r}$$

as a function of r ($0 \leqslant r \leqslant k \leqslant n$; r, k, n are integers).

8.5.25. Find the asymptotic behaviour of the quantity $g(n) = \min\limits_{0 \leqslant k \leqslant n} f(n, k)$ (k is an integer) for $n \to \infty$, if:

(1) $f(n, k) = 2^{n-k} + 2^{2^k}$;

(2) $f(n, k) = k2^k + \frac{1}{k} 2^{2n-k}$.

Solutions, Answers, and Hints

CHAPTER ONE

1.1.

1.1.1. (1) $\binom{n}{k}$; (2) 2^n. **1.1.2.** (1) 9, 13, 50; (2) (010011).

1.1.3. $\binom{n-1}{k-1}$. **1.1.5.** (1) $n2^{n-1}$; (2) $\binom{n}{k}2^{n-1}$. **1.1.6.** (1) 2^m;

(2)[1] $\binom{m}{\frac{k+m-r}{2}}\binom{n-m}{\frac{k-m+r}{2}}$; (3)[1] $\sum_{j=0}^{k}\binom{m}{\frac{j+m-r}{2}}\times$

$\binom{n-m}{\frac{j+r-m}{2}}$.

1.1.9. Hint. Prove that the number $\psi(n)$ of vectors $\tilde{\alpha}\in B_n^{2n}$ such that $\sum_{i=1}^{m}\alpha_i\leqslant m/2$ for all $m=\overline{1,\ 2n}$ satisfies the recurrence relation $\psi(n)=\sum_{i=1}^{n}\psi(i-1)\,\psi(n-i)$, $\psi(0)=\psi(1)=1$. Using this relation, prove that $\psi(6)=132$.

1.1.10. $\binom{n-r(k-1)}{k}$.

1.1.11. (1) **Hint.** Consider the stratum $B^n_{[n/2]}$. (2) **Hint.** Among $n+2$ tuples in B^n, there always exist two tuples of the same weight.

1.1.12. (1) $\binom{n-l}{k-l}$; (2) $\binom{k}{l}$; (3) $2^k+2^{n-k}-1$.

1.1.13. Assuming that $0\leqslant l<k\leqslant n$, we calculate the number $p(n,\ k,\ l)$ of pairs $(\tilde{\alpha},\ \tilde{\beta})$ such that $\tilde{\alpha}\in A$, $\tilde{\beta}\in B$. On the one hand, $p(k,\ l,\ n)=|A|\binom{n-l}{k-l}$, while on the other hand, $p(n,\ k,\ l)\leqslant$

---

[1] For a non-integral a, $\binom{n}{a}$ is assumed here to be equal to 0.

$|B|\binom{k}{l}$. Using the identity $\binom{n}{l}\binom{n-l}{k-l}=\binom{n}{k}\binom{k}{l}$, we obtain the required inequality. For $l \geqslant k$, the result will be the same.

1.1.14. (1) An ascending chain of length n contains one vertex in each stratum. In a stratum B_1^n, a vertex can be chosen in n different ways. After a vertex of the chain has been chosen in B_1^n, the vertex of the chain in the second stratum can be chosen in $n-1$ ways, and so on.

1.1.15. (1) Let A be a set of pairwise incomparable tuples in B^n, $A_i = A \cap B_i^n$ and $Z(\tilde{\alpha})$ be the set of ascending chains of length n, that contains the vertex $\tilde{\alpha}$. We have

$$n! \geqslant \left| \bigcup_{\tilde{\alpha} \in A} Z(\tilde{\alpha}) \right| = \sum_{\tilde{\alpha} \in A} |Z(\tilde{\alpha})| = \sum_{i=0}^{n} |A_i|\, i!\,(n-i)!$$

$$\geqslant \left[\frac{n}{2}\right]! \times \left]\frac{n}{2}\right[! \times \sum_{i=0}^{n} |A_i| = |A| \times \left[\frac{n}{2}\right]! \times \left]\frac{n}{2}\right[!.$$

Hence $|A| \leqslant \binom{n}{[n/2]}$. (2) **Hint.** For $i \leqslant k \leqslant n/2$, the inequality $i!\,(n-i)! \geqslant k!\,(n-k)!$ is valid.

1.1.16. Hint. We associate each number of the form $p_1^{\alpha_1} p_2^{\alpha_2} \ldots p_n^{\alpha_n}$ with a vector $(\alpha_1, \alpha_2, \ldots, \alpha_n)$ in B^n. Then the set A corresponds to the set $A' \subseteq B^n$ consisting of pairwise incomparable tuples. Using the solution of Problem 1.1.15 (1), we obtain the required statement.

1.1.17. Let $i_1, i_2, \ldots, i_k$ be the numbers of those coordinates at which $\tilde{\alpha}$ and $\tilde{\beta}$ are different. Any vector $\tilde{\gamma}$ such that $\tilde{\alpha} \leqslant \tilde{\gamma} \leqslant \tilde{\beta}$ can be obtained from $\tilde{\alpha}$ by replacing 0 by 1 in some of the above-mentioned coordinates. The number of ways in which such a substitution can be made is 2^k.

1.1.18. For the cube B^1, the partition consists of a single chain $Z = \{(0), (1)\}$. $Z_1 = \{(10)\}$ and $Z_2 = \{(00), (01), (11)\}$ can be taken as a partition chains for B^2. The properties (1) and (2) are satisfied for these partitions. We assume that the statement is proved for cubes with a dimensionality not exceeding n and then prove the statement for B^{n+1}. By hypothesis, there exists a partition of faces $B_0^{n+1,n+1}$ and $B^{n+1,n+1}$ into a chain, which satisfies the conditions (1) and (2). Partitions in $B_0^{n+1,n+1}$ and $B_1^{n+1,n+1}$ can be chosen in such a way that they are isomorphic, i.e. such that the chain $Z_0 = (\tilde{\alpha}_1, \tilde{\alpha}_2, \ldots, \tilde{a}_s)$ belongs to a partition of the cube $B_0^{n+1,n+1}$ if and only if the chain $Z_1 = \{\tilde{\alpha}_1', \tilde{\alpha}_2', \ldots, \tilde{\alpha}_s'\}$ obtained from Z_0 by substituting unity for zero at the $(n+1)$-th coordinate in each tuple $\tilde{\alpha}_i$ in Z_0 belongs to a partition of the face $B_1^{n+1,n+1}$. Let $Z_0 = \{\tilde{\alpha}_1, \tilde{\alpha}_2, \ldots, \tilde{\alpha}_s\}$ and $Z_1 = \{\tilde{\alpha}_1', \tilde{\alpha}_2', \ldots, \tilde{\alpha}_s'\}$ be two isomorphic chains in partitions of the faces $B_0^{n+1,n+1}$ and $B_1^{n+1,n+1}$. We construct two new chains $\hat{Z}_0 = \{\tilde{\alpha}_1, \tilde{\alpha}_2, \ldots, \tilde{\alpha}_s, \tilde{\alpha}'_s\}$ and $\hat{Z}_1 = \{\tilde{\alpha}'_1, \tilde{\alpha}'_2, \ldots, \tilde{\alpha}'_{s-1}\}$. The chain $\hat{Z}_0$ is obtained from Z_0 by adding the vertex $\tilde{\alpha}'_s \in Z_1$, while the chain $\hat{Z}_1$ is obtained from Z_1 by omitting the vertex $\tilde{\alpha}'_s$. We proceed in this way with each pair of isomorphic chains. As a result, we obtain a partition of the cube B^{n+1} into non-intersecting chains. The fulfilment of conditions (1) and (2) can be easily verified.

1.1.19. $2^k \sum\limits_{i=0}^{r} \binom{n-k}{i}$.

1.1.20. (2) We prove that the stratum B_1^n is a complete set for $n > 2$. Let $\tilde{\alpha}_i$ be a vector in B_1^n, in which the i-th coordinate is equal to unity. Let the distances $\rho(\tilde{\alpha}_i, \tilde{\beta}) = r_i$, $i = \overline{1, n}$ be specified. Let us demonstrate how $\tilde{\beta}$ can be reconstructed. If $\|\tilde{\beta}\| = k > 0$, we obviously have $r_i \in \{k-1, k+1\}$. Let $\max\limits_{1 \leqslant i \leqslant n} r_i \geqslant 2$ and $A(\tilde{\beta}) = \{i: r_i = \min\limits_{1 \leqslant j \leqslant n} r_j\}$. Then the i-th coordinate of the vector $\tilde{\beta}$ is equal to unity if $i \in A(\tilde{\beta})$ and equal to zero otherwise. If $\max\limits_{1 \leqslant j \leqslant n} r_j = 1$, then $\tilde{\beta} = \tilde{0}$. For $n = 5$, the stratum B_1^n is not a basis set since the set {(00001), (00010), (00100), (01000)} is complete. (3) For an even n and $k = n/2$, for any $n > 1$ and $k = 0, n$. (4) Let $A = \{\tilde{\alpha}_1, \ldots, \tilde{\alpha}_k\}$ be a complete set. Let us consider the set of equalities $\rho(\tilde{\alpha}_i, \tilde{\beta}) = r_i$, $i = \overline{1, k}$. Obviously, $r_i \in \{0, 1, \ldots, n\}$, $i = \overline{1, k}$. Each tuple $(r_1, r_2, \ldots, r_k)$ for which the set has a solution uniquely determines a certain vector $\tilde{\beta} \in B^n$. Hence it follows that $(n+1)^k \geqslant 2^n$. **Hint.** The upper bound for the number of vectors in the base system can be obtained from the fact that the set of equations $\rho(\tilde{\alpha}_i, \tilde{\beta}) = r_i$, $i = \overline{1, k}$, can be reduced to an ordinary set of linear equations. For $k > n$, some of the equations of this system can be expressed as linear combinations of remaining equations. Such equations, and hence the vectors corresponding to them, can be omitted.

1.1.21. Sufficiency. The commutation of coordinates of all vectors in B^n, as well as the inversion of the coordinates $i_1, i_2, \ldots, i_k$ of all vectors in B^n do not change the separation between the vertices. **Necessity.** Let φ be a mapping of the cube B^n into itself such that for any $\tilde{\alpha}$ and $\tilde{\beta}$, $\rho(\varphi(\tilde{\alpha}), \varphi(\tilde{\beta})) = \rho(\tilde{\alpha}, \tilde{\beta})$. Let us consider the set $D = B_0^n \cup B_1^n$ which is complete in B^n (see the previous problem). Let $\varphi(D) = \{\varphi(\tilde{0}), \varphi(\tilde{\alpha}_1), \ldots, \varphi(\tilde{\alpha}_n)\}$ be an image of the set D and $\tilde{\alpha}_i$ be a vector in B_1^n whose i-th coordinate is equal to unity. We put $\tilde{\beta}_i = \varphi(\tilde{0}) \oplus \varphi(\tilde{\alpha}_i)$, $i = \overline{1, n}$. We have $\| \tilde{\beta}_i \| = \rho(\tilde{0}, \tilde{\beta}_i) = \rho(\varphi(\tilde{0}) \oplus \varphi(\tilde{0}), \varphi(\tilde{0}) \oplus \varphi(\tilde{\alpha}_i)) = \rho(\varphi(\tilde{0}), \varphi(\tilde{\alpha}_i)) = \rho(\tilde{0}, \tilde{\alpha}_i) = 1$. Let $j(i)$ be the number of the coordinate of the vector $\tilde{\beta}_i$, which is equal to unity, $i = \overline{1, n}$. Then the mapping $\pi: i \to j(i)$ is a permutation on the set $\{1, 2, \ldots, n\}$. For an arbitrary $\tilde{\alpha} = (\alpha_1, \ldots, \alpha_n)$ in B^n, we put $\pi(\tilde{\alpha}) = (\alpha_{\pi(1)}, \ldots, \alpha_{\pi(n)})$. We show that for any $\tilde{\gamma} \in B^n$, the equality $\varphi(\tilde{\gamma}) = \pi(\tilde{\gamma}) \oplus \varphi(\tilde{0})$ is valid. Thus, the necessity is proved. Indeed, $\rho(\varphi(\tilde{\gamma}), \varphi(\tilde{0})) = \rho(\tilde{\gamma}, \tilde{0})$, $\rho(\varphi(\tilde{\gamma}), \varphi(\tilde{\alpha}_i)) = \rho(\tilde{\gamma}, \tilde{\alpha}_i)$. On the other hand, $\rho(\pi(\tilde{\gamma}) \oplus \varphi(\tilde{0}), \varphi(\tilde{0})) = \rho(\tilde{\gamma}, \tilde{0})$, $\rho(\pi(\tilde{\gamma}) \oplus \varphi(\tilde{0}), \varphi(\tilde{\alpha}_i)) = \rho(\tilde{\gamma}, \tilde{\alpha}_i)$. Obviously, the sets $\varphi(D)$ and D are complete in B^n. Hence it follows that $\varphi(\tilde{\gamma})$ and $\pi(\tilde{\gamma}) \oplus \varphi(\tilde{0})$ coincide.

1.1.22. $\binom{2^{n+1}-1}{2^n-1}$.

1.1.23. (1) For $n=1$ and $k=\overline{0, 1}$, the statement is true. $\left(\text{We assume that } \binom{0}{k}=0 \text{ for all } k=0, 1, \ldots.\right)$ Let us suppose that the statement is proved for all dimensionalities which are less than or equal to $n-1$ and for all $k=0, 1, \ldots$. Let $\tilde{\alpha}$ be $(\alpha_1, \ldots, \alpha_n)$, and $\tilde{\alpha}'$ be $(\alpha'_1, \ldots, \alpha'_{n-1}) = (\alpha_2, \ldots, \alpha_n)$. If $\alpha_1 = 0$, we obviously have $|M_k^n(\tilde{\alpha})| = |M_k^{n-1}(\tilde{\alpha}')| = 1 + \binom{n-1-(i_1-1)}{k} + \ldots + \binom{n-1-(i_k-1)}{1} = 1 + \binom{n-i_1}{k} + \ldots + \binom{n-i_k}{1}$. If $\alpha_1 = 1$, then $|M_k^n(\tilde{\alpha})| = \binom{n-1}{k} + |M_{k-1}^{n-1}(\tilde{\alpha}')| = \binom{n-1}{k} + 1 + \binom{n-1-(i_2-1)}{k-1} + \ldots + \binom{n-1-(i_k-1)}{1} = 1 + \binom{n-i_1}{k} + \ldots + \binom{n-i_k}{1}$. (2) For arbitrary k and $\tilde{\alpha} \in B_k^n$, the number $\mu_k(\tilde{\alpha}) = |M_k^n(\tilde{\alpha})|$ will be called the number of the tuple $\tilde{\alpha}$ in the stratum B_k^n. Let $\tilde{\alpha} \in B_k^n$ be a vector whose unit coordinates are $i_1, \ldots, i_k$, $1 \leqslant i_1 < \ldots < i_k \leqslant n$. Then in view of what has heen stated above, $\mu_k(\tilde{\alpha}) = 1 + \binom{n-i_1}{k} + \ldots + \binom{n-i_k}{1}$. Further, let the tuple $\tilde{\beta} \in B_l^n$ be obtained from $\tilde{\alpha}$ by substituting zeros for the last $k-l$ unit coordinates. We show that $Z_l^n(M_k^n(\tilde{\alpha})) = M_l^n(\tilde{\beta})$ or, which is the same, $Z_l^n(M_k^n(\tilde{\alpha})) = \{\tilde{\gamma} : \mu_l(\tilde{\gamma}) \leqslant \mu_l(\tilde{\beta})\}$. We first prove that for any $\tilde{\gamma} \in B_l^n$ such that $\mu_l(\tilde{\gamma}) < \mu_l(\tilde{\beta})$, there exists a tuple $\tilde{\delta} \in B_k^n$ such that $\tilde{\gamma} < \tilde{\delta}$ and $\mu_k(\tilde{\delta}) \leqslant \mu_k(\tilde{\alpha})$. We choose for $\tilde{\delta}$ a tuple obtained from $\tilde{\gamma}$ by substituting unities for the last $k - l$ zero coordinates. We must prove that $\mu_k(\tilde{\delta}) \leqslant \mu_k(\tilde{\alpha})$. If $\mu_k(\tilde{\delta}) > \mu_k(\tilde{\alpha})$, there exists a t such that $\alpha_i = \delta_i$ for $i < t$, $\alpha_t = 0$, $\delta_t = 1$. Considering the tuples $\tilde{\alpha}'$ and $\tilde{\delta}'$ obtained from $\tilde{\alpha}$ and $\tilde{\delta}$ respectively by substituting zeros for the last $k - l$ unit coordinates, we have $\mu_l(\tilde{\alpha}') \leqslant \mu_l(\tilde{\delta}')$, $\tilde{\alpha}' = \tilde{\beta}$, $\mu_l(\tilde{\delta}') \leqslant \mu_l(\tilde{\gamma})$. We arrive at a contradiction to the inequality $\mu_l(\tilde{\beta}) > \mu_l(\tilde{\gamma})$. Similarly, we can prove that for any $\tilde{\gamma} \in B_l^n$ such that $\mu_l(\tilde{\gamma}) > \mu_l(\tilde{\beta})$ there does not exist a tuple $\tilde{\delta} \in M_k^n(\tilde{\alpha})$ such that $\tilde{\gamma} < \tilde{\delta}$. (3) It follows from (1) and (2). (4) Let $A \subseteq B_k^n$ and A be other than the initial segment of the stratum B_k^n. We construct the set $F \subseteq B_k^n$ such that $|F| = |A|$, $\sum_{\tilde{\alpha} \in F} \nu(\tilde{\alpha}) < \sum_{\tilde{\alpha} \in A} \nu(\tilde{\alpha})$, $|Z_{k-1}^n(F)| \leqslant |Z_{k-1}^n(A)|$. Thus, the statement is proved. We denote by $K(\tilde{\alpha})$ the set of the numbers of unit coordinates of the tuple $\tilde{\alpha}$. Let $\tilde{\alpha} \in B_k^n \setminus A$ and $\tilde{\beta} \in A$ be the tuples on which $\min |K(\tilde{\sigma}) \setminus K(\tilde{\tau})|$ is attained, where the minimum is taken over all pairs $(\tilde{\sigma}, \tilde{\tau})$ such that $\tilde{\sigma} \in B_k^n \setminus A$, $\tilde{\tau} \in A$, $\nu(\tilde{\sigma}) < \nu(\tilde{\tau})$. Let $M = K(\tilde{\alpha}) \setminus K(\tilde{\beta})$, $N = K(\tilde{\beta}) \setminus K(\tilde{\alpha})$. Obviously, $M \cap N = \varnothing$, $|M| = |N|$. For arbitrary $\tilde{\sigma}$ and $\tilde{\tau}$ in B^n, we denote by $\tilde{\sigma} \setminus \tilde{\tau}$ the tuple of the form $\tilde{\sigma} \oplus (\tilde{\sigma} \cap \tilde{\tau})$, and assume that $\tilde{\alpha}' = \tilde{\alpha} \setminus \tilde{\beta}$ and $\tilde{\beta}' = \tilde{\beta} \setminus \tilde{\alpha}$. Clearly, $\nu(\tilde{\alpha}') < \nu(\tilde{\beta}')$. We put

$$D = \{\tilde{\gamma} : \tilde{\gamma} \in A, \ \tilde{\alpha}' \cap \tilde{\gamma} = \tilde{0}, \ \tilde{\beta}' < \tilde{\gamma}; \ (\tilde{\gamma} \setminus \tilde{\beta}') \cup \tilde{\alpha}' \notin A\},$$

$$E = \{\tilde{\delta} : \tilde{\delta} = (\tilde{\gamma} \setminus \tilde{\beta}') \cup \tilde{\alpha}', \ \tilde{\gamma} \in D\},$$

$$F = E \cup (A \setminus D).$$

Obviously, $F \subseteq B_k^n$, $|F| = |A|$ and $\sum_{\tilde{\alpha} \in F} \nu(\tilde{\alpha}) < \sum_{\tilde{\alpha} \in A} \nu(\tilde{\alpha})$, since $\nu((\tilde{\gamma} \setminus \tilde{\beta}') \cup \tilde{\alpha}') < \nu(\tilde{\gamma})$ for all $\tilde{\gamma} \in D$. It remains for us to prove that $|Z_{k-1}^n(F)| \leqslant |Z_{k-1}^n(A)|$. We shall just give a hint towards the solution. We must first prove that if $\tilde{\sigma} \in Z_{k-1}^n(F) \setminus Z_{k-1}^n(A)$, then $\tilde{\sigma} \cap \tilde{\beta}' = \tilde{0}$ and $\tilde{\alpha}' < \tilde{\sigma}$, and then show that the tuple $\tilde{\varepsilon} = (\tilde{\sigma} \setminus \tilde{\alpha}') \cup \tilde{\beta}'$ is contained in $Z_{k-1}^n(A) \setminus Z_{k-1}^n(F)$. (5) **Hint.** Prove by induction on $k - l$. (6) Let k be the highest integer for which $a_k > 0$. Then $l = k - 1$, the statement is reduced to (4). By taking an inductive step (induction is carried out on $k - l$), we assume that $A_i = A \cap B_i^n$, $A_i' = A \cap \left(\bigcup_{j=i}^{k} B_j^n \right)$ and note that

$$Z_l^n(A) = Z_l^n(Z_{l+1}^n(A_{l+1}')) \cup A_l.$$

It is clear that

$$\min_{C\subseteq B_l^n,\ |C|=m} |Z_l^n(C)| \leqslant \min_{C\subseteq B_l^n,\ |C|>m} |Z_l^n(C)|.$$

By inductive hypothesis, $\min |Z_{l+1}^n(A'_{l+1})|$ is attained when each $Z_i^n(A)$, $i > l+1$, is an initial segment of the stratum B_i^n. But then according to (2), the set $Z_{l+1}^n(A'_{l+1})$ is an initial segment. By an appropriate choice of A_l, the set $Z_l^n(A) = Z_l^n(Z_{l+1}^n(A'_{l+1})) \cup A_l$ can be made an initial segment of the stratum B_l^n. This leads to the statement.

1.1.24. For each $l=\overline{0, n}$, we suppose that $A_l = A \cap \beta_l^n$. We denote by $t(n, m, k)$ the number of pairs $(\tilde{\alpha}, \tilde{\beta})$ such that $\tilde{\alpha} \in A_m$, $\tilde{\beta} \notin A$, $\tilde{\beta} \cap \tilde{\alpha} = \tilde{0}$, or $\tilde{\beta} \cup \tilde{\alpha} = \tilde{\beta}$, $\tilde{\beta} \in B_k^n \cup B_{m+k}^n$. By hypothesis, $\tilde{\beta} \notin A$. On the one hand, $t(n, m, k) = a_m \binom{n-m}{k}$. On the other hand, $t(n, m, k) \leqslant \left(\binom{n}{m+k} - a_{m+k}\right)\binom{m+k}{m} + \left(\binom{n}{k} - a_k\right)\binom{n-k}{m}$. This leads to the statement.

1.1.25. (1) Let $\tilde{\alpha}$ be the center of the set A. Then $A' = \{\tilde{\alpha} \oplus \tilde{\beta}, \tilde{\beta} \in A\}$ is the required set. (2) We can assume that in view of (1), $\tilde{0}$ is the center of A. In order to construct A', it is sufficient for each $i = \overline{1, n}$ to substitute the tuple $\tilde{\alpha}^i$ for each tuple $\tilde{\alpha}$ in A_1^i for which $\tilde{\alpha}^i \notin A$. (3) **Hint.** Let $A \subseteq B^n$ have the center $\tilde{0}$ and possess the property (1). Let $S^{i,j}$ be a transformation consisting in that for each $\tilde{\alpha} \in A_{10}^{i,j}$ such that $\tilde{\alpha}^{i,j} \notin A$, the tuple $\tilde{\alpha}$ is replaced by $\tilde{\alpha}^{i,j}$. Let us arrange the pairs (i, j) such that $1 \leqslant i < j \leqslant n$, as follows:

$$(1, n), (1, n-1), \ldots, (1, 2), (2, n), (2, n-1), \ldots, (2, 3), \ldots, (n-1, n).$$

Applying to A consecutively all transformations $S^{i,j}$ starting from $(i, j) = (1, n)$ and finishing with $(i, j) = (n-1, n)$, we obtain the required A'.

1.1.26. (2) For A such that $|A| = 2^{n-1}$ we can take any $(n-1)$-dimensional face containing $\tilde{1}$. For an odd n, we can take the set $\{\tilde{\alpha}: \|\tilde{\alpha}\| \geqslant (n+1)/2\}$. On the other hand, if $|A| > 2^{n-1}$, there exist in A two opposite tuples whose intersection is $\tilde{0}$

1.1.27. (1) **Hint.** Consider the set $\left\{\tilde{\alpha}: \|\tilde{\alpha}\| \geqslant \left]\frac{n+k+1}{2}\right[, \tilde{\alpha} \in B^n\right\}$.

1.1.28. (1) The proof is similar to 1.1.25. (2) $A = \{\tilde{\alpha}: \tilde{\alpha} \in b^n, \|\tilde{\alpha}\|$ is even$\}$.

1.1.31. (1) Each face $B^{n, i_1, \ldots, i_k}_{\sigma_1 \ldots \sigma_k}$ of direction $I = \{i_i, \ldots, i_k\}$ can be put in one-to-one correspondence with its *code*, viz. a vector $(\gamma_1, \ldots, \gamma_n)$ with coordinates $\gamma_i \in \{0, 1, -\}$, $i = \overline{1, n}$, such that $\gamma_{i_1} = \sigma_1, \ldots, \gamma_{i_k} = \sigma_k$, the remaining coordinates being blanks. The fixed k places can be filled with zeros and unities in 2^k ways. (4) The required number is equal to the number of vectors $(\gamma_1, \ldots, \gamma_n)$ with k coordinates in the set $\{0, 1\}$ and $n - k$ blanks. The k places from n for significant symbols (i.e. for 0 and 1) can be chosen in $\binom{n}{k}$ ways, and the fixed k places can be filled by zeros and unities in 2^k ways. (6) The code of the k-dimensional face containing a given vertex $(\alpha_1, \ldots, \alpha_n)$ can be specified by arranging k blanks instead of certain k coordinates of the vector $(\alpha_1, \ldots, \alpha_n)$. (8) The intersection of two faces is a face. A j-dimensional face can be chosen in the face of dimensionality l in $\binom{l}{j} 2^{l-j}$ ways. Thus, l coordinates have already been fixed in the code of the k-dimensional face. Among the remaining $n - l$ coordinates, we must arrange $k - j$ blanks. This can be done in $\binom{n-l}{k-j}$ ways.

1.1.33. Let $\tilde{\gamma}_1, \tilde{\gamma}_2, \tilde{\gamma}_3$ be the codes of the faces g_1, g_2, g_3 respectively. Two faces do not intersect if and only if there exists a coordinate at which the codes of these faces have significant symbols, and these symbols are opposite. If, however, the faces g_1 and g_2 intersect, the code of their intersection contains significant symbols of coordinates at which at least one of the codes $\tilde{\gamma}_1$ and $\tilde{\gamma}_2$ has a significant coordinate, and significant coordinates of the code of the intersection coincide with the coorresponding coordinates of at least one of the codes $\tilde{\gamma}_1$ and $\tilde{\gamma}_2$. If $\tilde{\delta}$ is the code of the intersection of the faces g_1 and g_2 and the face g_3 has no vertices in common with this intersection, there exists an i such that the i-th coordinates of the vectors $\tilde{\delta}$ and $\tilde{\gamma}_3$ are significant and do not coincide. Then the i-th coordinate of the vector $\tilde{\gamma}_3$ does not coincide with the i-th significant coordinate of one of the vectors $\tilde{\gamma}_1$ or $\tilde{\gamma}_2$. This means that the corresponding faces do not intersect. We arrive at a contradiction.

1.1.34. The problem can be easily reduced to the case when $\sum_{i=1}^{s} 2^{n_i} = 2^n$. We shall carry out the proof by induction on n. For $n = 0, 1$, the statement is obviously true. We make a transition $n \to n + 1$. Let $\min_{1 \leqslant i \leqslant s} n_i \geqslant 1$. We put $n_i' = n_i - 1$. Then n_i'

are non-negative numbers, and $\sum_{i=1}^{s} 2^{n_i} = 2^n$. By inductive hypothesis, there exists a partition of each of the faces $B_0^{n+1,\, n+1}$ and $B_1^{n+1,\, n+1}$ of the cube B^{n+1} into the faces of dimensionalities $n'_1, \ldots, n'_s$. We choose these partitions so that they are identical, i.e. such that for each $i = \overline{1, s}$, the union of n'_i-dimensional faces forms an n_i-dimensional face g_i. We obtain the required partition. For $\min_{1 \leqslant i \leqslant s} n_i = 0$, let m be the number of those i for which $n_i = 0$. The condition $\sum_{i=1}^{s} 2^{n_i} = 2^{n+1}$ implies that m is even. Let us consider a new set $n'_1, n'_2, \ldots, n'_{s-m/2}$ obtained from the previous one by substituting $m/2$ unities for m zeros. As earlier, we partition the cube B^{n+1} into faces, and then split certain $m/2$ faces of dimensionality 1 into m faces of dimensionality 0.

1.1.36. (1) **Majorization. Hint.** Consider $N = \{\tilde{\alpha} : \tilde{\alpha} \in B^n, \|\tilde{\alpha}\|\}$ is even. **Minorization.** Let $N \subseteq B^n$, $|N| < 2^{n-1}$. There exists (see 1.1.31 and 1.1.34) a partition of the cube B^n into nonintersecting faces of dimensionality 1. The number of faces in such a partition is 2^{n-1}. Hence it follows that at least one of the faces does not contain vertices from N. (3) **Hint.** Consider $N = \{\tilde{\alpha} : \tilde{\alpha} \in B^n, \|\tilde{\alpha}\| \equiv 0 \pmod 3)\}$ or $N_1 = \{\tilde{\alpha} : \tilde{\alpha} \in B^n, \nu(\tilde{\alpha}) \equiv 3 \pmod 4)\}$. (4) **Minorization.** It should be noted that for a vertex $\tilde{\alpha} = (\alpha_1, \ldots, \alpha_n)$ to be contained in an $(n-2)$-dimensional face with a code $\tilde{\gamma} = (\gamma_1, \ldots, \gamma_n)$ (see solution of Problem 1.1.31) with significant coordinates in the places i and j, it is necessary that $\alpha_i = \gamma_i$ and $\alpha_j = \gamma_j$. Let $N \subseteq B^n$, $|N| = m$ be a set such that its each $(n-2)$-dimensional face contains the vertex $\tilde{\alpha}$ from N. Let us consider a binary matrix M whose lines are vectors in N. The above remark implies that for each pair of numbers (i, j), $1 \leqslant i < j \leqslant n$ and each pair (σ, τ), $\sigma, \tau \in \{0, 1\}$, there exists a line $\tilde{\alpha} = (\alpha_1, \ldots, \alpha_n)$ such that $\alpha_i = \sigma$, $\alpha_j = \tau$. Hence any two columns of the matrix M are pairwise incomparable. The number of pairwise incomparable binary sets of length m does not exceed $\binom{m}{[m/2]}$. Hence $\binom{m}{[m/2]} \geqslant n$. **Majorization.** Let m be the minimum integer such that $\binom{m}{[m/2]} \geqslant n$. We construct a binary matrix M with m lines and n pairwise incomparable columns. We add the lines $\tilde{0}$ and $\tilde{1}$ to the matrix M. Then the obtained set of vectors-lines is "piercing" for the set of all $(n-2)$-dimensional faces of the cube B^n.

1.1.37. For $n = 1, 2$, the statement is obvious. Let $\tilde{\alpha}_1, \tilde{\alpha}_2, \ldots, \tilde{\alpha}_{2^n}$ be a cycle in B^n. Let $\tilde{\beta}_\sigma$, where $\tilde{\beta} \in B^n$, $\sigma \in \{0, 1\}$, denote a vector of length $n + 1$, whose first n coordinates coincide with the corresponding coordinates of the tuple $\widetilde{\beta}$, and the $(n + 1)$-th coordinate is σ. Then the sequence $\widetilde{\alpha}_1 0, \widetilde{\alpha}_2 0, \ldots, \widetilde{\alpha}_{2^n} 0, \widetilde{\alpha}_{2^n} 1, \ldots, \widetilde{\alpha}_2 1, \widetilde{\alpha}_1 1$ is a cycle in B^{n+1}, containing all its vertices.

1.1.39. (1) Yes. (2) No. (3) Yes. (4) Yes.

1.1.43. Hint. It should be noted that if the tuples $\widetilde{\sigma} = (\sigma_1, \ldots \sigma_n)$ and $\widetilde{\sigma'} = (\sigma_1', \ldots, \sigma_n')$ are comparable, then one of the sums $\sum_{i=1}^{n} (-1)^{\sigma_i} a_i$ and $\sum_{i=1}^{n} (-1)^{\sigma_i'} a_i$ is greater than unity in absolute value.

1.2.

1.2.1. $2^{2^{n-1}}$ $(n \geqslant 1)$. **1.2.2.** 2 $(n \geqslant 1)$. **1.2.3.** $C_m^0 + C_m^1 + \ldots + C_m^{k-1}$, where $m = 2^n$ $(n \geqslant 1,\ k \geqslant 1)$.

1.2.7. (2) 2^{n+1} $(n \geqslant 1)$. **1.2.10.** (1) In five ways. (2) In nine ways.

1.2.12. (1) ***Basis of induction.*** If the complexity of a formula is unity (generated by the set of connectives $M = \{\neg, \&; \vee, \rightarrow\}$), this formula has one of the following forms: $(\neg x)$, $(x \& y)$, $(x \vee y)$ or $(x \rightarrow y)$ (within the notation of variables). Therefore, the statement is obvious for formulas of complexity 1. ***Induction step.*** Let the statement be valid for all formulas (generated by the set of connectives M) of complexity not exceeding a number l $(l \geqslant 1)$. We take an arbitrary formula $\mathfrak{A}$ (generated by M) of complexity $l + 1$. We assume for the sake of definiteness, that $\mathfrak{A} = (\mathfrak{B} \vee \mathfrak{C})$ (remaining cases are considered similarly). The connective $\vee$ between formulas $\mathfrak{B}$ and $\mathfrak{C}$ has an index equal to unity. Further, if the connective in formula $\mathfrak{B}$ (or $\mathfrak{C}$) treated as a connective of formula $\mathfrak{B}$ (resp. $\mathfrak{C}$) has an index greater than 1, the index of this connective in formula $\mathfrak{A}$ cannot be lower. If, however, the connective in $\mathfrak{B}$ (or $\mathfrak{C}$) has an index equal to unity, its index in formula $\mathfrak{A}$ will be equal to 2 (the index increases since formula $\mathfrak{A}$ contains a left parenthesis before subformula $\mathfrak{B}$).

1.2.15. (3) (0 1 0 0). **1.2.18.** (3) $\mathfrak{A} \sim \mathfrak{B}$. **1.2.19.** (2) $f_1(x, y) \equiv 0$, $f_2(x, y) = x$, $f_3(x, y) = y$, $f_4(x, y) = x \oplus y$. **1.2.20.** (1) $((x \downarrow y) \downarrow (x \downarrow y))$. (2) $((x \downarrow y) \downarrow ((x \downarrow x) \downarrow (y \downarrow y)))$. **1.2.21.** (1) $((x \mid y) \mid (x \mid y))$. (2) $((x \vee (x \vee y)) \sim y)$. (3) No, it cannot.

1.2.22. (2) The function represented by a formula generated by the set $\{\rightarrow\}$ assumes the value 1 on at least half the tuples of values of its arguments. The function xy does not satisfy this condition.

1.2.23. In general, it is impossible. For example, for $S = \{\neg\}$, the function $\bar{x}$ cannot be represented by a formula of even depth generated by S.

1.2.24. If only constant functions can be represented by a formula generated by the set S, the statement is obvious. Let a certain function $f(\widetilde{x^n}) \not\equiv \text{const}$ be represented by a formula generated by S. Identifying all the variables in the function $f(\widetilde{x^n})$ and in the formula representing it, we obtain a function $\varphi(x)$ (and the formula $\mathfrak{A}(x)$ corresponding to it). We must consider the following three cases: (a) $\varphi(x)$ is a constant, (b) $\varphi(x) = x$ and (c) $\varphi(x) = \bar{x}$. Having obtained (in cases (a) and (c)) a formula representing the function x (and having a depth not less than 1), we can construct a formula of an indefinitely large depth for each specific function represented by a formula generated by S. Let $\varphi(x) \equiv 0$. Then there exists a tuple $\tilde{\alpha} = (\alpha_1, \alpha_2, \ldots, \alpha_n)$ on which the function f is equal to 1. Substituting x for all x_i corresponding to $\alpha_i = 1$ and $\varphi(x) \equiv 0$ for the remaining x_i, we obtain a function $\psi(x) = x$. If, however, $\varphi(x) \equiv 1$, the function $\bar{x}$ can be obtained in a similar way.

1.2.26. (2) $x \vee y = (x \to y) \to y$. (3) $x \sim y = (x \to y) \,\&\, (y \to x)$.

1.2.30. (2) In general, it cannot. For example, $\mathfrak{A} = y_2 \to x$ is a formula which is not identically true. However, the formula $S^{x}_{y_1 \to y_2} \mathfrak{A} \mid = y_2 \to (y_1 \to y_2)$ is identically true.

1.2.31. (1) If we assume that $f(0, 0) = 0$, then putting $x = y = z = 0$, we obtain $f(f(0, f(0, 0)), f(f(0, 0), f(0, 0))) = f(f(0,0), f(0, 0)) = 0$, which contradicts the conditions of the problem. Thus, $f(0, 0) = 1$. Putting $f(0, 1) = 0$, we obtain (for $x = y = z = 0$) $(f(0, f(0, 0)), f(f(0, 0), f(0, 0))) = f(f(0, 1), f(1, 1)) = f(0, f(1, 1))$ and by hypothesis, the relation $f(1, 1) = 0$ must be satisfied. However, considering that $x = 0$, $y = z = 1$, we see that $f(f(0, f(1, 1)), f(f(0, 1), f(0, 1))) = f(f(0, 0), f(0, 0)) = 0$, which is in contradiction to the conditions of the problem. Thus, $f(0, 1) = 1$. Finally assuming that $f(1, 1) = 0$ and putting $x = y = 0$ and $z = 1$, we get $f(f(0, f(0, 1)), f(f(0, 0), f(0, 1))) = f(f(0, 1), f(1, 1)) = f(1, 0)$. Taking into account the condition of the problem, we obtain $f(1, 0) = 1$. But in this case $f(f(1, f(1, 1)), f(f(1, 1), f(1, 1))) = f(f(1, 0), f(0, 0)) = f(1, 1) = 0$, which contradicts the condition of the problem. Consequently, $f(1, 1) = 1$ as well. Thus, $f(0, 0) = f(0, 1) = f(1, 1) = 1$. Hence $f(x, y) = x \to y$ or $f(x, y) \equiv 1$. The equivalences (a)-(e) are verified directly (for example, by using the basic equivalences). (2) No, they do not. It is sufficient to consider the function $f(x, y) = x \sim y$.

1.3.

1.3.6. 2^n. **1.3.7.** (1) 2^{n-1}. (2) $2^n - 2$. (3) 2^{n-1}. **1.3.8.** (1) $k \times l$. (2) $k \times 2^m + l \times 2^n - k \times l$. (3) $k(2^m - l) + l(2^n - k)$. **1.3.9.** (2) $f_0^1(\tilde{x}^3) = x_2 x_3$. $f_0^2(\tilde{x}^3) = \tilde{x}_1$, $f_{01}^{1,3}(\tilde{x}^3) = x_2$. **1.3.11.** $2^{(3/4)\cdot 2^n}$. **1.3.12.** 12. **1.3.16.** 4. **1.3.18.** $\binom{n}{r}$. **1.3.19.** $2^{S_n^r} - 2^{S_n^{r-1}}$, $S_n^k = \sum_{i=0}^{k} \binom{n}{i}$.

1.3.20. 0 if k is odd and $\binom{2^n - 1}{k}$ if k is even. **1.3.23.** $\tilde{\beta}_P = (1101111011100110)$. **1.3.26.** $f(\tilde{x}^n) = \bar{x}_1 \bar{x}_2 \ldots \bar{x}_n$.

1.3.29. Hint. Using the decomposition formula, carry out the proof by induction on n.

1.3.30. (1) $2^{n-k}+2^k-2$; (2) $\frac{1}{3}(2^{n+1}+(-1)^n)$.

1.3.31. For $n=1$, the statement is obvious. We proceed by induction from $n-1$ to n. If $l \leqslant 2^{n-1}$, by inductive hypothesis there exists a polynomial $P(\tilde{x}^{n-1})$ of length $\leqslant n-1$, for which $|N_P|=l$. Then the polynomial $x_n P(\tilde{x}^{n-1})$ becomes equal to unity at l vertices of the cube B^n. If $2^{n-1}<l\leqslant 2^n$, we consider the polynomial $P(\tilde{x}^{n-1})$ which becomes equal to unity at 2^n-l vertices of the cube B^{n-1}. Then the polynomial $x_n P(\tilde{x}^{n-1}) \oplus 1$ is the sought one.

1.4.

1.4.4. Hint. Let $K=x_1^{\alpha_1}\ \dots\ x_n^{\alpha_n}$, $L=x_1^{\beta_1}\ \dots\ x_n^{\beta_n}$ and let $K\circ L$ be an e.c. obtained from K by deleting the letters $x_i^{\alpha_i}$ for which $\alpha_i \neq \beta_i$. Note that each implicant of the function $f(\tilde{x}^n)$ can be represented in the form $K\circ L$, where K and L are suitable elementary conjunctions from a perfect d.n.f. of the function $f(\tilde{x}^n)$, which may coincide.

1.4.5. (1) 2^{n-1}. (2) 2^{n-1}. (3) $3\times 2^{n-3}$. (4) 2^{n-2}. (5) $k+(n-k)\times(n-k-1)$.

1.4.6. (1) $\binom{l}{k}\binom{n-l}{k+m-l}$.

1.4.8. (2) $\bar{x}_1x_2x_3 \vee x_1\bar{x}_2 \vee x_1\bar{x}_3 \vee x_1x_4 \vee x_2x_3x_4$.

1.4.9. (1) $2^{2^n-2^{n-r}}$. (2) $2^{2^n-2^{n-r}}\sum_{i=0}^{r}(-1)^i\binom{r}{i}2^{-i2^{n-r}}$.

1.4.11. (1) $\bar{x}_2\bar{x}_3 \vee \bar{x}_1x_4 \vee x_2x_3 \vee x_1\bar{x}_4$.

1.4.12. (1) $\bar{x}_1x_3$, x_1x_2, $\bar{x}_1\bar{x}_2x_4$, $\bar{x}_2x_3x_4$.

1.4.13. Hint. Prove that there are no two proper tuples of two different kernel implicants which are adjacent, making use of the fact that the set $N\subset B^n$ such that $|N|>2^{n-1}$ always contains the edge (see Problem 1.1.44).

1.4.14. (1)2^{n-1}. (2) 2^{n-3}. (3) 0. (4) 2^{n-2}. (5) k.

1.4.15. $2^{2^n-2^{n-r}}(1-(1-2^{-2^r})^{n-r})$.

1.4.18. (1) 1. (2) $5^{2^{n-r}}$. (3) $5^{2^{n-1}}$. (4) 1.

1.4.20. One. Prove that all simple implicants are kernel implicants.

1.4.23. Hint. Carry out induction on n, using expansion (3) of the function in the variables.

1.4.24. (1) For two functions. (2) For 2^{n+1} functions.

1.4.25. For example, $k=n2^n-1$.

1.4.27. (1) 0 if $k \neq 0$, $k+m<n$; $\binom{n}{m}$ if $k=0$; $\binom{n}{k}$ if $k+m=n$. (2) On two functions for even n and on four functions for odd n.

1.5.

1.5.2. (1) x_3. (2) x_1, x_2. (3) x_4. (4) x_1. **1.5.4.** (1) x_2, x_3. (2) x_1, x_3. (3) x_3. **1.5.7.** m. **1.5.8.** (1) x_1, x_2, x_3, x_4. (2) x_1, x_2, x_3. (3) x_3, x_4. (4) x_1, x_2, x_4.

1.5.10. (1) $n \geqslant 3$. (2) No values of n. (1) $n \geqslant 3$. (4) Even n. (5) $n = 4k - 2$, $k = 1, 2, \ldots$.

1.5.12. (2) $|P^c(X^3)| = 218$.

1.5.13. Hint. Using the condition of the problem, prove that there exist tuples $\tilde{\alpha}'$, $\tilde{\beta}'$, $\tilde{\gamma}'$, such that $\rho(\tilde{\alpha}', \tilde{\beta}') = \rho(\tilde{\beta}', \tilde{\gamma}') = 1$. Let $\tilde{\alpha}'$ and $\tilde{\beta}'$ differ in the i-th and $\tilde{\beta}'$ and $\tilde{\gamma}'$ in the j-th place. Then the variables x_i and x_j are essential.

1.5.14. Hint. It is sufficient to prove that if x_i or $\bar{x}_i$ appear in a prime implicant K of the function f, then x_i is an essential variable. Deleting x_i^σ from K, we obtain an elementary conjunction K' that is not an implicant. Hence it follows that there exists a tuple $\tilde{\alpha} = (\alpha_1, \ldots, \alpha_{i-1}, \bar{\sigma}, \alpha_{i+1}, \ldots, \alpha_n)$ such that $K'(\tilde{\alpha}) = 1$ but $\tilde{f}(\alpha) = 0$. At the same time, $f(\tilde{\alpha}^i) = 1$ since $K(\tilde{\alpha}^i) = 1$. Hence x_i is essential.

1.5.15. Hint. If x_i explicitly enters the polynomial $P(\tilde{x}^n)$, the latter can be presented in the form $x_iQ \oplus R$, where Q and R are polynomials independent of x_i and $Q \not\equiv 0$. Let the tuple $\tilde{\alpha} = (\alpha_1, \ldots, \alpha_{i-1}, \alpha_{i+1}, \ldots, \alpha_n)$ be such that $Q(\tilde{\alpha})=1$. Then $P(\alpha_1, \ldots, \alpha_{i-1}, x_i, \alpha_{i+1}, \ldots, \alpha_n) = x_i \oplus R(\tilde{\alpha})$. The statement now follows from Problem 1.5.1 (1).

1.5.16. No, it does not. **Hint.** Consider $f(\tilde{x}^n) = x_1 \oplus x_2 \oplus \ldots \oplus x_n$.

1.5.17. Hint. Let $f(\tilde{x}^n)$ satisfy the condition of the problem. Then there exists an integer i ($1 \leqslant i \leqslant n$) such that for any $\tilde{\alpha}$ and $\tilde{\beta}$ (such that $\tilde{\alpha} \in B_i^n$, $\tilde{\beta} \in B_{i-1}^n$) $f(\tilde{\alpha}) \neq f(\tilde{\beta})$. The statement now follows from the fact that for any i in $B_{i-1}^n \cup B_i^n$, there exist edges of any direction of i.

1.5.18. Hint. See Problem 1.5.13.

1.5.19. If $f(\tilde{x}^n)$ is expressed through a polynomial of degree greater than unity, without loss of generality it can be presented in the form $x_1x_2P_1 \oplus x_1P_2 \oplus x_2P_3 \oplus P_4$, where $P_1 \not\equiv 0$ and P_1, P_2, P_3 and P_4 depend on the variables $x_3, \ldots, x_n$. Let $\tilde{\alpha} = (\alpha_3, \ldots, \alpha_n)$ be a tuple such that $P_1(\tilde{\alpha}) = 1$. Then the component $f_{\alpha_3 \ldots \alpha_n}^{3, \ldots, n}(\tilde{x}^n)$ has the form $x_1x_2 \oplus \lambda_1x_1 \oplus \lambda_2x_2 \oplus \lambda_3$, and hence this component essentially depends on x_1 and x_2. Hence it follows that the required three tuples will be found in the face $B^{n;\,3,\,\ldots,\,n}_{\alpha_3\ldots\alpha_n}$. If $f(\widetilde{x^n})$ is expressed by a first-degree polynomial and essentially depends on x_1 and x_2, then for any tuple $(\alpha_3, \ldots, \alpha_n)$ any three vertices in the face $B^{n;\,3,\,\ldots,\,n}_{\alpha_3\ldots\alpha_n}$ are the required vertices.

1.5.21. We assume the contrary. Without loss of generality, we can assume that the maximum number of essential variables among the components of the form $f^i_\alpha(\widetilde{x^n})$ belongs to the component $f^1_1(\widetilde{x^n})$. Then the assumption is equivalent to the statement that $f^1_1(\widetilde{x^n})$ fictitiously depends on a certain variable, say x_2, i. e. $f^1_1 = f^{1,\,2}_{11} = f^{1,\,2}_{10}$. Since x_1 is an essential variable, at least one of the relations $f^{1,\,2}_{10} \neq f^{1,\,2}_{00}$ and $f^{1,\,2}_{11} \neq f^{1,\,2}_{01}$ must be satisfied. Let, for example, $f^{1,\,2}_{11} \neq f^{1,\,2}_{01}$. Then the function

$$f^2_1 = \bar{x}_1 f^{1,\,2}_{01} \vee x_1 f^{1,\,2}_{11}$$

essentially depends on x_1, as well as on all essential variables of the subfunction $f^{1,\,2}_{11} = f^1_1$. This is in contradiction with the fact that f^1_1 has the maximum number of essential variables.

1.5.22. (1) The validity of the statement follows from the previous problem. (2) No, the statement is not valid. **Hint**. Consider $f(\widetilde{x^3}) = x_1x_3 \vee x_2\bar{x}_3$ and the variables x_1 and x_2.

1.5.23. (1) $\bar{x}_1$, x_2. (2) 1. (3) x_1, x_2. (4) $x_2 \to x_1$, $x_1 \sim x_2$. 1. **1.5.25.** 6. **1.5.26.** Yes, it can. **Hint.** Consider the function $x_1x_2x_3 \vee x_1x_2x_4 \vee x_1x_3x_4 \vee x_2x_3x_4$.

1.5.27. The numbers of coordinates of the tuples $\widetilde{\alpha}$, $\widetilde{\beta}$, $\widetilde{\gamma}$ can be divided into the following four groups:

$$A_1 = \{i\colon \alpha_1 = \beta_i = \gamma_i = 0\}, \quad A_2 = \{i\colon \alpha_i = \beta_i,\ \beta_i < \gamma_i\},$$
$$A_3 = \{i\colon \alpha_i < \beta_i,\ \beta_i = \gamma_i\}, \quad A_4 = \{i\colon \alpha_i = \beta_i = \gamma_i = 1\}.$$

Let $f(\widetilde{\alpha}) = f(\widetilde{\gamma}) = \sigma$, $f(\widetilde{\beta}) = \bar{\sigma}$. Then if $f(\widetilde{0}) = f(\widetilde{1}) = \sigma$, we put $x_i = x$ if $i \in A_1 \cup A_2$, and $x_i = y$ if $i \in A_3 \cup A_4$. If $f(\widetilde{0}) = \sigma$, $f(\widetilde{1}) = \bar{\sigma}$, we put $x_i = x$ for $i \in A_1$, $x_i = y$ for $i \in A_2$, $x_i = z$ for $i \in A_3 \cup A_4$. If $f(\widetilde{0}) = \bar{\sigma}$, we put $x_i = x$ for $i \in A_1 \cup A_2$, $x_i = y$ for $i \in A_3$, and $x_i = z$ for $i \in A_4$. The essential dependence on the obtained functions follows from Problem 1.5.13.

1.5.29. $x_1 \oplus x_2 \oplus \ldots \oplus x_n \oplus \sigma$, $\sigma \in \{0, 1\}$ for $n \geqslant 2$, $x_1^{\sigma_1}x_2^{\sigma_2} \vee x_2^{\sigma_2}x_3^{\sigma_3} \vee x_3^{\sigma_3}x_1^{\sigma_1}$, $\sigma_i \in \{0, 1\}$, $i = \overline{1, 3}$ for $n = 3$; $x_1^\sigma x_2^{\bar{\sigma}}$, $x_1^\sigma \vee x_2^{\bar{\sigma}}$ for $n = 2$.

1.5.30. The function obtained from $f(\widetilde{x^n})$ by identifying the variables x_i and x_j has the form $x_i f^{i,\,j}_{11} \vee \bar{x}_i f^{i,\,j}_{00}$. The conditions of the problem imply that at least one of the components $f^{i,\,j}_{11}$ and $f^{i,\,j}_{00}$ is not identically equal to zero.

1.5.33. Hint. At least one of the functions f and g becomes equal to unity on an odd number of tuples.

1.5.34. Hint. See Problem 1.5.30.

CHAPTER TWO

2.1.

2.1.2. No, it does not. We assume (by definition) that $[M] = P_2$ for any set M in P_2, i.e. introduce such a closure operation. Then relations (1)-(3) in Problem 2.11.1 are obviously satisfied, while relation (4) is not valid.

2.1.3. (1) Yes. (2) No. (3) No. (4) Yes. (5) Yes. (6) No. (7) No.

2.1.4. (1) $\{1\}$. (2) $\{0, 1, x_1, \tilde{x}_1\}$. (3) $\{x_1, x_1 \vee x_2, x_1 \vee x_2 \vee x_3\}$. (4) $\{1, x_1, x_1 \sim x_2, x_1 \oplus x_2 \oplus x_3\}$.

2.1.5. Let $f(\tilde{x}^n)$ be a function in a closed class, which essentially depends on all its variables ($n \geqslant 2$). Then the function

$$f_1(y_1, y_2, \ldots, y_n, x_2, \ldots, x_n) = f(f(y_1, y_2, \ldots, y_n), x_2, \ldots, x_n)$$

essentially depends on all $2n - 1$ variables (here $\{y_1, y_2, \ldots, y_n\} \cap \{x_2, \ldots, x_n\} \neq \varnothing$). The function $f(z_1, z_2, \ldots, z_n)$ can then be substituted for the variable y_1 in the function f_1 to obtain a function essentially depending on $3n - 2$ variables, and so on.

2.1.6. $[0]$, $[1]$, $[x]$, $[0, 1]$, $[0, x]$, $[1, x]$, $[x, \bar{x}]$, $[0, 1, x]$, $[0, 1, x, \bar{x}]$.

2.1.7. (1) Yes. (2) No, not always. (3) No, not always (the only exception being all P_2 and its complement, viz. a closed empty class $\varnothing$).

2.1.8. (3) $(x \to y) \to y = x \vee y$, $\bar{x} = \overline{x \oplus x \oplus x}$. (4) A negation is obtained from the second function by identifying all the variables. (5) $m(x, y, 0) = xy$, $x^0 \oplus 0 = \bar{x}$.

2.1.9. (1) $K_1 \subseteq K_2$, but the strict inclusion $K_1 \supset K_2$ is also possible. (2) $K_1 \not\supseteq\not\subseteq K_2$; for example, for $M_1 = \{xy, \bar{x}\}$ and $M_2 = \{\bar{x}\}$ we have $K_1 = [xy]$, and $K_2 = P_2 \setminus [\bar{x}]$. (5) See Problem 2.1.9(2).

2.1.10. (2) $M_1 = [x, 0]$, $M_2 = [x, 1]$. (3) $M_1 = [\bar{x}, 0]$, $M_2 = [0]$. (5) See Problem 2.1.10(2).

2.1.11. (1) $\{0, \bar{x}\}$. (2) $\{1, x \oplus y\}$. (3) $\{x \vee y, x \,\&\, y \,\&\, z\}$. (4) $\{x \oplus 1, m(x, y, z)\}$, since $x \oplus y \oplus z = m(m(x, y, \bar{z}), m(\bar{x}, \bar{y}, z), m(\bar{x}, y, z))$ and in the closed class $[m(x, y, z)]$ any function which essentially depends on one argument is equal to an identical function.

2.1.12. Let M be a precomplete class (in P_2). It means that $[M] \neq P_2$ but for any function $f \in P_2 \setminus M$ we have $[M \cup \{f\}] = P_2$. If $[M] \neq M$, there exists a function g in $[M] \setminus M$, such that $[M] = [M \cup \{g\}] = P_2$. We arrive at a contradiction.

2.1.13. If $M_2^1 = M^1$, the statement is obvious. Let $M_2^1 \neq M^1$. We assume that $M_1^1 = M_2^1$. Since M_1 and M_2 are closed classes, and $M_1^1 \subset M_1$ and $M_2^1 \subset M_2$, then irrespective of the form of a function f in M_1 (or in M_2), substituting for its variables the functions from M_1^1 $(= M_2^1)$ depending on the same variable x, we again obtain functions in M_1^1. But this means that by adding to M_1 an arbitrary function g in $M_2 \setminus M_1$ $(\neq \varnothing)$, we get $[M_1 \cup \{g\}] \cap M^1 = M_1^1 \neq M^1$, and this contradicts to the precompleteness of the class M_1 in the class M.

2.1.14. (1) $[0, 1, x]$, $[\bar{x}]$. (2) $[0]$, $[1]$. (4) $[0, x]$, $[x \vee y]$.

2.1.15. (1) No, it is not. (3) No, it is not.

2.1.16. If $x \in M$, the statement is obvious. Let $x \notin M$. By definition of the superposition operation, the substitution of an identical function can be considered as a transformation made only in the set M and carried out in accordance with the superposition operation. Therefore, $[M \cup \{x\}] \subseteq M \cup \{x\}$. The inverse inclusion follows from the properties of closure operation.

2.1.17. The statement follows from the fact that the function $x \mid y$ forms a complete system in P_2.

2.1.18. Let $f(\tilde{x}^n) \not\equiv \text{const}$. If by identifying all the variables in the function $f(\tilde{x}^n)$ we obtain an identical function or a negation, the statement is obvious. Let $f(x, \ldots, x) \equiv \text{const}$. For definiteness, we assume that $f(x, \ldots, x) \equiv 0$. Since $f(\tilde{x}^n) \not\equiv \text{const}$, there exists a tuple $\tilde{\alpha} = (\alpha_1, \ldots, \alpha_n)$ such that $f(\tilde{\alpha}) = 1$. We put $x_i = x$ for $\alpha_i = 1$ and $x_i = y$ for $\alpha_i = 0$. Then from the function f we obtain a function $g(x, y)$ satisfying the following conditions: $g(0, 0) = g(1, 1) = 0$, $g(1, 0) = 1$. If $g(0, 1) = 1$, then $g(x, y) = x \oplus y$; if $g(0, 1) = 0$, then $g(x, y) = \overline{x \to y}$. Obviously $[x \oplus y] \ni x$ and $[\overline{x \to y}] \ni \bar{x}$.

2.1.20. Since the set P_2 is countably infinite, the set of all its finite subsets is also countably infinite. Therefore, the power of all closed classes in P_2 which have finite complete systems in not more than countable.

2.1.21. If a closed class in P_2 differs from the sets $\{0\}$, $\{1\}$, $\{0, 1\}$, it contains an identical function (see Problem 2.1.18). Consequently, it cannot be expanded to the basis.

2.1.22. Let $\mathcal{P}$ be a complete system in M. We take in $\mathcal{P}$ an arbitrary function f_1. It does not belong to any precomplete classes $K_1, \ldots, K_{r_1}$ in M. Then we choose in $\mathcal{P}$ a function f_2 which does not belong to at least one of the remaining precomplete classes, and so on. The number of functions in the system $\mathcal{P}'$ obtained as a result of such a process does not exceed the number of precomplete classes in M. The system $\mathcal{P}'$ is complete in M, since otherwise it can be expanded to a precomplete class, while it follows from the method of construction that the system $\mathcal{P}'$ is not contained completely in any of precomplete classes in M. If $\mathcal{P}$ is a basis, all the functions in $\mathcal{P}$ must be used in the construction of the corresponding system $\mathcal{P}'$, otherwise we would arrive at a contradiction to the irreducibility of the system $\mathcal{P}$.

2.1.23. The proof can be carried out by induction on the number of occurrences of the connective $\to$ in the formulas representing the functions in the closed class $[x \to y]$. If there is one occurrence of the connective $\to$ in a formula, the formula (accurate to notation of variables) has one of the following forms: $x \to x$ or $x \to y$. In this case, the statement is obvious. Further, if $\mathfrak{A} = f_1 \to f_2$, we must consider two possible cases:

(a) The formula corresponding to the function f_2 has at least one occurrence of the connective $\to$. Then $f_2 = y_j \vee \varphi_2(\widetilde{y^m})$ and $\mathfrak{A} = \overline{f}_1 \vee f_2 = y_j \vee (\overline{f}_1 \vee \varphi_2)$.

(b) The formula corresponding to the function f_2 is a variable (say, y). Then $\mathfrak{A} = \overline{f}_1 \vee y$.

2.1.24. We carry out induction on the number of occurrences of the connective $\to$. If the connective $\to$ appears only once, we have $x \to y$, and the statement is obvious. Let $\mathfrak{A} = f_1(\widetilde{x^n}) \to f_2(\widetilde{x^n})$, and let the function f corresponding to $\mathfrak{A}$ essentially depend on at least two variables. If f_2 satisfies the conditions of the problem, then $|N_{f_2}| > 2^{n-1}$ and $|N_f| \geqslant |N_{f_2}| > 2^{n-1}$. Let us now suppose that f_2 essentially depends on only one variable. We assume for definiteness that $f_2 \equiv x_1$ (f_2 cannot be equal to the negation of the variable since $\overline{x} \notin [x \to y]$). Under such an assumption, $f = \overline{f}_1 \vee x_1$. Obviously, $f(1, x_2, \ldots, x_n) \equiv 1$. If we had $f(0, x_2, \ldots, x_n) \equiv 0$, then $f(x_1, x_2, \ldots, x_n) \equiv x_1$, which contradicts to the fact that the function $f(\widetilde{x^n})$ essentially depends on at least two variables. Consequently, $f(0, x_2, \ldots, x_n) \not\equiv 0$ and $|N_f| > 2^{n-1}$. Finally, if the function f_2 depends on all variables $x_1, \ldots, x_n$ inessentially, then $f_2 \equiv 1$, but in this case $f \equiv 1$, i.e. it depends on all the variables $x_1, \ldots, x_n$ inessentially.

2.1.25. Let K be a precomplete class in P_2. We assume that $\overline{x} \notin K$. In view of Problem 2.1.16, the addition of the function $\overline{x}$ to the class K does not lead to a complete system in P_2 (since $[K \cup \{\overline{x}\}] = K \cup \{\overline{x}\} \not\ni x \mid y$). This is in contradiction with the precompleteness of the class K.

2.2.

2.2.1. (2) No. (5) Yes. **2.2.3.** (1) $xz \vee yt$. (3) $xy \oplus y \oplus 1$.

2.2.4. Carry out induction on the number of occurrences of connectives in the set $\{0, 1, \neg, \&, \vee\}$ to the formula $\mathfrak{A}$.

2.2.6. Let $i = 1$, i. e. the function $f(\widetilde{x^n})$ essentially depends on the variable x_1. Then there exist tuples $\widetilde{\alpha} = (0, \alpha_2, \ldots, \alpha_n)$ and $\widetilde{\alpha}' = (1, \alpha_2, \ldots, \alpha_n)$ such that $f(\widetilde{\alpha}) \neq f(\widetilde{\alpha}')$. Since $f^*(1, \overline{\alpha}_2, \ldots, \overline{\alpha}_n) = \overline{f}(0, \alpha_2, \ldots, \alpha_n)$ and $f^*(0, \overline{\alpha}_2, \ldots, \overline{\alpha}_n) = \overline{f}(1, \alpha_2, \ldots, \alpha_n)$, we have $f^*(1, \overline{\alpha}_2, \ldots, \overline{\alpha}_n) \neq f^*(0, \overline{\alpha}_2, \ldots, \overline{\alpha}_n)$, i. e. x_1 is an essential variable of the function $f^*(\widetilde{x^n})$.

2.2.8. (1) 0. (2) 2^{n-1}. (3) 1. **2.2.9.** (1) Yes. (4) No. (5) Yes. **2.2.10.** Use the decomposition $f(\widetilde{x^n}) = \overline{x}_1 f(0, x_2, \ldots, x_n) \vee x_1 f(1, x_2, \ldots, x_n)$.

2.2.11. If $f(\tilde{x}^n)$ is a self-dual function, then it assumes opposite values on any two opposite tuples of variables. Consequently, the number of tuples on which the self-dual function $f(\tilde{x}^n)$ assumes the value 1 is equal to the number of pairs of opposite tuples (of length n), i.e. $|N_f| = 2^{n-1}$.

2.2.13. $\sum_{i=0}^{n-1} (-1)^i C_n^i 2^{2n-1-i}, \quad n \geqslant 1.$

2.2.14. $x \oplus y \oplus z$, $x \oplus y \oplus z \oplus 1$, $m(x, y, z)$, $m(x, y, z) \oplus 1$, $m(x, y, z) \oplus x \oplus y$, $m(x, y, z) \oplus x \oplus y \oplus 1$, $m(x, y, z) \oplus x \oplus z$, $m(x, y, z) \oplus x \oplus z \oplus 1$, $m(x, y, z) \oplus y \oplus z$, $m(x, y, z) \oplus y \oplus z \oplus 1$. It can be easily verified that $m(x, y, \bar{z}) = m(x, y, z) \oplus x \oplus y$, $m(x, \bar{y}, \bar{z}) = m(x, y, z) \oplus y \oplus z \oplus 1$, $m(\bar{x}, \bar{y}, \bar{z}) = m(x, y, z) \oplus 1$.

2.2.15. (1) For $n = 3$. (2) For $n = 4m + 3$, $m = 0, 1, 2, \ldots$. (3) The function is not self-dual for any $n \geqslant 2$ since $f(\tilde{0}) = f(\tilde{1})$. (4) For odd $n \geqslant 3$.

2.2.16. Only for odd n.

2.2.17. (3) $g(\tilde{y}^5) = y_1 \oplus y_2 \oplus y_3 \oplus y_4 \oplus y_5 \in S$, $m(\bar{x}_1, x_3, f) \in S$ and $m(x_2, x_3, f) \in S$ if $f \in S$. Substituting the function $m(\bar{x}_1, x_3, f) = \bar{x}_1 f \oplus x_3 f \oplus \bar{x}_1 x_3$ in $g(\tilde{y}^5)$ for y_1, the function $m(x_2, x_3, f) = x_2 f \oplus x_3 f \oplus x_2 x_3$ for y_2, the variable x_2 for y_3, and the variables x_4 and x_5 for y_4 and y_5 respectively, obtain $\bar{x}_1 f \oplus x_3 f \oplus \bar{x}_1 x_3 \oplus \bar{x}_2 f \oplus x_3 f \oplus x_2 x_3 \oplus x_2 \oplus x_4 \oplus x_5 = \bar{x}_1 f \oplus \bar{x}_1 x_3 \oplus x_2 \bar{f} \oplus x_2 x_3 \oplus x_4 \oplus x_5$.

2.2.18. (1) We have $\bar{f}(\tilde{x}^n) \vee f^*(\tilde{x}^n) = \overline{f(x_1, \ldots, x_n) \,\&\, f(\bar{x}_1, \ldots, \bar{x}_n)} \equiv \text{const}$. This constant cannot be equal to 0 since otherwise the relation $f(\tilde{x}^n) \equiv 1$ must be satisfied. Therefore, $f(x_1, \ldots, x_n) \,\&\, f(\bar{x}_1, \ldots, \bar{x}_n) \equiv 0$. Proceeding from the equality $|N_f| = |N_{f^*}|$, we conclude that in each of the functions $f(\tilde{x}^n)$, $f^*(\tilde{x}^n)$ and $f'(\tilde{x}^n) = f(\bar{x}_1, \ldots, \bar{x}_n)$, the number of zeros is equal to the number of unities, i.e. $|N_f| = |N_{f^*}| = |N_{f'}| = 2^{n-1}$. Consequently, in view of the relation $f \,\&\, f' \equiv 0$, $f(\tilde{\alpha}^n) = 0$ if and only if $f'(\tilde{\alpha}^n) = 1$. Therefore, $f(\tilde{x}^n) \equiv f^*(\tilde{x}^n)$.

2.2.19. (3) $(x \downarrow x) \to (x \oplus \bar{x}) = 1$.

2.2.20. Let $f(\tilde{x}^n)$ be a non-self-dual function, and let all its n variables ($n \geqslant 3$) be essential. From the non-self-duality of f it follows that there exist two opposite tuples $\tilde{\alpha} = (\alpha_1, \ldots, \alpha_n)$ and $\tilde{\alpha}' = (\bar{\alpha}_1, \ldots, \bar{\alpha}_n)$ on which the function f assumes identical values. We first assume that $f(\tilde{0}^n) \neq f(\tilde{1}^n)$, and hence $\tilde{\alpha} \neq \tilde{0}^n$ and $\tilde{\alpha} \neq \tilde{1}^n$. Let us consider the function $g(x, y)$ obtained from $f(\tilde{x}^n)$ as a result of substituting x for any variable x_i such that $\alpha_i = 0$ and as a result of substituting y for each of the remaining variables.

We have $g(0, 0) \neq g(1, 1)$ and $g(0, 1) = g(1, 0)$. Obviously, $g(x, y) \notin S$. Let us now suppose that $f(\tilde{0}^n) = f(\tilde{1}^n)$. Since $f(\tilde{x}^n) \not\equiv$ const (because it essentially depends on at least three variables), there exists a tuple $\tilde{\beta} = (\beta_1, \ldots, \beta_n)$ such that $f(\tilde{\beta}) \neq f(\tilde{0}^n)$. Starting with the tuple $\tilde{\beta}$, we construct a function $h(x, y)$ such that if $\beta_i = 0$, we substitute x for the variable x_i in $f(\tilde{x}^n)$, and if $\beta_i = 1$, we substitute y for x_i. The function $h(x, y)$ satisfies the following relations: $h(0, 0) = h(1, 1) \neq h(0, 1)$. Obviously, $h(x, y)$ essentially depends on two variables and is not self-dual.

2.2.21. See the previous problem.

2.2.22. It can be easily seen that (see Problem 2.2.14) $m(x^\alpha, x^\beta, x^\gamma) = x^{m(\alpha, \beta, \gamma)}$. If $m(\alpha, \beta, \gamma) = 0$, then $m(x^\alpha, x^\beta, x^\gamma) = \bar{x}$. Consequently, substituting $m(x^\alpha, y^\beta, z^\gamma)$ for x, y and z in the same function $m(x^\alpha, y^\beta, z^\gamma)$ we obtain either $m(x, y, z)$, or one of the functions $m(x, y, z) \oplus x \oplus y$, $m(x, y, z) \oplus x \oplus z$, $m(x, y, z) \oplus y \oplus z$. If we have one of the last three functions, the function $m(x, y, z)$ can be easily constructed. For example, $m(x, y, m(x, y, z) \oplus x \oplus y) \oplus x \oplus y = m(x, y, z)$. Let us now prove that the function $m(x, y, z)$ can be used to obtain a function which essentially depends on n ($\geqslant 4$) variables. If n is odd and $n \geqslant 5$, we have $g_5(\tilde{x}^5) = m(x_1, x_2, m(x_3, x_4, x_5)), \ldots, g_{2l+1}(\tilde{x}^{2l+1}) = g_{2l-1}(x_1, x_2, \ldots, x_{2l-2}, m(x_{2l-1}, x_{2l}, x_{2l+1})), \ldots$. It can easily be proved (by induction of l) that each of these functions essentially depends on all its arguments. For an even $n = 2l \geqslant 4$, the corresponding function can be obtained, for example, from the function g_{2l+1} by putting $x_{2l-1} = x_2$. While proving that the function $g_n(\tilde{x}^n)$ essentially depends on all its arguments, it should be borne in mind that the median $m(x, y, z)$ is equal to 1 only on the tuples containing at least two unities.

2.2.24. We must take an arbitrary tuple $(\alpha_1, \alpha_2, \ldots, \alpha_n)$ of values of variables $x_1, x_2, \ldots, x_n$ and compare on it the values of the left- and right-hand sides of the given relation. We should distinguish between the following two cases: (1) there are at least two zeros among the values α_1, α_2 and α_3; and (2) there are at least two unities among α_1, α_2 and α_3.

2.2.25. (1) In the solution to Problem 2.2.22 it was shown how the median $m(x, y, z)$ can be constructed from $m(x, y, z)$ or $m(x, \bar{y}, \bar{z})$. Further, $m(x, x, x) \equiv m(x, \bar{x}, \bar{x}) \equiv x$, and hence (see Problem 2.2.23(1)) the function $x \oplus y \oplus z$ can be constructed from $m(x, y, z)$ (or $m(x, \bar{y}, \bar{z})$). Self-dual functions $\bar{x}$ (and x) of one variable are obviously contained in $[m(x, y, z)]$ (and in $[m(x, \bar{y}, \bar{z})]$). It follows from Problem 2.2.12 that there exist no self-dual functions which essentially depend on two variables. Finally, using the previous problem, we conclude that in order to construct any self-dual function of $n \geqslant 3$ variables, it is sufficient to have all self-dual functions of $n - 1$ variables (as well as the functions $m(x, y, z)$ and $x \oplus y \oplus z$ if $n = 3$). If $n = 3$, we can use the result in Problem 2.2.14. (**Remark.** Problem 2.2.14 can be solved by using the solution of Problem 2.2.24 and the functions $\bar{x}$, $x \oplus y \oplus z$ and $m(x, y, z)$.) (2) Let $f(\tilde{x}^n)$ be an arbitrary self-dual function, and all its variables are essential ($n \geqslant 3$). We prove that if $n \geqslant 4$ and $f(\tilde{x}^n)$ is not a linear function (i.e. cannot be represented in the form $x_1 \oplus x_2 \oplus \ldots \oplus x_n \oplus \sigma$, where $\sigma \in \{0, 1\}$), a nonlinear self-dual function essentially depending on three variables can be obtained from $f(\tilde{x}^n)$ by identification of variables. (If f is a linear self-dual function, a similar statement for it is obvious.) Thus, let $f(\tilde{x}^n)$ be a nonlinear function essentially depending on $n \geqslant 4$ variables. The nonlinearity of the function f implies that in the Zhegalkin polynomial representing this function there exists at least one nonlinear term. For the sake of definiteness, let this term contain the variable x_1. Then $f = x_1\varphi_1(x_2, \ldots, x_n) \oplus \varphi_0(x_2, \ldots, x_n)$ and $\varphi_1 \equiv$ const. We take the tuple $(\alpha_2, \ldots, \alpha_n)$ on which φ_1 vanishes. We have $f(0, \alpha_2, \ldots, \alpha_n) = f(1, \alpha_2, \ldots, \alpha_n)$. Let us consider two cases. First we assume that $\alpha_2 = \ldots = \alpha_n = 1$. Since x_1 is an essential variable of the function f, there exist two tuples $\tilde{\beta}$ and $\tilde{\gamma}$ adjacent in the first component and such that $f(\tilde{\beta}) \neq f(\tilde{\gamma})$. Obviously, these two tuples differ from the tuples $\tilde{1}$ and $(0, 1, \ldots, 1)$. We substitute y for those variables x_i which correspond to zeros in the tuples $\tilde{\beta}$ and $\tilde{\gamma}$ and z for all the remaining variables except x_1. We get the function $g(x_1, y, z)$ satisfying the relations $g(0, 0, 1) = \sigma$, $g(1, 0, 1) = \bar{\sigma}$, and in view of the self-duality of the function g, $g(1, 1, 0) = \bar{\sigma}$ and $g(0, 1, 0) = \sigma$. If $g(1, 1, 1) = \sigma$ (and $g(0, 1, 1) = \sigma$), then $g(x_1, y, z) = m(\bar{x}_1, y, z) \oplus \sigma$. If, however, $g(1, 1, 1) = \bar{\sigma} = g(0, 1, 1)$, then $g(x_1, y, z) = m(x_1, y, z) \oplus \sigma$. Let us now consider the case when not all $\alpha_2, \ldots, \alpha_n$ are equal to 1. In view of the self-duality of the function f, not all α_i are equal to 0. Let us take the tuples $\tilde{1}$ and $\tilde{\delta} = (0, 1, \ldots, 1)$. If $f(\tilde{1}) = f(\tilde{\delta})$, the above-mentioned tuples $\tilde{\beta}$ and $\tilde{\gamma}$ differ both from $\tilde{\delta}$ and from $\tilde{1}$, and the problem is reduced to the first case. If, however, $f(\tilde{1}) \neq f(\tilde{\delta})$, the identification of the variables in the function $f(\tilde{x}^n)$ is carried out proceeding from the tuples $(0, \alpha_2, \ldots, \alpha_n)$ and $(1, \alpha_2, \ldots, \alpha_n)$. We obtain a self-dual function $h(x_1, y, z)$ satisfying the relations: $h(\tilde{1}) = \sigma$, $h(0, 1, 1) = \bar{\sigma}$, and $h(0, 0, 1) = h(1, 0, 1)$. If $h(0, 0, 1) = \sigma$, then $h(x_1, y, z) = m(x_1, \bar{y}, z) \oplus \bar{\sigma}$. If, however, $h(0, 0, 1) = \bar{\sigma}$, then $h(x_1, y, z) = m(x_1, y, \bar{z}) \oplus \bar{\sigma}$.

Since the superposition operation on a linear function can lead only to linear functions, any system $\mathcal{P}$ complete in S must contain a nonlinear self-dual function $f(\tilde{x}^n)$ from which, as was shown above, the function of the form $m(x^{\alpha}, y^{\beta}, z^{\gamma})$, where α, β, and γ belong to the set $\{0, 1\}$, can be constructed. If $f(x, x, \ldots, x) = \bar{x}$, then $[f(\tilde{x}^n)]$ contains $\overline{m(x, y, z)}$ (see Problem 2.2.14), and hence

$[f(\widetilde{x}^n)] = S$ (see the previous problem). Let us now suppose that $f(x, x, \ldots, x) = x$, i.e. f preserves 0 (and 1). Then there must be a function $f_1(\widetilde{x}^m)$ in $\mathcal{P}$ which does not preserve 0. If f_1 is a nonlinear function, then $[f_1] = S$. If, however, f_1 is a linear function, then $[f, f_1] = S$, and $\{f, f_1\}$ is a basis in S (the function f generates only the functions preserving 0 while f_1 generates only linear functions).

2.2.26. (1) No, it cannot since $f \in S$ and $g \notin S$. (2) Yes, it can.

2.2.27. The largest value of n is 4.

2.2.28. There are two such functions: $x_1 \oplus x_2 \oplus x_3 \oplus x_4 \oplus \sigma$, where $\sigma \in \{0, 1\}$.

2.2.29. (1) No. (2) Yes. (3) No. (4) Yes.

2.3.

2.3.2. Let $f(\widetilde{x}^n)$ assume opposite values on any two adjacent tuples. Let $f(\widetilde{0}) = \sigma$. Then $f(\widetilde{\alpha}) = \overline{\sigma}$ if the number $\|\widetilde{\alpha}\|$ is odd and $f(\widetilde{\alpha}) = \sigma$ if $\|\widetilde{\alpha}\|$ is even. Let $l_\sigma(\widetilde{x}^n) = x_1 \oplus x_2 \oplus \ldots \oplus x_n \oplus \sigma$. Then $N_f = N_{l_\sigma}$. In view of the uniqueness of the representation of the functions by polynomials, $f = l_\sigma$, and hence $f \in L$. The opposite is not true (see for example, $f(\widetilde{x}^3) = x_1 \oplus x_3$).

2.3.6. 2^n. **Solution.** Let $f(\widetilde{x}^n) = c_1x_1 \oplus \ldots \oplus c_nx_n \oplus c_{n+1}$. Since $f \in S$, we have $c_1(x_1 \oplus 1) \oplus \ldots \oplus c_n(x_n \oplus 1) \oplus c_{n+1} \oplus 1 = f(\widetilde{x}^n)$. Hence $\sum_{i=1}^{n} c_i = 1$, i.e. the number of coefficients c_i, $i = \overline{1, n}$, which are equal to unity is odd. The number of vectors $(c_1, \ldots, c_n)$ with an odd weight in B^n is 2^{n-1}. The coefficient c_{n+1} can be chosen arbitrarily. Hence the statement.

2.3.10. If $f(\widetilde{x}^n) \in L$ and $f(\widetilde{x}^n)$ essentially depends on all its variables, then $f(\widetilde{x}^n) = x_1 \oplus \ldots \oplus x_n \oplus \sigma$, where $\sigma \in \{0, 1\}$. Hence we easily obtain the necessary condition of the statement. **Sufficiency.** Let $f \notin L$. Then there exist two adjacent tuples $\widetilde{\alpha}$ and $\widetilde{\beta}$ such that $f(\widetilde{\alpha}) = f(\widetilde{\beta})$ (see Problem 2.3.2). Let i be the number of the coordinate at which $\widetilde{\alpha}$ and $\widetilde{\beta}$ differ. Then $f(\alpha_1, \ldots, \alpha_{i-1}, x_i, \alpha_{i+1}, \ldots, \alpha_n)$ fictitiously depends on x_i.

2.3.11. Let $P(\widetilde{x}^n)$ be a polynomial of degree $k \geqslant 3$. Without any loss of generality, we can assume that an e.c. $A = x_1x_2 \ldots x_k$ is an addend of the polynomial P. Putting $x_k = x_{k+1} = \ldots = x_n$ we obtain the polynomial Q in which A is the only e.c. of rank k. A few cases are possible here. (a) All the elementary conjunctions of rank $k - 1$ generated by the set X^k appear in the polynomial Q. We put $x_{k-1} = x_k$. (b) There exist i,j $(1 \leqslant i < j \leqslant k)$ such that e.c. $x_1 \ldots x_{i-1}x_{i+1} \ldots x_k$ and $x_1 \ldots x_{j-1}x_{j+1} \ldots x_k$ do not appear in Q. Then we put $x_i = x_j$. (c) The number of $(k - 1)$-rank terms is equal to $k - 1$. Let the e.c. $x_1 \ldots x_{k-1}$ do not appear in Q. Then we identify the variables x_1 and x_k. In all the cases, we obtain a polynomial of degree $k - 1$.

2.3.12. Let us consider the polynomial P representing the function f. If the degree of the polynomial P is greater than 2, by identification of variables we can get a polynomial Q of degree 2 from it (see Problem 2.3.11). Without any loss of generality, we can assume that x_1x_2 appears in Q. By identifying all the variables that differ from x_1 and x_2, we obtain a nonlinear function of three variables.

2.3.13. It follows from the solution to Problem 2.3.12 that by identifying some variables of the function f, it is possible to obtain a function of three variables, representing the polynomial of second degree. It remains for us to prove that if there are exactly two second-degree terms in this polynomial, this number can be reduced by unity upon identification of variables. If, for example, x_1x_2 and x_1x_3 appear in P and x_2x_3 does not appear in it, we put $x_1 = x_3$.

2.3.14. If $f(\widetilde{x^n}) \notin L$, there exist i and j such that $f(\widetilde{x^n}) = x_ix_jP_1 \oplus x_iP_2 \oplus x_jP_3 \oplus P_4$, where P_s are functions independent of x_i and x_j and $P_1 \not\equiv 0$. We put $i = 1$ and $j = 2$. Let $\widetilde{\alpha} = (\alpha_3, \ldots, \alpha_n)$ be a tuple of the values of the variables $x_3, \ldots, x_n$ such that $P_1(\widetilde{\alpha}) = 1$. Then $\varphi(x_1, x_2) = f(x_1, x_2, \alpha_3, \ldots, \alpha_n) = x_1x_2 \oplus l(x_1x_2)$, where $l(x_1, x_2)$ is a linear function. The function $\varphi(x_1, x_2)$ becomes equal to unity on an odd number of tuples. Hence it follows that the face $B^{n;\,3,\,\ldots,\,n}_{\alpha_3\ldots\alpha_n}$ is the required one.

2.3.18. Let $f \in L$. If f forms a basis in L, then $f(\widetilde{0}) \neq 0$ and $f(\widetilde{1}) \neq 1$, and hence f essentially depends on an odd number of variables. But then it is self-dual and cannot form a basis in L.

2.3.19. If Φ is a system of functions complete in L, there exists in Φ a function $f_1 \notin S$, i.e. a function essentially depending on an even number of variables. Then a constant can be obtained from it. Let σ be such a constant. The system of functions Φ must also contain the function $f_2 \notin T_\sigma$. Substituting the constant σ for the variables of the function f_2, we get $\bar{\sigma}$. The system Φ also contains a function f_3 essentially depending on more than one variable. It can be easily seen that $[0, 1, f_3] = L$. Thus, from any system complete in L we can always isolate a complete subsystem consisting of three functions. Hence the statement.

2.3.20. There are three classes $\{0\}$, $\{1\}$ and $\{0, 1\}$ consisting only of constants, two classes $\{x\}$ and $\{x, \bar{x}\}$ consisting of functions essentially depending on more than one variable, and four classes consisting of functions of not more than one variable and containing constants and functions differing from the constants: $[0, x]$, $[1, x]$, $[0, 1, x]$ and $[0, 1, x, \bar{x}]$. If a class of linear functions contains a function essentially depending on more than one variable we have one of the following possibilities: the class contains or does not contain a function of an even number of variables; the class contains or does not contain a function not preserving 0. Hence it follows that the classes $[x \oplus y \oplus z \oplus 1]$, $[x \oplus y \oplus z]$, $[x \oplus y]$, $[x \oplus y \oplus 1]$, L and only these classes are closed classes of linear functions, which contain a function essentially depending on more than one variable.

2.3.22. Hint. If $f(\tilde{x}^3)$ satisfies the condition of the problem and is not self-dual, then for any $\tilde{\alpha} \in B^3$, $f(\tilde{\alpha}) = f(\tilde{\bar{\alpha}})$, and hence f can be specified completely by specifying its values on the tuples of the form $(\alpha_1, \alpha_2, 0)$. Besides, condition (1) implies that the function f becomes equal to unity exactly on two tuples of this form. Thus, there exist not more than six non-self-dual functions satisfying conditions (1) and (2). It can be easily verified that all these functions are linear.

2.3.25. For $n \neq 2$, there exist no such functions. All functions of the form $x_1x_2 \oplus \alpha x_1 \oplus \beta x_2 \oplus \gamma$ $(\alpha, \beta, \gamma \in \{0, 1\})$, and only these functions satisfy conditions (1) and (2).

2.3.26. Two for an odd n and 0 for an even n. **2.3.27.** In five ways. **2.3.28.** $f(\bar{\alpha}_1, \ldots, \bar{\alpha}_{n-1}, \alpha_n) = 1$. **2.3.29.** For an odd n. **2.3.30.** See Problem 2.3.14. **2.3.31.** See Problem 2.3.13. **2.3.32.** xy, $x \vee y$, $x \mid y$, $x \downarrow y$.

2.3.33. Hint. At least one of the functions f_1, f_2 and f_3 is not linear.

2.4.

2.4.2. (2) For even n. (4) For $n \neq 4k - 1$, $k = 1, 2, \ldots$. **2.4.3.** (1) For all $n \geqslant 2$. (3) For $n = 3k$, $k = 1, 2, \ldots$. **2.4.4.** In seven ways. **2.4.5.** (2) $3 \times 2^{2^{n-2}}$. (5) $3 \times 2^{n-1}$. (8) $(2^{2^n} - 2^{2^{n-1}})/2$. (12) 0. (15) $2^{2^{n-1}} - 1$. **2.4.7.** (1) No, it cannot since $x \oplus y \notin T_1$ and $x \rightarrow y \in T_1$.

2.4.8. (1) If $f \in T_0$, its polynomial does not contain 1 as an addend. Therefore, f can be expressed in terms of xy and $x \oplus y$. The equality $T_0 = [x \vee y, x \oplus y]$ follows from the previous one and from the fact that $x \vee y = xy \oplus x \oplus y$.

2.4.9. Let Φ be a basis in T_0. Then there exists a function $f_1 \in \Phi \setminus T_1$. By identifying all its variables, we obtain the constant 0. Φ also contains a function f_2 which is not monotonic. The function f_2 is obviously nonlinear. By identifying the variables and substituting 0 we can obtain from f_2 a function $\varphi_1(x, y)$ of the form $xy \oplus \sigma x \oplus \tau y$, $\sigma, \tau \in \{0, 1\}$ (see Problem 2.3.13). Since f_2 is not monotonic, there exist tuples $\tilde{\alpha}$ and $\tilde{\beta}$ such that $\tilde{\alpha} < \tilde{\beta}$ and $f(\tilde{\alpha}) > f(\tilde{\beta})$. Let A (resp. B) be a set of coordinates of the tuple $\tilde{\alpha}$ $(\tilde{\beta})$ which are equal to unity. Let us consider the function $\varphi_2(x, y, z)$ obtained from f_2 by putting in it $x_i = x$ for $i \in A$, $x_i = y$ for $i \in B$ and $x_i = z$ for $i \notin B$. We have $\varphi_2(0, 0, 0) = \varphi_2(1, 1, 0) = 0$, $\varphi_2(1, 0, 0) = 1$. If in this case φ_2 is a self-dual function, then either $\varphi_2 = x \oplus y \oplus z$, or $\varphi_2 = x\bar{y} \vee xz \vee \bar{y}z$. In both cases, we have $[0, \varphi_1, \varphi_2] = T_0$. If φ_2 is not self-dual, we consider the function $\varphi_3(x, y) = \varphi_2(x, y, 0)$. We have either $\varphi_3 = x \oplus y$ or $\varphi_3 = x\bar{y}$. If $\varphi_3 = x \oplus y$, then $[\varphi_1, \varphi_2] = T_0$. When $\varphi_3 = x\bar{y}$ and $\varphi_1 \neq x \vee y$, all the obtained functions 0, φ_1 and φ_3 belong to the same closed class, namely, the class G consisting of all functions f such that for any two tuples $\tilde{\alpha} = (\alpha_1, \ldots, \alpha_n)$ and $\tilde{\beta} = (\beta_1, \ldots, \beta_n)$ such that $f(\tilde{\alpha}) = f(\tilde{\beta})$, there exists i for which $\alpha_i = \beta_i = 1$. In view of the completeness of Φ in T_0, there exists a function f_3 which does not belong to the class G. The function f_3 has the following property: there exist tuples $\tilde{\alpha}$ and $\tilde{\beta}$ such that $f(\tilde{\alpha}) = f(\tilde{\beta})$ and $\tilde{\alpha}$ and $\tilde{\beta}$ do not have unit coordinates in common. Let A (B) be a set of unit coordinates of the tuple $\tilde{\alpha}$ (resp. $\tilde{\beta}$). Let $\varphi_4(x, y)$ be a function obtained from f_3 by substituting x for x_i if $i \in A$, y for x_i if $i \in B$, and 0 for x_i if $i \notin A \cup B$. Then φ_4 is either $x \vee y$ or $x \oplus y$. In both cases, we have $[x\bar{y}, \varphi_4] = T_0$. Thus, the completeness of the system Φ in T_0 implies that it contains a subsystem containing not more than three functions. Hence follows the statement.

2.4.10. The solution is similar to that for the previous problem. Make use of the fact that a complete system contains a non-monotonic function and one of the functions f_σ, $\sigma \in \{0, 1\}$, which do not belong to classes G_σ consisting of all functions f such that for any two tuples $\tilde{\alpha}$ and $\tilde{\beta}$ with the property $f(\tilde{\alpha}) = f(\tilde{\beta}) = \sigma$, there exists i for which the i-th coordinates of the tuples $\tilde{\alpha}$ and $\tilde{\beta}$ equal to σ. An example of a basis consisting of one function is the set $\{xy \oplus z \oplus u\}$.

2.4.11. Hint. Consider the function $xy \oplus y \oplus z$.

2.4.12. (3) Let us prove the completeness of the system $\Phi = \{xy \vee xz \vee yz, x \oplus y \oplus z\}$ in $T_0 \cap S$ by induction. The functions in $T_0 \cap S$ which depend on not more than two variables can be obtained by identifying the variables y of the functions in Φ. Let all the functions in $T_0 \cap S$, which depend on less than n ($n \geqslant 3$) variables, be superpositions of functions in Φ. Let $f(\tilde{x}^n) \in T_0 \cap S$. Then $f(x_1, x_1, x_3, x_4, \ldots, x_n)$, $f(x_3, x_2, x_3, x_4, \ldots, x_n)$, and $f(x_1, x_2, x_2, x_4, \ldots, x_n)$ depend on less than n variables. We can now use the result of solution of Problem 2.2.24.

2.4.13. $f(1, 1, 1) = 0$ since $f \notin T_1$; $f(1, 0, 1) = 0$ since $f \notin M$, and $f(1, 0, 1) = f(0, 1, 0) = 0$ since $f \notin S$.

2.4.14. If $f \in L$, there exists $k \geqslant 1$ such that f is congruent to $x_1 \oplus x_2 \oplus \ldots \oplus x_{2k+1} \oplus 1$. If $f \notin L$, by identifying the variables we can obtain either a function φ of the form $xy \oplus \sigma(x \oplus y) \oplus 1$, $\sigma \in \{0, 1\}$, or $xy \oplus xz \oplus yz \oplus 1$. In both cases, $L \cap S \subseteq [\varphi]$.

2.4.18. Let us prove, for example, that $T_0 \cap L$ is a precomplete class in T_0. Let $f \in T_0 \setminus L$. By identifying the variables in f, we can obtain a function φ either in the form $xy \oplus l(x, y)$ or $xy \oplus xz \oplus yz \oplus l(x, y, z)$. In both cases, the system $\{x \oplus y, \varphi\}$ is complete in T_0, i.e. $[(T_0 \cap L) \cup \{f\}] = T_0$.

2.4.19. Hint. $xy \oplus z \in T_0 \setminus (T_1 \cup L \cup S)$ and $\{1, xy \oplus z\}$ is a system complete in P_2.

2.4.21. $x_1 \oplus x_2 \oplus x_4$. **2.4.22. Hint.** See Problem 2.3.20.

2.4.23. We prove that if $f \notin (T_0 \cap T_1) \cup S$, a constant can be obtained from it. If $f \in T_0 \setminus T_1$ or $f \in T_1 \setminus T_0$, then by identifying all the variables of the function f, we obtain a constant.

Let $f \notin T_0 \cup T_1$. Since $f \notin S$, the function f essentially depends on more than one variable and it is possible to identify the variables in order to obtain a function which essentially depends on two variables, viz. one of the functions $\bar{x}\bar{y}$ or $\bar{x} \vee \bar{y}$. In both cases, we can obtain a constant.

2.5.

2.5.2. (1) $n = 2, 3$. (2) For even n. (3) For odd n.

2.5.4. Use the expansion from Problem 2.5.3.

2.5.5. Five functions among which one has a fictitious variable. One of the functions is self-dual.

2.5.8. (1) It is not true. Consider $f(\tilde{x}^3) \equiv x_3$ and the tuples (110) and (001).

2.5.12. $\bar{x}$, n functions.

2.5.14. We assume that a prime implicant I of the function $f(\tilde{x}^n)$ contains a negation of variable. Let us suppose, for example, that $I = \bar{x}_1 L$, where L is an elementary conjunction. Let us consider $N_L = \{\tilde{\alpha} : L(\tilde{\alpha}) = 1\}$ and an arbitrary tuple $0, \alpha_1, \ldots, \alpha_n$ in N_L. In view of its monotonicity, $f(1, \alpha_1, \ldots, \alpha_n) = 1$. It follows that xL, and hence L are implicants of the function f. This is in contradiction to the fact that I is a prime implicant.

2.5.16. For $k = 4l + 2$ $\left(\frac{k-2}{4} \leqslant l \leqslant \frac{n-2}{4}\right)$ and for $k = 4l + 1$ $\left(\frac{k-1}{4} \leqslant l \leqslant \frac{n-1}{4}\right)$.

2.5.18. (1) Yes, there exists. (2) No, there does not exist. **Hint.** A contracted d.n.f. $\mathscr{D}$ of the function f has a length equal to three. The formula $\mathscr{D}^*$ dual to $\mathscr{D}$ is a contracted c.n.f. of the function f^*. The contracted d.n.f. of the function f is obtained from $\mathscr{D}^*$ by opening the parentheses and by applying the rules $AA = A$, $A \vee A = A$, $A \vee AB = A$. Consider all the cases when the contracted d.n.f. of the function f^* has the length 3.

2.5.19. All lower unities of a monotonic function are pairwise incomparable. The maximum number of pairwise incomparable functions in B^n is $\binom{n}{[n/2]}$ (see Problem 1.1.15).

2.5.21. All the functions $f(\tilde{x}^n)$ such that $f(\tilde{\alpha}) = 1$ for $\|\tilde{\alpha}\| > n/2$ and $f(\tilde{\alpha}) = 0$ for $\|\tilde{\alpha}\| < [(n-1)/2]$ are monotonic.

2.5.22. (1) 2. (3) $n + 2$. (5) $2^n - 2$.

2.5.23. (1) Let $f(\tilde{x}^n) \subset M \cap S$. We consider the expansion $f = \bar{x}_n f_0^n \vee x_n f_1^n$. Since f is self-dual, we have $(f_1^n)^* = f_0^n$. In view of the monotonicity of f, we have $f_0^n \in M$, $f_1^n \in M$, $f_0^n \vee f_1^n = f_1^n$. Thus, if the function f_1^n in M is specified, the function f is specified as well. It remains for us to verify that for $n \geqslant 1$, there exists a function $f \in M^{n-1}$ such that $f \vee f^* \neq f$. We can take, for example, the function $f(x_1, x_2, \ldots, x_{n-1}) \equiv 0$. (3) Let $f(\tilde{x}^n) \in M$. If the component $f_{10}^{1,2}$ is given, the function f is defined by monotonicity on at least 2^{n-2} tuples: if $(\alpha_3, \ldots, \alpha_n)$ is a tuple and $f_{10}^{1,2}(1, 0, \alpha_3, \ldots, \alpha_n) = 0$, then $f(0, 0, \alpha_3, \ldots, \alpha_n) = 0$. If $f_{10}^{1,2}(1, 0, \alpha_3, \ldots, \alpha_n) = 1$, then $f(1, 1, \alpha_3, \ldots, \alpha_n) = 1$. Thus, in order to specify $f(\widetilde{x}^n) \in M$ it is sufficient to specify the components $f_{10}^{1,2}$, $f_{01}^{1,2}$ and additionally define the function by monotonicity. Then f remains indeterminate on not more than 2^{n-2} tuples.

2.5.25. $m_e(1) = 1$, $m_e(2) = 2$, $m_e(3) = 9$, $m_e(4) = 114$.

2.5.26. We have $|M^4| = 168$ (see Problem 2.5.24) and $|S^4| = 256$. The statement can now be proved by induction considering that $|S^n| = |S^{n-1}|^2$, $|M^n| < |M^{n-1}|^2$. **2.5.27.** Four.

2.5.28. Let $f(\widetilde{x}^n) \in M$. For each ascending chain $Z \in B^n$, either $f(\widetilde{\alpha}) = 0$ for all $\widetilde{\alpha} \in B^n$, or there exists a tuple $\widetilde{\beta} \in Z$ such that $f(\widetilde{\beta}) = 1$ and $f(\widetilde{\alpha}) = 0$ for all $\widetilde{\alpha} \in Z$, $\widetilde{\alpha} < \widetilde{\beta}$. Let a partition be specified in B^n on $\binom{n}{[n, 2]}$ ascending chains. Then to specify the function $f(\widetilde{x}^n) \in M$, it is sufficient to indicate for each chain in the partition whether it contains a vertex $\widetilde{\alpha}$ such that $f(\widetilde{\alpha}) = 1$, and if it does, specify the vertex $\widetilde{\beta}$ for which $f(\widetilde{\beta}) = 1$ and $f(\widetilde{\alpha}) = 0$ for all $\widetilde{\alpha} \in Z$ such that $\widetilde{\alpha} < \widetilde{\beta}$. Since the length of each ascending chain does not exceed $n + 1$, there are not more than $n + 2$ possibilities for each chain.

2.5.29. Let us consider the partition of the cube B^n into non-intersecting ascending chains, considered in the solution of Problem 1.1.18. This partition has the following properties. (1) Chains of minimum length have not more than two vertices. (2) For each two-dimensional face containing three vertices of a chain with $k + 2$ vertices, the fourth vertex of the face is contained in the chain of the partition, which has k vertices. Property (1) implies that on each chain of minimum length, a monotonic function can be specified in not more than three ways. Property (2) implies that if a monotonic (partial) function is defined on all chains containing k unities, on each chain of length $k + 2$ this function is defined by monotonicity everywhere except, perhaps, at two vertices. Consequently, there exist not more than three ways to additionally define it on such a chain. Considering that the number of chains is $\binom{n}{[n/2]}$, we have $[M^n] \leqslant 3^{\binom{n}{[n/2]}}$.

2.5.32. $(x_1 \vee x_2)(x_3 \vee x_4) \ldots (x_{2n-1} \vee x_{2n})$.

2.5.33. Any function essentially depending on more than one variable has not more than one lower unity.

2.5.34. Hint. See solution to Problem 1.1.36(4).

2.5.39. (1) No, it is impossible. (2) No, it is impossible. (3) Yes, it is. **2.5.44.** No, it cannot. **2.5.45.** No, it is impossible.

2.5.46. The statement that any function $f(\widetilde{x}^n) \in M$ can be represented in terms of superpositions generated by the functions of the set $\{0, 1, xy, x \vee y\}$ can be proved by induction on n, using Problem 2.5.3. The statement then follows from the fact that $\{1, xy, x \vee y\} \subseteq T_1$, $\{0, xy, x \vee y\} \subseteq T_0$ and that the function xy cannot be obtained from the set $\{0, 1, x \vee y\}$ and the function $x \vee y$ cannot be obtained from $\{0, 1, xy\}$.

2.5.48. Any basis in M contains constants as well as a function essentially depending on at least two variables. On the other hand, let Φ be a basis in M containing more than four functions. Then at least three of them depend on more than one variable. Among these functions, there exists a function f_1 which is not an elementary conjunction. Then by substituting constants we obtain $x \vee y$. There also exists a function f_2 which **is** not an e.d. from which we get xy. Then $\{0, 1, f_1, f_2\}$ is a subsystem complete in M.

2.5.49. See Problem 2.5.48. **2.5.50.** (2) $\{1, x(y \vee z)\}$. **2.5.51.** See Problem 2.2.24.

2.5.53. Let us prove, for example, that $M \cap T_0$ is precomplete. The only function $f \in M \setminus T_0$ is the constant 1. In $M \cap T_0$, there are functions $0, xy, x \vee y$ which together with 1 constitute a system complete in M. Let there exist one more closed class Q which is not contained in any of the above conditions. Then there exist $f_1 \notin M \cap T_0$, $f_2 \notin M \cap T_1$, $f_3 \notin \mathscr{D} \cup \{0, 1\}$ and $f_4 \notin \mathscr{K} \cup \{0, 1\}$. In this case $f_1 \equiv 1$, $f_2 \equiv 0$, and we obtain xy from f_3 by substituting constants and identifying variables, and $x \vee y$ from f_4.

2.6.

2.6.1. Assume that the converse is true and use the completeness criterion.

2.6.2. (4) Complete. (6) Incomplete. (7) Complete. (8) Incomplete.

2.6.3. Prove that $\bar{f} \notin T_0 \cup T_1 \cup L \cup M$.

2.6.4. (1) The system can be either incomplete, for example, $f_1(\tilde{x}^3) = m(x_1, x_2, x_3) \oplus x_1 \oplus x_2$, $f_2(\tilde{x}^3) = (x_1 \sim x_2) \vee x_3$, or complete, for example, $f_1(\tilde{x}^3) = \overline{m(x_1, x_2, x_3)}$, $f_2(\tilde{x}^3) = x_1 \vee x_2x_3$. (2) Complete. (3) Incomplete.

2.6.5. Complete. Prove that $f_1 \notin T_1 \cup L \cup S \cup M$ and $f_3 \notin T_0$.

2.6.6. (1) $\{xy \oplus z, (x \oplus y) \sim z\}$, $\{(x \vee y)(\bar{x} \vee \bar{y}), (x \oplus y) \sim \bar{z}, m(x, y, z)\}$. (3) $\{0, (x \mid (xy)) \to \bar{z}\}$, $\{x \oplus y, (x \mid (xy)) \to \bar{z}\}$, $\{(x \to y) \downarrow (y \sim z), (x \mid (xy)) \to \bar{z}\}$.

2.6.7. $\{x \downarrow y\}$, $\{x \mid y\}$, $\{x \oplus yz \oplus z \oplus 1\}$; $\{\bar{x}, xy\}$, $\{\bar{x}, x \vee y\}$, $\{\bar{x}, x \to y\}$; $\{1, x \oplus y, xy\}$, $\{1, x \oplus y, x \vee y\}$, $\{0, 1, x \oplus y \oplus yz\}$; $\{0, 1, x \oplus y \oplus z, xy\}$, $\{0, 1, x \oplus y \oplus z, x \vee y\}$, $\{0, 1, x \oplus y \oplus z, m(x, y, z)\}$.

2.6.8. There are seven such bases: two of them contain one function each, three contain two functions each, and two more bases containing three functions each can be isolated from the complete system $\{xy, x \vee y, x \oplus y, x \sim y\}$.

2.6.9. (1)-(3) Yes, it can. (4) No, it cannot.

2.6.10. No, it does not. Let $f_1 \not\equiv 1$. Then $f_1 \notin M$, and either $f_1 \notin T_1$ or $f_1 \notin S$. The function f_2 can be "deleted" in the former case and the function f_4 in the latter case. The possibility connected with the relation $f_2 \not\equiv 0$ is analyzed in a similar way. Let us now assume that $f_1 \equiv 1$ and $f_2 \equiv 0$. If $f_3 \notin M$, the system $\{f_1, f_2, f_3\}$ is complete. Let $f_3 \in M$. Then the function f_4 must not belong to the class M. If $f_4 \notin L$, the system $\{f_1, f_2, f_4\}$ is complete. It remains for us to analyze the case when $f_4 \in L \setminus M$. Since $f_4 \notin S \cup M$ and $f_4 \in L$, f_4 essentially depends on an even number of variables (and differs from constants 0 and 1). Consequently, either $f_4 \notin T_1$ or $f_4 \notin T_0$. Therefore, one of the systems $\{f_1, f_3, f_4\}$ or $\{f_2, f_3, f_4,\}$ is necessarily complete.

2.6.11. (1) P_2. (2) $M \setminus (T_0 \cup L) = \varnothing$. (3) $M \setminus (T_0 \cap T_1) = \{0, 1\}$. (4) $M \cap T_0 \cap T_1$. (5) $T_0 \cap L$. (6) S. (7) $L \cap S$. (8) T_0. (9) $T_0 \cap T_1$.

2.6.12. (1) $K_1 \subset K_2$. (2) $K_1 \not\supseteq\not\subseteq K_2$. (3) $K_1 \not\supseteq\not\subseteq K_2$. (4) $K_1 \not\supseteq\not\subseteq K_2$. (5) $K_1 = K_2$. (6) $K_1 \not\supseteq\not\subseteq K_2$. **2.6.13.** (2) 56.

2.6.14. (1) If $f \notin T_0 \cup T_1 \cup S$, then $f \not\equiv$ const and $f(\tilde{0}) = 1$, while $f(\tilde{1}) = 0$. Therefore, $f \notin M$. If f belonged to L, in view of the relation $f \notin T_0 \cup T_1$, the function f would essentially depend on an odd number of variables (and have a free term equal to 1). However, in this case $f \in S$. (2) The number of Sheffer's functions in the set $P_2(X^n)$ is equal to the difference between the number of all functions $f(\tilde{x}^n)$, satisfying the conditions $f(\tilde{0}) = 1$ and $f(\tilde{1}) = 0$, and the number of self-dual functions subject to the same condition. Thus, $P_2(X^n)$ contains $2^{2^n-2} - 2^{2^{n-1}} - 1$ Sheffer's functions $(n \geqslant 2)$.

2.6.15. Let $f(\tilde{x}^n)$ be a Sheffer's function (here $n \geqslant 3$ and all the variables are essential). Since $f(\tilde{x}^n) \notin S$, there exist two opposite tuples $\tilde{\alpha} = (\alpha_1, \ldots, \alpha_n)$ and $\tilde{\alpha}' = (\bar{\alpha}_1, \ldots, \bar{\alpha}_n)$ on which the function f assumes the identical values $f(\tilde{\alpha}) = f(\tilde{\alpha}') = \sigma$. Obviously, $\tilde{\alpha} \neq \tilde{0}$ and $\tilde{\alpha} \neq \tilde{1}$ (since $f(\tilde{0}) = 1$ and $f(\tilde{1}) = 0$). Let us substitute x for variables x_i corresponding to zero components in the tuple $\tilde{\alpha}$ and y for the remaining variables. We obtain a function $g(x, y)$ satisfying the relations $g(0, 0) = 1$, $g(1, 1) = 0$, and $g(0, 1) = g(1, 0) = \sigma$. If $\sigma = 0$, then $g(x, y) = x \downarrow y$, and if $\sigma = 1$, then $g(x, y) = x \mid y$.

2.6.16. (1) For $n = 4m + 2$, $m = 0, 1, 2, \ldots$. (2)-(5) For no values of n. (6) For all $n \geqslant 2$.

2.6.18. Functions which possess the properties mentioned in the condition of the problem essentially depend on at least three variables.

2.6.21. See solution to Problem 2.2.25(2).

2.6.23. $x\bar{y}$. The function $x \vee y$, for example, cannot be obtained from it. Generally, the class $[x\bar{y}]$ does not contain functions assuming the value of 1 on more than half the tuples of values of the variables.

2.6.24. The functions satisfying the requirement in the condition of the problem essentially depend on at least four variables. An example of such a function is $x_1 \oplus x_2 \oplus x_2x_3 \oplus x_3x_4$.

2.6.25. The proof can be carried out by induction on n. The *basis of induction*: $n = 4$ (see Problem 2.3.12). *Inductive step.* Let us assume that the statement is valid for $n = s \geqslant 4$, and let us prove it for $n = s + 1$. Let the function $f(\widetilde{x^{s+1}})$ essentially depend on all its arguments and not belong to the class L. We represent the function f in the following form: $g(\widetilde{x^s}) \oplus x_{s+1}h(\widetilde{x^s})$. Since x_{s+1} is an essential variable of the function f, $h(\widetilde{x^s}) \not\equiv 0$. Let us consider several possibilities. (a) A fictitious variable is x_i in the function g and x_j $(i \neq j)$ in the function h. Then it is sufficient to put $x_i = x_j$. (b) Neither g nor h contain fictitious variables. If $h \notin L$, we apply the inductive assumption to the function h. If $h \in L$ but $g \notin L$, we apply the inductive assumption to g. If $\{g, h\} \subset L$, then $f = (x_1 \oplus \ldots \oplus x_s)\, \overline{x}_{s+1} \oplus c_0 \oplus c_1x_{s+1}$, where $c_0, c_1 \in \{0, 1\}$, and we can, for example, identify x_s with x_{s+1}. (c) The function h does not contain fictitious variables while the function g does. Let x_i be one of such variables. If $h \notin L$, we apply the inductive assumption to h. If $h \in L$, we identify x_i with x_{s+1}. (d) Finally, we assume that g does not contain fictitious variables and h does. Let x_1 be one of them. If $g \in L$ (and hence $h \not\equiv 1$), we identify x_1 with the variable on which h depends essentially (for the sake of definiteness, we assume that x_2 is such a variable). We obtain $f(x_2, x_2, x_3, \ldots, x_{s+1}) = x_3 \oplus \ldots \oplus x_s \oplus c \oplus x_{s+1}h(x_2, x_2, x_3, \ldots, x_s)$. Obviously, this function is nonlinear and $x_2, \ldots, x_{s+1}$ are essential variables in it. Let us now consider the case when $g \notin L$. By inductive hypothesis, there exist in g two variables (x_i and x_j) whose identification in g leads to a nonlinear function which essentially depends on $s - 1$ variables. If the function h does not degenerate to 0 as a result of the identification of x_i and x_j, then x_{s+1} remains an essential variable in the new function as well. Let us now suppose that the function h degenerates to 0 as a result of the identification of x_i and x_j. Then the two variables x_i and x_j are essential in h. For definiteness, we assume that $x_i = x_2$ and $x_j = x_3$. We have $h(\widetilde{x^s}) = x_2x_3h_1(x_4, \ldots, x_s) \oplus x_2h_2(x_4, \ldots, x_s) \oplus x_3h_3(x_4, \ldots, x_s)$ and $h_3 \equiv h_1 \oplus h_2$. In the function f, we put $x\ = x_{s+1}$. If in this case we obtain a function φ which essentially depends on s arguments, the proof is completed. (The nonlinearity of φ is obvious since $\varphi(x_1, x_2, x_2, x_4, \ldots, x_s) = g(x_1, x_2, x_2, x_4, \ldots, x_s) \notin L$.) However, it may happen that for $x_1 = x_{s+1}$, a certain variable in φ becomes fictitious. Such a variable can only be one of the variables x_2 and x_3 (since identifying the variables x_2 and x_3 in φ, we must obtain a function $g(x_1, x_2, x_2, x_4, \ldots, x_s)$ which essentially depends on $s - 1$ variables). Let x_2 be a fictitious variable of the function φ. Since $f = x_1(x_2x_3g_1 \oplus x_2g_2 \oplus x_3g_3 \oplus g_4) \oplus x_2x_3g_5 \oplus x_2g_6 \oplus x_3g_7 \oplus g_8 \oplus x_{s+1} \times (x_2x_3h_1 \oplus x_2h_2 \oplus x_3h_3)$, where each function of $g_1, \ldots, g_7, g_8, h_1, h_2, h_3$ depends on the variables $x_4, \ldots, x_s$ and $h_3 \equiv h_1 \oplus h_2$, under the above assumption concerning the fictitious nature of the variable x_2 in the function φ, we have $\varphi = f(x_1, x_2, \ldots, x_s, x_1) = x_1(x_3g_3 \oplus g_4) \oplus x_3g_7 \oplus g_8 \oplus x_1x_3h_3$, i.e. $g_1 \equiv h_1$, $g_2 \equiv h_2$ and $g_5 \equiv g_6 \equiv 0$. Consequently, $f(\widetilde{x^n}) = x_2(x_1 \oplus x_{s+1})(x_3h_1 \oplus h_2) \oplus x_1x_3g_3 \oplus x_1g_4 \oplus x_3g_7 \oplus g_8 \oplus x_{s+1}x_3h_3$. We put $x_2 = x_{s+1}$. This leads to the function $\psi = f(x_1, x_2, x_3, x_4, \ldots, x_s, x_2) = (x_1 \oplus$

1) $x_2(x_3h_1 \oplus h_2) \oplus x_1x_3g_3 \oplus x_1g_4 \oplus g_8 \oplus x_3g_7 \oplus x_2x_3h_3$. Since the function $f(x_1, x_2, x_2, x_4, \ldots, x_s, x_2) = g(x_1, x_2, x_2, x_4, \ldots x_s)$ essentially depends on all its arguments, only one of the variables x_2 and x_3 can be fictitious for the function ψ. But this is possible only for $h_1 \equiv 0$ (and under some additional conditions). Then $h_3 \equiv h_2$ and $f = x_2(x_1 \oplus x_{s+1}) h_2 \oplus g_8 \oplus x_1x_3g_3 \oplus x_1g_4 \oplus x_3g_7 \oplus x_{s+1}x_3h_2$. Since $h \not\equiv 0$, for $h_1 \equiv 0$ the function h_2 cannot "degenerate" to 0, i.e. $h_2 \not\equiv 0$. Consequently, the function $\psi = x_2(x_1 \oplus 1) h_2 \oplus x_1x_3g_3 \oplus g_8 \oplus x_1g_4 \oplus x_3g_7 \oplus x_2x_3h_2$ essentially depends both on x_2 and x_3 even for $h_1 \equiv 0$. Thus, if none of the identifications $x_2 = x_3$ or $x_1 = x_{s+1}$ leads to the required function, it is sufficient to put $x_2 = x_{s+1}$. This completes the analysis of all possibilities. The fulfilment of the inductive step is just proved.

2.6.28. (1)-(4). No, they are not true. In part (4) we can take $f(x, y) = x \to y$.

2.6.29. Since $|T_0^n| = |T_1^n| = 2^{2^n-1}$, $|L^n| = 2^{n+1} \leqslant 2^{2^n-1}$ (for $n \geqslant 2$), $|S^n| = 2^{2^n-1}$ and $|M^n| \leqslant 2^{2^n-2} + 2 \leqslant 2^{2^n-1}$ (for $n \geqslant 2$), the set $\mathfrak{A}$ is not contained completely in any class in T_0, T_1, S, L and M.

2.6.30. An identical function is not contained in any basis in P_2 (see Problem 2.1.18). Any of the remaining functions of one variable and of zero variables among the Boolean functions does not belong to any precomplete class in P_2. At the same time, the hereditary system of any such function is an empty set, and hence is contained in each precomplete class (in P_2). Let now $f(\tilde{x}^n)$ be a function of a certain simple base $\mathfrak{B}$ and $n \geqslant 2$ (all variables are essential). By the definition of a prime basis, the system $\mathfrak{C} = (\mathfrak{B} \setminus \{f\}) \cup \mathfrak{N}(f)$ is not complete in P_2, i.e. is contained completely in at least one precomplete class K. If f belonged to K, the system $\mathfrak{B}$ $(\subseteq \mathfrak{C} \cup \{f\})$ would not be complete.

2.6.31. (1) 1, $\overline{x}$. (2) 0, 1, xy, $x \vee y$, $x \downarrow y$, $x \mid y$, $m(x, y, z)$, $m(x, y, \overline{z})$, $m(x, \overline{y}, \overline{z})$, $m(\overline{x}, \overline{y}, \overline{z})$. (3) $\overline{x}$, $x \oplus y$, $x \oplus y \oplus 1$, xy, $x \vee y$, $x \to y$, $\overline{x \to y}$, $x \oplus y \oplus z$, $m(x, y, z)$, $m(x, y, \overline{z})$. (4) $x \oplus y$, $x \sim y$, xy, $x \vee y$, $x \to y$, $\overline{x \to y}$, $x \downarrow y$, $x | y$, $\overline{x} \oplus \overline{y} \oplus z$, $\overline{x \oplus y \oplus z}$, $m(x, y, z)$, $m(x, y, \overline{z})$, $m(x, \overline{y}, z)$, $m(\overline{x}, \overline{y}, z)$. While solving problems (2)-(4), it is expedient to use the results formulated in Problems 2.2.14, 2.2.20 and 2.3.12.

2.6.33. (2) Taking into account Problems 2.2.20, 2.3.12 and 2.5.38, we conclude that any prime basis in P_2 consists only of functions which essentially depend on not more than three variables. Let us first consider functions which essentially depend on two arguments and are not Sheffer's functions: $x \oplus y$, $x \sim y$, $x \to y$, $\overline{x \to y}$, xy, and $x \vee y$. Let us find out which of them can be contained in a four-element basis in P_2. It can be seen that $x \oplus y \notin T_1 \cup S \cup M$, $x \sim y \notin T_0 \cup S \cup M$, $x \to y \notin T_0 \cup S \cup L \cup M$ and $\overline{x \to y} \notin T_1 \cup S \cup L \cup M$. Consequently, all these functions are unsuitable for our purpose. We assume that there exists a four-element prime basis containing xy. Since $xy \in (T_0 \cap T_1 \cap M) \setminus (L \cup S)$, the remaining three functions in the basis must be as follows: $f_1 \in (T_1 \cap M) \setminus T_0$, $f_2 \in (T_0 \cap M) \setminus T_1$ and $f_3 \in$

$(T_0 \cap T_1)\setminus M$. Obviously, $f_1 \equiv 1$ and $f_2 \equiv 0$. If f_3 did not belong to L, the system $\{f_1, f_2, f_3\}$ would be complete. Consequently, $f_3 \in L$, and hence $f_3 = x \oplus y \oplus z$. The case with the function $x \vee y$ is analyzed in a similar way. Let us now find out which nonlinear functions that essentially depend on three arguments can be contained in a four-element prime basis. If $f(x, y, z) \notin L \cup S$, then identifying variables in f, we can obtain a non-self-dual function essentially depending on two arguments (see Problem 2.2.20). But in this case $f(x, y, z)$ cannot be contained in a prime four-element basis (see the possibilities with the functions of two arguments considered above). It remains for us to consider the case when $f(x, y, z) \in S \setminus L$. If $f \notin T_0 \cap T_1$, and hence $f \notin M$, then for any function $g \notin S$ the system $\{f, g\}$ is complete in P_2. Therefore, we can assume that $f \in T_0 \cap T_1$. If $f \notin M$, then (accurate to redesignation of variables) $f = m(x, y, \bar{z})$. If, however, $f \in M$, then $f = m(x, y, z)$. In the former case, $f \in (T_0 \cap T_1 \cap S)\setminus(L \cup M)$, and hence in the corresponding prime four-element basis (if it exists at all), the remaining three functions must satisfy the conditions $f_1 \in (T_1 \cap S)\setminus T_0$, $f_2 \in (T_0 \cap S)\setminus T_1$ and $f_3 \in (T_0 \cap T_1)\setminus S$. However, $(T_1 \cap S)\setminus T_0 = (T_0 \cap S)\setminus T_1 = \varnothing$, i.e. there exist neither f_1 nor f_2 with the above properties. Hence, the first version is impossible. In the latter subcase, $f \in (T_0 \cap T_1 \cap S \cap M)\setminus L$. Consequently, the remaining three functions in the basis must be linear (otherwise the function f could be deleted from the basis). Linear functions essentially depending on an even number of variables and differing from constants cannot belong to such a basis since each such function is contained neither in the set $T_0 \cup S \cup M$ nor in the set $T_1 \cup S \cup M$. A linear function essentially depending on an odd number of variables can belong to such a basis only if it preserves 0 and 1. Therefore, it can be only the function $x \oplus y \oplus z$. This leaves behind only constants.

CHAPTER THREE

3.1.

3.1.1. (1) Consider two cases: $x = k - 1$ and $x \neq k - 1$. (2)-(7) Consider two cases: $x \geqslant y$ and $x < y$. (8) Let $x \leqslant y$. Then $\sim x = k - 1 - x \geqslant y - x$ and $(\sim x) \dot{-} (y \dot{-} x) = k - 1 - x - (y - x) = k - 1 - y = \sim y$. If, however, $x > y$, we have $(\sim x) \dot{-} (y \dot{-} x) = \sim x$. (9) Consider two cases: $x \leqslant y$ and $x > y$. (10)-(11) The equalities are proved directly by using the relations $\sim x = k - 1 - x$ and $\bar{x} = x + 1$. (12) Consider two cases: $x = k - 2$ and $x \neq k - 2$. (13) Consider two cases: $x = k - 1$ and $x \neq k - 1$. (14) Consider five cases: (a) $x = y = k - 1$; (b) $x = k - 1$, $y \neq k - 1$; (c) $x \neq k - 1$, $y = k - 1$; (d) $x \neq k - 1$, $y \neq k - 1$, $x \leqslant y$, and (e) $x \neq k - 1$, $y \neq k - 1$, $x > y$. (15) Consider four cases: (a) $x = y = k - 1$; (b) $x = k - 1$, $y \neq k - 1$; (c) $x \neq k - 1$, $y = k - 1$, and (d) $x \neq k - 1$, $y \neq k - 1$. (16)-(17) Consider three cases: (a) $x = k - 1$, (b) $x \neq k - 1$, $x \geqslant y$, and (c) $x \neq$

$k - 1$, $x < y$. (18) Consider two cases: $x = k - 1$ and $x \neq k - 1$. (19) Consider two cases: $x = 0$ and $x \neq 0$. (20) Consider three cases: $x = 0$, $x = 1$ and $x \geqslant 2$. (21) Consider four cases: $x < i - 1$, $x = i - 1$, $x = i$ and $x > i$. (22) Consider two cases: $x = k - 1$ and $x \neq k - 1$. (23) Consider three cases: $x = k - 1$, $x = k - 2$, and $x \leqslant k - 3$. If $x = i$ in the last case, there are exactly $k - 2 - i$ unities among the values of the functions $j_0(x)$, $j_0(x \dot{-} 1)$, . . ., $j_0(x \dot{-} (k - 3))$.

3.1.2. Hint. Relations from Problem 3.1.1 can be useful for solving this problem. (1) $J_1(x) = J_0(\max(J_0(x), J_2(x)))$. (2) $\sim x = \max(J_0(x), \min(x, J_1(x)))$. (3) $f_1(x) = \max(1, J_1(x^2))$, $f_2(x) = J_1(f_1(x))$, $f_2^2(x) = j_0(x)$ and $\bar{x} = \max(j_0(x), J_1(x))$. (4) $j_0(x) = (x^2 - 1)^2$ for $k = 3$ and $j_0(x) = (x^4 - 1)^2$ for $k = 5$. (5) Let $xy + x - y^2 + 1 = \varphi(x, y)$. We have $\varphi(x, x) = x + 1$, $\varphi(x, x + 1) = 0$. Further we obtain all the constants and (a) $\varphi(2, x) = j_1(x)$ for $k = 3$, (b) $\varphi(0, x) = 1 - x^2$ for $k = 5$, and hence we can construct the functions $-x^2$, x^4, $x + 4$ $(= x - 1)$ and $j_1(x) = 1 - (x - 1)^4$. (6) We put $x\bar{y} = \psi(x, y)$. This gives: (a) $\sim x = \psi(1, \psi(1, \psi(x, 1)))$ for $k = 3$, (b) $\psi_1(x) = \psi(\psi(x, 1), 1) = 4x = -x$, $\psi_2(x) = \psi(1, x) = x + 1$, $\psi_3(x) = \psi_2(\psi_2(x)) = x + 2$, $\psi_4(x) = \psi_3(\psi_3(x)) = x + 4$ and $\sim x = \psi_4(\psi_1(x)) = 4 - x$ for $k = 5$. (7) $x \dot{-} x = 0$, $j_0(0) = 1$, $3 \dot{-} 1 = 2$, $j_0(x \dot{-} 1) \dot{-} j_0(x) = j_1(x)$, $j_0(x \dot{-} 2) \dot{-} j_0(x \dot{-} 1) = j_2(x)$, $1 \dot{-} j_0(x \dot{-} 2) = j_3(x)$, $\varphi_1(x) = (3 \dot{-} j_0(x)) \dot{-} j_0(x)$, $\varphi_2(x) = \varphi_1(x) \dot{-} j_1(x)$, $\varphi_3(x) = ((\varphi_2(x) \dot{-} j_3(x)) \dot{-} j_3(x)) \dot{-} j_3(x) = \bar{x}$. (8) $\sim x = \max(J_0(x), J_0^2(x + 2), 2J_1(x))$. (9) $j_4(x) = (((J_2((x \dot{-} 1) \dot{-} 1) \dot{-} 1) \dot{-} 1) \dot{-} 1) \dot{-} 1$. (10) $j_5(x) = J_3^2(x + 4)$. (11) $j_1(x) = -J_{k-1}(x - 2)$. (12) $J_{k-1}(x) = (\ldots((x \dot{-} \underbrace{(\sim x)) \dot{-} (\sim x)) \dot{-} \ldots \dot{-} (\sim x)) \dot{-} (\sim x)}_{k-2 \text{ times}}$. (13) $J_{k-2}(x) =$

$$(\ldots((k - 1) \dot{-} \underbrace{(x + 2)) \dot{-} \ldots \dot{-} (x + 2)) \dot{-} (x + 2)}_{k-1 \text{ times}}. \qquad (14)$$

$1 \dot{-} 2 \times 1 = 0$, $\sim 0 = k - 1$, $(\sim x) \dot{-} 2(k - 1) = j_0(x)$. (15) $\bar{x} = (\sim((\sim x) \dot{-} 1)) \dot{-} J_{k-1}(x)$, and $J_{k-1}(x)$ can be represented by a formula generated by the set $\{\sim x, x \dot{-} y\}$ (see 12)).

3.1.3. If α and k are mutually prime, there are no numbers among α, 2α, . . ., $i\alpha$, . . ., $(k - 1)\alpha$ which are multiple to k or mod k comparable. Consequently, the division of these numbers by k gives different remainders. Since we have $k - 1$ given numbers, there exists among them a number which is mod k comparable with 1. Let this number be $l_0\alpha$. Then we have $\underbrace{(x+\alpha) + \ldots + \alpha}_{l_0 \text{ times}} = x + 1$. To complete the proof, we note that $J_i(x) = J_{k-1}(x + k - 1 - i))$, $i = 0, 1, \ldots, k - 2$.

3.1.4. (1) Take $y = m - 1$. (2) Consider the function $1 \dot{-} 2x$.

3.1.5. In this case, we can use the relation $x \supset y = \min(k - 1, y - x + k - 1)$ with ordinary (and not mod k) addition and subtraction in $y - x + k - 1$. By induction in i, we can prove that $h_i(x) = \min(k - 1, i(k - 1 - x))$, $i = 1, 2, \ldots, k - 1$. Indeed, $h_1(x) = \sim x = k - 1 - x = \min(k - 1, k - 1 - x)$. Further, if $h_{i-1}(x) = \min(k - 1, (i - 1)(k - 1 - x))$, $i \geqslant 2$, then $h_i(x) = \min(k - 1, h_{i-1}(x) - x + k - 1) = \min(k - 1,$

$\min(k-1, (i-1)(k-1-x)) + (k-1-x)) = \min(k-1, (k-1) + (k-1-x), i(k-1-x)) = \min(k-1, i(k-1-x))$. Consequently, $h_{k-1}(x) = \min(k-1, (k-1)(k-1-x)) = \begin{cases} 0 & \text{if } x = k-1, \\ k-1 & \text{if } x \neq k-1, \end{cases}$ and hence $\sim h_{k-1}(x) = J_{k-1}(x)$.

3.1.6. Obviously, a comparison of $x^2 \equiv x^3 \pmod{k}$, and $x^3 \equiv x^4 \pmod{k}$ cannot be carried out identically for any $k \geqslant 3$ since for $x = k-1$ we have $x^2 = (k-1)^2 = 1$, $x^3 = (k-1)^3 = k-1$ and $x^4 = (k-1)^4 = 1$. Let us now consider the relation $x^2 \equiv x^4 \pmod{k}$. We put $x = 2$. This gives $4 \equiv 16 \pmod{k}$. Trying the values of k from 3 to 15, we see that the comparison can be carried out only for $k = 3, 4, 6$ and 12. Hence for each of these values of k, we must find out whether the comparison $x^2 \equiv x^4 \pmod{k}$ is valid for any values of x in E_k. It turns out that for any $k = 3, 4, 6$ and 12 the comparison $x^2 \equiv x^4 \pmod{k}$ is fulfilled identically in E_k. For example, for $k = 6$ we have

$$x^2 = \begin{pmatrix} 0 & 1 & 2 & 3 & 4 & 5 \\ 0 & 1 & 4 & 3 & 4 & 1 \end{pmatrix} = x^4.$$

3.1.7. For $k = 3, 5$ and 7, the number of different functions of the given form is $k-1$, while for $k = 4, 6, 8, 9, 10$ it is equal to 3, 2, 4, 7, 4 respectively. For example, for $k = 4$ we have

$$x = \begin{pmatrix} 0 & 1 & 2 & 3 \\ 0 & 1 & 2 & 3 \end{pmatrix}, \quad x^2 = \begin{pmatrix} 0 & 1 & 2 & 3 \\ 0 & 1 & 0 & 1 \end{pmatrix}, \quad x^3 = \begin{pmatrix} 0 & 1 & 2 & 3 \\ 0 & 1 & 0 & 3 \end{pmatrix}$$

and $x^{2l} = x^2$, $x^{2l+1} = x^3$ $(l \geqslant 1)$.

3.1.8. $j_{k-1}(x) = J_{k-1}(x) + \ldots + J_{k-1}(x)$ ($k-1$ terms); $j_i(x) = j_{k-1}(x - i - 1)$, $0 \leqslant i \leqslant k-2$. If $g(x) \in P_k^{(1)}$, then $g(x) = g(0) j_0(x) + \ldots + g(i) j_i(x) + \ldots + g(k-1) j_{k-1}(x)$. It remains for us to take into account $l j_i(x) = j_i(x) + \ldots + j_i(x)$ (l terms).

3.1.9. Let us consider the functions $f(x) - c$ $(c \in \{1, 2\})$, $f(x+1)$, $f(x+2)$ and $f(h_{12}(x))$, where $h_{12}(x) = x + j_1(x) - j_2(x)$. Each of them assumes the same number of values as the function $f(x)$ has. Therefore, we can assume that $f_1(x) = \begin{pmatrix} 0 & 1 & 2 \\ 0 & 0 & a \end{pmatrix}$ and $f_2(x) = \begin{pmatrix} 0 & 1 & 2 \\ 0 & b_1 & b_2 \end{pmatrix}$, where $a, b_1, b_2 \in \{1, 2\}$. Then $g(x) = \begin{pmatrix} 0 & 1 & 2 \\ 0 & b_1 & a + b_2 \end{pmatrix}$. If $b_1 = 1$, then $b_2 = 2$ and $a + b_2 = 2$, and hence $a = 0$, which contradicts to the condition of the problem (since $a \in \{1, 2\}$). If, however, $b_1 = 2$, we have $b_2 = 1$ and $a + b_2 = 1$. Consequently, $a = 0$, and we again arrive at a contradiotion.

3.1.10. Since the function $f(x) - c$ $(c \in \{1, 2\})$ assumes the same number of values as the function $f(x)$ does, the coefficient a_2 can assume any value in E_3. We can assume that $a_2 = 0$. Then $f(x) = a_0 x^2 + a_1 x = \begin{pmatrix} 0 & 1 & 2 \\ 0 & a_0 + a_1 & a_0 + 2a_1 \end{pmatrix}$. The following three cases are assumed to be favourable: (a) $a_0 + a_1 = a_0 + 2a_1$,

(b) $a_0 + a_1 = 0$, but $a_0 + 2a_1 \neq 0$, (c) $a_0 + a_1 \neq 0$, but $a_0 + 2a_1 = 0$. In case (a), we have $a_1 = 0$, and a_0 is arbitrary. In case (b): $a_1 \neq 0$, $a_0 = - a_1 = 2a_1$. Case (c): $a_1 = a_0$ and $2a_1 \neq 0$ (hence $a_1 \neq 0$). **Answer:** a_2 is arbitrary and (1) if $a_1 = 0$, then a_0 is arbitrary, (2) if $a_1 \neq 0$, then either $a_0 = a_1$ or $a_0 = 2a_1$ (i.e. if $a_1 \neq 0$, then $a_0 \neq 0$).

3.1.11. (1) $\bar{x} = \max(\min(1, J_0(x)), J_1(x)) = j_0(x) + 2j_1(x)$. (2) $\sim x = \max(J_0(x), \min(2, J_1(x)), \min(1, J_2(x))) = 3j_0(x) + 2j_1(x) + j_2(x)$. (3) $-j_0(x) = J_0(x) = 4j_0(x)$. (4) $2J_1(x) = \min(4, J_1(x)) = 4j_1(x)$. (5) $J_2(x^2 + x) = \max(J_1(x), J_3(x)) = 4j_1(x) + 4j_3(x)$. (6) $(\sim x)^2 + x = \max(\min(1, J_0(x)), \min(1, J_1(x)), J_2(x), J_3(x)) = j_0(x) + j_1(x) + 3j_2(x) + 3j_3(x)$. (7) $3j_1(x) - j_3(x) = \max(J_1(x), J_3(x)) = 3j_1(x) + 3j_3(x)$. (8) $x + 2y = \max(\min(1, J_1(x), J_0(y)), \min(J_2(x), J_0(y)), \min(J_0(x), J_1(y)), \min(1, J_2(x), J_1(y)), \min(1, J_0(x), J_2(y)), \min(J_1(x), J_2(y))) = j_1(x)\, j_0(y) + 2j_2(x)\, j_0(y) + 2j_0(x)\, j_1(y) + j_2(x)\, j_1(y) + j_0(x)\, j_2(y) + 2j_1(x)\, j_2(y)$. (9) $\max(x, y) = \max(\min(1, J_1(x), J_0(y)), \min(1, J_1(x), J_1(y)), \min(1, J_0(x), J_1(y)), \min(J_0(x), J_2(y)), \min(J_1(x), J_2(y)), \min(J_2(x), J_2(y)), \min(J_2(x), J_1(y)), \min(J_2(x), J_0(y))) = j_1(x) j_0(y) + j_1(x) j_1(y) + j_0(x)\, j_1(y) + 2j_0(x)\, j_2(y) + 2j_1(x)\, j_2(y) + 2j_2(x)\, j_2(y) + 2j_2(x)\, j_1(y) + 2j_2(x)\, j_0(y)$. (10) $x \dot{-} y^2 = \max(\min(1, J_1(x), J_0(y)), \min(1, J_2(x), J_1(y)), \min(1, J_2(x), J_2(y)), \min(J_2(x), J_0(y))) = j_1(x)\, j_0(y) + j_2(x)\, j_1(y) + j_2(x)\, j_2(y) + 2j_2(x)\, j_0(y)$. (11) $x^2y = \max(\min(1, J_1(x), J_1(y)), \min(1, J_2(x), J_1(y)), \min(J_1(x), J_2(y)), \min(J_2(x), J_2(y))) = j_1(x)\, j_1(y) + j_2(x)\, j_1(y) + 2j_1(x)\, j_2(y) + 2j_2(x)\, j_2(y)$. (12) $xy = \max(\min(1, J_1(x), J_0(y)), \min(1, J_3(x), J_2(y)), \min(2, J_2(x), J_0(y)), \min(2, J_1(x), J_1(y)), \min(2, J_3(x), J_1(y)), \min(2, J_2(x), J_2(y)), \min(J_3(x), J_0(y)), \min(J_1(x), J_2(y))) = j_1(x) j_0(y) + j_3(x)\, j_2(y) + 2j_2(x)\, j_0(y) + 2j_1(x)\, j_1(y) + 2j_3(x)\, j_1(y) + 2j_2(x)\, j_2(y) + j_3(x)\, j_0(y) + j_1(x)\, j_2(y)$.

3.1.12. Putting $\widetilde{x^n} = (\alpha_1, \alpha_2, \ldots, \alpha_n) = \widetilde{\alpha^n}$, we get $f(\widetilde{\alpha^n}) = \min\limits_{\widetilde{\sigma}} \{\max(f(\widetilde{\sigma}), \sim J_{\sigma_1}(\alpha_1), \sim J_{\sigma_2}(\alpha_2), \ldots, \sim J_{\sigma_n}(\alpha_n))\}$. Let us prove that the right-hand side of this relation coincides with the left-hand side. If at least one $\sigma_i \neq \alpha_i$, for example, for $i = i_0$, then $\max(f(\widetilde{\sigma}), \sim J_{\sigma_1}(\alpha_1), \ldots, \sim J_{\sigma_{i_0-1}}(\alpha_{i_0-1}), \sim J_{\sigma_{i_0}}(\alpha_{i_0}), \sim J_{\sigma_{i_0+1}}(\alpha_{i_0+1}), \ldots, \sim J_{\sigma_n}(\alpha_n)) = \max(f(\widetilde{\sigma}), \sim J_{\sigma_1}(\alpha_1), \ldots, \sim J_{\sigma_{i_0-1}}(\alpha_{i_0-1}), \sim 0, \sim J_{\sigma_{i_0+1}}(\alpha_{i_0+1}), \ldots, \sim J_{\sigma_n}(\alpha_n)) = k - 1$.

If, however, $\widetilde{\sigma} = \widetilde{\alpha}$, then $\max(f(\widetilde{\sigma}), \sim J_{\sigma_1}(\alpha_1), \ldots, \sim J_{\sigma_n}(\alpha_n)) = \max(f(\widetilde{\alpha}), \sim (k-1), \ldots, \sim (k-1)) = f(\widetilde{\alpha})$, and hence $\min\limits_{\widetilde{\sigma}} \{\max(f(\widetilde{\sigma}), \sim J_{\sigma_1}(\alpha_1), \ldots, \sim J_{\sigma_n}(\alpha_n))\} = \min(k-1, \ldots, k-1, f(\widetilde{\alpha}), k-1, \ldots, k-1) = f(\widetilde{\alpha})$.

3.2.

3.2.1. (1) In order to simplify the notation, we put $T(\{0, 2\}) = T$ and $\mathcal{U}(\{0, 1\}, \{2\}) = \mathcal{U}$. (a) $\sim x \in T$ and $\notin \mathcal{U}$; (b) (b) $j_1(x) \in T \cap \mathcal{U}$; (c) $J_2(x) \in T \cap \mathcal{U}$; (d) $x \dot{-} y \in T$, and $\notin \mathcal{U}$; (e) $x + y \notin T$ and $\notin \mathcal{U}$; (f) $\min(x, y) \in T \cap \mathcal{U}$.

(2) We put $T(\{1, 3\}) = T$, $\mathcal{U}(\{0, 1\}, \{2\}, \{3\}) = \mathcal{U}_1$, $\mathcal{U}(\{0, 3\}, \{1, 2\}) = \mathcal{U}_2$. (a) $\overline{x} \notin T$, $\notin \mathcal{U}_1$ and $\notin \mathcal{U}_2$; (b) $\sim x \notin T$ and $\in \mathcal{U}_1 \cap \mathcal{U}_2$ (c) $j_0(x) \notin T$, $\in \mathcal{U}_1$ and $\notin \mathcal{U}_2$; (d) $x + 2y \in T$, $\notin \mathcal{U}_1$ and $\notin \mathcal{U}_2$; (e) $\max(x, y) \in T \cap \mathcal{U}_1$ and $\notin \mathcal{U}_2$; (f) $x^2y \in T$, $\notin \mathcal{U}_1$ and $\notin \mathcal{U}_2$.

3.2.2. (1) The subset $\{2\}$ and the partition $\{0\} \cup \{1, 2\}$ are suitable. (2) $\mathcal{E} = \{0, 2\}$ and $D = \{\{0, 2\}, \{1\}\}$. (3) $\mathcal{E} = \{1\}$ and $D = \{\{0\}, \{1, 2\}\}$. (4) $\mathcal{E} = \{2\}$. There is no suitable partition. (5) $\mathcal{E} = \{0\}$ and $D = \{\{0, 1\}, \{2\}, \{3\}\}$. (6) $\mathcal{E} = \{0\}$ and $D = \{\{0, 2\}, \{1, 3\}\}$. (7) $\mathcal{E} = \{1, 3\}$ and $D = \{\{0, 2\}, \{1, 3\}\}$. (8) $\mathcal{E} = \{0\}$ and $D = \{\{0, 1, 4\}, \{2, 3\}\}$. (9) $\mathcal{E} = \{4\}$ and $D = \{\{0, 1, 2, 4\}, \{3\}\}$. (10) $\mathcal{E} = \{0, 2\}$ and $D = \{\{0, 2, 4\}, \{1, 3, 5\}\}$. (11) $\mathcal{E} = \{0, 1\}$ and $D = \{\{0, 1\}, \{2, \ldots, k-1\}\}$. (12) $\mathcal{E} = \{0, k-1\}$; $D = \{\{0, 2\}, \{1\}\}$ for $k = 3$ and $D = \{\{0, k-1\}, \{1, 2, \ldots, k-2\}\}$ for $k \geqslant 4$.

3.2.3. (1) If $\mathcal{E} = \varnothing$ or $\mathcal{E} = E_k$, any function in P_k preserves this set (for $\mathcal{E} = \varnothing$, it is assumed by definition that $f(\varnothing, \ldots, \varnothing) = \varnothing$). Let $\mathcal{E} \neq \varnothing$ and $\mathcal{E} \neq E_k$. Then there exists $a \in E_k$ such that $a \notin \mathcal{E}$. Obviously, $f(x) \equiv a \notin T(\mathcal{E})$. Hence $T(\mathcal{E}) \neq P_k$.

(2) **Hint.** It should be first proved that different classes $T(\mathcal{E})$ correspond to different proper subsets in E_k. The number of all non-empty subsets in E_k is $2^k - 1$.

(3) If $\mathcal{E} = \varnothing$, then $|(T(\mathcal{E}))^n| = k^{k^n}$. Let $1 \leqslant m = |\mathcal{E}| \leqslant k-1$. Then $|(T(\mathcal{E}))^n| = m^{m^n}k^{k^n-m^n}$ (since the values from $\mathcal{E}$ should be assumed on $\mathcal{E}^n = \underbrace{\mathcal{E} \times \ldots \times \mathcal{E}}_{n \text{ times}}$, while on the remaining $k^n - m^n$ tuples, the values of the functions can be arbitrary).

3.2.4. (1) If $D = \{\{0\}, \{1\}, \ldots, \{k-1\}\}$, i.e. $s = k$, or $D = \{\{0, 1, \ldots, k-1\}\}$ i.e. $s = 1$, then any function in P_k preserves this partition. Let $s \neq 1$ and $s \neq k$. In the partition $D = \{\mathcal{E}_1, \ldots, \mathcal{E}_s\}$ we take a subset $\mathcal{E}_{i_1}$ such that $|\mathcal{E}_{i_1}| \geqslant 2$, and another subset $\mathcal{E}_{i_2}$, differing from $\mathcal{E}_{i_1}$. Let $a_1, a_2 \in \mathcal{E}_{i_1}$ and $b \in \mathcal{E}_{i_2}$. Let us consider the function $f(x) = \begin{cases} a_1 & \text{for } x = a_1 \\ b & \text{for } x \neq a_1. \end{cases}$ Obviously, $f(a_2) = b$, and hence f does not preserve the partition D, and $f \notin \mathcal{U}(D)$.

(2) **Hint.** We must first prove that for $s \neq 1$ and $s \neq k$ different classes $\mathcal{U}(D)$ correspond to different partitions of E_k. For $k = 3$, the set has the following partitions: $D_1 = \{\{0\}, \{1\}, \{2\}\}$, $D_2 = \{\{0, 1\}, \{2\}\}$, $D_3 = \{\{0, 2\}, \{1\}\}$, $D_4 = \{\{0\}, \{1, 2\}\}$, $D_5 = \{\{0, 1, 2\}\}$. $\mathcal{U}(D_1) = \mathcal{U}(D_5) = P_3$. The remaining three classes $\mathcal{U}(D_2)$, $\mathcal{U}(D_3)$ and $\mathcal{U}(D_4)$ differ from one another and from P_3. Consequently, there exist four different classes of the type $\mathcal{U}(D)$ in P_3 (if we also take into account the entire set P_3 as well). For $k = 4$, the set E_k has the following partitions: $D_1 = \{\{0\}, \{1\},$

$\{2\}, \{3\}\}$, $D_2 = \{\{0, 1\}, \{2\}, \{3\}\}$, $D_3 = \{\{0, 2\}, \{1\}, \{3\}\}$, $D_4 = \{\{0, 3\}, \{1\}, \{2\}\}$, $D_5 = \{\{1, 2\}, \{0\}, \{3\}\}$, $D_6 = \{\{1, 3\}, \{0\}, \{2\}\}$, $D_7 = \{\{2, 3\}, \{0\}, \{1\}\}$, $D_8 = \{\{0, 1\}, \{2, 3\}\}$, $D_9 = \{\{0, 2\}, \{1, 3\}\}$, $D_{10} = \{\{0, 3\}, \{1, 2\}\}$, $D_{11} = \{\{0\}, \{1, 2, 3\}\}$, $D_{12} = \{\{1\}, \{0, 2, 3\}\}$, $D_{13} = \{\{2\}, \{0, 1, 3\}\}$; $D_{14} = \{\{3\}, \{0, 1, 2\}\}$, $D_{15} = \{\{0, 1, 2, 3\}\}$. Consequently, there are 14 different classes of type $\mathcal{U}(D)$ in P_4. In P_5, there are 51 classes of the type $\mathcal{U}(D)$.

(3) Let $(i_1, i_2, \ldots, i_n)$ be a tuple of numbers in the set $\{1, 2, \ldots, s\}$. The number of such tuples is s^n. Arrange them in the alphabetic order and assign them numbers from 1 to s^n. The tuple $(1, 1, \ldots, 1, 1)$ has the number 1, the tuple $(1, 1, \ldots, 1, 2)$ the number 2, the tuple $(1, 1, \ldots, 1, 2, 1)$ the number $s + 1$, and so on. If a tuple $(i_1, i_2, \ldots, i_n)$ has a number m, the number $|\mathscr{E}_{i_1}|\,|\mathscr{E}_{i_2}| \ldots |\mathscr{E}_{i_n}|$ is denoted by d_m. The number of functions in the set $P_k(X^n)$, which preserve the partition $D = \{\mathscr{E}_1, \mathscr{E}_2, \ldots, \mathscr{E}_s\}$ is equal to $\sum\limits_{(j_1, j_2, \ldots, j_r)} |\mathscr{E}_{j_1}|^{d_1} |\mathscr{E}_{j_2}|^{d_2} \ldots |\mathscr{E}_{j_r}|^{d_r}$, where $r = s^n$ and the sum is taken over all possible tuples of length s^n, consisting of numbers from the set $\{1, 2, \ldots s\}$.

3.2.5. We denote the set $\mathscr{E}$ by $\mathscr{E}_1$ and $E_k \setminus \mathscr{E}$ by $\mathscr{E}_2$. Let $|\mathscr{E}| = l$ and, as in Problem 3.2.4(3), let d_m stand for the number $|\mathscr{E}_{i_1}| \ldots |\mathscr{E}_{i_n}|$, where $(i_1, \ldots, i_n)$ is the tuple of length n, consisting of numbers belonging to the set $\{1, 2\}$, where m is the number of the tuple in the alphabetic order. Obviously, $d_1 = l^n$.

(1) Since $T(\mathscr{E}) \setminus \mathcal{U}(D) = T(\mathscr{E}) \setminus (T(\mathscr{E}) \cap \mathcal{U}(D))$, the number of functions in $P_k(X^n)$ contained in the set $T(\mathscr{E}) \setminus \mathcal{U}(D)$, i.e. $|(T(\mathscr{E}) \setminus \mathcal{U}(D))^{(n)}|$ is $|(T(\mathscr{E}))^{(n)}| - |(T(\mathscr{E}) \cap \mathcal{U}(D))^{(n)}| = l^{l^n} \cdot k^{k^n - l^n} - \sum\limits_{\substack{j_1=1 \\ (j_2, \ldots, j_r)}} |\mathscr{E}_{j_1}|^{d_1} |\mathscr{E}_{j_2}|^{d_2} \ldots |\mathscr{E}_{j_r}|^{d_r} = l^{l^n} k^{k^n - l^n} - \sum\limits_{(j_2, \ldots, j_r)} |\mathscr{E}_{j_2}|^{d_2} \ldots |\mathscr{E}_{j_r}|^{d_r}$, where $r = 2^n$, and the last sum is taken over all such tuples of length $2^n - 1$, which consist of numbers belonging to the set $\{1, 2\}$. The above formula is valid for $n \geqslant 1$. If $n = 0$, then $|(T(\mathscr{E}) \setminus \mathcal{U}(D))^{(0)}| = 0$.

(2) For $n \geqslant 1$, we have $|(\mathcal{U}(D) \setminus T(\mathscr{E}))^{(n)}| = |(\mathcal{U}(D))^{(n)}| - |(T(\mathscr{E}) \cap \mathcal{U}(D))^{(n)}| = \sum\limits_{(j_1, j_2, \ldots, j_r)} |\mathscr{E}_{j_1}|^{d_1} |\mathscr{E}_{j_2}|^{d_2} \ldots |\mathscr{E}_{j_r}|^{d_r} - \sum\limits_{\substack{j_1=1 \\ (j_2, \ldots, j_r)}} |\mathscr{E}_{j_1}|^{d_1} |\mathscr{E}_{j_2}|^{d_2} \ldots |\mathscr{E}_{j_2}|^{d_r} = \sum\limits_{\substack{j_1=2 \\ (j_2, \ldots, j_r)}} |\mathscr{E}_{j_1}|^{d_1} |\mathscr{E}_{j_2}|^{d_2} \ldots |\mathscr{E}_{j_r}|^{d_r} = (k-l)^{l^n} \sum\limits_{(j_2, \ldots, j_r)} |\mathscr{E}_{j_2}|^{d_2} \ldots |\mathscr{E}_{j_r}|^{d_r}$ (here, as before, $r = 2^n$ and $j_i \in \{1, 2\}$, $i = 2, \ldots, r$). For $n = 0$, $|(\mathcal{U}(D) \setminus T(\mathscr{E}))^{(0)}| = |\mathscr{E}_2| = k - l$.

(3) $|(T(\mathscr{E}) \cup \mathcal{U}(D))^{(n)}| = |(T(\mathscr{E}))^{(n)}| + |(\mathcal{U}(D) \setminus T(\mathscr{E}))^{(n)}| =$

$l^{l^n} k^{k^n - l^n} + (k-l)^{l^n} \sum\limits_{(j_2, \ldots, j_r)} |\mathscr{E}_{j_2}|^{d_2} \ldots |\mathscr{E}_{j_r}|^{d_r}$. (For $n = 0$, we have $|(T(\mathscr{E})| \cup \mathscr{U}(D))^0| = k$.)

3.2.6. (1) If $f_1(x) = \begin{pmatrix} 0 & 1 & \ldots & k-1 \\ a_0 & a_1 & \ldots & a_{k-1} \end{pmatrix}$ and $f_2(x) = \begin{pmatrix} 0 & 1 & \ldots & k-1 \\ b_0 & b_1 & \ldots & b_{k-1} \end{pmatrix}$ are the functions in S_k, they have different values (i. e. $a_i \neq a_j$ and $b_i \neq b_j$ for $i \neq j$), and hence $f(x) = f_1(f_2(x))$ $\begin{pmatrix} 0 & 1 & \ldots & k-1 \\ a_{b_0} & a_{b_1} & \ldots & a_{b_{k-1}} \end{pmatrix}$ and if $i \neq j$, then $b_i \neq b_j$, and therefore $a_{b_i} \neq a_{b_j}$. Consequently, $f(x) \in S_k$. Let $f_1(x)$ and $f_2(x)$ be functions in CS_k. Then there exist i, j ($i \neq j$) such that $f_2(i) = f_2(j)$. But under this condition, the equality $f_1(f_2(i)) = f_1(f_2(j))$ is also observed. Therefore, $f(x) = f_1(f_2(x)) \in CS_k$.

(2) For $k = 3$, we have $|S_3 \cap \mathscr{U}(\{0, 1\}, \{2\})| = 2$. (This class contains the following functions: $\begin{pmatrix} 0 & 1 & 2 \\ 0 & 1 & 2 \end{pmatrix}$ and $\begin{pmatrix} 0 & 1 & 2 \\ 1 & 0 & 2 \end{pmatrix}$.) If $k \geqslant 4$, then $|S_k \cap \mathscr{U}(\{0, k-2\}, \{k-1\}, \{1, \ldots, k-3\})| = 2!(k-3)!$.

3.2.7. (1) $2x^4 + x^3 + 2x^2 + x$. (2) $3x^4 + 3x^3 + 3x^2 + 2x$. (3) $x + x^3 - x^4$. (4) $5x^6 + 2x^5 + 4x^4 + x^3 + x^2 + 4x$. (5) $6x^6 + 3x^5 + 5x^4 + 3x^2 + 5x$. (6) x^2y^2. (7) $xy^2 + xy + 2x$. (8) $x^2y^2 + x^2y + xy^2 + 2xy + x$. (9) $(k-1)(1-(x+2)^{k-1}) = \sum\limits_{i=0}^{q-2} \binom{k-1}{i} 2^i x^{k-1-i}$. (10) $1 - (x - x^2 - 2)^{k-1} = \sum\limits_{i=0}^{k-2} (-1)^{i+1} \binom{k-1}{i} 2^i x^{k-1-i} (1-x)^{k-1-i}$.

3.2.8. (1) It is sufficient to represent the function $2j_0(x)$ through a polynomial since $2j_i(x) = 2j_0(x-i)$, $i = 1, 2, 3$. We write the required polynomial as an expression with indeterminate coefficients: $2j_0(x) = a_0 + a_1x + a_2x^2 + a_3x^3$ (the remaining powers are not needed since $x^{2l+2} = x^2$ and $x^{2l+3} = x^3 \bmod 4$ for $l \geqslant 1$). Assigning the values 0, 1, 2 and 3 to x, we obtain the following system of equations:

$$\begin{cases} a_0 = 2, \\ a_0 + a_1 + a_2 + a_3 = 0, \\ a_0 + 2a_1 = 0, \\ a_0 + 3a_1 + a_2 + 3a_3 = 0. \end{cases}$$

Solving this system, we obtain (a) $a_0 = 2$, $a_1 = 1$, $a_2 = 0$, $a_3 = 1$; (b) $a_0 = 2$, $a_1 = 3$, $a_2 = 0$, $a_3 = 3$; (c) $a_0 = 2$, $a_1 = 1$, $a_2 = 2$, $a_3 = 3$; (d) $a_0 = 2$, $a_1 = 3$, $a_2 = 2$, $a_3 = 1$. Consequently, $2j_0(x) = 2 + x + x^3 = 2 + 3x + 3x^3 = 2 + x + 2x^2 + 3x^3 = 2 + 3x + 2x^2 + x^3$.

(2) If the function $f(x)$ in P_4 is such that $f(E_4) \subseteq \{0, 2\}$ it can

be represented in the form $f(x) = \sum_{i=0}^{3} b_i 2 j_i(x)$, where $b_i \in \{0, 1\}$, $i = 0, 1, 2, 3$. Then we must use the result of Problem 3.2.8(1). If, however, $f(E_4) \subseteq \{1, 3\}$, we can consider the function $g(x) = f(x) - 1$.

(3) Assuming that the function $f(x, y)$ can be represented by a mod 4 polynomial, we have $f(x, y) = a_{00} + a_{01}y + a_{02}y^2 + a_{03}y^3 + a_{10}x + a_{11}xy + \ldots + a_{32}x^3y^2 + a_{33}x^3y^3$. Considering this relation for the pairs (0, 0), (0, 2), (2, 0) and (2, 2), we get a system which has no solution:

$$\begin{cases} a_{00} = 2, \\ a_{00} + 2a_{01} = 0, \\ a_{00} + 2a_{10} = 0, \\ a_{00} + 2a_{01} + 2a_{10} = 0. \end{cases}$$

3.2.9. Hint. It is sufficient to prove that the functions $f(x) - f^2(x)$ and $f(x) - f^3(x)$ assume the values only in the set $\{0, 2\}$ and then to make use of the result of Problem 3.2.8(2).

3.2.10. If the function $f(x)$ in P_4 can be presented by a mod 4 polynomial, it can be written in the form $a_0 + a_1x + a_2x^2 + a_3x^3$, where $a_i \in E_4$, $i = 0, 1, 2, 3$ (see Problem 3.2.8(1)). But one and the same function can be represented by different polynomials. Besides, each such function is represented by four different polynomials (see, for example, the solution to Problem 3.2.8(1)). The latter fact can be easily proved by considering that the following relations are valid in P_4: $2x^3 = 2x^2 = 2x$ and $3x^3 = 2x + x^3$. The number of corresponding functions is equal to 64.

3.2.11. (1) The functions x, x^2 and x^3 in P_6 should be compared.

(2) Since $3x^2 = 3x \pmod 6$, each function that can be represented by a mod 6 polynomial is represented by two different polynomials. Therefore, the number of functions in P_6 which can be represented by mod 6 polynomials is $6^3/2 = 108$.

(3) The necessary and sufficient conditions for representing the function $(a + bj_0(x))^2$ by a mod 6 polynomial consists in that the function $(2a + b)\, bj_0(x)$ must be represented by a mod 6 polynomial. It remains for us to find the values $c \in E_6$ for which the function $cj_0(x)$ can be represented by a mod 6 polynomial, and after determining from each such value c the corresponding pairs of values (a, b) choose the pairs for which the function $a + bj_0(x)$ cannot be represented by a mod 6 polynomial. **Answer:** either $b = 2$ and $a = 2,5$, or $b = 4$ and $a = 1,4$.

3.2.12. (1), (2), (4) No, it cannot. (3) and (5) Yes, it can.

3.2.13. (1) $T(\{0, k-1\})$. (2) $T(\{1, 2\})$. (3) $\mathcal{U}(\{0, 1\}, \{2, \ldots, k-1\})$. (4) $T(\{0, 1\})$. (5) $T(\{0, k-1\})$. (6) $T(\{0\})$. (7) $T(\{1\})$. (8) $T(\{k-1\})$. (9) $T(\{1, k-1\})$. (10) $T(\{0, 2\})$. (11) $T(\{k-2\})$. (12) $T(\{0, 1\})$. (13) $\mathcal{U}(\{0\}, \{1, 2, \ldots, k-1\})$. (14) $T(\{1, 2, \ldots, k-1\})$. (15) $T(\{1, k-2\})$. (16) $\mathcal{U}(\{0, k-1\}, \{1, \ldots, k-2\})$. (17) $T(\{0, 1\})$.

3.2.14. (1) The set $\{j_0(x),\ x + y\}$ is such a subsystem (see Problem 3.1.8 and use the Słupecki criterion). The completeness of this system can be proved in a different way. We have $j_0(x) + j_0(j_0(x)) \equiv 1$; $x + 1 = \overline{x}$, $\overline{x} + 1 = x + 2, \ldots, (x + (k-2)) + 1 = x + (k-1)$; $j_i(x) = j_0(x - i) = j_0(x + (k - i))$, $i = 1, 2, \ldots, k-1$; $j_2(j_0(x) + j_0(y)) = j_{0,0}(x, y) = j_0(x) j_0(y)$, $j_{l,m}(x, y) = j_{0,0}(x - l, y - m)$, $l = 0, 1, \ldots, k-1$ and $m = 0, 1, \ldots, k-1$; $j^{(s)}_{l,m}(x, y) = j_{l,m}(x, y) + \ldots + j_{l,m}(x, y)$ (s terms), $s = 1, 2, \ldots, k-1, k$. If $f(x, y)$ is an arbitrary function of two arguments, then $f(x, y) = \sum\limits_{(l,m)} f(l, m)\, j^{f(l,m)}_{l,m}(x, y)$, where the summation is carried out over all pairs $(l, m) \in E_k \times E_k$. It remains for us to take into account the fact that the system of all functions in P_k which depend on two arguments is complete in P_k.

(2) $\{0, x + y, xy\} \subset T(\{0\})$, $\{j_0(x), j_1(x), \ldots, j_{k-1}(x)\} \subset T(\{0, 1\})$, $\{1, 2, \ldots, k-1\} \subset T(\{1, 2, \ldots, k-1\})$.

3.2.15. (1) $A_1 \setminus \{i\} \subset T(E_k) \setminus \{i\})$, $i = 1, \ldots, k-2$. (2) $A_1 \setminus \{0, k-1, J_0(x), J_{k-1}(x)\}$. **Hint:** $J_1(J_1(x)) = 0$, $J_1(1) = k - 1$ (see also Problems 3.1.1(18) and (19)).

3.2.16. (1) The subsystem is complete: $0 = J_0(1)$, $2 = J_0(0)$, $J_1(x) = J_0(\max(J_0(x), J_2(x)))$. (2) The subsystem is not complete. Consider the partition $D = \{\{0, 1\}, \{2\}\}$. (3) The subsystem is not complete and preserves the partition $\{\{0\}, \{1\}, \{2, 3\}\}$. (4) The subsystem is not complete and preserves the partition $\{\{0\}, \{1, 2\}, \{3\}\}$.

3.2.17. (1) The completeness of the given system in S_k can be proved as follows. By induction on i ($i \geqslant 1$), we prove that any function g in S_k satisfying the condition $g(x) \equiv x$ for $x > i$ is generated by functions in the set $\{h_{01}(x), h_{02}(x), \ldots, h_{0i}(x)\}$. Putting then $i = k - 1$, we obtain the required result. The inductive step is substantiated as follows. Let $g(x) \equiv x$ for $x > i + 1$ and $g(i + 1) = j$, $g(l) = i + 1$ (here $j \neq i + 1$ and $0 \leqslant l \leqslant i$). Let us take a function $h(x)$ coinciding with $g(x)$ everywhere except at $x = l$ and $x = i + 1$. At these two values of the argument, the function h is: $h(l) = j$ and $h(i + 1) = i + 1$. Obviously, the function $h(x)$ has already been constructed (in view of the inductive hypothesis). It can be easily seen that $g(x) = h(h_{0(i+1)}(h_{0i}(h_{0(i+1)}(x))))$.

(2) Since $h_{0i}(h_{i,(i+1)}(h_{0i}(x))) = h_{0(i+1)}(x)$, $i = 1, 2, \ldots, k-2$, the given system generates each function in Problem 3.2.17(1).

(3) Consider that $h_{i,(i+1)}(x + (k-1)) + 1 = h_{(i+1),(i+2)}(x)$, $i = 0, 1, 2, \ldots, k-3$.

3.2.18. It follows from Problem 3.2.17(1) that the given system generates the set S_k. Using the function $x + j_0(x)$, we can easily construct any function not assuming exactly one value (in E_k). Then given that we have all the functions assuming not more than i values ($1 \leqslant i \leqslant k-2$), we must show how an arbitrary function in $P^{(1)}_k$, which does not assume $i + 1$ values, can be constructed. Let $g(x)$ be an arbitrary function not assuming only one value. Then there exist exactly two values of the argument, viz. l_0 and l_1, for which the values of the function $g(x)$ are equal, i.e. $g(l_0) = g(l_1)$. The elements of the set $E_k \setminus \{l_0, l_1\}$ will be denoted by $l_2, l_3, \ldots, l_{k-1}$. Let us take two functions, $g_1(x)$ and $g_2(x)$ in the set S_k such that $g_1(l_r) = r$, $r = 0, 1, \ldots,$

$k-1$, and $g_2(s)=l_s$, $s=1, 2, \ldots, k-1$, while $g_2(0)=l\in E_k\setminus g(E_k)$. We denote the function $x+j_0(x)$ by $h_0(x)$. We have $g(x)=g_2(h_0(g_1(x)))$. Let us now take an arbitrary function $g'(x)$ not assuming $i+1$ values $(1\leqslant i\leqslant k-2)$. Let one of these values be l_0'. We assume that $g'(b_0)=g'(b_1)$. Let us consider the function $f_1(x)$ coinciding with $g'(x)$ everywhere except at $x=b_0$, and $f_1(b_0)=l_0'$. Obviously, the function $f_1(x)$ has already been constructed (by the inductive hypothesis). Let $a_1, \ldots, a_s$ be all different values assumed by the function $g'(x)$ (here $s=k-i-1\geqslant 1$), and $E_k\setminus\{a_1, \ldots, a_s\}=\{l_0', l_1', \ldots, l_i'\}$. We take the function $f_2(x)$ in the following form:

$$f_2(x)=\begin{cases} x & \text{if } x\neq l_0', \\ g'(b_0) & \text{if } x=l_0'. \end{cases}$$

It can be easily seen that the function $f_2(x)$ does not assume only one value, and hence we know how to construct it. We have $g'(x)=f_2(f_1(x))$.

3.2.19. (1) $x\dot{-}x=0$, $J_0(0)=k-1$, $(J_0(0))^2=1$. Further, we obtain the remaining constants (for $k\geqslant 4$): $(k-1)\dot{-}1=k-2$, $(k-2)\dot{-}1=k-3, \ldots$. Then we construct $\max(x, y)$ and $\min(x, y)$: $x\dot{-}(x\dot{-}y)=\min(x, y)$, $(k-1)\dot{-}x=\sim x$, $\sim\min(\sim x, \sim y)=\max(x, y)$. Thus, the initial system generates the Rosser-Turquette system.

(2) $(x\dot{-}y)+y=\max(x, y)$, $x+(k-1)+\ldots+(k-1)=x+1$ (here $k-1$ is added $k-1$ times). Consequently, we have constructed the system $\{x+1, \max(x, y)\}$ known to be complete beforehand.

(3) If $k=2l+1$ $(l\geqslant 1)$, we proceed as follows: $((x+2)+\ldots+2)+2=x+1$ (2 is added $l+1$ times); $x\dot{-}(x\dot{-}y)=\min(x, y)$, $\sim\min(\sim x, \sim y)=\max(x, y)$. We have obtained a Post's system. Let us now suppose that $k=2l$ $(l\geqslant 2)$. Then we construct a Rosser-Turquette system: $\min(x, y)$ and $\max(x, y)$ are obtained in the same way as for an odd k. Then we construct the constants $x\dot{-}x=0$, $0+2=2$, $2+2=4, \ldots, (k-4)+2=k-2$, $\sim 0=k-1$, $\sim 2=k-3, \ldots, \sim(k-2)=1$, and then the functions $J_i(x)$: $\underbrace{((k-1)\dot{-}x)\dot{-}\ldots\dot{-}x}_{k-1 \text{ times}}=J_0(x)$,

$J_0(x+k-2m)=J_{2m}(x)$, $m=1, 2, \ldots, l-1$, $J_{2m}(\sim x)=J_{2m}(k-1-x)=J_{k-1-2m}(x)$, $m=0, 1, \ldots, l-1$.

(4) $x\dot{-}x=0$, $1-0^2=1$, $-1=k-1$, $(k-1)\dot{-}x=\sim x$, $x\dot{-}(x\dot{-}y)=\min(x, y)$, $\sim\min(\sim x, \sim y)=\max(x, y)$. Using $k-1$, 1 and $x\dot{-}y$, we obtain the constants $k-2=(k-1)\dot{-}1$, $k-3=(k-2)\dot{-}1, \ldots, 2=3\dot{-}1$. Then we obtain $\underbrace{((k-1)\dot{-}x)\dot{-}\ldots\dot{-}x}_{k-1 \text{ times}}=J_0(x)$, $J_{k-1}(x)=J_0(\sim x)$,

$J_{k-2}(x)=J_0((k-2)\dot{-}x)\dot{-}J_{k-1}(x)$, $J_{k-3}(x)=J_0((k-3)\dot{-}x)\dot{-}J_0((k-2)\dot{-}x)$ and so on. Thus, the initial system generates a Rosser-Turquette system.

(5) $x+\ldots+x=0$ (k terms), $(\sim 0)\dot{-}2\times 0=k-1$, $(\sim x)\dot{-}2(k-1)=j_0(x)$. We obtained the system $\{j_0(x), x+y\}$ which is known to be complete beforehand (see Problem 3.2.14(1)).

(6) The completeness of this system can be proved by a method similar to that used for proving the completeness of Post's system (i.e. the system $\{\bar{x}, \max(x, y)\}$). We shall give a fragment of the proof, concerning the construction of the function $\sim x$. We have $\min(x, x + 1, x + 2, \ldots, x + (k - 1)) \equiv 0$. Further, in the ordinary way, we obtain all the constants and the functions $j_0(x)$, $j_1(x), \ldots, j_{k-1}(x)$: $j_i(x) = \min(x + (k - 1), x + (k - 2), \ldots, x + (k - (i - 1)), x + (k - (i + 1)), \ldots, x + 1, x)$, $i = 0, 1, \ldots, k - 1$. Then we construct the functions $g_{s,i}(x)$ such that $g_{s,i}(i) = s$ and $g_{s,i}(x) = k - 1$ for $x \neq i$: $g_{s,i}(x) = \min(j_i(x) + (k - 1), k - 1 - s) + s$. Finally, we get the Lukasevich negation: $\sim x = \min(g_{k-1,0}(x), g_{k-2,1}(x), \ldots, g_{k-l,l-1}(x), \ldots, g_{0,k-1}(x))$.

(7) $\min(x, x) - 1 = x - 1$, $(x - 1) - 1 = x - 2, \ldots, (x - (k - 2)) - 1 = \bar{x}$, $\overline{\min}(x, y) - 1 = \min(x, y)$. We have constructed the system $\{\bar{x}, \min(x, y)\}$.

(8) From $\bar{x} + y$, we construct $x + y$: $\bar{x} + \ldots + \bar{x} + x = k - 1$ ($\bar{x}$ is taken $k - 1$ times), $(k - 1)(k - 1) = 1$, $J_{k-1}(1) = 0$, $\bar{x} + 0 = \bar{x}$, $\bar{\bar{x}} + y = x + y + 2, \ldots, \bar{x} + y + (k - 1) = x + y$. Further, we have $J_{k-1}(x) J_{k-1}(x) = j_{k-1}(x)$ and $j_{k-1}(x + k - 1) = j_0(x)$. We have constructed the system $\{j_0(x), x + y\}$.

(9) $1^2 \dot{-} x = j_0(x)$, $1^2 + x = \bar{x}$, $j_i(x) = j_0(x + (k - i))$, $i = 1, 2, \ldots, k - 1$. We have $x + y = y + \sum_{i=1}^{k-1} \varphi_i(x)$, where $\varphi_i(x)\,(j_i(x))^2 + \ldots + (j_i(x))^2$ (i terms), i.e. $\varphi_i(x) = i j_i(x)$, $i = 1, \ldots, k - 1$. We have obtained the system $\{j_0(x), x + y\}$, which is known to be complete beforehand.

(10) $x + \ldots + x \equiv 0$ (k terms), $J_0(0) = k - 1$, $(k - 1) + \ldots + (k - 1) = 1$ ($k - 1$ terms), $1 \times J_0^2(x) = j_0(x)$. We have constructed the system $\{x + y, j_0(x)\}$.

(11) $(\sim (k - 2)) \dot{-} (k - 2) = 1 \dot{-} (k - 2) = 0$, $(\sim x) \dot{-} 0 = \sim x$, $(\sim (\sim x)) \dot{-} y = x \dot{-} y$, $x \dot{-} (x \dot{-} y) = \min(x, y)$, $\sim \min(\sim x, \sim y) = \max(x, y)$, $\sim (k - 2) = 1$, $x \times 1 + 1 = \bar{x}$. We have constructed the Post system $\{\bar{x}, \max(x, y)\}$.

(12) $(k - 1)^2 \dot{-} (k - 1) = 0$, $0^2 - x = -x$, $x^2 - (-y) = x^2 + y$, $-(k - 1) = 1$. We have constructed the complete system from Problem 3.2.19(9).

(13) $x \dot{-} x = 0$, $2 \times 1 + x = x + 2$. Then we obtain all the constants: $0 + 2 = 2$, $2 + 2 = 4, \ldots, (k - 4) + 2 = k - 2$, $1 + 2 = 3$, $3 + 2 = 5, \ldots, (k - 3) + 2 = k - 1$. Then we have $(k - 1) \dot{-} x = \sim x$, $x \dot{-} (x \dot{-} y) = \min(x, y)$, $\sim \min(\sim x, \sim y) = \max(x, y)$, $\underbrace{((k \dot{-} 1) \dot{-} x) \dot{-} \ldots \dot{-} x}_{k-1 \text{ times}} = J_0(x)$. Then we construct the remaining functions $J_i(x)$ (see Problem 3.2.19(4)). Thus, the initial system generates a Rosser-Turquette system.

(14) $1^2 - 1 = 0$, $0^2 - x = -x$, $1^2 - (-x) = \bar{x}$. The complete system $\{\bar{x}, \min(x, y)\}$ has been constructed (see Problem 3.2.19(6)).

(15) The function $x_1 + x_2 + \ldots + x_l + 2(l - 1)$, $l \geqslant 2$, can be easily constructed from $x + y + 2$. If $k = 2m$ ($m \geqslant 2$),

for $l = m + 1$ we obtain the function $x_1 + \ldots + x_{m+1}$. Using then the constant $0 = 1^2 \dot{-} 1$, we get $x_1 + x_2 + 0 + \ldots + 0 = x_1 + x_2$ (which contains $m - 1$ zeros). Then we construct $j_0(x)$: $1^2 \dot{-} x = j_0(x)$. Thus, we have constructed the system $\{j_0(x), x + y\}$, which is known to be complete beforehand (see Problem 3.2.14(1)). If $k = 2m + 1$ $(m \geqslant 1)$, for $l = m + 2$ we obtain the function $x_1 + \ldots + x_{m+2} + 1$, and from it, the function $\overline{x} = x + 0 + \ldots + 0 + 1$ $(m + 1$ zeros) and $x_1 + \overline{x_2} + 0 + \ldots + 0 + 1 = x_1 + x_2$ (m zeros). Then we proceed in the same way as for an even k.

(16) $\overline{xj_0(x)} = j_0(x)$, $\min(x, j_0(x)) = 0$, $\overline{xj_0(0)} = \overline{x}$. We have obtained the complete system $\{\overline{x}, \min(x, y)\}$ (see Problem 3.2.19(6)).

3.2.20. While solving these problems, it is expedient to compile tables with two inputs (in x and y) for the functions of two variables under consideration.

(1) $x - y + 2$ is an essential function. Consequently, it is sufficient to construct the functions $\overline{x}$, $x + j_0(x)$ and $h_{01}(x) = x + j_0(x) - j_1(x)$, see S. Picard's theorem. We have $(k - 1)^2 \dot{-} (k - 1) = 0$, $(k - 1)^2 \dot{-} x = j_0(x)$, $j_0(0) = 1$, $x - 1 + 2 = \overline{x}$; $x - \overline{y} + 2 = x - y + 1$, $x - \overline{y} + 1 = x - y$, $0 - x = -x$, $x - (-y) = x + y$. From $x + y$ and $j_0(x)$ we construct $x + j_0(x)$; $h_{01}(x) = (x + j_0(x)) - j_0(x + (k - 1)) = x + j_0(x) - j_1(x)$.

(2) The functions $x + y^2$ and $xy + 1$ are essential. We construct the functions $\overline{x}$, $h_{01}(x)$ and $x + j_0(x)$: $j_2(j_2(x)) \equiv 0$, $x \times 0 + 1 \equiv 1$, $x \times 1 + 1 = \overline{x}$, $j_2(\overline{x}) = j_2(x + 2) = j_0(x)$, $x + (j_0(x))^2 = x + j_0(x)$, $\underbrace{((x + 1) + 1) + \ldots + 1}_{k-1 \text{ times}} = x - 1$, $j_0(x - 1) = j_1(x)$,

$$h_{01}(x) = x + j_0(x) + \underbrace{(j_1(x))^2 + \ldots + (j_1(x))^2}_{k-1 \text{ times}}.$$

(3) $(\sim x) - y$ and $x \dot{-} y$ are essential functions, $x \dot{-} x = 0$, $(\sim x) - 0 = \sim x$, $(\sim(\sim x)) - y = x - y$, $0 - x = -x$, $x - (-y) = x + y$, $\sim 0 = k - 1$, $x + (k - 1) = x - 1$, $x - (k - 1) = \overline{x}$, $1 \dot{-} x = j_0(x)$, $j_0(x - 1) = j_1(x)$. From $x + y$ and $j_0(x)$, we construct $x + j_0(x)$; $h_{01}(x) = (x + j_0(x)) - j_1(x)$.

(4) $\overline{x} - y$ and $x^2 - y$ are essential functions; $(j_1(x))^2 - j_1(x) = 0$, $\overline{x} - 0 = \overline{x}$, $\overline{x} - \overline{y} = x - y$, $0 - x = -x$, $x - (-y) = x + y$, $j_1(\overline{x}) = j_0(x)$. From $x + y$ and $j_0(x)$, we construct $x + j_0(x)$; $h_{01}(x) = (x + j_0(x)) - j_1(x)$.

(5) xy is an essential function. From $\overline{x}$ and $j_0(x)$, we construct all the functions $j_i(x)$, $i = 1, 2, \ldots, k - 1$: $j_i(x) = j_0(x + (k - i))$. Further we have $(x + s - i)\, j_i(x) = g_{s,i}(x)$, $s = 0, 1, \ldots, k - 1$; $i = 0, 1, \ldots, k - 1$ ($g_{s,i}(i) = s$ and $g_{s,i}(x) = 0$ for $x \neq i$). If $g(x)$ is an arbitrary function in $P_k^{(1)}$, then $g(x) = (g_{g(0)-1,0}(x) + 1)\ (g_{g(1)-1,1}(x) + 1) \ldots (g_{g(i)-1,i}(x) + 1) \ldots (g_{g(k-1)-1,k-1}(x) + 1)$.

(6) From $x - 1$, we construct, as usual, $\overline{x}$: $(x - 1) - 1 = x - 2, \ldots, (x - (k - 2)) - 1 = x - (k - 1) = x + 1 = \overline{x}$. Let us denote the second function by $f(x, y)$. Since $1 \dot{-} x = j_0(x)$,

$f(x, y) = (x + j_0(x))\, j_0(y) + j_0(x)(y - j_1(y))$. We have $f(x, x) = x\, j_0(x) + j_0(x) + j_0(x)\, x - j_0(x)\, j_1(x) = j_0(x)$, $j_0(\overline{j_0(x)}) \equiv 0$, $f(x, 0) = x + j_0(x)$, $f(0, x) = j_0(x) + x - j_1(x) = h_{01}(x)$. Thus, we have constructed the system $\{\overline{x}, h_{01}(x), x + j_0(x)\}$ generating all functions of one variable in P_k (see Picard's theorem). It follows from the form of the functions $f(x, 0)$ and $f(0, x)$ that $f(x, y)$ is an essential function. Applying the Słupecki criterion, we conclude that the initial system is complete.

(7) Obviously, the given function is essential. We denote it by $\varphi(x, y)$. We have $\varphi(x, y) = j_0(x)\, y + \overline{x}\, j_0(y)$, $\varphi(x, x) = j_0(x)\, x + \overline{x} j_0(x) = j_0(\overline{x})$, $\varphi(x, j_0(x)) = j_0(x)\, j_0(x) + \overline{x} j_0(j_0(x)) = j_0(x) + (\overline{x} - j_0(x)) = \overline{x} = x + 1$; $j_0(j_0(x) + 1) \equiv 0$, $\varphi(0, x) = x + j_0(x)$; $(x + \underbrace{1) + \ldots + 1}_{k-1 \text{ times}} = x - 1$, $j_0(x - 1) = j_1(x)$,

$$\varphi(j_1(x), x) = j_0(j_1(x))\, x + (j_1(x) + 1)\, j_0(x) = \begin{pmatrix} 0 & 1 & 2 \ldots k-1 \\ 0 & 0 & 2 \ldots k-1 \end{pmatrix} + \begin{pmatrix} 0 & 1 & 2 \ldots k-1 \\ 1 & 0 & 0 \ldots \quad 0 \end{pmatrix} = h_{01}(x).$$

We have constructed the system $\{\overline{x}, x + j_0(x), h_{01}(x)\}$ which is complete in $P_k^{(1)}$.

(8) The initial function is obviously essential. We denote it by $\psi(x, y)$. We have $\psi(x, x) = \overline{x}$, $\psi(\overline{x}, x) = \overline{\overline{x}} j_0(\overline{x} - x) + (\overline{x} - j_1(\overline{x}))\, j_0(x) + x j_0(\overline{x}) = 0 + 0 + J_{k-1}(x)$, $J_{k-1}(\overline{J_{k-1}(x)}) \equiv 0$, $\psi(0, x) = j_0(-x) + 0 j_0(x) + x \times 1 = j_0(x) + x$, $\psi(x, 0) = \overline{x} j_0(x) + (x - j_1(x)) \times 1 + 0 \times j_0(x) = j_0(x) + x - j_1(x) = h_{01}(x)$.

(9) Obviously, the given function is essential. We denote it by $f(x, y)$. We have $f(x, x) = \begin{pmatrix} 0 & 1 & 2 \ldots k-1 \\ 1 & 2 & 1 \ldots \quad 1 \end{pmatrix} = h_1(x)$,
$h_1(h_1(x)) = \begin{pmatrix} 0 & 1 & 2 \ldots k-1 \\ 2 & 1 & 2 \ldots \quad 2 \end{pmatrix} = h_2(x)$; $f(h_1(x), h_2(x)) = h_3(x) = \begin{cases} 0 & \text{for } k = 3, \\ 3 j_1(x) & \text{for } k \geqslant 4. \end{cases}$ If $k \geqslant 4$, we also consider the superposition $h_3(h_3(x)) \equiv 0$. Further, $h_1(0) = 1$, $f(x, 0) = x + j_0(x)$, $f(x, 1) = \overline{x}$, $f(1, x) = \begin{pmatrix} 0 & 1 & 2 \ldots k-1 \\ 1 & 2 & 0 \ldots \quad 0 \end{pmatrix} = h_4(x)$,
$f(x, h_4(x)) = \begin{pmatrix} 0 & 1 & 2 \ldots k-1 \\ 1 & 0 & 2 \ldots k-1 \end{pmatrix} = h_{01}(x)$.

(10) The initial function is essential. We denote it by $\varphi(x, y)$. We have $\varphi(x, x) = j_0(x) + 2 j_1(x) = h_1(x)$, $\varphi(x, h_1(x)) = h_2(x) = \begin{cases} J_2(x) & \text{for } k = 3, \\ x + 2 j_1(x) & \text{for } k \geqslant 4. \end{cases}$ If $k = 3$, then $h_2 = (h_1(h_2(x))) \equiv 0$. If, however, $k \geqslant 4$, $h_1(h_2(x)) = j_0(x)$, $j_0(h_1(j_0(x))) \equiv 0$. Further, we obtain $h_1(0) = 1$, $\varphi(0, x) = h_{01}(x)$, $\varphi(1, x) = \overline{x}$, $\varphi(x, 0) = x + j_0(x)$.

(11) Obviously, the given function is essential. We denote it by $\psi(x, y)$. We have $\psi(x, x) = \overline{x} j_0(x) + j_1(x)(x + j_0(x)) + j_1(x)(j_2(x) - j_1(x)) = j_0(x) + j_1(x) - j_1(x)\, j_1(x) = j_0(x)$, $\psi(j_0(j_0(x)), j_0(x)) = 1 + j_0(x)$, $j_0(1 + j_0(x)) \equiv 0$, $j_0(0) = 1$,

$\psi(0, x) = \bar{x}$, $\psi(1, x) = h_{01}(x)$. From $\bar{x}$ we obtain $x - 1$, and then $j_1(x) : j_0(x - 1) = j_1(x)$; $\psi(1 + j_1(x), x) = x + j_0(x)$.

3.2.21. (1) If $k = 2m + 1$ $(m \geqslant 1)$, the system is complete since $(x + 2) + \ldots + 2 = x + 1$ (2 is added here $m + 1$ times), and hence the initial system generates the complete system $\{\bar{x}, \max(x, y)\}$. If, however, $k = 2m$ $(m \geqslant 2)$, the system is incomplete since it preserves the subset $\mathscr{E}$ of all odd numbers in E_k, i.e. $\mathscr{E} = \bigcup_{l=0}^{m-1} \{2l + 1\}$.

(2) The system is incomplete and preserves the set $\{1, 2\}$.

(3) For $k = 2m + 1$ $(m \geqslant 1)$, the system is complete since it generates the complete system $\{\bar{x}, \min(x, y)\}$ (see Problem 3.2.19(6)). The function $\bar{x}$ can be constructed as follows: $\underbrace{(x - 2) - \ldots - 2}_{m \text{ times}} = x - 2m = x + 1$. If $k = 2m$ $(m \geqslant 2)$, the system is incomplete since it preserves the set $\mathscr{E} = \bigcup_{l=0}^{m-1} \{2l\}$.

(4) The system is incomplete and preserves the set $\{0, 1\}$.

(5) For $k = 2m + 1$ $(m \geqslant 1)$, the system is complete since it generates the complete system $\{j_0(x), x+y\}$ (see Problem 3.2.14(1)). Indeed, $\underbrace{2x + \ldots + 2x}_{m+1 \text{ times}} + y = x + y$, $\underbrace{(x + 2) + \ldots + 2}_{m+1 \text{ times}} = x + 1$, $\underbrace{x + \ldots + x}_{k \text{ times}} = 0$, $0 + 1 = 1$, $1^2 \dot{-} x = j_0(x)$. If $k = 2m$ $(m \geqslant 2)$, the system is incomplete since it preserves the set $\mathscr{E} = \sum_{l=0}^{m-1} \{2l\}$.

(6) The system is incomplete and preserves a certain $(k - 1)$-element subset in E_k.

(7) The system is incomplete and preserves a certain one-element subset in E_k.

(8) For odd k, the system is complete (generates, for example, the system $\{x + y, j_0(x)\}$). For even k, it is incomplete since it preserves the subset of all even numbers in E_k.

(9) The system is incomplete since it preserves a certain two-element subset in E_k.

(10) If $k = 2m + 1$ $(m \geqslant 1)$, the system is complete: $\underbrace{(x + 2) + \ldots + 2}_{m+1 \text{ times}} = \bar{x}$, $(x \dot{-} y) + y = \max(x, y)$. For $k = 2m$ $(m \geqslant 2)$, the system is incomplete since it preserves the subset $\mathscr{E} = \bigcup_{l=0}^{m-1} \{2l\}$.

(11) For $k = 2m + 1$ $(m \geqslant 1)$, the system is incomplete since it preserves the subset $\mathscr{E} = \bigcup_{l=0}^{m} \{2l\}$. If $k = 2m$ $(m \geqslant 2)$, the system is complete. Indeed, $x \dot{-} (x \dot{-} y) = \min(x, y)$, $\sim\min(\sim x, \sim y) = \max(x, y)$, $x \dot{-} x = 0$, $\sim 0 = k - 1$, $2j_0(0) = 2$, $(k - 1) \dot{-} 2 = k - 3, \ldots, 5 \dot{-} 2 = 3$, $3 \dot{-} 2 = 1$, $(k - 1) \dot{-}$

$1 = k - 2$, $(k - 3) \dot{-} 1 = k - 4$, ..., $5 \dot{-} 1 = 4$, $J_1(x \dot{-} i) = J_{i+1}(x)$, $i = 1, \ldots, k-2$, $J_{k-1}(\sim x) = J_0(x)$. Thus, the Rosser-Turquette system has been constructed.

(12) For $k \geqslant 4$, the system is incomplete. For $k = 4$, it preserves the partition $D = \{\{0, 1\}, \{2, 3\}\}$, and for $k \geqslant 5$, it preserves, for example, the partition $D = \{\{0, 1\}, \{2\}, \ldots, \{k-3\}, \{k-2, k-1\}\}$. For $k = 3$, the system is complete. Indeed, $\sim(J_0(x) + J_1(x)) = J_2(x)$, $J_2(\sim x) = J_0(x)$, $J_0(\max(J_0(x), J_2(x))) = J_1(x)$, $J_1(1) = 2$, $\sim 2 = 0$, $\sim\max(\sim x, \sim y) = \min(x, y)$, i.e. we have constructed a Rosser-Turquette system.

(13) For $k \geqslant 4$, the system is incomplete. For $k = 4$, the partition $D = \{\{0, 1\}, \{2, 3\}\}$ is preserved, while for $k \geqslant 5$ it preserves, for example, the partition $D = \{\{0, 1\}, \{2, \ldots, k-3\}, \{k-2, k-1\}\}$. For $k = 3$, the system is complete: $\sim\min(\sim x, \sim y) = \max(x, y)$, $\min(2 - j_0(x) - 2j_1(x), x) = J_2(x)$, $J_2(\sim x) = J_0(x)$, $J_1(x) = J_0(\max(J_0(x), J_2(x)))$, $\sim 0 = 2$. Thus, we have constructed the Rosser-Turquette system.

(14) The system is incomplete. It preserves, for example, the partition $D = \{\{0, 1, \ldots, k-2\}, \{k-1\}\}$.

(15) The system is complete. It is not difficult to construct, for example, the following Rosser-Turquette system: $x \dot{-} (x \dot{-} y) = \min(x, y)$, $\sim\min(\sim x, \sim y) = \max(x, y)$, $x \dot{-} x = 0$, $\sim 0 = k - 1$, $(k-1) \dot{-} (k-2) = 1$, $(k-2) \dot{-} 1 = k-3$, ..., $3 \dot{-} 1 = 2$, $\underbrace{((k-1) \dot{-} x) \dot{-} \ldots \dot{-} x}_{k-1 \text{ times}} = J_0(x)$, $J_0(x \dot{-} i) \dot{-} J_0(x \dot{-} (i-1)) = J_i(x)$, $i = 1, 2, \ldots, k-1$.

3.2.22. The functions appearing in each of the given systems are polynomials. Consequently, for composite k these systems are incomplete. If, however, k is a prime number not less than 3, the completeness of a system can be conveniently proved by constructing one of the following systems: (a) the system $\{j_0(x), x + y\}$ complete in P_k (for any k, see Problem 3.2.14(1)) or (b) the system $\{1, x + y, xy\}$ complete in the class of polynomials.

(1) It is sufficient to construct the functions $x + y$ and xy. We have $1 + 1 + 1 \times 1 = 3$, $1 + 3 + 1 \times 1 = 5$, ..., $1 + (k-2) + 1 \times 1 = k = 0$, $x + y + x \times 0 = x + y$, $\underbrace{(x + 1) + \ldots + 1}_{k-1 \text{ times}} = x - 1$, $x + 0 + x(y-1) = xy$.

(2) It is sufficient to construct the system $\{j_0(x), x + y\}$. $x - x + 1 = 1$, $1^2 - 1 = 0$, $x - 0 + 1 = x + 1$, $x - (y + 1) + 1 = x - y$, $1 \times x^2 = x^2$, $x^2 x^2 = x^4$, ..., $x^{k-3} x^2 = x^{k-1}$ ($k - 1$ is an even number since k is a prime number not less than 3), $1 - x^{k-1} = j_0(x)$, $0 - x = -x$, $x - (-y) = x + y$.

(3) We construct the system $\{j_0(x), x + y\}$: $x + \ldots + x = 0$ (k terms), $0 - 1 = -1$, $x^2(-1) = -x^2$, $x^2(-x^2) = -x^4$, ..., $x^2(-x^{k-3}) = -x^{k-1}$, $(-1) + \ldots + (-1) = 1$ ($k - 1$ terms), $1 + (-x^{k-1}) = 1 - x^{k-1} = j_0(x)$.

(4) We construct the functions 1, $x + y$ and xy: $x(k-1) + x - (k-1) + z = z + 1$, $(k-1) + 1 = 0$, $0 + 1 = 1$, $x \times 0 + x - 0 + y = x + y$, $(x + 1)y + x + 1 - y + 0 = xy + x + 1$, $x + (k-1) = x - 1$, $x(y-1) + x + 1 = xy + 1$, $(xy + 1) - 1 = xy$.

(5) We construct the system $\{j_0(x), x+y\}$. We have $(x + 2y) + \ldots + 2y = x - y$ (the number of $2y$ terms is $(k-1)/2$), $x - x = 0$, $0 - x = -x$, $x - (-y) = x + y$. (The function $x + y$ can be obtained more rapidly: $\underbrace{((x + 2y) + 2y) + \ldots + 2y}_{(k+1)/2 \text{ times}} = x + y$.) Further, $-(k-2) = 2$, $(k-2) - 2 = k - 4, \ldots, 3 - 2 = 1$, and from functions 1 and $x - y$ we obtain $1 - x$; $1 \times x^2 = x^2$, $x^2x^2 = x^4, \ldots, x^{k-3}x^2 = x^{k-1}$, $1 - x^{k-1} = j_0(x)$.

(6) We construct the system $\{j_0(x), x + y\}$: $x - x = 0$, $0 - x = -x$, $x - (-y) = x + y$, $\sim 0 = k - 1$, $-(k-1) = 1$, $x^2 \times 1 = x^2$, $x^2x^2 = x^4, \ldots, x^2x^{k-3} = x^{k-1}$, $1 - x^{k-1} = j_0(x)$.

(7) We construct the functions 1, $x + y$ and xy. We have $(x + 2y + 1) + 2z + 1 = x + 2y + 2z + 2$, $(x - 2) + 2y + 2z + 2 = x + 2y + 2z$, $((x + (2y + 2z)) + (2y + 2z)) + \ldots + (2y + 2z) = x + y + z$ (here the terms $2y + 2z$ are contained $(k+1)/2$ times), $x + y + y = x + 2y$, $((x + 2y) + 2y) + \ldots + 2y = x - y$ (the terms $2y$ are contained $(k-1)/2$ times), $x - x = 0$, $0 - x = -x$, $x - (-y) = x + y$, $x + 2 \times 0 + 1 = x + 1$, $0 + 1 = 1$, $(x + 1)(y + 1) - (x + 1) - (y + 1) = xy - 1$, $(xy - 1) + 1 = xy$.

(8) We construct the functions $x + y$ and xy. We have $((x + 2y) + 2y) + \ldots + 2y = x - y$ (the number of terms $2y$ is $(k-1)/2$), $x - x = 0$, $0 - x = -x$, $x - (-y) = x + y$, $(x(y + 1)^2 - x + y + 1) - (x(y - 1)^2 - x + y - 1) = 4xy + 2$, $((4xy + 2) - 1) - 1 = 4xy$. Further, if $k = 4m + 1$ $(m \geqslant 1)$, we consider the sum $4xy + 4xy + \ldots + 4xy$ of m terms $4xy$. It is equal to $-xy$. But since we have the function $-x$, we can easily obtain the function xy: $-(-xy) = xy$. If $k = 4m + 3$ $(m \geqslant 0)$, we construct the sum $4xy + 4xy + \ldots + 4xy$ of $m + 1$ terms. It is equal to xy.

(9) We construct the functions 1, $x + y$ and xy. We have $xx - x^2 = 0$, $0 + 0 + 1 = 1$, $x + 0 + 1 = x + 1$, $\underbrace{((x + 1) + 1) + \ldots + 1}_{k-1 \text{ times}} = x - 1$, $x + (y - 1) + 1 = x + y$, $x(x + y) - x^2 = xy$.

(10) $\underbrace{((x - 2y) - 2y) - \ldots - 2y}_{(k+1)/2 \text{ times}} = x - y$, $x - x = 0$, $0 - x = -x$, $x - (-y) = x + y$, $0 \times 0 + 0 + 1 = 1$, $x(y - 1) + x + 1 = xy + 1$, $(xy + 1) - 1 = xy$.

(11) We construct the functions 1, $x + y$ and xy. Since k is a prime number, $x^k = x$, and hence we have $1 + x - x + \underbrace{xx \ldots x}_{k \text{ times}} = 1 + x$ and $(x + 1) + 1 = x + 2, \ldots, (x + (k - 2)) + 1 = x + (k - 1)$. Further, we obtain $1 + x - (x + 1) + x(x + 1)(x + 2) \ldots (x + (k - 1)) \equiv 0$, $1 + 0 = 1$, $1 + x - y + x \times y \times 0 \times \ldots \times 0 = 1 + x - y$, $1 + 0 - x = 1 - x$, $1 + x - (1 - y) = x + y$, $1 + x - y + x \times y \times 1 \times \ldots \times 1 = xy + x - y + 1$, $(x + 1)(y + (k - 1)) + x + 1 - (y + (k - 1)) + 1 = xy + 2$, $(xy + 2) + k - 2 = xy$.

3.2.23. (1) Take into account the fact that $\{j_0(x), x + y\}$ is a complete system and that $j_i(x + i) = j_0(x)$, $i = 1, \ldots, k - 1$.

(2) Consider Post's system. (3) Adding, for example, the functions $\sim x$ and $\bar{x}$, we can construct Post's system. (4) Post's system can easily be constructed by adding, for example, the functions $\sim x$ and $\bar{x}$. (5) By adding, for example, the function $\bar{x}$, we can easily obtain Post's system. (6) It is sufficient to add the functions $\sim x$ and $\bar{x}$ to construct Post's system. (7) Apply directly the theorem on completeness and incompleteness of a polynomial class in P_k.

3.2.24. (1) For $k = 4$ and 6, we can consider the function $f(x + 1, x)$. (2) For $k = 4$, analyze the function $f(2, x)$ and for $k = 6$, the function $f(4, x)$. (3) For $k = 4$, we can consider any of the functions $(f(0, x))^2$, $(f(x + 1, x))^2$ and $1 - f(2, x)$, while for $k = 6$, the function $1 - f(4, x)$ should be considered. (4) For $k = 4$, any of the functions $(f(x, 0))^2$, $(f(x, x))^2$, $(f(x + 2, x))^2$ or $f(2, x) - 2$ is suitable. For $k = 6$, we can consider the function $f(4, x) - 2$.

3.2.25. (1) The subsystem $B = \{k - 1, j_0(x), x \dot{-} y\}$ is a basis. Let us prove this. We have $j_0(k - 1) = 0$, $j_0(0) = 1$, $(k - 1) \dot{-} 1 = k - 2$, $\ldots$, $3 \dot{-} 1 = 2$, $j_0((k - 1) \dot{-} x) = j_{k-1}(x)$, $j_{k-2}(x) = j_0((k - 2) \dot{-} x) - j_{k-1}(x)$, $j_{k-3}(x) = j_0((k - 3) \dot{-} x) \dot{-} j_0((k - 2) \dot{-} x)$, $\ldots$, $j_1(x) = j_0(1 \dot{-} x) \dot{-} j_0(2 \dot{-} x)$. Let $g(x)$ be an arbitrary function in $P_k^{(1)}$. It can be constructed as follows. At first we construct the function $g_0(x) = ((((k - 1) \dot{-} j_0(x)) \dot{-} j_0(x)) \dot{-} \ldots \dot{-} j_0(x))$, where $j_0(x)$ is subtracted from $k - 1$ such a number of times r_0 that $g_0(0) = g(0)$, i.e. $r_0 = k - 1 - g(0)$. (For $g(0) = k - 1$, we put $g_0(x) = k - 1$.) Then we subtract $r_1 = k - 1 - g(1)$ times the function $j_1(x)$ from $g_0(x)$, and so on. (In particular, the function $\bar{x}$ can be constructed in this way.) In order to prove the completeness of the system B, it remains for us to use the Słupecki criterion (or recall that $\min(x, y) = x \dot{-} (x \dot{-} y)$, $\max(x, y) = \sim\min(\sim x, \sim y)$, $\sim x = (k - 1) \dot{-} x$ and that the system $\{\bar{x}, \max(x, y)\}$ is complete in P_k). Further, we must prove that B does not contain complete proper subsystems. The function $x \dot{-} y$ cannot be removed from B since the functions $k - 1$ and $j_0(x)$ essentially depend on not more than one variable. Further, we obviously have $\{j_0(x), x \dot{-} y\} \subset T(\{0, 1\})$ and $\{k - 1, x \dot{-} y\} \subset T(\{0, k - 1\})$.

(2) $B = \{J_0(x), x \dot{-} y^2\}$. Obviously, $x \dot{-} y^2$ is an essential function. We shall show how all the functions of one variable can be constructed. We have $\underbrace{((J_0(x) \dot{-} J_0^2(x)) \dot{-} J_0^2(x)) \dot{-} \ldots \dot{-} J_0^2(x)}_{k-1 \text{ times}} \equiv 0$, $J_0(0) = k - 1$, $\underbrace{((k-1) \dot{-} (k-1)^2) \dot{-} \ldots \dot{-} (k-1)^2}_{i \text{ times}} = k - 1 - i$ $(i = 1, 2, \ldots, k-2)$, $\underbrace{((x \dot{-} 1^2) \dot{-} 1^2) \dot{-} \ldots \dot{-} 1^2}_{l \text{ times}} = x \dot{-} l$ $(l = 1, 2, \ldots, k-2)$, $J_i(x) = \underbrace{(J_0(x \dot{-} i) \dot{-} J_0^2(x \dot{-} (i-1))) \dot{-} \ldots \dot{-} J_0^2(x \dot{-} (i-1))}_{k-1 \text{ times}}$, $i = 1, 2, \ldots, k-1$. Let $g(x)$ be an arbitrary function in $P_k^{(1)}$. It can be constructed as follows. We first construct the function $g_0(x)$: if $g(0) = k-1$, then $g_0(x) \equiv k-1$. If however, $g(0) \neq$

$k-1$, then $g_0(x)=\underbrace{((k-1) \dot{-} J_0^2(x)) \dot{-} \ldots \dot{-} J_0^2(x)}_{r_0 \text{ times}}$, where $r_0=k-1-g(0)$. Then we construct the function $g_1(x)$: if $g(1)=k-1$, then $g_1(x)=g_0(x)$, but if $g(1) \neq k-1$, then $g_1(x)=\underbrace{(g_0(x) \dot{-} J_1^2(x)) \dot{-} \ldots \dot{-} J_1^2(x)}_{r_1 \text{ times}}$, where $r_1=k-1-g(1)$, and so on. On the k-th step, we obtain the function $g_{k-1}(x)$: if $g(k-1)=k-1$, then $g_{k-1}(x)=g_{k-2}(x)$, but if $g(k-1) \neq k-1$, then $g_{k-1}(x)=\underbrace{(g_{k-2}(x) \dot{-} J_{k-1}^2(x)) \dot{-} \ldots \dot{-} J_{k-1}^2(x)}_{r_{k-1} \text{ times}}$, where $r_{k-1}=k-1-g(k-1)$. Obviously, the function $g_{k-1}(x)$ coincides with the function $g(x)$. In order to complete the proof of the completeness of the system B, it is sufficient to use the Słupecki criterion. It can be easily seen that B does not contain complete proper subsystems since the function $J_0(x)$ depends only on one argument, and the function $x \dot{-} y^2$ preserves the set $\{0\}$.

Remark. For odd k, the subsystem $\{x-2, \max(x, y)\}$ is also a basis.

(3) $B=\{\sim x, \min(x, y), x+y\}$. We have $x+(\sim x)=k-1$, $x+k-1=x-1$, $\underbrace{(x-1)-\ldots-1}_{k-1 \text{ times}}=x$, $\sim\min(\sim x, \sim y)=\max(x, y)$. We have obtained Post's system. It should be also noted that $\{\sim x, \min(x, y)\} \subset T(\{0, k-1\})$, $\{\sim x, x+y\} \subset L$ and $\{\min(x, y), x+y\} \subset T(\{0\})$.

(4) If k is odd, for example, the subsystems $\{x+2, \max(x, y)\}$ and $\{x+2, x \dot{-} y\}$ are bases. This fact can be substantiated by using the following relations: $(x+2)+\ldots+2=x+1$ (the term 2 is taken $(k+1)/2$ times) and $x \dot{-} (x \dot{-} y)=\min(x, y)$. If k is even ($k \geqslant 4$), the basis is the following subsystem: $B=\{k-1, x+2, x \dot{-} y\}$. Let us prove this. We have $(k-1)+2=1$, $(k-1) \dot{-} 1=k-2, \ldots, 3 \dot{-} 1=2, 1 \dot{-} 1=0$, $(k-1) \dot{-} x=\sim x$, $x \dot{-} (x \dot{-} y)=\min(x, y)$, $\sim\min(\sim x, \sim y)=\max(x, y)$, $\underbrace{((k-1) \dot{-} x) \dot{-} \ldots \dot{-} x}_{k-1 \text{ times}}=J_0(x)$, $J_i(x)=J_0(x \dot{-} i) \dot{-} J_0(x \dot{-} (i-1))$, $i=1, 2, \ldots, k-1$. Thus, the subsystem B generates the Rosser-Turquette system which is known to be complete. Consequently, B is a complete system. Let us prove that B is a basis. Obviously, $\{k-1, x \dot{-} y\} \subset T(\{0, k-1\})$, the subsystem $\{k-1, x+2\}$ consists only of functions which essentially depend on not more than one variable, and for $k=2m$ the subsystem $\{x+2, x \dot{-} y\}$ is contained in the class $T(\{0, 2, \ldots, k-2\})$. Consequently, B does not contain complete proper subsystems.

(5) The basis is the subsystem $B=\{j_0(x), x+y^2\}$. In order to prove this, we must take into account that $j_0(j_0(x))+(j_0(x))^2 \equiv 1$ and that the function $x+y^2$ can be used for constructing all the functions of one variable (by using $j_0(x), j_1(x), \ldots, j_{k-1}(x)$) almost in the same way as the function $x+y$ was used in the solution of Problem 3.1.8.

3.2.26. Take into account the fact that $\{0, 1, \ldots, k-1, J_{k-1}(x), \min(x, y), \max(x, y)\} \subset \mathfrak{U}(\{0, 1, \ldots, k-2\}, \{k-1\})$, $\min(J_i(x), J_{k-1}(x)) \equiv 0$, $0 \leqslant i \leqslant k-2$, and $\max(J_0(J_0(x)), J_0(x)) \equiv k-1$.

3.2.27. If the variables of the functions belong to the set $\{x_1, x_2, \ldots, x_n, \ldots\}$, the number of different functions in any closed class in P_k (including P_k as a particular case) is not more than countable. Let a closed class A ($\subseteq P_k$) have a finite complete system. Then any basis in it is finite. Consequently, the power of the set of its bases does not exceed the power of the set of all its finite subsystems, which (as is known from the theory of sets) is not more than countable.

3.2.28. (1) Any precomplete class is closed. Consequently, B is a closed class, and the set A is also a closed class in $P_k^{(1)}$ as an intersection of closed classes. It should be also noted that the class B, and hence the set A contain an identity function. Let us consider a closed class $\mathfrak{P}(A)$ consisting of all the functions of the system P_k, which preserve the set A (the function $f(\widetilde{x^n})$, $n \geqslant 0$, preserves the set A if for any functions $\varphi_1(x), \ldots, \varphi_n(x)$ in A the superposition $f(\varphi_1(x), \ldots, \varphi_n(x))$ is a function in A). Obviously, $B \subseteq \mathfrak{P}(A)$ and $\mathfrak{P}(A) \neq P_k$ (since, for example, any function in $P_k^{(1)} \setminus A$ is not contained in $\mathfrak{P}(A)$). But since B is a precomplete class, it cannot be exactly contained in any closed class differing from the entire P_k. Therefore, $B = \mathfrak{P}(A)$. The class $\mathfrak{P}(A)$ is uniquely defined by the set A. Consequently, the class B having the properties specified in the conditions of the problem is unique.

(2) It follows from Problem 3.2.27(1) that the number of such precomplete classes in P_k is not more than the number of all subsets of the set $P_k^{(1)}$. Since $|P_k^{(1)}| = k^k$, the power of the set of all subsets in $P_k^{(1)}$ is 2^{k^k}.

3.2.29. (1) Let us prove that the class K_1 contains only one (accurate up to redesignation of variables) function of one variable $g(x) = x^2$. We denote the function x^2y^2 by $f(x, y)$. We have $f(x, x) = x^4 = x^2$, $f(x^2, x) = f(x, x^2) = x^4x^2 = x^2$, $f(x^2, x^2) = x^4x^4 = x^2$. Obviously, $x^2 \neq x$ and $x^2 \neq \text{const}$.

(2) We denote the function $j_1(x)\, j_2(y)$ by $\varphi(x, y)$. We have $\varphi(x, x) = 0$, $\varphi(0, x) = \varphi(x, 0) = \varphi(0, 0) = 0$. Therefore, K_2 contains only one function depending on not more than one variable, and this function is an identical zero.

3.2.30. We denote the function $j_2(x)\, j_2(y)$ by $\psi(x, y)$. Let us consider the general form of the superposition generated by the set $\{\psi(x, y)\}$: $\mathfrak{A}(x_1, \ldots, x_m, y_1, \ldots, y_n) = \psi(\mathfrak{A}_1(x_1, \ldots, x_m), \mathfrak{A}_2(y_1, \ldots, y_n))$, where $\mathfrak{A}_i$ is either a variable or a superposition generated by the set $\{\psi(x, y)\}$, $i = 1, 2$. Let us prove that if at least one $\mathfrak{A}_i$ differs from a variable, the superposition $\mathfrak{A}$ is equal to an identical zero. Indeed, any function represented by the superposition generated by the set $\{\psi(x, y)\}$ does not assume the value of 2. Therefore, if for example, $\mathfrak{A}_1$ differs from a variable, then $\psi(\mathfrak{A}_1(E_3, \ldots, E_3), \mathfrak{A}_2(E_3, \ldots, E_3)) \subseteq \psi(E_3 \setminus \{2\}, E_3) = \{0\}$. Thus the class K_3 contains only three pairwise incongruent functions: 0, $j_2(x)$ and $j_2(x)\, j_2(y)$, which essentially depend on zero, one and two variables respectively.

3.2.31. If in the function $f_n(\widetilde{x^n})$ $(n \geqslant 2)$ we identify any l variables $(2 \leqslant l \leqslant n)$, we obtain a function congruent to $f_{n-l+1}(\widetilde{x^{n-l+1}})$ (since $j_2(x)\; j_2(x) = j_2(x)$). Further, like in the previous problem, we can prove that if at least one $\mathfrak{A}_i$ in the superposition $\mathfrak{A} = f_n(\mathfrak{A}_1(x_{11}, \ldots, x_{1m_1}), \mathfrak{A}_2(x_{21}, \ldots, x_{2m_2}), \ldots, \mathfrak{A}_n(x_{n_1}, \ldots, x_{mn_n}))$ differs from a variable, then $\mathfrak{A}$ is equal to an identical zero. Hence K_4 contains, in addition to the functions $f_1, f_2, \ldots, f_n, \ldots$, only an identical zero, and any function f_n generates any function f_l for $l < n$, but is not generated by the set $\{f_1, f_2, \ldots, f_{n-1}\}$. Hence it follows that the set of closed classes in K_4 consists of K_4 itself, the class $\{0\}$ and the classes $[f_l]$, $l = 1, 2, \ldots,$ and $\{0\} \subsetneqq [f_1] \subsetneqq [f_2] \subsetneqq \ldots \subsetneqq [f_l] \subsetneqq \ldots \subsetneqq K_4$. Obviously, none of these classes is precomplete in K_4 (since any class in this chain differing from K_4 is strictly contained in some other class differing from K_4).

3.2.32. Let us calculate the number of non-essential functions in $P_k^{(n)}$ and subtract the obtained expression from k^{k^n}. A function is not essential if it assumes less than k different values (in E_k) or assumes all the k values (in E_k), but essentially depends only on one variable. The number of the first type functions can be determined by the inclusion-exclusion method. It is $C_k^{k-1} \times (k-1)^{k^n} - C_k^{k-2}(k-2)^{k^n} + \ldots + (-1)^i C_k^{k-i}(k-i)^{k^n} + \ldots + (-1)^{k-2} C_k^2 2^{k^n} + (-1)^{k-1} C_k^1$. The number of the second-type functions is $(k!)n$.

3.2.33. Let us establish a one-to-one correspondence between all rational numbers of the segment [0, 1] and the set of functions $\{f_2, f_3, \ldots, f_m, \ldots\}$. We take an arbitrary real number $\gamma \in [0, 1]$ and denote by C_γ the subset of all such functions in $\{f_2, f_3, \ldots, f_m, \ldots\}$ which given the chosen correspondence are in compliance with all rational numbers less than γ. For $\gamma_1 < \gamma_2$, we have $C_{\gamma_1} \subsetneqq C_{\gamma_2}$. Let $B_\gamma = [C_\gamma]$. Obviously, $B_{\gamma_1} \subsetneqq B_{\gamma_2}$ for $\gamma_1 < \gamma_2$.

CHAPTER FOUR

4.1.

4.1.1. It follows from the calculation of the type (v, x) pairs, where $v \in V$, $x \in X$, and v and x are incident.

4.1.2. (2) None.

4.1.5. (1) Let $O(v)$ be the set of vertices adjacent to v and $O'(v) = \{v\} \cup O(v)$. By hypothesis, $|O'(v)| \geqslant (n+1)/2$ and $|V \setminus O'(v)| \leqslant (n-1)/2$. Hence it follows that each vertex from $V \setminus O'(v)$ is adjacent to a certain vertex from $O'(v)$, and hence the graph is connected. (2) It is possible for an even n and impossible for an odd n.

4.1.10. Let $[v, u]$, $[w, t]$ be two chains of maximum length, which have no common vertices in a graph G. The graph G is connected. Consequently, there exists a chain Z connecting, for example, vertices v and w. Let v_1 be the last vertex of the chain $[v, u]$ which is met on the path from v to w along the chain Z and w_1 be the first vertex of the chain $[w, t]$, met on this path after v_1. The vertex v_1 of the chain $[v, u]$ divides it into two parts: $[v, v_1]$ and $[v_1, u]$. Let $[v, v_1]$ be not shorter than $[v_1, u]$. Similarly, the vertex w_1 divides the chain $[w, t]$ into two parts. Let $[w, w_1]$ be not shorter than $[w_1, t]$. Then the chain $[v, v_1, w_1, w]$ is longer than each of the chains $[v, u]$ and $[w, t]$. We arrive at a contradiction.

4.1.18. Hint. Consider the complements of the graphs G and H.

4.1.20. $2p$. **4.1.22.** Six.

4.1.24. (1) Let v, u and w be three vertices of the same power of a graph G in R_n. We assume that v and u are adjacent, and v and w are not. For vertices u and w, at least one of the inclusions $O(u) \subseteq O(w)$ or $O(w) \subseteq O(u)$ is valid. It follows hence from this and from the equality $|O(u)| = |O(w)|$ that $O(u) = O(w)$ and therefore $v \in O(w)$. Let us now consider the pair v and w. It follows from what has been proved above that v and w are adjacent. But the inclusion $O'(v) \subseteq O'(w)$ does not take place since $u \in O'(v) \setminus O(w)$. The inclusion $O'(w) \subset O'(v)$ is not valid either since $|O'(w)| = |O'(v)|$. We arrive at a contradiction.

4.1.26. (1) The number of edges of the graph G is equal to the sum of the numbers of edges of graphs H_i, divided by $n - 1$. (2) This follows from 4.1.26 (1). **4.1.28.** Two.

4.1.34. (1) Yes, there exists. **4.1.35.** Yes, it is. **4.1.39.** No, it is not.

4.2.

4.2.5. Let G be a plane 2-connected graph with at least two internal faces. If there exists an internal face separated from an external face by a single simple chain, then the deletion of this chain leads to a 2-connected graph (prove this statement). Suppose that there is no such a face. Then any internal face having a common chain with the external face has at least one vertex in common with it, which does not lie on this chain. Let us prove that this case is impossible. We number all the internal faces. We mark the connected pieces of the boundary of the external face which simultaneously belong to an internal face with the number i by an index i. Then there exist such indices i and j which are met during the circumvention of the boundary of the external face in the sequence i, j, i, j. We denote the corresponding pieces of the boundary by $\Gamma_{i1}, \Gamma_{j1}, \Gamma_{i2}, \Gamma_{j2}$. Let us choose on these pieces points a_1, b_1, a_2, b_2 on the plane (one point on each piece). Then the points a_1 and a_2 can be connected by a curve whose all points except a_1 and a_2 are internal points of the face with the number i, while points b_1 and b_2 can be connected by a curve whose internal part lies in the face with the number i. The curves intersect at a certain point d which is hence an internal point of two different faces. We arrive at a contradiction.

4.2.9. Let G be a 2-connected planar graph and G' be its plane representation. Let n be the number of vertices, m the number of edges, and r the number of faces of the graph G' (including the internal and external faces). According to Problem 4.2.6, we have $r = m - n + 2$. The number of pairs of the form (v, x), where v is a vertex incident to the edge x is equal to $2m$. On the other hand, this number is equal to the sum of the powers of the graph vertices. Since each power by hypothesis is not less than 6, we can write $2m \geqslant 6n$, i.e. $m \geqslant 3n$. The boundary of each face has at least two edges, and each edge belongs to the boundary of not more than two faces. Hence $3r \leqslant 2m$. It can be easily seen that the system $r = m - n + 2$, $m \geqslant 3n$, $3r \leqslant 2m$ is incompatible.

4.2.14. We assume that the graph K_5 is planar. Let K_5' be its plane representation. Then the number r of the faces of the graph K_5' is $r = m - n + 2$, where m is the number of edges, and n the number of vertices. As in Problem 4.2.9, we have $3r \leqslant 2m$. Hence $m \leqslant 3n - 6$. But $m = 10$ and $n = 5$. We arrive at a contradiction. While proving the non-planarity of the graph $K_{3,3}$, we note that each its cycle contains at least four edges. Assuming that $K_{3,3}$ is a planar graph, use Problem 4.2.11.

4.2.22. Each vertex of a 6-connected graph has a power not less than 6 (see Problem 4.2.9).

4.2.27. (3) $\chi(B^n) = 2$, $\chi'(B^n) = n$.

4.2.28. Two.

4.2.45. (1) Let $\tau(m)$ be the number of terminal coverings of a chain of length m. Then $\tau(m) = \tau(m-2) + \tau(m-3)$, $\tau(1) = \tau(2) = \tau(3)$. The solution of recurrence relations of this type is given in Sec. 8.3.

4.2.49. (1) $\bar{\nu}_k(G)$ is the average number of vertices which are not adjacent to vertices of the subset $U \subseteq V$ of power k. Consequently, there exists a subset U_0, $|U_0| = k$, such that $\nu(U_0) \leqslant \bar{\nu}_k(G)$. If W is the set of vertices which are not adjacent to vertices from U_0, then $W \cup U_0$ is the vertex covering of power not exceeding $k + \bar{\nu}_k(G)$. (2) Let $e(v, u) = 1$ if the vertices v and u are not adjacent and $e(v, u) = 0$ if they are. Let $S_k(v)$ be the number of subsets $U \subseteq V$ composed of k vertices none of which is adjacent to v, and let $d(v)$ be the power of the vertex v. Then

$$\sum_{U \subseteq V,\ |U| = k} \nu(U) = \sum_{U \subset V,\ |V| = k} \ \sum_{v \in V} e(v, u)$$

$$= \sum_{v \in V} S_k(v) = \sum_{v \in V} \binom{n - d(v)}{k} \leqslant |V| \binom{n - d_0}{k}. \text{ Hence}$$

$$\bar{\nu}_k(G) \leqslant |V| \prod_{i=0}^{|V|-1} \left(1 - \frac{d_0}{|V| - i}\right).$$

4.3.

4.3.2. The proof can be carried out by induction on the number of vertices of the digraph. On an inductive step, it is expedient to delete one of the vertices (v_1 or v_2) from the digraph.

4.3.5. Use induction on the number k.

4.3.8. The statement can be proved by induction on the number of vertices in the tournament.

4.3.10. For $n = 1$, the inequality is obvious. Let it be valid for such n that $1 \leqslant n \leqslant m$. Let us consider the tournament T with $m + 1$ vertices. Since any tournament with $m + 1$ vertices has exactly C^2_{m+1} arcs, there exists a vertex v_0 with an out-degree $\geqslant [(m + 1)/2]$. Having deleted the vertex v_0 in T, we get a tournament T' which has m vertices and contains a set S consisting of at least $f(m)$ of compatible arcs. The arcs emerging from v_0 and the arcs belonging to S are compatible. Using the inductive hypothesis, we obtain

$$f(m+1) \geqslant \left[\frac{m+1}{2}\right] + \left[\frac{m}{2}\right]\left[\frac{m+1}{2}\right] = \left[\frac{m+1}{2}\right]\left[\frac{m+2}{2}\right].$$

4.3.12. Any even-order group contains at least one element inverse to itself and differing from the unit element of the group. If the group of the tournament T were even, it would contain an element α satisfying the above property. Since α is not a unit element, there exist two vertices v_1 and v_2 such that $\alpha(v_1) = v_2$ and $\alpha(v_2) = v_1$. Let T contain the arc (v_1, v_2). Then the arc $(\alpha(v_1), \alpha(v_2))$ should also belong to T. We arrive at a contradiction.

4.3.15. The proof can be carried out by induction on the number of strong components of the tournament. If there is only one vertex in condensation or there are two vertices in it, the condensation is by definition a transitive digraph.

4.3.19. Carry out induction on the number of vertices in the digraph. On an inductive step, one of the vertices with zero out-degree should be deleted, after establishing beforehand the existence of such a vertex in the given digraph.

4.3.21. Proof can be carried out by induction on the number of vertices in the digraph.

4.3.25. Carry out induction over the length of the contour.

4.3.29. Proof can be carried out by induction on the determinant order.

4.4.

4.4.2. Starting from an arbitrary vertex of the tree, we construct a chain, adding a new vertex on each step as long as possible. At the moment when it becomes impossible, the end vertices of the chain will become pendant vertices of the tree. The process of construction terminates since the set of vertices is finite, and there are no cycles in the tree.

4.4.4. (1) According to Problem 4.1.26, the number of edges in the graph G and its connectiveness can be reconstructed from $F(G)$.

4.4.9. The induction can be carried out on the magnitude of the tree radius.

4.4.12. The power is continual. **4.4.14.** 2. **4.4.23.** Carry out induction over the length of the vector. **4.2.28.** It is true. **4.4.30.** It is true. **4.4.31.** Generally, it is not true. **4.4.32.** Generally, it is not true. **4.4.33.** $n-1$.

4.4.34. An irreducible network does not have parallel edges and is not p-reducible. Hence $m \leqslant \binom{n}{2} - 1$. Each internal vertex in an irreducible network has a power not lower than 3 and poles have a power not lower than 2 for $n > 2$.

4.4.40. (1) (a), (b) and (c). Yes, it can. (3) (a) and (b). No, it cannot. **4.4.43.** Consider Γ_m^p, $m > 3$. **4.4.52.** No, it is not.

4.5.

4.5.2. (1) $2^{\binom{n}{2}}$. (2) $\binom{n^2}{m}$. **4.5.3.** (2) Use the inclusion-exclusion relation.

4.5.5. (1) Make use of the fact that in a connected graph, $m \leqslant \binom{n}{2}$ and $m \geqslant n-1$. (2) A connected graph with m edges has not more than $m+1$ vertices. The number of pairs of different vertices is hence not larger than $\binom{m+1}{2}$. Hence $\psi(m) \leqslant \binom{\binom{m+1}{2}}{m}$. Then Stirling's formula should be used.

4.5.6. Note that the number of vertices in a graph does not exceed $2m$. The further line of reasoning is the same as in Problem 4.5.5 (2).

4.5.7. See Problems 4.5.5 (2) and 4.5.6.

4.5.9. (1) The code of a tree with m edges is a dual vector of length $2m$ with m unit coordinates. (2) See Problem 4.4.24.

4.5.10. Use the solution of Problem 4.5.4 and Stirling's formula. **4.5.14.** See Problems 8.3.18 and 8.3.19.

4.5.16. The network $\Gamma(a, b)$ having properties I and II is a result of substituting the networks of the type Γ_k^p, $k = \overline{1, m+1}$ for the edges of the network $\Gamma_{n-1}^s(a, b)$. The number of such networks is equal to the number of arrangements of m objects in $n-1$ boxes so that none of the boxes is empty, and is $\binom{m-1}{n-2}$.

4.5.17. (2) Use the solution of Problem 4.5.14.

4.5.19. If a graph is not connected, the set of its vertices can be divided into two parts that there are no edges connecting the vertices of different parts. The number of vertices in one of the parts lies between 1 and $[n/2]$. The graph with numbered vertices is completely determined by the choice of the edges. The number of edges which are forbidden is equal to $k(n-k)$, where k is the number of vertices in one of the parts.

4.5.22. Any subgraph of the cube B^n is completely defined by specifying the set of its vertices. The set of vertices of a connected subgraph is defined by specifying its certain spanning set (i.e. the tree containing all the vertices). In order to specify the tree which is a subgraph of the cube B^n, we can choose any vertex of the cube belonging to the tree (there are not more than 2^n ways of doing so) and a tree with k vertices (there are not more than 4^{k-1} ways of doing so). For each edge of a tree, its direction as that of an edge of the cube B^n can be specified in not more than n ways. Hence we obtain the required estimate.

4.5.26. $\binom{n}{3}\dfrac{\binom{\binom{n}{2}-3}{m-3}}{\binom{\binom{n}{2}}{m}}$. **4.5.29.** $\binom{n}{k}2^{\left(-\binom{k}{2}\right)}$.

4.5.31. Let $\delta(n)$ be a submultiple of graphs for which $p(G) \geqslant 1$. In view of inequality (1), we have $\delta(n) \leqslant -\overline{p}(n) \to 0$ as $n \to \infty$. Hence we obtain the result.

4.5.32. (2) $Z(\Gamma(K_{2,3}), t_1, t_2, t_3, t_4, t_5) = \dfrac{1}{2!\,3!}(t_1^2+t_2)(t_1^3+3t_1t_2+2t_3)$.

4.5.33. (1) Let us consider an arbitrary permutation $a_1a_2\ldots a_n$ of numbers of the set $\{1, 2, \ldots, n\}$ and arrange parentheses in it so as to obtain a substitution with a cyclic structure $(j) = (j_1, j_2, \ldots, j_n)$. First we have j_1 cycles of length 1, then j_2 cycles of length 2, and so on (i.e. the substitution has the following form: $(a_1)(a_2)\ldots(a_{j_1})(a_{j_1+1}a_{j_1+2})\ldots$). Let us now assume that two substitutions π_1 and π_2 with a cyclic structure (j) are constructed from two different permutations of elements of the set $\{1, 2, \ldots, n\}$ in the way described above. When does π_1 coincide with π_2? The coincidence is possible for two reasons: (1) identical cycles in the substitutions π_1 and π_2 are in different positions, (2) although the cycles are identical (as cycles of a substitution), in the above construction they start with the different elements (say, 123 and 231). The first reason leads to a repetition of the same substitution $\prod_{k=1}^{n} j_k!$ times, while the second reason to a repetition $\prod_{k=1}^{n} k^{j_k}$ times, the reasons being mutually independent. (4) The proof can be carried out by induction on n using the relations from parts (1) and (2) of this problem.

4.5.34. Use the following obvious fact: a cycle is an even substitution if and only if its length is odd.

4.5.36. This follows from the Polya theorem.

4.5.37. Interpret the coefficients of the binomial $1+x$ appropriately and use the Polya theorem.

4.5.40. Associate each graph with a tuple of connected graphs, i.e. the tuple of all its connective components. Then apply the Polya theorem.

4.5.42. Any tournament is uniquely determined by its condensation (with numbered vertices) and the tuple of strong components. The numbering of the vertices is required only for associating them with the corresponding strong components of the tournament.

4.6.

4.6.3. 14. **4.6.4.** See Problem 4.6.2. **4.6.6.** This follows from the fact that any function $f(\widetilde{x^n})$ has a d.n.f. whose complexity does not exceed $n2^{n-1}$. **4.6.16.** See Problem 4.6.15.

4.6.20. This follows from the fact that the complexity of a scheme dual to a given one is equal to the complexity of the initial scheme, and from Problem 4.6.18.

4.6.21. If a scheme contains not more than seven contacts, it can be plotted on a plane so than the addition of an edge between the poles leaves it plane. Consequently, there exists a scheme dual to it. The substitution in the latter scheme $x^{\bar{\sigma}}$ for all contacts of the type x^{σ} leads to a scheme representing a negation of the function which is represented by the initial scheme.

4.6.22. Let us choose an arbitrary contact of a scheme without repetition representing the Boolean function f and consider a chain passing through the contact. The values of variables differing from those in the chosen chain can be fixed in such a way that the contacts not contained in the chain will be disconnected. The obtained scheme represents an e.c. depending on the chosen variable. Thus, it turns out that a certain component of the function f, and hence the function f itself, essentially depends on the chosen variable.

4.6.25. The scheme Σ in Fig. 21 has the sets $\{x, y\}$, $\{r, w\}$, $\{x, r, v, z\}$ and $\{x, w, t, u\}$ among its sections. Then if there exists a scheme without repetition Σ_1 representing the function f^*, it contains chains with conductivities xy, rw, $xrvz$, $xwtu$. Without any loss of generality, we can assume that the contact x adjoins the pole a of the network Σ_1. Then either contact r or contact w also adjoins this pole. In the former case, Σ_1 does not contain the chain $xrvz$, and in the latter, $xwtu$.

4.6.26. No, it is not. Make use of the fact that $\bar{f}(x_1, \ldots, x_n) = f^*(\bar{x}_1, \ldots, \bar{x}_n)$ and the result of Problem 4.6.25.

CHAPTER FIVE

5.1.

5.1.1. Let $\rho(v, w) \leqslant 2t$ for some v and w in C. Then $S_t^n(v) \cap S_t^n(w) \neq \varnothing$. Consequently, any mapping ψ: $B^n \to C$ such that $S_t^n(u) \subseteq \psi^{-1}(u)$ for any $u \in C$ is not single-valued.

5.1.2. (1) Generally, it is not true. (2) It is true. (3) Generally, it is not true.

5.1.3. The sets $C_0 = \{\tilde{\alpha} \in C, \|\alpha\| \text{ is even}\}$ and $C_1 = \{\tilde{\alpha} \in C, \|\tilde{\alpha}\| \text{ is odd}\}$ are codes detecting a single error. At least one of them contains not less than half the number of words in C.

5.1.4. (1) It detects one and corrects 0 mistakes. (2) It detects $n - 1$ and corrects $[(n - 1)/2]$ mistakes.

5.1.7. (2) 16. **5.1.8.** 2^{n-1}. **5.1.9.** 2. **5.1.12.** No, there does not.

5.1.13. Let for a certain $n > 7$ there exist a densely packed $\langle n, 3\rangle$-code. Then in view of 5.1.11, the number $\sum_{i=0}^{3} \binom{n}{i}$ for this n is a power of two. Consequently, for a certain k the equality $(n + 1)(n^2 - n - 6) = 3 \times 2^k$ is valid. Then either $n + 1$ is a power of two, or $n + 1$ has the form 3×2^r for a certain natural r. If $n + 1 = 2^r$, then $n^2 - n + 6 = 3 \times 2^{k-r}$. Substituting $n = 2^r - 1$ into the last equality, we obtain $2^{2r-3} - 3 \times 2^{r-3} + 1 = 3 \times 2^{k-r-3}$. For $r > 3$, the left-hand side is an odd number exceeding 3, while the right-hand side is either an even number or 3. The second case can be considered similarly.

5.1.14. Let the vertices $\tilde{\alpha} = (\alpha_1, \ldots, \alpha_n)$, $\tilde{\beta} = (\beta_1, \ldots, \beta_n)$ and $\tilde{\gamma} = (\gamma_1, \ldots, \gamma_n)$ form an $\langle n, d\rangle$-code. Without loss of generality, $\tilde{\alpha} = \tilde{0}$ and $\rho(\tilde{\alpha}, \tilde{\beta}) = d$. We put $A_{\sigma\tau} = \{i: \beta_i = \sigma, \gamma_i = \tau\}$, $\sigma, \tau \in \{0, 1\}$. Let us consider a vertex $\tilde{\delta}$ such that $\delta_i = 0$ for $i \in A_{11}$ and $\delta_i = 1$ for $i \notin A_{11}$. We have $\rho(\tilde{\alpha}, \tilde{\delta}) = \|\tilde{\delta}\| = |A_{01} \cup A_{10} \cup A_{00}| \geqslant |A_{01} \cup A_{10}| = \rho(\tilde{\beta}, \tilde{\gamma}) \geqslant d$, $\rho(\tilde{\delta}, \tilde{\beta}) \geqslant \|\tilde{\gamma}\| \geqslant d$, $\rho(\tilde{\delta}, \tilde{\gamma}) \geqslant \|\tilde{\beta}\| = d$.

5.1.16. Without loss of generality, we assume that $\tilde{0} \in C$. For any $\tilde{\alpha} \in B^n_{d+1}$, there exists a unique vertex $\tilde{\gamma} \in C$ such that $\rho(\tilde{\alpha}, \tilde{\gamma}) \leqslant d$. Since the weight of any non-zero code word is not less than $2d + 1$, $\tilde{\gamma} \in B^n_{2d+1}$. Let $A(\tilde{\gamma}) = \{\tilde{\alpha}: \tilde{\alpha} \in B^n_{d+1}, \rho(\tilde{\alpha}, \tilde{\gamma}) = d\}$. Then $\bigcup_{\tilde{\gamma} \in C \cap B^n_{2d+1}} A(\tilde{\gamma}) = B^n_{d+1}$ and for any $\tilde{\gamma}_1, \tilde{\gamma}_2$ in $C \cap B^n_{2d+1}$ we have $A(\tilde{\gamma}_1) \cap A(\tilde{\gamma}_2) = \varnothing$.

5.1.17. The code distance of any equidistant code of power higher than 2 is even.

5.1.20. (1) In each of the faces $B^{n+d,\, 1, \ldots, d}_{0 \ldots 0}$ and $B^{n+d,\, 1, \ldots, d}_{1 \ldots 1}$ we construct codes C_0 and C_1 of power $m(n, d)$. Then $C_0 \cup C_1$ is an $\langle n + d, d\rangle$-code of power $2m(n, d)$. (3) **Hint.** If C is an $\langle n, d\rangle$-code in B^n, for $(n - 1)$-dimensional face g the set $C \cap g$ is an $\langle n - 1, d\rangle$-code.

5.1.23. Let $n < 2d$ and let C be an arbitrary $\langle n, d\rangle$-code of power $m \geqslant 2$. We construct a matrix M whose lines are code words. Let R be the sum of pairwise distances between (unordered) pairs of code words. On the one hand, $R \geqslant \binom{m}{2} d$. On the other hand, $R = \sum_{i=1}^{n} h_i (m - h_i)$, where h_i is the number of unities in the i-th row of the matrix M. Since $h(m-h) \leqslant m^2/4$, we have $\frac{m(m-1)}{2} d \leqslant n \frac{m^2}{4}$. Hence $m \leqslant \frac{2d}{2d-n}$.

5.1.25. (3) Let C be the maximum $\langle n, k, d\rangle$-code. Then the set $C_i = \{\tilde{\alpha}: \tilde{\alpha} \in C, \alpha_i = 0\}$ is an $\langle n-1, k, d\rangle$-code. The number of pairs of the form $(i, \tilde{\beta})$ such that $1 \leqslant i \leqslant n$, $\tilde{\beta} \in C_i$, does not exceed $n \max |C_i| \leqslant nm(n-1, k, d)$. On the other hand, each vector $\tilde{\alpha} \in C$ generates $n-k$ such pairs. Hence

$$(n-k)\, m(n, k, d) \leqslant nm(n-1, k, d).$$

5.1.26. Let $\tilde{\alpha} \in B^n$, $C \subseteq B^n$ and let $C_{\tilde{\alpha}} = \{\tilde{\gamma}: \tilde{\gamma} = \tilde{\alpha} \oplus \tilde{\beta}, \tilde{\beta} \in C\}$. Then, if C is an $\langle n, d\rangle$-code and $\rho(\tilde{\alpha}, \tilde{\beta}) < d$, $C_{\tilde{\alpha}} \cap C_{\tilde{\beta}} = \varnothing$. Indeed, let $\tilde{\gamma} \in C_{\tilde{\alpha}}$, $\tilde{\delta} \in C_{\tilde{\beta}}$ and $\tilde{\delta}' = \tilde{\delta} \oplus \tilde{\beta}$, $\tilde{\gamma}' = \tilde{\gamma} \oplus \tilde{\alpha}$. Then $\rho(\tilde{\gamma}, \tilde{\delta}) = \rho(\tilde{\alpha} \oplus \tilde{\gamma}', \tilde{\beta} \oplus \tilde{\delta}') = \| \tilde{\alpha} \oplus \tilde{\gamma}' \oplus \tilde{\beta} \oplus \tilde{\delta}' \| = \| (\tilde{\alpha} \oplus \tilde{\beta}) \oplus (\tilde{\gamma}' \oplus \tilde{\delta}') \| \neq 0$, since otherwise $\tilde{\alpha} \oplus \tilde{\beta} = \tilde{\gamma}' \oplus \tilde{\delta}'$, and hence $\rho(\tilde{\alpha}, \tilde{\beta}) = \rho(\tilde{\gamma}', \tilde{\delta}')$. But $\rho(\tilde{\alpha}, \tilde{\beta}) < d$ and $\rho(\tilde{\gamma}', \tilde{\delta}') \geqslant d$ since $\tilde{\gamma}', \tilde{\delta}' \in C$. Hence follows the statement.

5.1.27. This follows from Problem 5.1.26 if we take into account that in any $(d-1)$-dimensional face of the cube B^n, the pairwise distances between vertices do not exceed $d-1$, and the number of vertices is 2^{d-1}.

5.2.

5.2.2. A linearly independent system is, for example, B_1^n. If s vectors in B^n are linearly independent, all their combinations of the form (1) are pairwise different. If there existed a subset consisting of $n+1$ linearly independent vectors, we would have the equality $|B^n| = 2^{n+1}$.

5.2.4. This follows from the fact that an (n, k)-code is a k-dimensional subspace of B^n, i.e. the maximum number of linearly independent vectors in it is k, and any linear combination of code vectors belongs to the code.

5.2.5. If there is a vector with an odd weight in the code, half the number of code words have odd weights and the other half even weights. Otherwise, all the vectors are of an even weight. The former statement follows from the fact that the number of linear combinations containing the given vector with the odd weight is equal to the number of combinations which do not contain it. All the combinations are divided into pairs in which exactly one combination has an odd weight.

5.2.6. A non-zero vector in B^n can be chosen in $2^n - 1$ ways. If i linearly independent vectors are chosen, the corresponding subspace has 2^i vectors. Any vector in the complement to this space constitutes with i chosen vectors a linearly independent set, and any vector in the subspace can be represented as a linear combination of the chosen vectors. Thus, the $(i + 1)$-th vector can be chosen in $2^n - 2^i$ ways.

5.2.7. See Problem 5.2.6. **5.2.8.** 2^{n-1}. **5.2.10.** It is true.

5.2.11. (1) $m(C(H)) = 8$, $d(C(H)) = 2$; (5) $m(C(H)) = 32$, $d(C(H)) = 7$.

5.2.14. The matrix $M(C)$ generating the (n, k)-code C can be reduced to the form $(I_k P)$ by substituting for rows their linear combinations and transposing the columns. If a vector $\tilde{\alpha}$ is orthogonal to each row of the matrix $M(C)$, then a vector $\tilde{\beta}$ obtained from $\tilde{\alpha}$ by an appropriate transposition of coordinates is orthogonal to each row of the matrix $M(C)$, and vice versa. One of generating matrices of the code C dual to the code C^* generated by the matrix $(I_k P)$ has the form $(P^T I_{n-k})$, i.e. C^* is an $(n, n - k)$-code. Hence, the code C^* dual to the code C is also an $(n, n - k)$-code.

5.2.15. The problem is similar to 5.2.5.

5.2.16. The code distance d is obviously not smaller than the minimum weight of the non-zero code vector. If $\rho(\tilde{\alpha}, \tilde{\beta}) < d$ for some code words $\tilde{\alpha}$ and $\tilde{\beta}$, we arrive at a contradiction to the condition of the problem since $\tilde{\alpha} \oplus \tilde{\beta}$ is also a code vector and $\rho(\tilde{0}, \tilde{\alpha} \oplus \tilde{\beta}) < d$.

5.2.17. The solution follows from Problems 5.2.15 and 5.2.16.

5.2.18. The number of unities in the matrix of the code C is equal to $\frac{1}{2} |C| n$. On the other hand, this number is not smaller than $d(|C| - 1)$.

5.2.20. Let $\tilde{\alpha} = (\alpha_1, \ldots, \alpha_n)$ be a weight-d vector orthogonal to the matrix H. We shall denote by h_i the i-th column of the matrix H. It follows from the orthogonality of $\tilde{\alpha}$ to each row of the matrix H that $\sum_{i=1}^{n} \alpha_i h_i = \tilde{0}$. This leads to a linear dependence between the rows h_i which appear in the linear combination. Consequently, to each vector $\tilde{\alpha}$ of weight d in the null space of the matrix H, there correspond d linearly dependent columns of this matrix. Thus, if each $d - 1$ columns of the matrix H are linearly independent, the minimum weight of the code vector is not smaller than d, and conversely, if there exists a set of $d - 1$ linearly dependent rows, there exists a vector of weight smaller than d in the orthogonal subspace.

5.2.21. The problem is a corollary of Problem 5.2.20.

5.2.24. (1) It follows from 5.2.18 that $g(9, 5) \leqslant 10$. We can easily construct a linear $\langle 9, 5\rangle$-code of power 4. Therefore, $g(9, 5) \in \{4, 8\}$. Let us assume that C is a linear $\langle 9, 5\rangle$-code of power 8. Then four code vectors have an odd weight. No two of these four vectors lie outside B_5^9. But among three vectors in B_5^9, there are two the distance between which does not exceed 4.

5.3.

5.3.4. Consider the code $\{a, aabb, bb\}$.

5.3.8. See Problem 5.3.7.

5.3.10. The number of words of length smaller than l in the k-letter alphabet is $\sum_{i=0}^{l-1} k^i = \frac{k^l - 1}{k - 1}$. Hence if $k^l - 1 < m(k - 1)$, there exists in M a word of a length not smaller than $\log_k(1 + m(k - 1))$.

5.3.11. (2) For any divisible code, there exists a prefix code with the same tuple of code word lengths (see Ref. 6, Part 5).

5.3.18. (1) Prove by induction on m.

5.3.19. This follows from 5.3.18.

5.3.22. (1) Let $C = \{w_1, \ldots, w_m\}$ be a dual prefix code, and let the maximum length of a code word be n. Let $w = \alpha_1 \ldots \alpha_{\lambda(w)}$ be a word in C. Let $\tilde{\gamma} = (\gamma_1, \ldots, \gamma_n)$ be a vector with $\gamma_i = \alpha_i$, $i = \overline{1, \lambda(w)}$ and with blanks in the remaining coordinates. Then the vector $\tilde{\gamma}$ is a code of an $(n - \lambda(w))$-dimensional face of the cube B. It follows from the prefix form of the code that faces corresponding to different code words do not intersect. Hence $\sum_{j=1}^{m} 2^{n-\lambda(w_j)} \leqslant 2^n$.

(2) It follows from the completeness of the code that for any $\tilde{\alpha} \in B^n$ there exists a code word w which is the prefix of $\tilde{\alpha}$. This means that the tuple $\tilde{\alpha}$ is contained in the face corresponding to the word w. It follows from the prefix form of the code that the faces corresponding to different code words are disjoint. Thus, the set of faces corresponding to the words of the complete prefix code define a partition of the cube B^n into disjoint faces. This leads to the required equality. (3) The proof is similar to that in Problem 1.1.34.

5.3.23. (1) It is true. The statement follows from the fact that any optimal code is a complete prefix code.

5.3.24. Prove by induction on m.

5.3.25. A prefix code with code word lengths $\lambda_1, \ldots, \lambda_m$ exists if and only if $\sum_{i=1}^{m} 2^{\lambda_i} \leqslant 1$ (see Problem 5.3.22). Therefore, $L_m = \min \sum_{i=1}^{m} \lambda_i$, where the minimum is taken over all sets $\{\lambda_1, \ldots, \lambda_m\}$ of natural numbers satisfying the condition $\sum_{i=1}^{m} 2^{-\lambda_i} \leqslant 1$. The minimum $\sum_{i=1}^{m} \lambda_i$ is attained on the sets $\{\lambda_1, \ldots, \lambda_m\}$ such that $|\lambda_i - \lambda_j| \leqslant 1$, $1 \leqslant i, j \leqslant m$. Indeed, if there exist λ_i and λ_j such that $\lambda_i - \lambda_j \geqslant 2$, after the substitution of $\lambda_i - 1$ for λ_i and $\lambda_j + 1$ for λ_j the quantity $\sum_{i=1}^{m} \lambda_i$ does not change, and the condition $\sum_{i=1}^{m} 2^{-\lambda_i} \leqslant 1$ remains fulfilled. Let $\{\lambda_1, \ldots, \lambda_m\}$ be a set of numbers, such that $\lambda_i = \lambda$ for $i = 1, m-r$, $\lambda_i = \lambda + 1$ for $m - r < i \leqslant m$, where λ is an integer. Then $\sum_{i=1}^{m} \lambda_i = m\lambda + r$, and the condition assumes the form $m2^{-\lambda} - r2^{-\lambda-1} \leqslant 1$. It follows from the condition that $\lambda > (\log_2 m) - 1$, and hence $\sum_{i=1}^{m} \lambda_i \geqslant m\,[\log_2 m]$.

5.3.27. (1) For $m = 3$, assuming that $p_1 \geqslant p_2 \geqslant p_3 > 0$, we have $L(P) = p_1 + 2p_2 + 2p_3 = 1 + p_2 + p_3 > 1$. The statement then follows from the fact that upon expansion operation the value of the code does not decrease. (2) For given $\varepsilon > 0$ and $m \geqslant 2$, consider the distribution

$$P = \left(1 - \delta, \frac{\delta}{m-1}, \ldots, \frac{\delta}{m-1}\right), \text{ where } \delta = \frac{\varepsilon}{|\log_2 m|}.$$

CHAPTER SIX

6.1.

6.1.1. (1) No, it is not since the output signal at the instant t depends on the input signal at the subsequent instant. (3) Yes, it is.

6.1.2. (3) No, it is not. (4) Yes, it is.

6.1.3. (2) Yes, it is. (4) Yes, it is.

6.1.4. (1) Yes, it is. (2) No, it is not. (4) No, it is not.

6.1.5. (1) It is impossible to supplement the definition to a deterministic function. (2) The function φ admits the supplement to the definition of deterministic function. (4) The supplement to a definition is impossible.

6.1.6. (1) It is sufficient to consider the function

$$\varphi(X_1, X_2) = \begin{cases} \tilde{1}^{\omega} & \text{if } X_2 = \tilde{0}^{\omega}, \\ \tilde{x}_1^{\omega} & \text{if } X_2 \neq \tilde{0}^{\omega} \text{ and } X_1 = \tilde{x}_1^{\omega}. \end{cases}$$

6.1.9. (2) The weight of the function φ is equal to 2. **6.1.10.** (1) The operators are equivalent. (2) No, they are not. **6.1.11.** (2) Yes, it is. For example, $\varphi_1 = \varphi_{\tilde{x}^s}$ for $s = 1$ and $\tilde{x}^s = 0$. (3) No, it is not.

6.1.12. (1) Yes, it is. The weight is 4. (3) Yes, it is. The weight is 7.

6.1.13. (2) If $r = 3$, for a suitable function we can take $\varphi(\tilde{x}^{\omega}) = \langle 3/4 \rangle$.

6.1.16. For $r = 3$, for such a function we can take

$$\varphi: \begin{cases} y(1) = x(1), \\ y(t) = (x(t) \oplus x(t-1))\, y(t-1), \quad t \geqslant 2. \end{cases}$$

6.1.18. See hint to Problem **6.1.6** (1).

6.1.20. (2) In order to solve the problem, it is sufficient to consider the function

$$\varphi(\tilde{x}^{\omega}) = 00x(2)\,x(3)\,\ldots\,x(t)\,\ldots.$$

6.1.22. It is a generated operator.

6.1.24. (2) If $|A| = 1$ and $|B| \geqslant 2$, then $|\Phi_{A,B}| = \mathfrak{c}$. If $|B| = 1$ and $|A| \geqslant 1$, then $|\Phi_{A,B}| = 1$.

6.1.25. (1) If $|A| > 1$, the power of each class $K_j(s)$ is $\mathfrak{c}$. 2) The number of different classes is $|A|^s$.

6.1.26. (2) If $|B| = 1$ and $|A| \geqslant 1$, then $|\hat{\Phi}_{A,B}| = 1$.

6.2.

6.2.1. (4) The canonical equations have the following form:

$$\begin{cases} y(t) = F(x(t) \vee \bar{x}(t))\, \bar{q}(t-1), \\ q(t) = (x(t) \vee \bar{x}(t))\, \bar{q}(t-1), \\ q(0) = 0. \end{cases}$$

The input signal $x(t)$ can be omitted since the function φ depends on it inessentially.

6.2.2. (1) The operator definition can be supplemented so that we obtain a b.d.-operator of weight 4, described by the following canonical equations:

$$\begin{cases} y(t) = \bar{x}(t)\, q_1(t-1)\, q_2(t-1) \oplus q_1(t-1) \oplus q_2(t-1), \\ q_1(t) = m(x(t),\ q_1(t-1),\ q_2(t-1)) \oplus x(t) \oplus q_1(t-1), \\ q_2(t) = \bar{x}(t)\, (\bar{q}_1(t-1) \vee \bar{q}_2(t-1)) \oplus q_1(t-1) \oplus q_2(t-1), \\ q_1(0) = q_2(0) = 0. \end{cases}$$

6.2.3. (4) The weight of the operator is 3 (the states corresponding to pairs $(q_1, q_2) = (1, 0)$ and $(q_1, q_2) = (1, 1)$ can be identified).

6.2.4. To each function of weight w (in the indicated set of b.d.-functions) there corresponds a canonical array containing $n + 1$ "input" columns (including the column describing the state of the function) and $m + 1$ "output" columns (taking into account the transition function). This array contains wk^n "input" tuples. At the "output" of the array, there must be some tuples of the set containing wk^m elements ("output" tuples). In order to obtain the majorant, we must assume that each "input" tuple can be associated with any "output" tuple.

6.2.8. (2) The canonical equations of the operator ψ have the form

$$\begin{cases} y(t) = q(t-1), \\ q(t) = \overline{q}(t-1), \\ q(0) = 0, \end{cases}$$

i.e. ψ is an autonomous operator.

6.2.9. (1) The canonical equations of the operator obtained from the operator φ by introducing the feedback in variables x_2 and y_1 have the form

$$\begin{cases} y'(t) = 1, \\ q'(t) = x_1(t)\,\overline{x_3}(t)\,q'(t-1), \\ q'(0) = 0. \end{cases}$$

6.2.10. (1) The operator weight is 2.

6.2.11. (1) The following b.d.-operator can be taken for φ:

$$\varphi: \begin{cases} y(t) = x_1(t) \vee \overline{q}(t-1), \\ q(t) = x_1(t) \oplus x_2(t), \\ q(0) = 0. \end{cases}$$

6.2.14. (1) and (2) Consider the operator $\varphi_d(\varphi_d)$. (5) It is expedient to analyze the operator $\varphi_d(\varphi_{\equiv 0})$ where $\varphi_{\equiv 0}$ is an operator generated by the constant 0.

6.2.15. (3) The weight of the superposition is 4. **6.2.16.** (2) The operator is autonomous. (3) The operator is not autonomous.

6.2.19. (2) Such a scheme exists.

6.2.22. First draw all possible three-vertex digraphs with numbered vertices, which satisfy condition (c) and are such that the out-degree of each vertex is equal to 2. The digits 0, 1 and 2 can be taken as the numbers for vertices. The vertex marked by 0 is convenient to be taken as the initial one. Then the arcs of each of the constructed digraphs must be "loaded" in all possible ways so as to obtain Moore's diagrams of some b.d.-operators.

6.3

6.3.1. (1) M is a closed class. (2) The set M is not a closed class.

6.3.3. Prove by induction on the number of delays that any operator ψ in $[\varphi_0, \varphi_3]_{\mathcal{O}}$, which has one input, transforms the word $\tilde{0}^{\omega}$ either into the word of the form $y(1) \ldots y(n_0)\,[0]^{\omega}$, or into the word of the form $y(1) \ldots y(n_0)\,[1]^{\omega}$, where n_0 is the length of the preperiod (which depends on the choice of the operator ψ).

6.3.7. (1) The system is incomplete. (2) The system is complete.

6.3.8. (2) $\{\varphi_3(X), \varphi_{j_0(x)}(X), \varphi_{x_1+x_2}(X_1, X_2)\}$.

6.3.10. The statement can be proved as follows. Let M be a closed class in $\hat{\Phi}_{(k)}$, which differs from the entire set $\hat{\Phi}_{(k)}$. We assume that M is not a precomplete class and consider the totality of all subsets M' in $\hat{\Phi}_{(k)} \setminus M$, which satisfy the condition $[M \cup M']_{\mathcal{O}} \neq \hat{\Phi}_{(k)}$. This totality is not empty. We choose in it a maximum chain (by inclusion) whose existence can be established either with the help of Zermelo's axiom of choice, or by directly taking into account the countability of the set $\hat{\Phi}_{(k)}$ (by arranging the set $\hat{\Phi}_{(k)} \setminus M$ in the form of a natural scale). The union of all the sets of the chosen maximum chain will be denoted by M_0. Then $M \cup M_0$ is a precomplete class in $\hat{\Phi}_{(k)}$. In the proof, we essentially use the fact that there exists in the set $\hat{\Phi}_{(k)}$ a finite system complete with respect to the set of operations $\Theta = \{O_1, O_2, O_3, O_4, S\}$.

6.3.13. This fact can be proved almost in the same way in which the validity of a similar statement was established in Boolean algebra (see Problem 2.1.16).

6.3.14. Yes, there exists such a function (see Problems 6.3.13 and 2.1.25).

6.3.15. Cf. Problem 2.1.17.

6.3.16. For $k \geqslant 3$, the statement directly follows from the corresponding result in P_k. In the general case (for $k \geqslant 2$), the subsets of autonomous operators can be used.

6.3.17. It is countable. **6.3.19.** This power is continual.

6.3.20. It is expedient to use the "power maps", i.e. to compare the powers of the corresponding sets.

CHAPTER SEVEN

7.1

7.1.1. (1) (a) $T(P) = 1^3 0^2 1^2$. (b) The machine T is not applicable to the word $1^3 0 1^3$. (c) $T(P) = 10\,[01]^2 1$.

7.1.2. (4) The program of one of possible machines has the form

	q_1	q_2	q_3
0	$q_0 0 S$	$q_2 0 S$	$q_3 0 S$
1	$q_2 1 R$	$q_3 1 R$	$q_1 1 R$

7.1.3. (2) (a) $1^2 0^2 1 q_0 0 1$; (c) $[10]^2\, 0 q_0 1^2$.

7.1.4. (3) One of Turing's machines transforming the configuration K_1 into K_0 is specified by the following program:

$$\begin{array}{ll} q_1 0 q_2 0 R & q_4 1 q_4 1 L \\ q_1 1 q_1 1 R & q_5 0 q_6 0 R \\ q_2 0 q_3 1 R & q_5 1 q_5 1 L \\ q_2 1 q_2 1 R & q_6 1 q_7 0 R \\ q_3 0 q_4 1 L & q_7 0 q_0 0 R \\ q_4 0 q_5 0 L & q_7 1 q_1 1 R \end{array}$$

7.1.5. (1) This can be done as follows: each command of the form $q_i\alpha q_j\beta S$ (where α and β belong to the external alphabet A) in the program of the Turing machine is replaced by $|A|+1$ commands $q_i\alpha q'_j\beta R$, $q'_j\gamma q_j\gamma L$ (γ runs through the alphabet A), where q'_j is a new state (each state is characterized by its own q_j).

7.1.7. In order to construct the machine T_m, it is sufficient to add m additional (new) states $q'_1, \ldots, q'_m$ and "supplement" the program of the machine T, for example, with the following commands: $q'_1aq'_1aS, \ldots, q'_maq'_maS$, where a is a fixed character of the external alphabet.

7.1.9. (1) (a) The composition T_1T_2 is inapplicable to the word $1^30^21^2$. (b) T_1T_2 is applicable to the word 1^401; as a result, we obtain 1010^31^2.

7.1.10. (1) (a) This iteration is inapplicable to words of the form 1^{3k} $(k \geqslant 1)$. (b) It is inapplicable to words of the form 1^{3k+1} $(k \geqslant 1)$ either. (c) The iteration is applicable to any word of the form 1^{3k+2} $(k \geqslant 1)$. As a result, we obtain the word 1.

7.1.11. (1) (a) $T(P) = 10^41$. (b) $T(P) = 1^501$.

7.1.14. (3) One of the possible Turing machines is specified by

$$\begin{array}{ll} q_10q_00R & q_40q_51L \\ q_11q_20R & q_41q_41R \\ q_20q_01S & q_50q_60L \\ q_21q_30R & q_51q_51L \\ q_30q_40R & q_60q_10R \\ q_31q_31R & q_61q_61L \end{array}$$

7.1.16. (3) An example of the possible machines is

$$\begin{array}{ll} q_10q_01L & q_31q_40R \\ q_11q_20R & q_40q_40S \\ q_20q_20S & q_41q_50R \\ q_21q_30R & q_50q_11R \\ q_30q_30S & q_51q_51S \end{array}$$

7.1.17. (1) $f(x) = x + 1$, $f(x, y) = x + y + 2$.

7.1.18. If we assume, as usual, that machines start to operate in the state q_1, and the extreme left unity of the code of the number x is scanned at the initial instant, the machines mentioned in the problem can compute only one of the following three functions: x, $x - 1$, and a function defined nowhere.

7.1.19. Yes, it is true.

7.1.20. (1) For a fixed (finite!) set of states, there exists only a finite number of pairwise non-equivalent Turing machines (with a given external alphabet). (2) There exists an l such that for any $n \geqslant 1$, the subset of all functions of n variables in M contains not more than l elements.

7.2

7.2.1. (2) $x_1 + x_2 - \operatorname{sgn} x_2$.

7.2.2. (1) At first, we can prove the primitive recursion of the functions $x_1 \dot{-} x_2$ and x^2 and then apply the superposition operation. The "direct proof" of the primitive recursion of the function $g(x) = x^2$ is as follows:

$$\begin{cases} g(0) = 0, \\ g(x+1) = h_1(x, g(x)) = x^2 + 2x + 1, \end{cases}$$

i.e. $h_1(x, y) = 2x + y + 1$;

$$\begin{cases} h_1(x, 0) = 2x + 1 = g_1(x), \\ h_1(x, y+1) = s(h_1(x, y)) = (2x + y + 1) + 1; \end{cases}$$

$$\begin{cases} g_1(0) = 1, \\ g_1(x+1) = h_2(x, g_1(x)) = 2x + 3, \end{cases}$$

i.e. $h_2(x, y) = y + 2$;

$$\begin{cases} h_2(x, 0) = 2, \\ h_2(x, y+1) = s(h_2(x, y)) = (y + 2) + 1. \end{cases}$$

7.2.5. $f(x, y) = \overline{\text{sgn}}\, x \cdot g_1(y) + g_2(x \dot{-} 1) \times \text{sgn}\, x \times \overline{\text{sgn}}\, y + g_3(x \dot{-} 1, y \dot{-} 1) \times \text{sgn}\, x \times \text{sgn}\, y$.

7.2.7. (2) $\mu_{x_1}([x_1/2]) = 2x_1$. (4) $\mu_{x_2}(x_1 \dot{-} x_2) = (x_1 + x_2) \times \text{sgn}\, x_1$, $\mu_{x_2}(x_1 \dot{-} x_2) = x_1 - x_2$.

7.2.8. (4) $f(x_1, x_2) = x_1(1 \dot{-} x_2)$.

7.2.9. No, it is not correct. **7.2.10.** (1) No, it cannot. (2) $1 + \text{sgn}\, x$.

7.2.12. All these classes are countably infinite.

7.2.17. No, not always: the function $f_2(x, y)$ can be identically equal to 0.

7.2.18. (1) No, not always.

7.2.19. (3) This relation is valid. **Hint.** This function resulting from the minimization operation is $2x \dot{-} 1$. (4) Yes, it is. **Hint.** This function resulting from the minimization operation is

$$\left(\left[\frac{x+3}{2}\right]^2 \dot{-} \left(\left[\frac{x+1}{2}\right] \dot{-} \left[\frac{x}{2}\right]\right)\right) \dot{-} 1.$$

7.2.20. (1) Yes, it can. (2) This statement is false for any function f_1 in $K_{g.r} \setminus K_{pr.r}$. (3) The statement is correct for some functions f_1 and f_2 in the set $K_{g.r} \setminus K_{pr.r}$.

7.2.21. (1) No, they are not always satisfied. (2) This inclusion is false for any function $f(x)$ in $K_{g.r} \setminus K_{pr.r}$. (3) This relation is valid for any function $f(x)$ in $K_{g.r} \setminus K_{pr.r}$.

7.2.24. (2) No, it cannot. **7.2.26.** Yes, it is true.

7.3

7.3.1. (a) This follows from the fact that there exists a function in $K_{pr.r}$ which assumes all the values.

7.3.2. Any primitive recursive function can be associated with an infinite set of terms reflecting the way of obtaining the function from the simplest ones.

7.3.3. If there exists a partially recursive universal function $F(x_0, x_1, \ldots, x_n)$, it is defined everywhere. Then the function $F(x_1, x_1, x_2, \ldots, x_n) + 1$ is general recursive and has a number y in the numbering corresponding to the universal function $F^{(n+1)}$. But in this case $F(y, y, \ldots, y) = F(y, y, \ldots, y) + 1$.

7.3.6. Carry out the proof by "diagonalization". **7.3.7.** (1)-(4). No, it is not. (5) Yes, it is. **7.3.8.** The solution is similar to that of Problem 7.3.3.

7.3.11. Consider the function

$$g(x) = \begin{cases} f(x) & \text{if} \quad h(x) = 0, \\ 1 & \text{if} \quad h(x) = 1. \end{cases}$$

7.3.12. Consider the sequence $\alpha_0, \alpha_1, \ldots$, such that $\alpha_i = 1$ if $\varphi_i(i)$ is defined and $\alpha_i = 0$ if $\varphi_i(i)$ is not defined, $i = 0, 1, \ldots$.

7.3.13. (1) For any autonomous finite automaton, its output sequence is quasi-periodic.

7.3.15. For example, the inversion of a word.

7.3.21. (2) This follows from the fact that there exist not more than $c^{s_T(P)}$ different configurations of length $s_T(P)$ for a certain c which depends on the alphabet of states and on the external alphabet of the machine T. If a configuration is repeated, T operates an infinitely long time.

CHAPTER EIGHT

8.1.

8.1.1. (1) The number of ways is $(20)_3 = 20 \times 19 \times 18$. The tickets are not equivalent. The first ticket can be distributed in 20 ways, the second in 19 ways, and the third in 18 ways. (2) 20^3. (3) $C(20, 3)$.

8.1.2. (1) $9!$. (2) $C(9, 3) \times C(6, 3)$. The first rank can be chosen in $C(9, 3)$ ways, and then the second in $C(6, 3)$ ways. Using the multiplication rule, we obtain the result.

8.1.3. (1) Each term of a permutation with repetitions can be chosen independently in n ways. Using the multiplication rule, we obtain $\hat{P}(n, r) = n^r$. (2) From each (n, r)-combination without repetition, we can obtain $r!$ different (n, r)-permutations, and each (n, r)-permutation can be obtained in this way. Hence $r!\,C(n, r) = P(n, r)$. Since (see Example 3) $P(n, r) = (n)_r$, we have $C(n, r) = (n)_r/r! = \binom{n}{r}$. (3) Each (n, r)-combination A with repetitions composed of elements of the set $U = \{a_1, \ldots, a_n\}$ we associate with a vector $\widetilde{\alpha}(A)$ of length $n + r - 1$ consisting of r unities and $n - 1$ zeros and such that the number of zeros between the $(i - 1)$-th and i-th unities is equal to the number of elements a_i in the combination A, $i = 2, \ldots, n$, while the number of zeros before the first unity (after the $(n - 1)$-th unity) is equal to the number of elements a_1 (resp. elements a_n) constituting the combination A. There is a one-to-one correspondence between the combinations and the vectors. On the other hand, the number of vectors with $n-1$ unities and r zeros is $C(n+r-1,\ n-1)$ since each such vector can be put in one-to-one correspondence with a combination of $n+r-1$ elements taken $n-1$ at a time. Therefore, taking into account the result of the previous problem, we find that $\hat{C}(n, r) = \binom{n+r-1}{n-1}$.

8.1.4. (1) k^n. (2) $k_1 k_2 \ldots k_n$. (3) $\binom{n}{r}$.

8.1.5. (1) 2^{mn}. (2) $(2^m)_n$.

8.1.6. $\binom{m}{r}\binom{n}{s}$. The multiplication rule is used.

8.1.7. (1) $\binom{r}{h}\binom{r-h}{k}(n-\alpha-\beta)_{r-h-k}$. At first we choose the places for the letter a $\left(\text{in } \binom{r}{h} \text{ ways}\right)$, then we choose the places for the letter b $\left(\text{in } \binom{r-h}{k}\right)$ ways. We still have $r-h-k$ places the first of which can be occupied by one of the remaining $n-\alpha-\beta$ letters, the second by one of $n-\alpha-\beta-1$ letters, and so on, $(n-\alpha-\beta)_{r-h-k}$ ways in all.

8.1.8. (1) $4(n-4)$. The suite can be chosen in four ways, after which the smallest value of a card can be chosen in $n-4$ ways. (2) $4n(n-1)$. The number of four cards can be chosen in n ways, after which the remaining card can be chosen in $4(n-1)$ ways. (3) $12n(n-1)$. (4) $4\binom{n}{5}$. (5) $4^5(n-4)$. (6) $4n\binom{4n-4}{2}$. (7) $12\binom{n}{2}^2 n + 4\binom{n}{2} n^3$.

8.1.9. (1) 147. The number which appears on the two chips can be chosen in seven ways. It may turn out that this number appears on a chip twice. Then the second chip can be chosen in six ways. Otherwise, we must choose two different numbers from the six numbers that appear on the chips. The number of ways in which this can be done is equal to $\binom{6}{2}$. Thus, $7 \times \left(6 + \binom{6}{2}\right) = 147$. (2) 26. The number of occurrences of identical faces is 6, and the number of occurrences of pairwise different faces is $\binom{6}{3} = 20$. (3) $300 + 300^2 + 300^3$. A newlyborn baby can be given one, two or three names. The number of ways in which k names can be given is 300^k.

8.1.10. (1) $\binom{n-1}{k-1}$. We associate each division $n = n_1 + \ldots + n_k$ with a binary vector with $k-1$ unities and

$n - k$ zeros as follows: the number of zeros preceding the first unity is $n_1 - 1$, the number of zeros between the i-th and $(i + 1)$-th zero is $n_{i+1} - 1$, $i = 1, \ldots, k - 2$, the number of zeros after the $(k - 1)$-th unity is $n_k - 1$. This is a one-to-one correspondence. The number of such binary vectors is $\binom{n-1}{k-1}$. (2) $\binom{n+2}{2}$. **Hint.** The solution should be reduced to the problem of decomposition of n into three addends, after which the result of Problem 8.1.4 should be applied. (3) $\frac{1}{6}\binom{n+2}{2} + \frac{1}{2}([n/2] + 1)$. The representation of the number 7^n in the form of three multipliers corresponds to the representation of n in the form of the sum of three non-negative addends: $n = n_1 + n_2 + n_3$. Among these addends, two can be identical. In this case, the number of disordered partitions is $[n/2] + 1$. When all the addends are different, the number of ordered partitions is six times as large as the number of disordered partitions, which is equal to $\frac{1}{6}\left(\binom{n}{2} - 3\left(\left[\frac{n}{2}\right]+1\right)\right)$. Using the summation rule, we obtain the required result.

8.1.11. (1) The answer is $\binom{n-k(m-1)+1}{k}$. Let $\tilde{\alpha} = (\alpha_1, \ldots, \alpha_{n+k})$ be a tuple of zeros and unities, in which there are at least m zeros between two consecutive unities. Let $\tilde{\beta}(\tilde{\alpha}) = (\beta_1, \ldots, \beta_{n-k(m-1)+1})$ be a tuple obtained from $\tilde{\alpha}$ by deleting m zeros following the first, second, ..., $(k - 1)$-th unity. The tuple $\tilde{\beta}(\tilde{\alpha})$ contains k unities and is uniquely determined from $\tilde{\alpha}$. On the other hand, the tuple $\tilde{\alpha}$ is uniquely determined from the tuple $\tilde{\beta}(\tilde{\alpha})$ and the number m. Thus, the number of initial tuples is equal to the number of tuples of length $n - (m - 1)k + 1$ containing k unities. (2) $\binom{n+9}{9}$. We associate each number A satisfying the condition of the problem with a binary vector with n unities and ten zeros. The number of unities preceding the first zero is equal to the difference between n and the number of digits of the number A. The number of unities between the i-th and $(i + 1)$-th zeros, $1 \leqslant i \leqslant 8$ is equal to the number of digits equal to i in the number A, and the number of unities following the ninth zero is equal to the number of nines in A. Then we apply the result of Example 4 for $k = 10$. (3) $\binom{n+k}{k}$. Any of the shortest walks consists of n southward and k westward segments. The length of a segment is equal to the side of the square. Such a walk can be assigned a binary vector of length $n + k$, in which the i-th coordinate is equal to zero if the i-th segment is directed southwards, and equal to 1 if it has a westward direction. The number of such vectors is $\binom{n+k}{k}$.

8.1.12. (1) $(1+\alpha_1)(1+\alpha_2)\ldots(1+\alpha_n)$. Each divisor $p_1^{\beta_1}\ldots p_r^{\beta_r}$, $0\leqslant\beta_i\leqslant\alpha_i$, $i=1,\ldots,r$, can be put in a one-to-one correspondence with a vector $(\beta_1,\ldots,\beta_r)$. Then see Problem 8.1.4. (2) $2/2^r$. Each divisor $p_1^{\beta_1}\ldots p_r^{\beta_r}$ of the number n, which cannot be divided into the square of an integer, can be put in correspondence with a binary vector $(\beta_1,\ldots,\beta_r)$. (3) $\prod_{k=1}^{r}(p_k^{\alpha_k-1}-1)(p_k-1)^{-1}$. **Hint.** Note that after opening the parentheses in the expression $(1+p_1+\ldots+p_1^{\alpha_1})\ldots(1+p_r+\ldots+p_r^{\alpha_r})$ each divisor appears exactly once as a summand.

8.1.13. (1) $\binom{n}{k}=\frac{(n)_k}{k!}=\frac{n!}{k!\,(n-k)!}=\frac{(n)_{n-k}}{(n-k)!}=\binom{n}{n-k}$.

Another proof. The number $\binom{n}{k}$ is equal to the number of ways in which a k-element subset can be chosen from $U=\{a_1,\ldots,a_n\}$. But each k-element subset we can put in a one-to-one correspondence with its complement in an $(n-k)$-element set U. (2) $\binom{n}{k}\binom{k}{r}=\frac{n!\,k!}{k!\,(n-k)!\,r!\,(k-r)!}=\frac{n!\,(n-r)!}{(n-r)!\,r!\,(n-k)!\,(k-r)!}=\binom{n}{r}\binom{n-r}{k-r}$. (3) $\binom{n-1}{k}+\binom{n-1}{k-1}=\frac{(n-1)!}{k!\,(n-k-1)!}+\frac{(n-1)!}{(k-1)!\,(n-k)!}=\frac{(n-1)!}{k!\,(n-k)!}(n-k+k)=\binom{n}{k}$. (4) $\binom{n}{k-r}\Big/\binom{n}{k}=\frac{k!\,(n-k)!}{(k-r)!\,(n-k+r)!}=\frac{(k)_r}{(n-k+r)_r}$. (5) Use the fact that $\binom{n-r-1}{k-r}=\binom{n-r}{k-r}-\binom{n-r-1}{k-r-1}$. Then $\sum_{r=0}^{k}\binom{n-r-1}{k-r}=\sum_{r=0}^{k}\left(\binom{n-r}{k-r}-\binom{n-r-1}{k-r-1}\right)=\binom{n}{k}$. (6), (7) Directly. (8) Induction on n. For $n=k$, we have $\sum_{r=k}^{k}\binom{r}{k}=1=\binom{k+1}{k+1}$. Inductive transition $n\to n+1$: $\sum_{r=k}^{n+1}\binom{r}{k}=\sum_{r=k}^{n}\binom{r}{k}+\binom{n+1}{k}$. By inductive hypothesis, we have $\sum_{r=k}^{n}\binom{r}{k}=\binom{n+1}{k+1}$. Hence we obtain from this and the previous equality $\sum_{r=k}^{n+1}\binom{r}{k}=\binom{n+1}{k+1}+\binom{n+1}{k}=\binom{n+2}{k+1}$, Q.E.D.

8.1.14. (1) Let us consider the ratio $\binom{n+1}{k}\Big/\binom{n}{k}=(n+1)/(n-k+1)$. This ratio is greater than unity for all $0<k\leqslant n$. Hence the statement. (2) **Hint.** The ratio $\binom{n-r-1}{k-r-1}\Big/\binom{n-r}{k-r}<1$ for $k<n$. (3) The ratio $\binom{n}{k+1}\Big/\binom{n}{k}>1$ for $k\leqslant[n/2]$ and less than unity for $k>]n/2[$. (4) This follows from (3). (5) If the sum contains two binomial coefficients $\binom{n_i}{k}$ and $\binom{n_j}{k}$ such that $n_i-n_j>1$, then replacing them with $\binom{n_i-1}{k}$ and $\binom{n_j+1}{k}$, we obtain a sum smaller than the initial one. For this reason, the superscripts of any two binomial coefficients in the minimal sum differ by not more than 1 and assume not more than two values. If q is the smaller of them, we assume that the superscript of r binomial coefficients $(0\leqslant r<s)$ is $q+1$, and that of the remaining $s-r$ coefficients is q. Then from the condition $\sum_{i=1}^{s} n_i=n$ we obtain $(q+1)r+q(s-r)=n$, whence $q=[n/s]$, $r=n-s[n/s]$, while the minimum value of the sum is $(s-r)\binom{q}{k}+r\binom{q+1}{k}$.

(6) **Hint.** Note that $\binom{n}{j}\geqslant\binom{n}{i}$ if $\left|\frac{n}{2}-i\right|>\left|\frac{n}{2}-j\right|$. (7) We have $(p)_k=\binom{p}{k}k!$. The number $(p)_k$ is divisible and $k!$ is not divisible by p for $k<p$. Consequently, $\binom{p}{k}$ is divisible by p.

(8) We have $(2n)!=\binom{2n}{n}(n!)^2$. The number $(2n)!$ is divisible by any prime number p_i, where $n<p_i\leqslant 2n$, and the number $(n!)^2$ is not divisible by p_i. Consequently, $\binom{2n}{n}$ is divisible by p_i.

8.1.15. (1) Solution. We put $\alpha_{n-1}=\left[\frac{m}{(n-1)!}\right]$. Suppose that the coefficients $\alpha_{n-1}, \alpha_{n-2}, \ldots, \alpha_{n-i}$ have been determined. Then

$$\alpha_{n-(i+1)}=\left[\frac{m-\alpha_{n-1}(n-1)!-\alpha_{n-2}(n-2)!-\ldots-\alpha_{n-i}(n-i)!}{(n-i-1)!}\right].$$

The uniqueness of the representation will be proved by contradiction. Let two vectors $\tilde{\alpha}(m) = (\alpha_1, \ldots, \alpha_{n-1})$ and $\tilde{\beta}(m) = (\beta_1, \ldots, \beta_{n-1})$ correspond to a certain m. Let j be the largest of the subscripts in which the vectors differ from each other. Without any loss of generality, we can assume that $\alpha_j < \beta_j$. We have

$$0 = \sum_{i=1}^{n-1} \beta_i i! - \sum_{i=1}^{n-1} \alpha_i i! \geqslant (\beta_j - \alpha_j)\, j! - \sum_{i=1}^{j-1} \alpha_i i! \geqslant j! - \sum_{i=1}^{j-1} i \times i! > 0.$$

We arrive at a contradiction. (2) $\tilde{\alpha}(4) = (0, 2)$; $\tilde{\alpha}(15) = (1, 1, 2)$; $\tilde{\alpha}(37) = (1, 0, 2, 1)$. (3) $\mu(0, 2, 0, 4) = 100$; $\mu(0, 2, 1) = 10$; $\mu(1, 2, 3, 2) = 72$. (4) $\nu(2, 3, 1, 4) = 12$; $\nu(3, 5, 2, 1, 4) = 85$; $\nu(1, 4, 3, 5, 2) = 20$. (5) Let the number m be specified. We present it in the form $m = \alpha_1 \times 1! + \alpha_2 \times 2! + \ldots + \alpha_{n-1} \times (n-1)!$, where $\alpha_i \leqslant i$, $1 \leqslant i < n$. The construction of the required permutation π is equivalent to that of the vector $(\pi(1), \pi(2), \ldots, \pi(n))$. The coordinates $\pi(j)$ will be given by induction. We put $\pi(1) = \alpha_{n-1} + 1$. If the coordinates $\pi(1), \ldots, \pi(j-1)$ are specified, we put $\pi(j) = \alpha_{n-j} + 1 + s(j)$, where $s(j)$ is the number of $\pi(k)$, $1 \leqslant k < j$, for which $\pi(k) < j$. (6) $\pi_7 = (2, 1, 4, 3)$; $\pi_{18} = (2, 3, 4, 1)$; $\pi_{28} = (2, 1, 4, 5, 3)$.

8.1.16. (1) The coordinates of the vector $\tilde{\beta}(m) = (\beta_1, \ldots, \beta_k)$ are defined as follows: β_1 is the largest integer such that $m \geqslant \binom{\beta_1}{k}$. If $\beta_1, \ldots, \beta_i$ have already been specified, then β_{i+1} is the maximum integer such that $m - \binom{\beta_1}{k} - \binom{\beta_2}{k-1} - \ldots - \binom{\beta_i}{k-i+1} \geqslant \binom{\beta_{i+1}}{k-i}$. The uniqueness is proved as in Problem (8.1.15 (1). (2) (a) $\tilde{\beta}(19) = (6, 4, 1, 0)$ since $19 = \binom{6}{4} + \binom{4}{3} + \binom{1}{2} + \binom{0}{1}$; (b) $\tilde{\beta}(25) = (6, 3, 2)$; (c) $\tilde{\beta}(32) = (4, 3, 2, 1)$. (3) (a) $m = 23$ since $m = \binom{6}{3} + \binom{3}{2} + \binom{0}{1} = 23$; (b) $m = 13$; (c) $m = 10$. (4) First we put in correspondence to the vector $\tilde{\alpha} = (\alpha_1, \ldots, \alpha_n)$ in B_k^n a vector $\tilde{\beta} = (\beta_1, \ldots, \beta_k)$ in which $\beta_j + 1$ is the number of the coordinate of the $(k-j+1)$-th unity on the left in the tuple $\tilde{\alpha}$, $1 \leqslant j \leqslant k$. Then we put $\nu(\tilde{\alpha}) = \mu(\tilde{\beta}) + 1$. For example, if $\tilde{\alpha} = (1, 0, 0, 1, 0)$, then $\tilde{\beta} = (3, 0)$, $\nu(\tilde{\alpha}) = \mu(\tilde{\beta}) + 1 = \binom{3}{2} + \binom{0}{1} + 1 = 4$.

8.1.17. For $n=1$, equality (1) can be verified directly. Let the equality be proved for a certain $n \geqslant 1$. We have

$$\sum_{k=0}^{n+1} \binom{n+1}{k} t^k = \sum_{k=0}^{n+1} \left(\binom{n}{k} + \binom{n}{k-1} \right) t^k = \sum_{k=0}^{n} \binom{n}{k} t^k + \sum_{k=1}^{n+1} \binom{n}{k-1} t^k = \sum_{k=0}^{n} \binom{n}{k} t^k + \sum_{k=0}^{n} \binom{n}{k} t^{k+1} = (1+t)^n +$$

$$t(1+t)^n = (1+t)^{n+1}.$$

8.1.18. (1) Put $t=1$ in (1). (2) Put $t=-1$ in (1). (3) Differentiate (1) with respect to t and put $t=1$. (4) Differentiate (1) twice and put $t=1$. (5) Using (1) and (3), we obtain

$$\sum_{k=0}^{n} (2k+1) \binom{n}{k} = 2 \sum_{k=0}^{n} k \binom{n}{k} + \sum_{k=0}^{n} \binom{n}{k}$$

$$= n2^n + 2^n = (n+1)\, 2^n.$$

(6) Integrate identity (1) with respect to t between 0 and 1. (7) Integrate identity (1) with respect to t between -1 and 0. (8) Carry out induction on n by using Problem 8.1.13 (2) and 8.1.17 (7). (9) Compare the coefficients of t^k on the left- and right-hand sides of the identity $(1+t)^n (1+t)^m = (1+t)^{n+m}$.

(10) Put in (9) $k=n=m$. (11) Dividing by $\binom{2n}{n}$, reduce the problem to (10). (12) We have $\sum_{k=0}^{n} \sum_{r=0}^{n-k} \binom{n}{k} \binom{n-k}{r} =$

$$\sum_{k=0}^{n} \binom{n}{k} \sum_{r=0}^{n-k} \binom{n-k}{r} = \sum_{k=0}^{n} \binom{n}{k} 2^{n-k} = 3^n.$$ (13) $\sum_{r=k}^{n} (-1)^{k-r} \binom{n}{r} =$

$$\sum_{m=0}^{n-k} (-1)^m \binom{n}{k+m} = \sum_{m=0}^{n-k} (-1)^m \binom{n}{n-k-m} =$$

$$\sum_{r=0}^{n-k} (-1)^{n-k-r} \binom{n}{r}.$$ (14) The problem is reduced to (9) by substituting $n-k$ for k. (15) Note that $\sum_{k=n}^{m} (-1)^{k-n} \binom{k}{n} \binom{m}{k} =$

$$\binom{m}{n} \sum_{j=0}^{m-n} (-1)^j \binom{m-n}{j}.$$

8.1.19. (1) $2\sum_k \binom{n}{2k} = (1+1)^n + (1-1)^n = 2^n$. (2) $4\sum_k \binom{n}{4k} =$ $(1+1)^n + (1+i)^n + (1+i^2)^n + (1+i^3)^n = 2^n + (1+i)^n + (1-i)^n =$ $2^n + (\sqrt{2})^n \left(\cos\frac{\pi}{4} + i\sin\frac{\pi}{4}\right)^n + (\sqrt{2})^n \left(\cos\frac{\pi}{4} - i\sin\frac{\pi}{4}\right)^n =$ $2^n + 2^{n/2}\, 2\cos\frac{\pi n}{4}$. (3) $\sum_{\nu=0}^{m-1} e^{-\frac{2\pi i r}{m}\nu}\left(1 + e^{\frac{2\pi i \nu}{m}}\right)^n =$ $\sum_{\nu=0}^{m-1} e^{-\frac{2\pi i r}{m}\nu} \sum_{s=0}^{r-1}\sum_k \binom{n}{mk+s} e^{\nu\frac{2\pi i(lk+s)}{m}} = \sum_k \sum_{s=0}^{r-1} \binom{n}{mk+s} \times$ $\sum_{\nu=0}^{m-1} e^{\frac{2\pi i(lk+s-r)}{m}\nu} = m\sum_k \binom{n}{mk+r}$. In the latter case, we have used the identity

$$\sum_{\nu=0}^{m-1} e^{\frac{2\pi i n}{m}\nu} = \begin{cases} m \text{ if } n \text{ is multiple of } m, \\ 0 \text{ otherwise.} \end{cases}$$

8.1.20. (1) The solution is similar to that of Problem 8.1.19. (3). We have $\sum_{\nu=0}^{m-1} e^{-\frac{2\pi i r\nu}{m}}\left(1 + \alpha e^{\frac{2\pi i \nu}{m}}\right)^m =$

$$\sum_{\nu=0}^{m-1} e^{-\frac{2\pi i r\nu}{m}} \sum_{k=0}^{n} \binom{n}{k} \alpha^k e^{-\frac{2\pi i \nu k}{m}} =$$

$$\sum_{k=0}^{n} \binom{n}{k} \alpha^k \sum_{\nu=0}^{m-1} e^{-\frac{2\pi i(k-r)\nu}{m}} = m\sum_{s=0}^{n} \binom{n}{ms+r} \alpha^{ms+r}.$$

In the latter case, we have used the fact that the sum $\sum_{\nu=0}^{m-1} \exp\{2\pi i l\nu/m\}$ is equal to 0 for $1 < l < m$ and is equal to m for $l=0$. (2) $\frac{1}{2}\left((1+\sqrt{3})^n + (1-\sqrt{3})^n\right)$. Use (1) for $m=2$, $r=0$, $\alpha=\sqrt{3}$. (3) $\left((1+\sqrt[4]{-2})^n + i\,(1+i\sqrt[4]{-2})^n + (1-\sqrt[4]{-2})^n - i\,(1-i\sqrt[4]{-2})^n\right)/4\sqrt[4]{-2}$. Use (1) for $m=4$, $r=1$, $\alpha=\sqrt[4]{-2}$. (4) $(-3)^{\frac{r-1}{2}}\binom{n}{r}\left((1+\sqrt{-3})^{n-r} - (1-\sqrt{-3})^{n-r}\right)/2$. Using Problem 8.1.13 (2), we obtain $\sum_k (-1)^k\, 3\binom{n}{2k+1}\binom{2k+1}{r} =$

$$\binom{n}{r}\sum_k(-3)^k\binom{n-r}{2k+1-r}=(-3)^{\frac{r-1}{2}}\binom{n}{r}\sum_k(-3)^{\frac{2k+1-r}{2}}\times$$

$\binom{n-r}{2k+1-r}$. Further, we use (1) for $m=2$, $\alpha=\sqrt{-3}$ and $r=1-r$.

8.1.21. (1) 4. The term of the expansion $C_{20}^k 2^{k/2}3^{(20-k)/3}$ is rational if and only if k is even and $20-k$ is a multiple of three, i.e. for $k=2, 8, 14, 20$. (2) 13. (3) 9. (4) 6.

8.1.22. (1)—536. The general form of the k-th term of the expansion has the form $\binom{8}{k}A^k$, where $A=t(2-3t)$, and the expansion A^k contains from the k-th to $(2k)$-th powers of t. The power t^9 appears in A^5, A^6, A^7, and A^8 with the coefficients $\binom{5}{4}2(-3)^4$, $\binom{6}{3}2^3(-3)^3$, $\binom{7}{2}2^5(-3)^2$, $\binom{8}{1}2^7(-3)^1$ respectively. The coefficient of t^9 is the sum of these quantities and is equal to -536. (2) -91. (3) 13. (4) 0.

8.1.23. (1) We put $t=m-n$. Then the identity is transformed as follows:

$$\sum_{k=0}^{n}\frac{(n)_k}{(n+t)_k}=\frac{n+t+1}{t+1}\quad\text{or}\quad\sum_{k=0}^{n}\frac{(n)_k}{(n+t+1)_k}=\frac{1}{t+1}.\qquad(*)$$

The latter relation can be proved by induction. Assuming that $(0)_0=1$, we can easily verify the validity of $(*)$ for $n=0$. We assume that $(*)$ is valid for a certain n and prove that $\sum_{k=0}^{n+1}\frac{(n+1)_k}{(n+t+2)_k}=1/(t+1)$. We have

$$\sum_{k=0}^{n+1}\frac{(n+1)_k}{(n+2+t)_k}=\frac{1}{n+2+t}$$

$$+\sum_{k=1}^{n+1}\frac{(n+1)(n)_{k-1}}{(n+2+t)(n+1+t)_{k-1}}=\frac{1}{n+2+t}$$

$$+\frac{n+1}{n+2+t}\sum_{k=0}^{n}\frac{(n)_k}{(n+t)_k}=\text{(by using }(*))$$

$$=\frac{1}{n+2+t}+\frac{n+1}{n+2+t}\,\frac{1}{t+1}=\frac{1}{t+1}.$$

(2) $\sum_{k=0}^{n}\binom{m+k-1}{k}=\sum_{k=0}^{n}\binom{m+k-1}{m-1}=\sum_{k=0}^{n}\left(\binom{m+k}{m}-\binom{m+k-1}{m}\right)=\binom{m+n}{n}=\binom{m+n}{m}$. Similarly, $\sum_{k=0}^{m}\binom{n+k-1}{k}=\binom{m+n}{n}$.

8.1.24. (1) $\frac{(a)_k}{k!}+\frac{(a)_{k-1}}{(k-1)!}=\frac{(a)_k+k\,(a)_{k-1}}{k!}=\frac{(a+1)_k}{k!}$.

(2) The series $A(t)=\sum_{k=0}^{\infty}\binom{a}{k}t^k$ is a series of the function $f(t)=(1+t)^a$ since $f^{(k)}(0)/k!=\binom{a}{k}$. The series $A(t)$ converges for $|t|<1$ and for all a according to the d'Alembert criterion:

$$\binom{a}{k+1}t^{k+1}\Big/\binom{a}{k}t^k=\frac{a-k}{k+1}\,t,\quad \lim_{k\to\infty}\left|\frac{a-k}{k+1}t\right|=t<1.$$

Let the term $r_k(t)$ be the remainder of the series (in Cauchy's form). We shall demonstrate that $r_k(t)\to 0$ as $k\to\infty$ thus proving the equality. We have $r_k(t)=f^{(k+1)}(\Theta t)(1-\Theta)^k t^{k+1}/k!=(a)_{k+1}(1+\Theta t)^{a-k-1}(1-\Theta)^k t^{k+1}/k!=\binom{a-1}{k}t^k a(1+\Theta t)^{a-1}\left(\frac{1-\Theta}{1+\Theta t}\right)^k$. Here $\binom{a-1}{k}t^k\to 0$ as $k\to\infty$, the expression $a(1+\Theta t)^{a-1}$ is bounded, and $0<\left(\frac{1-\Theta}{1+\Theta t}\right)^k<1$ for $0<\Theta<1$, $|t|<1$. Hence it follows that $\lim_{k\to\infty} r_k(t)=0$. (3) $\binom{-a}{k}=(-a)(-a-1)\ldots(-a-k+1)/k!=(-1)^k\binom{a+k-1}{k}$. (4) $\sum_{k=0}^{n}\binom{a-k}{r}=\sum_{k=0}^{n}\left(\binom{a-k+1}{r+1}-\binom{a-k}{r+1}\right)=\binom{a+1}{r+1}-\binom{a-n}{r+1}$. (5) Use the identity $(1+t)^a(1+t)^b=(1+t)^{a+b}$. (6) It follows from (5) if we put $b=-1$. (7) It follows from (3) and (5). (8)

By using (5), we obtain $\sum\limits_{0\leqslant k,\ r\leqslant n}\binom{a}{k}\binom{b}{r}\binom{c}{n-k-r}=$ $\sum\limits_{0\leqslant k\leqslant n}\binom{a}{k}\sum\limits_{0\leqslant r\leqslant n-k}\binom{b}{r}\binom{c}{n-k-r}=\sum\limits_{0\leqslant k\leqslant n}\binom{a}{k}\binom{b+c}{n-k}=$ $\binom{a+b+c}{n}$. (9) By using (2), and the identity $\binom{2k}{k}=$ $(-4)^k\binom{-1/2}{k}$, we obtain $\sum\limits_{k=0}^{\infty}(-1)^k\binom{2k}{k}\left(\frac{1}{s}\right)^k=$ $\sum\limits_{k=0}^{\infty}\binom{-1/2}{k}\left(\frac{1}{2}\right)^k=(1+1/2)^{-1/2}=\sqrt{\frac{2}{3}}$. (10) It follows from (2) that $\frac{\sqrt{5}}{2}=\left(1+\frac{1}{4}\right)^{1/2}=\sum\limits_{k=0}^{\infty}\binom{1/2}{k}\left(\frac{1}{4}\right)^k$; $\frac{\sqrt{3}}{2}=$ $\left(1-\frac{1}{4}\right)^{1/2}=\sum\limits_{k=0}^{\infty}\binom{1/2}{k}\left(-\frac{1}{4}\right)^k$. Hence $2\sum\limits_{k=0}^{\infty}\binom{1/2}{2k+1}\times$ $\left(\frac{1}{4}\right)^{2k+1}=\frac{\sqrt{5}-\sqrt{3}}{2}$. (11) See 8.1.20 (1).

8.1.25. (1) $(mn)!/(m!)^n$. (2) $\binom{n}{k_1,\ k_2,\ \ldots,\ k_s}$. (3) Let us calculate (in two ways) the number $A_n(k_1,\ \ldots,\ k_s)$ of orderings of n objects among which there are k_1 objects of the first kind, k_2 objects of the second kind, etc., and k_s objects of the s-th kind. First method. We first choose k_1 places among n for arranging the objects of the first kind. This can be done in $\binom{n}{k_1}$ ways. Then we choose k_2 places for arranging the objects of the second kind. This can be done in $\binom{n-k_1}{k_2}$ ways, and so on. Using the multiplication rule, we find that

$$A_n(k_1,\ \ldots,\ k_s)=\binom{n}{k_1}\binom{n-k_1}{k_2}\ \cdots\ \binom{n-k_1-\ldots-k_{s-1}}{k_s}.$$

Second method. Let us calculate the number n of orderings of pairwise different objects. We shall assume that the objects are divided into s groups so that the i-th group contains k_i objects $(i=\overline{1,\ s})$. The choice of k_1 places for objects of the first kind, k_2 places for objects of the second kind, etc., can be made in

$A_n(k_1, \ldots, k_s)$ ways. In the group i, the objects can be arranged in k_i ways $(i = \overline{1, s})$. Hence $A_n(k_1, \ldots, k_s)\, k_1! \ldots k_s! = n!$
(4) For $s = 2$, the identity follows from (1) by substituting (t_2/t_1) for t. Suppose that this identity has been proved for a certain $s \geqslant 2$. We shall prove that

$$(t_1+t_2+\ldots+t_{s+1})^n = \sum_{\substack{k_1 \ldots k_{s+1} \\ k_1+\ldots+k_{s+1}=n}} \frac{n!}{k_1!k_2!\ldots k_{s+1}!} t_1^{k_1} t_2^{k_2} \ldots t_{s+1}^{k_{s+1}}.$$

We put $T = (t_1 + \ldots + t_s)$. Then $(t_{s+1} + T)^n =$

$$\sum_{k=0}^{n} \binom{n}{k} t_{s+1}^k T^{n-k} = \sum_{k=0}^{n} \binom{n}{k} (t_1+\ldots+t_s)^{n-k} t_{s+1}^k =$$

$$\sum_{k=0}^{n} \binom{n}{k} t_{s+1}^k \sum_{\substack{k_1 \ldots k_s \\ k_1+\ldots+k_s=n-k}} \frac{(n-k)!}{k_1! \ldots k_s!} t_1^{k_1} \ldots t_s^{k_s} =$$

$$\sum_{k=0}^{n} \sum_{\substack{k_1 \ldots k_s \\ k_1+\ldots+k_s=n-k}} \frac{n!}{k_1! \ldots k_s! k!} t_1^{k_1} \ldots t_{k_s}^{k_s} t_{k_{s+1}}^{k} =$$

$$\sum_{\substack{k_1, \ldots, k_s, k_{s+1} \\ k_1+\ldots+k_{s+1}=n}} \frac{n!}{k_1! \ldots k_{s+1}!} t_1^{k_1} \ldots t_{s+1}^{k_{s+1}}.$$

8.2

8.2.1. (1) Proof. For $n = 1$, $\hat{N}_0 = N - N_1 = S_0 - S_1$, and formula (2) is obviously valid. Let the formula be valid for $n - 1$ properties, and let $N_{\bar{i}_1, \ldots, \bar{i}_k}$ be the number of objects which do not possess any of the properties $i_1, \ldots, i_k$. We have

$$\hat{N}_0 = N_{\bar{i}_1, \ldots, \bar{i}_{n-1}} = N - \sum_{i=1}^{n-1} N_i + \sum_{1 \leqslant i < j \leqslant n-1} N_{i,j}$$

$$- \sum_{1 \leqslant i < j < k \leqslant n-1} N_{i,j,k} + \ldots + (-1)^{n-1} N_{1, \ldots, n-1}. \qquad (*)$$

This formula is also valid for a set of objects having the property n:

$$N_{\bar{1}, \ldots, \overline{n-1}, n} = N_n - \sum_{1 \leqslant i \leqslant n-1} N_{i,n}$$

$$+ \ldots + (-1)^{n-1} N_{1, \ldots, n-1, n} \qquad (**)$$

where $N_{\bar{1},\dots,\overline{n-1},n}$ is the number of objects having the property n and none of the properties $1, \dots, n-1$. Obviously,

$$N_{\bar{1},\dots,\overline{n-1},\bar{n}} = N_{\bar{1},\dots,\overline{n-1}} - N_{\bar{1},\dots,\overline{n-1},n}.$$

Subtracting (**) from (*), we obtain formula (2). (3) By definition, formula (3) can be written in the form

$$\hat{N}_m = \sum_{k=0}^{n-m} (-1)^k \binom{m+k}{m} \sum_{1\leqslant i_1<\dots<i_{m+k}\leqslant n} N_{i_1,\dots,i_{m+k}}.$$

Let us prove that any object possessing m properties will be taken into account by this formula only once, and all the other objects will not be taken into account at all. Indeed, the elements having $s < m$ properties are obviously not taken into account. The elements having $s = m + t$ given properties will be taken into account $\binom{m+t}{m+k}$ times in the second sum. But

$$\sum_{k=0}^{n-m} (-1)^k \binom{m+k}{m}\binom{m+t}{m+k}$$

$$= \binom{m+t}{m} \sum_{k=0}^{t} (-1)^k \binom{t}{k} = \begin{cases} 1 & \text{for } t=0, \\ 0 & \text{for } t>0. \end{cases}$$

Thus, the elements having exactly m properties are taken into account in (3) just once, while the remaining objects are not taken into account at all. (3) It should be noted that $\check{N}_{m+1} = \check{N}_m - \hat{N}_m$; then (4) can be proved by induction on m. By definition, $\check{N}_0 = N = S_0$. Then $\check{N}_1 = S_0 - (S_0 - S_1 + S_2 - \dots) = \sum_{k=1}^{n} (-1)^{k-1} S_k$. Thus, (4) holds for $m = 1$. Let identity (4) be valid for a certain $m \geqslant 1$. Then

$$\check{N}_{m+1} = \check{N}_m - \hat{N}_m = \sum_{k=0}^{n-m} (-1)^k \left(\binom{m-1+k}{m-1} - \binom{m+k}{m} \right) S_{m+k} =$$

$$\sum_{k=1}^{n-m} (-1)^{k-1} \binom{m-1+k}{m} S_{m+k} = \sum_{k=0}^{n-(m+1)} (-1)^k \binom{m+k}{m} S_{m+1+k}.$$

(4) Let us derive, for example, formula (5) from (3). By putting $m=n$ in (3), we obtain $\hat{N}_n = S_n$ in accordance with (5). Let formula (5) be valid for all $k \geqslant n-\nu+1$, $\nu \geqslant 1$. Substituting

$n-\nu$ for m in (3) and using the inductive hypothesis, we substitute the right-hand side of equality (5), in which k is replaced by $n-\nu+k$, for $S_{n-\nu+k}$ for any $k \geqslant 1$. This gives $S_{n-\nu} = \hat{N}_{n-\nu} - \sum_{k=1}^{\nu} (-1)^k \binom{n-\nu+k}{n-\nu} \times \sum_{m=n-\nu+k}^{n} \binom{m}{n-\nu+k} \hat{N}_m = \hat{N}_{n-\nu} - \sum_{m=n-\nu+1}^{n} \binom{m}{n-\nu} \hat{N}_m \times \sum_{k=1}^{m-n+\nu} (-1)^k \binom{m-n+\nu}{k} = \sum_{m=n-\nu}^{n} \binom{m}{n-\nu} \hat{N}_m$. Thus, formula (5) is proved. Formula (6) is proved similary. Let us prove (7). For this purpose, it is sufficient to verify that $R(n, r) = \sum_{k=r}^{n} (-1)^{k-r} \binom{k}{m} S_k \geqslant 0$ for all r. Let us use formula (5). We have $R(n, r) = \sum_{k=r}^{n} (-1)^{k-r} \binom{k}{m} \times \sum_{\nu=k}^{n} \binom{\nu}{k} \hat{N}_\nu = \sum_{\nu=r}^{n} \hat{N}_\nu \sum_{k=r}^{\nu} (-1)^{k-r} \binom{k}{m} \binom{\nu}{k} = \sum_{\nu=r}^{n} \hat{N}_\nu \binom{\nu}{m} \times \sum_{k=r}^{\nu} (-1)^{k-r} \binom{\nu-m}{k-m} = \sum_{\nu=r}^{n} \hat{N}_\nu \binom{\nu}{m} \binom{\nu-m-1}{r-m-1} \geqslant 0$. (5) Let us prove that $S_m > \hat{N}_m$. We put $R(n, m) = S_m - \hat{N}_m$. In view of (3), we have $R(n, m) = \sum_{k=1}^{n-m} (-1)^{k+1} \binom{m+k}{m} S_{m+k} =$ (according to (5)) $= \sum_{k=1}^{n-m} (-1)^{k+1} \binom{m+k}{m} \sum_{r=k+m}^{n} \binom{r}{k+m} \hat{N}_r = \sum_{r=m+1}^{n} \hat{N}_r \sum_{k=1}^{n-m} (-1)^{k+1} \binom{m+k}{m} \binom{r}{k+m} = \sum_{r=m+1}^{n} \hat{N}_r \binom{r}{m} \times \sum_{k=1}^{r-m} (-1)^{k+1} \binom{r-m}{k} = \sum_{r=m+1}^{n} \hat{N}_r \binom{r}{m} \geqslant 0$. Let us prove that $\hat{N}_m \geqslant S_m - (m+1) S_{m+1}$. We put $R'(n, m) = \hat{N} - S_m + (m+1) S_{m+1}$. In view of (3) we have $R'(n, m) = \sum_{k=r}^{n-m} (-1)^k \binom{m+k}{m} S_{m+k} =$

(according to (5)) $= \sum_{k=r}^{n-m} (-1)^k \binom{m+k}{m} \sum_{r=k+m}^{n} \binom{r}{k+m} \hat{N}_r =$

(see estimate $R(n, m)$) $= \sum_{r=m+2}^{n} \hat{N}_r \binom{r}{m} \sum_{k=2}^{r-m} (-1)^k \binom{r-m}{k} =$

$$\sum_{r=m+2}^{n} \hat{N}_r \binom{r}{m} (r-m-1) \geqslant 0.$$

8.2.2. The total number of ways in which the hats can be returned is $4! = 24$. The probability that exactly m persons obtain their own hats is $\check{N}_m/4!$, where $\check{N}_m$ is defined by formula (3) for $n = 4$. We have $\check{N}_0 = S_0 - S_1 + S_2 - S_3 + S_4 = 4! - 4\times 3! + 6 \times 2! - 4 \times 1! + 1 = 9$, $p_0 = 3/8$; $\check{N}_1 = S_1 - 2S_2 + 3S_3 - 4S_4 = 4 \times 3! - 2 \times 6 \times 2! + 3 \times 4 - 4 \times 1 = 8$, $p_1 = 1/3$, $p_2 = 1/4$, $p_3 = 0$, $p_4 = 1/24$.

8.2.3. (1) The number of distributions of the objects for which the given k boxes remain empty is $(n-k)^r$, $S_k = \binom{n}{k} (n-k)^r$. It remains for us to apply formula (2). (2) $S_{m+k} = \binom{n}{m} (n-m-k)^r$, $E(r, n, m) = \hat{N}_m = \sum_{k=0}^{n-m} (-1)^k \binom{m+k}{m} \times \binom{n}{m+k} (n-m-k)^r = \binom{n}{m} \sum_{k=0}^{n-m} (-1)^k \binom{n-m}{k} (n-m-k)^r$.
(3) Use formula (4) in Problem 8.2.1 (3).

8.2.4. Hint: Use Venn's circles. (1) 20%. (2) 60%. (3) 70%.

8.2.5. (1) 2. (2) 6. (3) 3.

8.2.6. (1) Each positive integer m divisible by n and not exceeding x satisfies the inequalities $1 \leqslant nm \leqslant x$, i.e. $0 < m \leqslant x/n$. Since m is an integer, $1 \leqslant m \leqslant [x/n]$. Hence we obtain the statement. (2) 457. Using formula (1), we obtain $S_1 = 675$, $S_2 = 141$, $S_3 = 9$, $\hat{N}_0 = 457$. (3) 734. (4) We have $N = S_0 = 30m$, $S_1 = 10m$, $S_2 = 3m$, $S_3 = m$, $\hat{N}_0 = S_0 - S_1 + S_2 - S_3 = 22m$. (5) The number p, such that $\sqrt{n} < p \leqslant n$, is prime if and only if it is not divisible by any prime number smaller than $\sqrt{n}$. The number $N_{i_1, \ldots, i_k}$ of numbers indivisible by any prime number from $p_{i_1}, \ldots, p_{i_k}$ is $[n/p_{i_1} \times \ldots \times p_{i_k}]$,

$$S_k = \sum_{1 < p_{i_1}, \ldots, p_{i_k} \leqslant \sqrt{n}} N_{i_1, \ldots, i_k}$$

for $k \geqslant 1$, $S_0 = n$. Applying formula (2), we find that the required number is $-1 + \sum\limits_{0 \leqslant k \leqslant \sqrt{n}} (-1)^k S_k = n - 1 + \sum\limits_{i \leqslant k \leqslant \sqrt{n}} (-1)^k S_k$. The unity is subtracted since 1 is not a prime number. (6) 25.

8.2.7. $1/3^n$. Irrespective of the distribution of other elements, each element can belong to one of the sets[1] $X\overline{Y}$, $\overline{X}Y$ or $\overline{X}\overline{Y}$. (2) $n\, 2^{n-1}$. One element in U belongs to the set $X\overline{Y} \cup \overline{X}Y$. The remaining elements belong to the set XY or $\overline{X}\overline{Y}$. (3) 3^n. The equality $\overline{X} \cup Y\overline{Z} = \overline{X} \cup \overline{Y}$ is equivalent to the equality $X\overline{Y} \cup \overline{X}Y\overline{Z} = U$. Therefore, each element in U is contained in exactly one of the sets $X\overline{Y}Z$, $X\overline{Y}\overline{Z}$, $\overline{X}Y\overline{Z}$. (4) $3^n - (n^2 + 7n + 16)\, 2^{n-3} + (1/2)\, n\,(n^2 + 1) + 1$. Solution. From among the total number 3^n of pairs (X, Y) satisfying the condition of Problem 8.2.7 (1), we must exclude the pairs belonging to the set $C = A_0 \cup A_1 \cup B_0 \cup B_1 \cup B_2$, where $A_0 = \{(X, Y)\colon |X| = 0\}$, $A_1 = \{(X, Y)\colon |X| = 1\}$, $B_0 = \{(X, Y)\colon |Y| = 0\}$, $B_1 = \{(X, Y)\colon |Y| = 1\}$, $B_2 = \{(X, Y)\colon |Y| = 2\}$. Obviously, $A_0 \cap A_1 = \varnothing$, $B_i \cap B_j = \varnothing$ $(0 \leqslant i < j \leqslant 2)$. Therefore $|C| = |A_0| + |A_1| + |B_0| + |B_1| + |B_2| - |A_0 \cap B_0| - |A_0 \cap B_1| - |A_0 \cap B_2| - |A_1 \cap B_0| - |A_1 \cap B_1| - |A_1 \cap B_2|$ (see the figure). It can easily be seen that $|A_0| = 2^n$ (n elements of the set U are distributed among the squares $\overline{X}Y$ and XY. The squares XY and $X\overline{Y}$ are empty since $X = \varnothing$). Similarly, $|A_1| = n2^{n-1}$, $|B_0| = 2^n$, $|B_1| = n2^{n-1}$, $|B_2| = \binom{n}{2} 2^{n-2}$. Further, we have $|A_0 \cap B_0| = 1$, $|A_0 \cap B_1| = |A_1 \cap B_0| = n$, $|A_1 \cap B_1| = n\,(n - 1)$, $|A_0 \cap B_2| = \binom{n}{2}$, $|A_1 \cap B_2| = n \binom{n-1}{2}$. Hence $|C| = (n + 2)\, 2^n + n\,(n - 1)\, 2^{n-3} - \dfrac{n\,(n^2 + 3)}{2} - 1$, and the number of required pairs is $3^n - |C|$. (5) $n\,(2^n - 2n)$. From among the pairs (X, Y) satisfying the conditions of problem (2) we must exclude those for which the power of a set X or Y having a lower power is smaller than 2. Considering that the smaller subset is composed of $n - 1$ elements, their number is n. (6) $(n + 2)\,(2^{n-1} - 1)$. It follows from Problem 8.2.7 (3) that the elements of the set U can be contained only in the squares $X\overline{Y}Z$, $X\overline{Y}\overline{Z}$ and $\overline{X}Y\overline{Z}$. Since $|Z| \leqslant 1$, two cases are possible here: (a) $Z = \varnothing$. Then the square $X\overline{Y}Z$ is empty, and there are $2^n - 2$ ways in which n element in U can be distributed among the squares $\overline{X}\overline{Y}\overline{Z}$ and $\overline{X}Y\overline{Z}$ so that $|X| \geqslant 1$ and $|Y| \geqslant 1$, i.e. so that the squares are not empty. (b) $|Z| = 1$. Then the square $X\overline{Y}Z$ contains one element which

---

[1] The expression "the element v belongs to the set $X\overline{Y}$" is understood here and below in the sense $v \in X \cap (U \setminus Y)$, and $v \in \overline{X}\overline{Y}$ is equivalent to $v \in (U \setminus X) \cap (U \setminus Y)$.

can be chosen in n ways. The remaining $n-1$ elements must be distributed among the squares $XY\bar{Z}$ and $\bar{X}Y\bar{Z}$ so that $|X| \geqslant 1$ and $|Y| \geqslant 1$. The condition $|X| \geqslant 1$ has already been satisfied since the square $XY\bar{Z}$ is not empty. Therefore, we must fulfil the only condition that the square $\bar{X}Y\bar{Z}$ is not empty. The number of such ways of distribution of $n-1$ elements is $2^{n-1}-1$. Therefore, in case (b) we have $n(2^{n-1}-1)$ ways, and the total number of ways is $(n+2)(2^{n-1}-1)$.

8.2.8. We calculate the number $N_{n,k}$ of seating arrangements for which k given pairs of warring knights are not separated. We combine each of the given pairs into a single "object". We obtain $2n-k$ objects which can be arranged in $(2n-k-1)!$ ways. In each of the given k pairs, the adversaries can change positions. Thus, we obtain $N_{n,k} = 2^k(2n-k-1)!$, and $S_k = \binom{n}{k} N_{n,k}$. Applying formula (2), we find that the required number is

$$\sum_{k=0}^{n} (-1)^k \binom{n}{k} 2^k (2n-k-1)!.$$

8.2.9. We calculate the number $N_{n,k}$ of seating arrangements of n married couples if k given couples are not separated. We combine each of such k couples into a single "object". Then we have $2n-k$ places for arranging k "objects", the number of arrangements being $2(2n-k-1)_{k-1}$ (the seats are assumed to be numbered). After k isolated couples are arranged, we presume the evenness of the seats occupied by ladies (these seats must be all either even or odd, otherwise two ladies will be seated side by side). The seating arrangement of ladies (gentlemen) remaining after the arrangement of married couples can be realized in $(n-k)!$ ways. Using the multiplication rule, we obtain $N_{n,k} = 2(2n-k-1)_{k-1}((n-k)!)^2$. Further, $S_k = \binom{n}{k} N_{n,k}$. It remains for us to apply formula (2).

8.3

8.3.1. (1) Induction. The numbers $a_0, a_1, \ldots, a_{k-1}$ are specified by hypothesis. If all the terms a_i have already been defined for $i \leqslant n$, by using (9) we then obtain $a_{n+1} = -p_1a_n - p_2a_{n-1} - \ldots - p_ka_{n-k+1}$. (2) We must prove that $c\lambda^{n+k} + p_1c\lambda^{n+k-1} + \ldots + p_kc\lambda^n = 0$ or, which is the same, $c\lambda^n(\lambda^k + p_1\lambda^{k-1} + \ldots + p_k) = 0$. Since λ is a root of polynomial (10), the expression in the parentheses vanishes. (3) The fact that $a_n = c_1\lambda_1^n + \ldots + c_k\lambda_k^n$ satisfies relation (9) follows from (2) and from the fact that if two sequences a_n and b_n satisfy relation (9), the sequence $d_n = \alpha a_n + \beta b_n$ satisfies this relation for all α and β.

Let us now prove that any sequence a_n satisfying (9) can be represented in the form $a_n = c_1\lambda_1^n + \ldots + c_k\lambda_k^n$, where c_i are appropriate constants. In view of Problem 8.3.1 (1), any sequence a_n satisfying (9) is completely defined by its first terms $a_0, a_1, \ldots, a_{k-1}$. Therefore, it remains for us to demonstrate that for any $a_0, \ldots, a_{k-1}$ there exist $c_1, \ldots, c_k$ such that

$$\begin{cases} c_1 + c_2 + \ldots + c_k = a_0, \\ c_1\lambda_1 + c_2\lambda_2 + \ldots + c_k\lambda_k = a_1, \\ \ldots\ldots\ldots\ldots\ldots\ldots \\ c_1\lambda_1^{k-1} + c_2\lambda_2^{k-1} + \ldots + c_k\lambda_{k-1}^{k-1} = a_{k-1}. \end{cases} \qquad (*)$$

The determinant of (*) is a Vandermonde determinant (see, for example, Richard Bellman, Introduction to Matrix Analysis). It is equal to $\prod_{1 \leqslant i < j \leqslant k} (\lambda_i - \lambda_j)$ and does not vanish if $\lambda_i \neq \lambda_j$ for $i \neq j$. Consequently, the set of equations (*) has a unique solution. (4) In order to prove that any sequence of the form specified in the problem satisfies (9), it is sufficient to demonstrate that the sequence $a_n = n^m\lambda^n$, where λ is a root of multiplicity $r > m$ of polynomial (10), satisfies (9). Substituting $a_s = s^m\lambda^s$, $s = n, \ldots, n + k$, into the left-hand side of (9) and putting $p_0 = 1$, we obtain $L(n) = (n+k)^m \lambda^{n+k} + p_1 (n+k-1)^m\lambda^{n+k-1} + \ldots$

$$\ldots + p_k n^m\lambda^n = \lambda^n \left(\sum_{j=0}^{k} (n+k-j)^m \lambda^{k-j} \right) =$$

$$\lambda^n \left(\sum_{j=0}^{k} \lambda^{k-j} p_j \sum_{i=0}^{m} \binom{n}{i} (k-j)^i n^{m-i} \right) =$$

$$\lambda^n n^m \left(\sum_{i=0}^{m} \binom{m}{i} n^{-i} \sum_{j=0}^{k} p_j \lambda^{k-j} (k-j)^i \right) =$$

$$\lambda^n n^m \left(\sum_{i=0}^{m} \binom{m}{i} n^i P_i(\lambda) \right), \text{ where } P_i(x) = \sum_{j=0}^{k} p_j (k-j)^i x^{k-j}.$$ It should be noted that $P_0(x) = P(x)$, where P is polynomial (10). Since λ is an r-tuple root of the polynomial P_0, $P_0(x) = (x - \lambda)^r Q_0(x)$, where Q_0 is a polynomial. It can be easily verified that $P_{i+1}(x) = x \dfrac{d}{dx} P_i(x)$. Therefore, $P_i = (x - \lambda)^{r-i} Q_i(x)$, where Q_i is a polynomial. Hence it follows that $P_i(\lambda) = 0$ for all $i < r$. Consequently, $L(\lambda) = 0$. Thus, (9) is fulfilled for the sequence $n^m \times \lambda^n$ for $m < r$, and thereby for the sequences of the form specified in the condition of the problem.

Let us now prove that any sequence a_n satisfying (9) (provided that λ_i is a root of multiplicity r_i, $i = 1, \ldots, s$, of polynomial (10)) has the form mentioned in the formulation of the problem. Here, without loss of generality we assume that $p_k \neq 0$, i.e. zero is not a root of the characteristic polynomial. (If $p_k = 0$, Eq. (9) can be simplified.) In order to prove this, it is sufficient to show (see solution of Problem 8.3.1 (3)) that for any $a_0, a_1, \ldots, a_{k-1}$ the system

$$\begin{cases} c_{1,1} + \ldots + c_{s,1} = a_0, \\ (c_{1,1} + c_{1,2} + \ldots + c_{1,r_1-1})\lambda_1 + \ldots + (c_{s,1} \\ + \ldots + c_{s,r_s-1})\lambda_s = a_1, \\ (c_{1,1} + c_{1,2}(k-1) + \ldots + c_{1,r_1-1}(k-1)^{r_1-1})\lambda_1^{k-1} \\ + \ldots + (c_{s,1} + c_{s,2}(k-1) + \ldots + c_{s,r_s-1} \\ \times (k-1)^{r_s-1})\lambda_s^{k-1} = a_{k-1} \end{cases}$$

has a solution. For this purpose, it is sufficient to prove that its determinant differs from zero or, which is the same, that the vectors $\vec{\Lambda}_0, \ldots, \vec{\Lambda}_{k-1}$, where $\vec{\Lambda}_i = (\lambda_1^i, i\lambda_1^i, \ldots, i^{r_1-1}\lambda_1^i, \ldots, \lambda_s^i, i\lambda_3^i, \ldots, i^{r_s-1}\lambda_s^i)$, $i = 0, \ldots, k-1$, are linearly independent. We assume that the converse is true. Then there exist constants $d_0, d_1, \ldots, d_{k-1}$, which are not all equal to zero and such that $\vec{\Lambda} = d_0\vec{\Lambda}_0 + \ldots + d_{k-1}\vec{\Lambda}_{k-1} = \vec{0}$. Let $Q(x) = \sum_{i=0}^{k-1} d_i x^i$ and let $\triangle$ be an operator such that $\triangle f(x) = x\dfrac{df}{dx}$. We put $\triangle^k f = \triangle(\triangle^{k-1}f)$. Then $\vec{\Lambda} = (Q(\lambda_1), \triangle Q(\lambda_1), \ldots, \triangle^{r_1-1}Q(\lambda_1), \ldots, Q(\lambda_s), \ldots, \triangle^{r_s-1}Q(\lambda_s))$. It should be noted that $Q(\lambda_1) = \triangle Q(\lambda_1) = \ldots = \triangle^{r-1}Q(\lambda)$ for $\lambda = 0$ if and only if λ is an r-multiple root of the polynomial $Q(x)$. Thus, $\vec{\Lambda} = \vec{0}$ means that λ_1 is a root of multiplicity r_1, λ_2 is a root of multiplicity r_2, and finally λ_s is a root of multiplicity r_s of the polynomial $Q(x)$. But $r_1 + r_2 + \ldots + r_s = k$ and $Q(x)$ is a polynomial of a power lower than k, and hence it cannot have k roots. We arrive at a contradiction. Therefore, the set of equations (*) has a unique solution.

8.3.2. (1) $c_1 + c_2 3^n$. The characteristic polynomial $x^2 - 4x + 3$ has the roots $\lambda_1 = 1$ and $\lambda_2 = 3$. Using Problem 8.3.1 (3), we obtain the general solution $c_1\lambda_1^n + c_2\lambda_2^n$. (2) $c_1(-3)^{n/2} + c_2(-1)^n(-3)^{n/2}$. (3) $c_1((+\sqrt{5})/2)^n + c_2((1-\sqrt{5})/2)^n$. (4) $(-1)^n(c_1 + c_2 n)$. (5) $(c_1 + c_2 n)(-4)^n + c_3(-2)^n$. (6) $(-1)^n(c_1 + c_2 n + c_3 n^2)$.

8.3.3. (1) $7 + 3^n$. The general solution has the form $C_1 + C_2 3^n$ (see Problem 8.3.2 (1)). From the initial conditions, we obtain $c_1 + 3c_2 = 10$, $c_1 + 9c_2 = 16$. Hence $c_1 = 7$, $c_2 = 1$, $a_n = 7 + 3^n$. (2) $3^n + (\sqrt{-1})^n + (-\sqrt{-1})^n$. (3) $c_1 + c_2 n + c_3(-2)^n$, where $c_1 = (14 - b - 4c)/9$, $c_2 = (b + c - 2a)/3$, $c_3 = (2b - c - a)/18$. (4) $\cos \alpha n$. The roots of the characteristic polynomial are $\lambda_{1,2} = e^{\pm i\alpha} = \cos\alpha \pm i \sin\alpha$. The general solution is $a_n =$

$c_1\lambda_1^n + c_2\lambda_2^n$. From the equalities $a_1 = \cos\alpha$, $a_2 = \cos 2\alpha$ we obtain $c_1 = c_2 = 1/2$. Hence $a_n = (\lambda_1^n + \lambda_2^n)/2 = \cos\alpha n$. (5) $1 - (-1)^n$. (6) $3^n(1+n)$.

8.3.4. (1) Substituting $an + b$ for a_n in (11), we obtain $a(n+2) + b + p(a(n+1)+b) + q(an+b) = \alpha n + \beta$. Comparing the coefficients of n on the left- and right-hand sides, as well as the free terms, we obtain $a = \alpha/(1+p+q)$, $b = ((1+p+q)\beta - (2+p))/(1+p+q)^2$. (2) It follows from the fact that $x = 1$ is a root of the polynomial $x^2 + px + q$ that $q = -1 - p$. Substituting $n(an+b)$ for a_n in the equality $a_{n+2} + pa_{n+1} - (p+1)a_n = \alpha n + \beta$, we find that $a = \alpha/[2(p+2)]$; $b = (2\beta(p+2) - \alpha(p+4))/(2(p+2)^2)$. (3) Since $x = 1$ is a multiple root of the polynomial $x^2 + px + q$, $p = 2$ and $q = 1$. Substituting $n^2(an+b)$ for a_n in the equality $a_{n+2} + 2a_{n+1} + a_n = \alpha n + \beta$ and comparing the coefficients of n^3, n^2, n and n^0, we find that the coefficients of n^3 and n^2 are equal to 0, and $a = \alpha/6$, $b = (\beta - \alpha)/2$. (4) For (1), the general solution has the form $a_n = c_1\lambda_1^n + c_2\lambda_2^n + \alpha n/(1+p+q) + (\beta(1+q+p) - \alpha(p+2))/(1+p+q)^2$, where $\lambda_{1,2} = (-p \pm \sqrt{p^2 - 4q})/2$; for (2) $a_n = c_1(-p-1)^n + c_2 + \alpha n/[2(p+2)] + (2\beta(p+2) - \alpha(p+4))/2(p+2)^2$; for (3) $a_n = n^2((\alpha n/6) + (\beta - 2)/2) + c_1 n + c_2$.

8.3.5. (1) $a_n = 1 + \binom{n}{2}$. The general solution of the recurrence relation $a_{n+1} - a_n = 0$ is an arbitrary constant c. A particular solution of equation (1) is sought in the form $a_n^* = n(an+b)$. Substituting this expression into (1), we obtain $a_n^* = n(n-1)/2$. The general solution of equation (1) has the form of $a_n = a_n^* + c$. From the condition $a_1 = 1$, we find that $c = 1$, and hence $a_n = 1 + n(n-1)/2$. (2) $a_n = 2(-4)^n - 3 \times 2^n + 5^n$. The general solution of the homogeneous equation $a_{n+2} + 2a_{n+1} - 8a_n = 0$ has the form $c_1(-4)^n + c_2 2^n$. A particular solution of equation (2) is sought in the form $a_n^* = d \times 5^n$. Substituting this relation, we obtain $d = 1$. The general solution of the inhomogeneous equation (2) has the form $c_1(-4)^n + c_2 2^n + 5^n$. From the initial conditions, we find that $c_1 = 2$, $c_2 = -3$. (3) $1 + n + 2^n$; (4) $n(n-5) + (-2)^n$. (5) $2^{n-3}(n^2 + 8)$; (6) $(-3)^n + 2^n + n2^n$.

8.3.6. (1) A non-degenerate case is when either $q_1 \neq 0$ or $p_2 \neq 0$. If $q_1 = p_2 = 0$, we obviously have $a_n = c_1 p_1^n$, $b_n = c_2 q_2^n$. Let $q_1 \neq 0$. Then $b_n = 1/q_1(a_{n+1} - p_1 a_n)$, $b_{n+1} = 1/q_1(a_{n+2} - p_1 a_{n+1})$. Substituting b_{n+1} and b_n into the second relation, we obtain $a_{n+2} + (-p_1 - q_2)a_{n+1} + (p_1 q_2 - p_2 q_1)a_n = 0$. The problem is reduced to Problem 8.3.1. (2) $a_n = (5 + 2n)2^n$, $b_n = -(1+2n)2^n$. (3) $a_n = c_1 + (-1)^{n+1}c_2 + 5.5n$, $b_n = c_1 + 0.5 + (-1)^n c_2 + 5.5n$.

8.3.7. (1) Induction on n. For $n = 2$, the relation $F_{2+m} = F_1F_m + F_2F_{m+1} = F_m + F_{m+1}$ holds for all $m \geqslant 1$. The inductive step $n \to n+1$: $F_{n+1+m} = F_{n+m} + F_{n+m-1} = F_{n-1}F_m + F_nF_{m+1} + F_{n-2}F_m + F_{n-1}F_{m+1} = F_nF_m + F_{n+1}F_{m+1}$. (2) Carry out induction on k. (3) If F_{n+1} and F_n had a common divisor $d > 1$, then F_{n-1} and F_n would have the same common divisor since $F_{n-1} = F_{n+1} - F_n$. It follows by induction that F_1 and F_2 should have the divisor d. (4) Method of representation. If $N = 2$, then

$N = F_2 + F_1$. If $N > 2$, we choose the largest n_1 such that $F_{n_1} \leqslant N$, then the largest n_2 such that $F_{n_2} \leqslant N - F_{n_1}$, and so on. Then $N = F_{n_1} + F_{n_2} + \ldots$. Since $F_{n+1} > F_n$ for $n > 1$, the representation cannot contain two numbers with the same index $n \geqslant 2$. The representation cannot contain two adjacent numbers F_n and F_{n+1} since $F_n + F_{n+1} = F_{n+2}$ and hence F_{n+2} must be chosen at the step on which F_{n+1} is chosen. (5) The general solution of the recurrence relation $F_{n+2} = F_{n+1} + F_n$ is given in Problem 8.3.2 (3). Using the initial conditions, we obtain the required result. (6) and (7) The proof is carried out by induction on n. (8) Applying twice the identity of Problem 8.3.7 (1), we obtain $F_{3n} = F_{n-1}F_{2n} + F_nF_{2n+1} = F_{n-1}\ (F_{n-1}F_n + F_nF_{n+1}) + F_n\ (F_n^2 + F_{n+1}^2) = F_{n-1}^2F_n + F_{n-1}F_nF_{n-1} + F_n^3 + F_{n+1}^2\ (F_{n+1} - F_{n-1}) = F_{n+1}^3 + F_n^3 + F_{n-1}^2F_n + F_{n+1}F_{n-1}\ (F_n - F_{n+1}) = F_{n+1}^3 + F_n^3 + F_{n-1}^2F_n - F_{n+1}F_{n-1}^2 = F_{n+1}^3 + F_n^3 - F_{n-1}^3$.

8.3.8. (1) $(1-t)^{-1}$. The generating function for the sequence $a_n = 1$, $n = 0, 1, \ldots$, is by definition the series $A(t) = \sum_{n=0}^{\infty} t^n$ which is an expansion of the function $f(t) = (1-t)^{-1}$ into a series for $|t| < 1$. (2) $1 + t + \ldots + t^N = (t^{N+1} - 1)/(t-1)$. (3) $(1-\alpha t)^{-1}$. (4) $e^{\alpha t}$. (5) $(1+t)^{-1}$. (6) $t(1-t)^{-2}$. We have $A(t) = \sum_{n=0}^{\infty} nt^n = t\sum_{n=0}^{\infty}(n+1)t^n$. The series $\sum_{n=0}^{\infty}(n+1)t^n$ is obtained by the termwise differentiation of the series $\sum_{n=0}^{\infty} t^n$ uniformly converging to $\varphi(t) = (1-t)^{-1}$ for $|t| < 1$. Thus, the series $\sum_{n=0}^{\infty} nt^n$ converges to the function $f(t) = t\varphi'(t) = t(1-t)^{-2}$. (7) $2t^2(1-t)^{-3}$. (8) $(1+t)^m$. (9) $(1+t)^{\alpha}$. (10) $t(t+1) \times (1-t)^{-3}$. (11) $t \sin\alpha\,(1 - 2t\cos\alpha + t^2)^{-1}$. We put $\varphi(\alpha) = e^{i\alpha} = \cos\alpha + i\sin\alpha$ and use the fact that $\varphi^n(\alpha) = e^{i\alpha n} = \cos\alpha n + i\sin\alpha n$, $\varphi^n(-\alpha) = e^{-i\alpha n} = \cos\alpha n - i\sin\alpha n$. Let us consider the generating functions $A_1(t) = \sum_{n=0}^{\infty}\varphi^n(\alpha)\,t^n = (1 - \varphi(\alpha)\,t)^{-1}$, $A_2(t) = \sum_{n=0}^{\infty}\varphi^n(-\alpha)\,t^n = (1 - \varphi(-\alpha)\,t)^{-1}$. Since $\sin\alpha n = (2i)^{-1}(\varphi\ (\alpha n) - \varphi(-\alpha n))$, we have $A(t) = \sum_{n=0}^{\infty}\sin\alpha n t^n = (2i)^{-1}(A_1(t) - A_2(t)) = (2i)^{-1}((1 - \varphi(\alpha)\,t)^{-1} + (1 - \varphi(-\alpha)\,t))^{-1} = (2i)^{-1}t\,(\varphi(\alpha) - \varphi(-\alpha))\quad(1 - (\varphi(\alpha) + \varphi(-\alpha))\,t + t^2)^{-1} = t\sin\alpha\,(1 - 2t\cos\alpha + t^2)^{-1}$. (12) $(1 - t\cos\alpha)(1 - 2z\cos\alpha + t^2)^{-1}$.

8.3.9. (1) e^t. By definition, $E(t) = \sum_{n=0}^{\infty} a_n t^n/n! = \sum_{n=0}^{\infty} t^n/n!$.

This series converges to e^t. (2) $e^{\alpha t}$. (3) te^t. (4) t^2e^t. (5) $(1+t)^m$. By definition, $E(t)=\sum_{n=0}^{\infty}(m)_n t^n/n! = \sum_{n=0}^{m}\binom{m}{n} t^n=(1+t)^n$. (6) $e^t(t^2+t)$.

8.3.10. (1)-(6) Compare the coefficients of t^k.

8.3.11. (1) $\binom{m}{n} q^{m-n}p^n$. (2) 1. (3) $(-1)^n\binom{1/2}{n}$. (4) $(-1)^{n-m}\binom{m}{n-m}$. (5) $(-1)^{n-m}\sum_k(-1)^{k(r-1)}\binom{m}{k}\times\binom{-m}{n-m-kr}$. It should be noted that $(t+t^2+\ldots+t^r)=t^m(1-t^r)^m(1-t)^{-m}$. The coefficient of t^{kr} in the expansion of $(1-t^r)^m$ is $(-1)^k\binom{m}{k}$. The coefficient of t^{n-m-kr} in the expansion of $(1-t)^{-m}$ is $\binom{-m}{n-m-kr}$. (6) $(1/2)^{n/2}\binom{-m}{n/2}$ for even n and 0 for odd n. (7) $(-2)^{-n}\sum_k\binom{2k}{k}\binom{-m}{n-k}$. The coefficient of t^k in the expansion of $(1+2t)^{-1/2}$ is $\binom{-1/2}{k}2^k$. The coefficient of t^{n-k} in the expansion of $(1-t/2)^{-m}$ is $\binom{-m}{n-k}(-1/2)^{n-k}$. Therefore, the coefficient of t^n in the expansion of $f(t)$ is $\sum_k\binom{-1/2}{k}2^k(-1/2)^{n-k}\binom{-m}{n-k}=(-1/2)^n\times$ $\sum_k\binom{-1/2}{k}(-4)^k\binom{-m}{n-k}=(-1/2)^n\sum_h\binom{2k}{k}\binom{-m}{n-k}$. (8) $\binom{-m}{n-2}2^{n-2}-\binom{-m}{n-3}2^{n-3}$. (9) $(-1)^{n-1}n^{-1}$. (10) $(-1)^{n-1}/(2n-1)$. We shall use the fact that $\int_0^t\frac{dx}{1+x^2}=\arctan t$. We have $(1+x^2)^{-1}=\sum_{n=0}^{\infty}(-1)^n x^{2n}$. Integrating the left- and right-hand sides between 0 and t, find that $\arctan t=\sum_{n=0}^{\infty}\frac{(-1)^n t^{2n+1}}{2n+1}$.

(11) $\frac{1\times3\times\ldots\times(2n-1)}{2\times4\times\ldots 2n}\,\frac{1}{2n+1}$. Hint: $\int_0^t\frac{dx}{\sqrt{1-x^2}}=\arcsin t$.

(12) $(-2)^{n/2}/(n/2)!$ for even n and 0 for odd n. (13) $(-1)^{n/2}/(n/2+1)!$ for even n and 0 for odd n. (14) $(-1)^n\binom{n-1}{m-1}$.

8.3.12. (1) Let us compare the coefficients of t^{n-1} in the identity $(1+t)^n(1+t)^{-m-2}=(1+t)^{n-m-2}$. On the one hand, this coefficient is $\sum\limits_s\binom{n}{n-s}\binom{-m-2}{s-1}=\sum\limits_s(-1)^{s-1}\binom{n}{s}\binom{m+s}{m+1}$. On the other hand, $\binom{n-m-2}{n-1}=(-1)^{n-1}\binom{m}{n-1}$. (2) Consider the identity $(1+t)^m(1-t)^m=(1-t^2)^m$ and compare the coefficients of t^{2n}. (3) Consider the identity $(1-t^{-1})^m(1-t)^{-n-1}=(-1)^m t^{-m}(1-t)^{m-n-1}$ and coefficients of t^{k-m}. (4) Consider the identity $((1+t)^n+(1-t)^n))^2=(1+t)^{2n}+2(1-t^2)^n+(1-t)^{2n}$ and coefficients of t^{2m}. (5) Consider the identity $((1+t)^{-n-1}+(1-t)^{-n-1})((1+t)^{-n-1}-(1-t)^{-n-1})=(1+t)^{-2n-2}-(1-t)^{-2n-2}$ and the coefficients of t^{2n+1}. (6) Consider the identity $(1-t)^{2n}(1+2t(1-t)^{-2})=(1+t^2)^n$ and the coefficients of t^{2m}.

8.3.13. (1) $A(t)=\sum\limits_{n=0}^{\infty}a_nt^n=\sum\limits_{n=0}^{\infty}\frac{a_n}{n!}t^n\int\limits_0^{\infty}e^{-x}x^n\,dx=\int\limits_0^{\infty}e^{-x}\times\sum\limits_{n=0}^{\infty}\frac{a_n}{n!}(xt)^n\,dx=\int\limits_0^{\infty}e^{-x}E(xt)\,dx$. (2) $\int\limits_0^{\infty}e^{-x}e^{xt}\,dx=\frac{1}{1-t}\int\limits_0^{\infty}e^{-u}\,du=(1-t)^{-1}$. (3) For the sequence a_n, we have $A(t)=\sum\limits_{j\leqslant n}(n)_j\,t^n$, $\dot{E}(t)=\sum\limits_{0\leqslant j\leqslant n}\frac{(n)_j}{n!}t^n=\sum\limits_{0\leqslant j\leqslant n}\frac{t^n}{(n-j)!}$. Further, $\int\limits_0^{\infty}e^{x}E(xt)\,dx=\int\limits_0^{\infty}e^{-x}\sum\limits_{j=0}^{n}\frac{(xt)^n}{(n-j)!}\,dx=\sum\limits_{j=0}^{n}\frac{t^n}{(n-1)!}\int\limits_0^{\infty}e^{-x}x^n\,dx=\sum\limits_{j=0}^{n}(n)_j\,t^n=A(t)$.

8.3.14. (1) We use the identity $(1+t)^{a+b}=(1+t)^a(1+t)^b$. Comparing the coefficients of t^n, we find that $\binom{a+b}{n}=\sum\limits_k\binom{a}{n-k}\binom{b}{k}=\sum\limits_k\frac{1}{k!\,(n-k)!}(a)_{n-k}(b)_k$. Multiplying both sides by $n!$, we obtain the required equality. (2) We put $a'=a/h$, $b'=b/h$. Applying the identity proved in 8.3.14 (1) to a' and b', we find that $(a'+b')_n=\sum\limits_k\binom{n}{k}(a')_{n-k}(b')_k$. Multiplying both sides by h^n, we arrive at the required identity (since $(a')_s h^n=(a)_{s,h}$).

8.3.15. (1) We multiply the equality $a_n=b_n-b_{n-1}$ by t^n and take the sum over n. In the region of convergence of the series $\sum\limits_{n=0}^{\infty}a_nt^n$ and $\sum\limits_{n=0}^{\infty}b_nt^n$, the following identities are valid:

$$\sum_{n=0}^{\infty} a_n t^n = \sum_{n=0}^{\infty} (b_n - b_{n-1}) t^n = \sum_{n=0}^{\infty} b_n t^n - \sum_{n=1}^{\infty} b_{n-1} t^n = B(t)(1-t).$$

(2) We multiply the equality $a_n = b_{n+1} - b_n$ by t^{n+1} and take the sum over n between 0 and ∞. We obtain $tA(t) = B(t) - b_0 - tB(t)$. (3) We note that $b_n = a_{n-1} - a_n$. Hence $B(t) = -A(t)(1-t) + a_0$, where $a_0 = B(1)$. (4) Multiplying the equality $a_n = nb_n$ by t^n and taking the sum, we obtain $A(t) = \sum_{n=1}^{\infty} nb_n t^n = t\frac{d}{dt}B(t)$. (5) The proof is similar to that in 8.3.15 (4). (6) Let us compare the coefficients of t^n in the equality $A(t) = (1-t)^{-k} B(t)$. By definition, this coefficient on the left-hand side is a_n and on the right-hand side, $\sum_{j=0}^{n} (-1)^j \binom{-k}{j} b_{n-j} = \sum_{j=0}^{n} \binom{k+j-1}{j} b_{n-j} = S^k(b_n)$. (7) We have $B(t^{1/2}) = \sum_{n=0}^{\infty} b_n t^{n/2}$, $B(-t^{1/2}) = \sum_{n=0}^{\infty} (-1)^n b_n t^{n/2}$. Multiplying the sum of these series by 1/2, we obtain $\frac{1}{2}(B(t^{1/2}) + B(-t^{1/2})) = \sum_{n=0}^{\infty} b_{2n} t^n = \sum_{n=0}^{\infty} a_n t^n A(t)$. (8) Note that $b_n = a_{n+1} - a_n$. The equality $A(t) = B(t)\, t\, (1-t)^{-1}$ now follows from 8.3.15 (2).

8.3.16. (1) First method. Let $C(t)$ be a generating function of the sequence 1, 0, 0, ... By hypothesis, $C(t) = A(t)B(t)$. Consequently, the equalities $1 = a_0 b_0$, $0 = a_0 b_1 + a_1 b_0$, ..., $0 = \sum_{i=0}^{n} a_{n-i} b_i$, ... must hold. Since $a_n = \binom{m}{n}$, $\sum_{i=0}^{n} \binom{m}{n-i} b_i = 0$ for all $n = 1, 2, \ldots$ and $\binom{m}{0} b_0 = 1$ for $n = 0$. Consecutively, we obtain $b_0 = 1$, $b_1 = -m$, $b_2 = m(m+1)/2$, $b_3 = -m(m+1) \times (m+2)/6$. By induction on n we can easily prove that $b_n = \binom{-m}{n}$. Thus, $B(t) = (1+t)^{-m}$. Second method. $A(t) = (1+t)^m$, $B(t) = [A(t)]^{-1} = (1+t)^{-m}$. Hence $b_n = \binom{-m}{n}$. (2) $b_0 = 1$, $b_1 = -a$, $b_n = 0$ for $n > 1$, $B(t) = 1 + at$. We have $A(t) = \sum_{n \geqslant 0} (at)^n = (1-at)^{-1}$, and since $A(t)B(t) = 1$, we obtain $B(t) = 1 - at$. (3) $b_0 = b_2 = 1$, $b_1 = -2$, $b_n = 0$, $n > 2$, $B(t) = (1-t)^2$. (4) $b_{2n} = (-1)^n$, $b_{2n+1} = 0$, $n \geqslant 0$, $B(t) = (1+t^2)^{-1}$. (5) $b_0 = b_1 = 1$, $B(t) = 1 + t$. (6) $b_n = \binom{1/2}{n}$, $B(t) = \sqrt{1+t}$.

8.3.17. (1) Multiplying by t^{n+2} and taking the sum, we get $A(t) = a_1 t - a_0 + ptA(t) - pa_0 t + qA(t)t^2$. (2) We represent $A(t)$ in the form $\frac{c_1}{1-\lambda_1 t}+\frac{c_2}{1-\lambda_2 t}$. Having determined c_1 and c_2, we find that $A(t)=\frac{1}{\lambda_1-\lambda_2}\left(\frac{a_1+pa_0+\lambda_1 a_0}{1-\lambda_1 t}-\frac{a_1+pa_0+\lambda_2 a_0}{1-\lambda_2 t}\right)$. Having determined the coefficient of t^n in the expansion of $A(t)$ into a series in t, we obtain an expression for a_n. (3) Let us represent $A(t)$ in the form $\frac{c_1}{1-\lambda t}+\frac{c_2}{(1-\lambda t)^2}$. From $\frac{c_1}{1-\lambda t}+\frac{c_2}{(1-\lambda t)^2}=\frac{a_0+(a_1-2\lambda a_0)t}{(1-\lambda t)^2}$, we find that $c_1=-\frac{a_1}{\lambda}+2a_0$, $c_2=\frac{a_1}{\lambda}-a_0$. Expanding $A(t)$ into a series, we obtain $A(t)=\sum_{n=0}^{\infty}(c_1\lambda^n+c_2(n+1)\lambda^n)t^n$, $a_n=\left(a_0+n\left(\frac{a_1}{\lambda}-a_0\right)\right)\lambda^n$.

8.3.18. (1) **Hint.** Use the identity in Problem 8.1.15 (3). (2) We multiply each relation of 8.3.18 (1) by t^{n+1} and take the sum over n between 0 and ∞. Using the initial conditions, we obtain the relations between generating functions. (3) $A(t)=\frac{1-t}{1-3t+t^2}$, $B(t)=\frac{t}{1-3t+t^2}$. (4) The roots of the equation $1-3t+t^2=0$ are $\lambda_1=(3+\sqrt{5})/2$ and $\lambda_2=(3-\sqrt{5})/2$. We express $A(t)$ in the form $A(t)=\frac{a}{1-\lambda_1 t}+\frac{b}{1-\lambda_2 t}$. Since $A(t)=(1-t)/(1-\lambda_1 t)\times(1-\lambda_2 t)$, equating the right-hand sides, we get $a(1-\lambda_2 t)+b(1-\lambda_1 t)=1-t$. Hence $a=(\lambda_1-1)/(\lambda_1-\lambda_2)=(1+\sqrt{5})/(2\sqrt{5})$, $b=1-a=(\sqrt{5}-1)/(2\sqrt{5})$. Thus, $A(t)=\frac{1}{2\sqrt{5}}\left(\frac{1+\sqrt{5}}{1-\lambda_1 t}+\frac{\sqrt{5}-1}{1-\lambda_2 t}\right)$. Similarly, we obtain $B(t)=\frac{1}{\sqrt{5}}\left(\frac{1}{1-\lambda_1 t}-\frac{1}{1-\lambda_2 t}\right)$. Expanding $A(t)$ and $B(t)$ into series in t, we find that $a_n=(2\sqrt{5})^{-1}((1+\sqrt{5})\lambda_1^n+(\sqrt{5}-1)\lambda_2^n)$, $b_n=(\sqrt{5})^{-1}(\lambda_1^n-\lambda_2^n)$. Taking into account the inequalities $0<\lambda_2<\lambda_1$, we obtain $\lim_{n\to\infty} a_n\lambda_1^{-n}=(1+\sqrt{5})/2\sqrt{5}$, $\lim_{n\to\infty} b_n\lambda_1^{-n}=1/\sqrt{5}$.

8.3.19. (1) Multiplying by t^n and taking the sum of the initial relation, we get $A(t)-a_0=tA^2(t)$. (2) We have $A(t)=(2t)^{-1}(1-(1-4t)^{1/2})=(2t)^{-1}\left(1-\sum_{n=0}^{\infty}(-1)^n\binom{1/2}{n}(4t)^n\right)=2\sum_{n=1}^{\infty}(-1)^{n-1}\times\binom{1/2}{n}(4t)^{n-1}=\sum_{n=0}^{\infty}\frac{1}{n+1}\binom{2n}{n}t^n$. (3) Multiplying by t^n and taking the sum over n, we obtain for the generating function

$A(t) = \sum_{n=0}^{\infty} a_n t^n$ the equation $A^2(t) = A(2t)$. We seek its solution in the form $A(t) = e^{\alpha t}$. This function obviously satisfies the equation. Considering that $a_1 = 1$, we find that $\alpha = 1$ whence $a_n = 1/n!$. The uniqueness of the solution follows from the initial relations.

8.3.20. (1) We number the vertices of the $(n+2)$-gon by $1, 2, \ldots, n+2$ in the clockwise direction. Two cases are possible here. The first case. None of the diagonals passes through the vertex $n+2$. Then there must be a diagonal between the vertices 1 and $n+1$, and the number of ways of dividing the polygon is a_{n-1}. The second case. There exists a diagonal emerging from the vertex $n+2$. Let k be the smallest number such that the $(k+1)$-th vertex is joined to the vertex $n+2$ by a diagonal. If $k \geqslant 2$, there exists a diagonal of the form $(1, k+1)$. Then the number of divisions of the initial $(k+1)$-gon is equal to the number of divisions of the $(k+1)$-gon with the vertices $1, 2, \ldots, k+1$, multiplied by the number of divisions of the $(n-k+2)$-gon with the vertices $k+1, k+2, \ldots, n+2$, i.e. is $a_{k-1}a_{n-k}$ $(2 \leqslant k \leqslant n)$. It is natural to assume that $a_0 = 1$. This gives $a_n = \sum_{k=1}^{n} a_{k-1}a_{n-k}$. As in Problem 8.3.19, we find that $a_n = \frac{1}{n+1}\binom{2n}{n}$. (2) $a_n = \frac{1}{n+1}\binom{2n}{n}$. The problem is solved in the same way as 8.3.20 (1).

8.3.21. (1) $n!$. (2) $(-1)^{n-1}/(n-1)!$. (3) $1^2 \times 3^2 \times \ldots \times (n-2)^2/n!$ for odd n, 0 for even n. (4) 0 for odd n, $(-1)^{n/2}2^{-n}/((n/2)!)^{-1}$ for even n. (5) n^2 for odd n and 0 for even n. (6) $a_n = F_n$, where $\{F_n\} = 1, 1, 2, 3$ is a Fibonacci's sequence. We first prove that if $\{a_n\}$ satisfies the relation

$$a_{n+2} = a_{n+1} + a_n, \quad a_0 = a_1 = 1, \tag{*}$$

it also satisfies the relation

$$a_{n+1}^2 - a_n a_{n+2} = (-1)^{n-1}, \quad a_0 = a_1 = 1. \tag{**}$$

For $n = 0$ and 1, the statement is valid. Let us prove that (*) leads to the identity

$$a_{n+1}^2 - a_n a_{n+2} = a_{n+3}^2 - a_{n+2}a_{n+4}. \tag{***}$$

We have $a_{n+1}^2 - a_n a_{n+2} = (a_{n+3} - a_{n+2})^2 - (a_{n+2} - a_{n+1})\, a_{n+2} = a_{n+3}^2 - 2a_{n+3}a_{n+2} + a_{n+1}a_{n+2} = a_{n+3}^2 - 2\,(a_{n+4} - a_{n+2})\, a_{n+2} + a_{n+1}a_{n+2} = a_{n+3}^2 - a_{n+2}a_{n+4} - a_{n+2}a_{n+4} + 2a_{n+2}^2 + a_{n+2}a_{n+2} = a_{n+3}^2 - a_{n+2}a_{n+4} + a_{n+2}\,(-a_{n+4} + a_{n+2} + a_{n+2} + a_{n+1}) = a_{n+3}^2 - a_{n+2}a_{n+4} + a_{n+2}\,(-a_{n+3} + a_{n+3}) = a_{n+3}^2 - a_{n+2}a_{n+4}$.

Identity (***) is proved. Relation (**) follows from (***) by induction. Thus, $\{F_n\}$ satisfies (**). In view of uniqueness (which can be proved by induction), relation (**) has no other solutions.

8.3.22. (1) Multiplying the recurrence relation by t^k and taking the sum over k, we obtain $(1-t)A_n(t) = A_{n-1}(t)$. (2) Since $a(0, 0) = 1$ and $a(0, k) = 0$ for all $k > 0$, $A_0(t) = 1$. Carrying out induction on n and using 8.3.22 (1), we find that $A_n(t) = (1-t)^{-n}$. (3) Expanding $a(n, k)$ into a series, we obtain $a(n, k) = \binom{-n}{k}(-1)^k = \binom{n+k-1}{k}$.

8.3.23. (1) 20. (2) 6. (3) 9.

8.3.24. (1) $A(t) = (1-t^2)^{-1}(1-t^3)^{-1}(1-t^5)^{-1}$. Let us prove that $A(t)$ is the required generating function. We note that $(1-t^k) = \sum_{s=0}^{\infty} t^{ks}$ for $k = 2, 3, 5$, i.e. all non-zero expansion coefficients are equal to 1. The coefficient of t^n in the expansion of the product $(1-t^2)^{-1}(1-t^3)^{-1}(1-t^5)^{-1}$ is equal to the sum of products of the form $a_k b_l c_m$, where a_k is the coefficient of t^{2k} in the expansion of $(1-t^2)^{-1}$, b_l is the coefficient of a_l in the expansion $(1-t^3)^{-1}$, and c_m is the coefficient of t^{5m}, and $2k + 3l + 5m = n$. It should be noted that $a_k = b_l = c_l = 1$, and the number of summands is exactly equal to a_n. (2) $A(t) = (1+t^2)(1+t^3)(1+t^5)$. (3) $A(t) = (1+t^2+\ldots+t^{2p})(1+t^3+\ldots+t^{3r})(1+t^5+\ldots+t^{5s})$.

8.3.25. (1) It should be noted that $A(qt) = \prod_{k=1}^{\infty}(1+q^{k+1}t) = A(t)(1+qt)^{-1}$, and hence $A(t) = A(qt)(1+qt)$. The coefficient of t^n on the left-hand side is, by definition, a_n, and on the right-hand side, $a_n q^n - a_{n-1}q^n$. By induction on n, we hence obtain $a_n = q^{n(n+1)/2}\prod_{k=1}^{n}(q^k-1)^{-1}$, $n = 1, 2, \ldots$, $a_0 = 1$. (2) See solution to Problem 8.3.24 (1).

8.3.26. (1), (2) $S(n, k, l+1) = \sum_{\nu=0}^{n}(-1)^{n-\nu}\left[\binom{n+1}{\nu+1} - \binom{n}{\nu+1}\right](n+1+l)^k = -\sum_{\nu=1}^{n+1}(-1)^{n+1-\nu}\binom{n+1}{\nu}(\nu+l)^k + \sum_{\nu=1}^{n}(-1)^{n-\nu}\times\binom{n}{\nu}(\nu+l)^k = -S(n+1, k, l) + S(n, k, l)$. (3) $S(n, k+1, l) = \sum_{\nu=0}^{n}(-1)^{n-\nu}\binom{n}{\nu}(\nu+l)^k(\nu+l) = \sum_{\nu=0}^{n}(-1)^{n-\nu} n\binom{n-1}{\nu-1}(\nu+l)^k + lS(n, k, l) = n\sum_{\nu=0}^{n}(-1)^{n-\nu}\left(\binom{n}{\nu} - \binom{n-1}{\nu}\right) + lS(n, k, l) = (n+l)S(n, k, l) + nS(n-1, k, l)$. (4) The proof is carried out by induction on k and by using 8.3.26 (3). For any l and $n > 0$, we have $S(n, 0, l) = \sum_{\nu=0}^{n} (-1)^{n-\nu} \binom{n}{\nu} = 0$. Let the statement be correct for a certain $k \geqslant 0$, any $n > k$ and any l. Let $k + 1 < n$. From 8.3.26 (3), we have $S(n, k+1, l) = (n+l) S(n, k, l) + nS(n-1, k, l)$. Since $k < n - 1$, in view of inductive hypothesis $S(n, k, l) = S(n-1, k, l) = 0$, and hence $S(n, k+1, l) = 0$. (5) Induction on n. We have $S(0, 0, l) = l^0 = 1 = 0!$. Let $S(n, n, l) = n!$ for a certain $n \geqslant 0$ and any l. Using (3) and (4) of this problem, we obtain $S(n+1, n+1, l) = (n+1+l) \times S(n+1, n, l) + (n+1) S(n, n, l) = (n+1) S(n, n, l) = (n+1)!$. (6) Induction on k. In view of 8.3.26 (5), we have $S(n, n, l) = n! > 0$ for all n and all l. Let $S(n, k, l) \geqslant 0$, for a certain $k > n$ and any n and l. Using 8.3.26 (3), we get $S(n, k+1, l) = (n+l) S(n, k, l) + nS(n-1, k, l) > 0$. (7) This follows from parts (1), (2) and (6) of this problem. (8) This follows from (3) and (5). (9) This can be derived from 8.3.26 (3) and (4) by induction on k.

8.3.27. (1) In view of 8.3.26 (9), we have $\sigma_1(t) = \sum_{k=0}^{\infty} S(1, k, 0) t^n = \sum_{k=1}^{\infty} t^n = t(1-t)^{-1}$. Further, in view of 8.3.26 (3) we have $S(n, k+1) = nS(n, k) + nS(n-1, k)$. Multiplying both sides of this expression by t^{k+1} and taking the sum over k, we obtain $\sigma_n(t)(1-nt) = nt\sigma_{n-1}(t)$. The statement can now be proved by induction on n. (2) For $n=1$, the right-hand sides of formulas in (1) and (2) for σ_n coincide. If we prove that for $\sigma_n(t) = t \sum_{k=1}^{n} (-1)^{n-k} k \binom{n}{k} (1-kt)^{-1}$ the recurrence relation $\sigma_{n-1}(t) = ((1-nt)/nt)\sigma_n(t)$ is valid, the statement will follow from it by induction. We have $\frac{1-nt}{nt} \sigma_n(t) = \frac{1-nt}{nt} \times$

$$t \sum_{k=1}^{n} (-1)^{n-k} k \binom{n}{k} (1-kt)^{-1} = \sum_{k=1}^{n} (-1)^{n-k} \binom{n-1}{k-1} \frac{1-nt}{1-kt} =$$

$$\sum_{k=1}^{n} (-1)^{n-k} \binom{n-1}{k-1} \left(1 - \frac{(n-k)t}{1-kt}\right) = \sum_{k=1}^{n-1} (-1)^{n-k+1} \binom{n-1}{k-1} \times$$

$$\frac{(n-k)t}{1-kt} = t \sum_{k=1}^{n-1} (-1)^{n-k-1} k \binom{n-1}{k} (1-kt)^{-1} = G_{n-1}(t).$$

8.4.

8.4.1. (1) (0, 0, 0, 1). (2) (2, 1, 0, 0). (3) (0, 0, 2, 0, 0, 0). (4) (2, 0, 0, 0, 0, 1, 0, 0). **8.4.2.** The transposition of the elements i and j will be denoted by (i, j). Then $(2, 3, 4, 1) = (1, 2) \times$

(2, 3) (3, 4); (4, 2, 3, 1) = (1, 4) (1, 2) (1, 3); (3, 4, 5, 6, 1, 2) = (1, 5) (2, 6) (3, 5) (4, 6); (8, 2, 1, 7, 4, 6, 3, 5) = (1, 3) (3, 7) × (4, 5) (5, 8) (8, 7).

8.4.3. (1) There exist four rotations of the square in the plane, which transform the former into itself: by 0°, by 90°, by 180° and by 270°. These rotations correspond to permutations (1, 2, 3, 4), (2, 3, 4, 1), (3, 4, 1, 2), (4, 1, 2, 3). The first permutation is of the (4, 0, 0, 0) type, while the remaining ones are of the (0, 0, 0, 1) type. The cycle index has the form $P_G(t_1, t_2, t_3, t_4) = (t_1^4 + 3t_4)/4$. (2) In addition to the permutations mentioned in 8.4.3 (1), we also have four permutations (1, 4, 3, 2), (3, 2, 1, 4), (2, 1, 4, 3), (4, 3, 2, 1) corresponding to the rotations of the square about the diagonals. Two of them are of the type (2, 1, 0, 0), while the other two are of the type (0, 0, 0, 1). Hence $P_G = (t_1^4 + 5t_4 + 2t_1^2t_2)/8$. (3) There exist 12 rotations of the tetrahedron: identical, eight rotations by 120° about the axis passing through its vertex and the centre of the opposite face, and three rotations about an axis passing through the midpoints of opposite edges. Hence $P_G = (t_1^4 + 8t_1t_3 + 3t_2^2)/12$. (4) $P_G = (t_1^4 + 8t_3^2 + 3t_1^2t_2^2)/12$. (5) $P_G = (t_1^4 + 8t_1t_3 + 3t_2^2)/12$. (6) $P_G = (t_1^3 + 3t_1t_2 + 2t_3)/6$. (7) $P_G = (t_1^4 + 8t_1t_3 + 3t_2^2)/12$. (8) From 24 rotations of the cube, one is identical, three are rotations by 180° and six are rotations by 90° about the straight lines passing through the centres of opposite faces, six are rotations by 180° about the straight lines passing through the midpoints of opposite edges, and eight are rotations by 120° about axes connecting opposite vertices. The identical permutation gives six cycles of length 1, the three permutations corresponding to the rotations by 180° give two cycles of length 1 and two cycles of length 2, the six permutations corresponding to the rotations by 90° give two cycles of length 1 and one cycle of length 4, and the other six permutations corresponding to the rotations by 180° give three cycles of length 2. The eight permutations corresponding to the rotations by 120° give two cycles of length 3. Hence $P_G = (t_1^6 + 3t_1^2t_2^2 + 6t_1^2t_4 + 6t_2^3 + 8t_3^2)/24$. (9) We have five types of rotation: by 0° (state of rest) corresponding to the identity permutation, the rotation by 90° about the diagonal, the rotation by 180° about the diagonal, the rotation by 180° about the straight line passing through the midpoints of opposite faces, and the rotations by 120° about the straight line passing through the midpoints of opposite faces. The identity permutation corresponding to the state of rest gives the contribution to Z_G equal to t_1^6, the second-type permutation gives $t_1^2t_4$, the third-type permutation $t_1^2t_2^2$, the fourth-type permutation t_2^3, and the fifth-type permutation t_3^2. Each of these types corresponds to 1, 6, 3, 6 and 8 rotations respectively. Hence $P_G = (t_1^6 + 6t_1^2t_4 + 3t_1^2t_2^2 + 6t_2^3 + 8t_3^2)/24$.

8.4.4. (1) We prove this statement directly by using Bernside's lemma. The group G consists of permutations $\pi_1 = (1, 2, 3, 4)$, $\pi_2 = (2, 3, 4, 1)$, $\pi_3 = (3, 4, 1, 2)$, $\pi_4 = (4, 1, 2, 3)$. It is immediately seen that for any pair $1 \leqslant i, j \leqslant 4$ there exists a permutation π such that $\pi_i = j$. Thus, all the elements are equivalent, and we have one equivalence class.

Let us obtain the result by using Bernside's lemma. We have $|G| = 4$, $b_1(\pi_1) = 4$, $b_1(\pi_i) = 0$ for $i = 2, 3, 4$. Hence $\nu(G) =$

$(4 + 0 + 0 + 0)/4 = 1$. (2) Note that the elements 1 and 2 are transformed into each other by permutation π_3, while permutation π_2 transforms elements 3 and 4 into each other, but none of the permutations transforms the elements of the set $\{1, 2\}$ into the elements of the set $\{3, 4\}$. Thus, we have two orbits. Applying Bernside's lemma, we obtain $|G| = 4$, $b_1(\pi_1) = 4$, $b_1(\pi_2) = b_1(\pi_3) = 2$, $b_1(\pi_4) = 0$, $\nu(G) = (4 + 2 + 2 + 0)/4 = 2$.

8.4.5. We must prove that $|G|\,\nu(G) = \sum_{\pi \in G} b_1(\pi)$, where $|G|$ is the order (number of elements) of the group G, $\nu(G)$ is the number of classes of G-equivalence (orbits) on the set Z_n, and $b_1(\pi)$ is the number of elements which do not change places upon permutation π. We put $G_{y\to x} = \{\pi \in G\colon \pi y = x\}$. If $M \subseteq Z_n$ is a certain orbit and $x \in M$, then $G = \bigcup_{y \in M} G_{y\to x}$. In this case, we obviously have $G_{v\to x} \cap G_{y\to x} = \varnothing$ for $v \neq y$. Note that $|G_{y\to x}| = |G_{x\to x}|$ if $y \sim x$, i.e if y belongs to the same orbit as x does. Indeed, if $\sigma \in G_{y\to x}$ and $G_{x\to x} = \{\pi_1, \ldots, \pi_m\}$, then $\{\sigma\pi_1, \ldots, \sigma\pi_m\} \in G_{y\to x}$ where $\sigma\pi_i \neq \sigma\pi_j$ for $i \neq j$. On the other hand, if $G_{y\to x} = \{\sigma_1, \ldots, \sigma_k\}$ and $\sigma \in G_{y\to x}$, then $\{\sigma^{-1}\sigma_1, \ldots, \sigma^{-1}\sigma_k\} \subseteq G_{x\to x}$ and $\sigma^{-1}\sigma_i \neq \sigma^{-1}\sigma_j$ for $i \neq j$. Hence it follows that $|G_{y\to x}| = |G_{y\to x}|$. Let now $M_1, \ldots, M_{\nu(G)}$ be orbits and $x_i \in M_i$. The

$$\sum_{x \in G} b_1(\pi) = \sum_{x \in Z_n} |G_{x\to x}| = \sum_{i=1}^{\nu(G)} \sum_{y \in M_i} |G_{y\to x_i}| = \sum_{i=1}^{\nu(G)} |G| = |G|\,\nu(G).$$

8.4.6. (1) Each permutation π in S_n of the type $\vec{b} = (b_1, b_2, \ldots, b_n)$ can be represented in the form of the product of cycles so that the length of the cycles does not decrease: $\pi = (i_1)(i_2)\ldots(i_{b_1})(i_{b_1+1}, i_{b_1+2})\ldots$. Such two notations can generally lead to different permutations. This can be in two cases: (a) when identical cycles occupy different places in these notations, and (b) when the cycles are equivalent (as the cycles of a permutation) but start with different elements (e.g. (1 2 3) and (2 3 1)). The first reason leads to the repetition of the same permutation $\prod_{k=1}^{n} b_k!$ times, while the second reason, to the repetition of $\prod_{k=1}^{n} k^{b_k}$ times. These reasons are independent. (2) By definition

$$P_{S_n}(t_1, \ldots t_n) = (n!)^{-1} \sum_{\pi \in S_n} t_n^{b_1(\pi)} \ldots t_n^{b_n(\pi)} = (n!)^{-1} \times \sum_{\vec{b}} \sum_{\pi \in H(\vec{b})} t_1^{b_1} \ldots t_n^{b_n} = (n!)^{-1} \sum_{\vec{b}} h(\vec{b}) \prod_{i-1}^{n} t_i^{b_i}.$$ (3) We have

$\exp\left(t_1 x + t_2 \frac{x_2}{2} + t_3 \frac{x^3}{3} + \ldots\right) = \prod_{k=1}^{\infty} \exp\ \{t_k x^k/k\} =$

$\prod_{k=1}^{\infty}\left(\sum_{b_k=0}^{\infty} x^{kb_k} t_k^{b_k} k^{-b_k} (b_k!)^{-1}\right)$. The coefficient of x^n is equal to the sum over possible non-negative integers $b_1, b_2, \ldots,$ such that $b_1 + 2b_2 + \ldots + nb_n = n$, of terms of the form $\prod_{k=1}^{n} t_k^{b_k} (b_k! k^{b_k})^{-1}$.

8.4.7. It should be noted that a cycle of an even length is an odd permutation, while a cycle of an odd length is an even permutation. Each odd permutation of the type $(b_1, \ldots, b_n)$ makes to the expression $P_{S_n}(t_1, \ldots, t_n)$ a contribution $t_1^{b_1} \ldots t_n^{b_n}$ and to the expression $P_{S_n}(t_1, -t_2, \ldots, (-1)^{n-1}t_n)$ a contribution $t_1^{b_1} \ldots t_n^{b_n}$. Therefore, the terms on the right-hand side of the equality to be proved, which correspond to odd permutations, are cancelled out, while the terms corresponding to even permutations are doubled. Besides, it can be easily seen that $|A_n| = |S_n|/2 = (n!)/2$. Hence follows the statement.

8.4.8. (1) We must prove that the set $\{\pi \times \sigma\}$ with the multiplication operation forms a group, that is

(1) there exists a unit element;
(2) each element has its inverse;
(3) the operation is associative.

The unit element is obviously $\pi_0 \times \sigma_0$ where π_0 (σ_0) is a unit permutation in the group G (respectively, in the group H). The inverse element to $\pi \times \sigma$ is a permutation $\pi^{-1} \times \sigma^{-1}$, where π^{-1} and σ^{-1} are the corresponding inverse elements. It remains to prove that $(\pi_1 \times \sigma_1) \times ((\pi_2 \times \sigma_2) \times (\pi_3 \times \sigma_3)) = ((\pi_1 \times \sigma_1) \times (\pi_2 \times \sigma_2)) \times (\pi_3 \times \sigma_3)$. Let us consider the action of a permutation on an element $v \in X \cup Y$. Without loss of generality, we can assume that $v \in X$. Then $(\pi_1 \times \sigma_1) \times ((\pi_2 \times \sigma_2) \times (\pi_3 \times \sigma_3))\, v = \pi_1 (\pi_2\pi_3)\, v$, $((\pi_1 \times \sigma_1) \times (\pi_2 \times \sigma_2)) \times (\pi_3 \times \sigma_3)\, v = (\pi_1\pi_2)\, \pi_3 v$, and the associativeness follows from the fact that the multiplication operation in the group G is associative. The order of the group is obviously equal to $|G|\,|H|$ since the elements π and σ of the permutation $\pi \times \sigma$ are chosen independently, and the two permutations $\pi_1 \times \sigma_1$ and $\pi_2 \times \sigma_2$ are obviously different if $\pi_1 \neq \pi_2$ or $\sigma_1 \neq \sigma_2$. (2) If b_i (c_i) is the number of cycles of length i of the permutation π (σ) on the set X (Y), then the number of cycles of length i of the permutation $\pi \times \sigma$ is equal to $b_i + c_i$. (3) We have

$$P_{G\times H}(t_1, \ldots t_n) = \frac{1}{|G \times H|} \sum_{\tau \in G\times H} t_1^{b_1(\tau)} \ldots t_n^{d_n(\tau)} = \frac{1}{|G|\,|H|} \times$$

$$\sum_{\pi\times\sigma} t_1^{b_1(\pi)+c_1(\sigma)} \ldots t_n^{b_n(\pi)+c_n(\sigma)} = \left(\frac{1}{|G|} \sum_{\pi \in G} t_1^{b_1(\pi)} \ldots t_n^{b_n(\pi)}\right) \times$$

$$\left(\frac{1}{|H|} \sum_{G\in H} t_1^{c_1(\sigma)} \ldots t_n^{c_n(\sigma)}\right) = P_G \times P_H.$$

8.4.9. (1) 20. The necklace of seven beads can be coloured in two colours in $2^7 = 128$ ways. We have a set G of seven different rotations $\pi_1, \ldots, \pi_7$ transforming the necklace into itself. The type of identity permutation is (7, 0, 0, 0, 0, 0, 0), while any other permutation is of the type (0, 0, 0, 0, 0, 0, 1). The cycle index is $P_G(t_1, \ldots, t_7) = (t_1^7 + 6t_7)/7$. According to Polya's theorem, the number of different equivalence classes is $P_G(2, 2, 2, 2, 2, 2, 2) = 20$. (2) The cycle index is $P(t_1, \ldots, t_n) = (t_1^n + (n-1)\,t_n)/n$. The number of necklaces is $P(k, \ldots, k) = (k^n + (n-1)\,k)/n$.

8.4.10. (1) The cycle index is (see Problem 8.4.3 (3)) $P_G = (t_1^4 + 8t_1t_3 + 3t_2^2)/12$. According to Polya's theorem, the number of colourings is $P_G(2, 2, 2, 2) = 5$. (2) The cycle index P_G can be taken from Problem 8.4.3 (9). The number of colourings is $P_G(3, 3, 3, 3, 3, 3) = 54$. (3) Let M be the set of cube faces, G the rotation group and N the set consisting of three colours: red, blue and white. We ascribe the weight x to red, y to blue and z to white colour. The cycle index (see Problem 8.4.3 (8)) is $P_G = (t_1^6 + 3t_1^2t_2^2 + 6t_1^2t_4 + 6t_2^3 + 8t_3^2)/24$. According to Polya's theorem, the function counting series is $P_G(f_1, f_2, f_3, f_4)$, where $f_k = x^k + y^k + z^k$, $k = 1, 2, 3, 4$. Therefore $P_G(f_1, f_2, f_3, f_4) = \frac{1}{24}((x + y + z)^6 + 3(x + y + z)^2(x^2 + y^2 + z^2)^2 + 6(x + y + z)^2 \times (x^4 + y^4 + z^4) + 6(x^2 + y^2 + z^2)^3 + 8(x^3 + y^3 + z^3)^2)$. The number of different colourings for which three faces are red, two are blue and one is white is equal to the coefficient $c_{3,2,1}(P_G)$ of $x^3y^2z^1$ in $P_G(f_1, f_2, f_3, f_4)$. We have $c_{3,2,1}P_G(f_1, f_2, f_3, f_4) = \frac{1}{24}c_{3,2,1}((x + y + z)^6 + 3(x + y + z)^2(x^2 + y^2 + z^2)^2) = \frac{1}{24}\left(\frac{6!}{3!2!1!} + 3 \times 2 \times 2\right) = 3$.

8.4.11. $P_G(N, N, \ldots, N)$. It follows from Polya's theorem.

8.4.12. (1) If a rooted tree has $k > 1$ vertices, it has $1 \leqslant n < k$ edges incident with the root. The subtrees "planted" on edges which are incident with the root will be referred to as branches of the tree. If the branches are isomorphic to one another, by transposing them we can obtain a tree isomorphic to the initial one. The permutation group of branches is a symmetry group of order n, and the figure counting series (branches) coincides with the rooted trees counting series $T(x)$. The result follows from Polya's theorem.

8.4.13. The solution is similar to that of the previous problem. If the connectivity components of a graph are isomorphic to one another, they can be transposed to give a graph isomorphic to the initial one. For a graph with n components, the permutation group of components coincides with S_n. The figure counting series is $l(n)$. The result follows from Polya's theorem.

8.4.14. (1), (2). Let π be a permutation of n elements, having k cycles. The element n in it may form a unit cycle. Then such a permutation can be put in a one-to-one correspondence with a permutation of $n - 1$ elements with $k - 1$ cycles (the number of such permutations is equal to $P(n - 1, k - 1)$). If the element n does not form a unit cycle, the permutation π can be obtained from a certain permutation of $n - 1$ elements having k cycles each by including the element into a certain cycle. For a fixed permutation of $n-1$ elements with k cycles, such an inclusion can be carried out in $n-1$ ways (the element n cannot be placed first since n is the largest element). Hence it follows that $P(n, k) = P(n-1, k-1) + P(n-1, k)(n-1)$. Multiplying the equality by t^k and taking the sum over k, we obtain the relation $\mathcal{P}_n(t) = (t+n-1)\mathcal{P}_{n-1}(t)$ between the generating functions. By induction taking into account the equality $P_1(t) = t$, we obtain

$$\mathcal{P}_n(t) = t(t+1)\ldots(t+n-1).$$

8.5.

8.5.1. (1) Note that $n \leqslant (i+1)(n-i) < ((n+1)/2)^2$ for $0 \leqslant i < n$. Hence $n^{n/2} \leqslant \prod_{1\leqslant i\leqslant n/2} (i+1)(n-i) \leqslant n! \leqslant \frac{n+1}{2} \prod_{1\leqslant i<n/2} (i+1)(n-i) \leqslant ((n+1)/2)^n$. (2) Note that $(i+1)(2n-i) \leqslant n(n+1)$. (3) We have $\left(1+\frac{1}{n}\right)^n = \sum_{i=0}^{n} \binom{n}{k} n^{-k} \leqslant \sum_{k=0}^{n} \frac{1}{k!} \leqslant 2 + \sum_{k=1}^{n} \frac{1}{2^k} < 3$. (4) Induction on n. For $n=1$, the inequality is valid. If $(n/3)^n < n!$ then $((n+1)/3)^{n+1} = (n/3)^n((n+1)/3)((n+1)/n)^n \leqslant$ (see (3)) $\leqslant (n/3)^n(n+1) \leqslant (n+1)!$ (5) Note that $(n!)^2 \leqslant (2n)!\,2^{-n}$ and use the result of Problem 8.5.1 (2). (6) Use Cauchy's inequality $\sqrt[k]{a_1a_2\ldots a_k} \leqslant (a_1+a_2+\ldots+a_k)/k$, $a_i \geqslant 0$ and the fact that the arithmetic mean of the multipliers on the left-hand side of the inequality is $\frac{2}{n(n+1)}\sum_{i=1}^{n} i^2 = \frac{2n+1}{3}$. (7) Putting $a_n = (2n-1)!!\times\sqrt{3n+1}\,(2n!!)^{-1}$, prove that $a_{n+1}^2/a_n^2 < 1$. Further, carry out induction on n. (8) Note that $i(2n-i) < n^2$ for $i < n$. (9) $e^n = \sum_{k=0}^{\infty} n^k/k! > n^n/n!$. (10) For $n=1$, the inequality is obviously valid. We assume that $(1+\alpha)^n > 1+\alpha n$ for all $-1 \leqslant \alpha$ and a certain $n \geqslant 1$. Then $(1+\alpha)^{n+1} = (1+\alpha)(1+\alpha)^n \geqslant (1+\alpha)(1+\alpha n) = 1+\alpha(n+1)+\alpha^2 n \geqslant 1+\alpha(n+1)$. (11) Expanding both sides of the inequality according to Newton's binomial theorem, we compare the terms of the expansion with the same numbers. We have $\binom{n}{k} n^{-k} = \binom{n+1}{k}(n+1)^{-k}$ for $k=0, 1$. Further, the ratio $\binom{n}{k} n^{-k} \Big/ \binom{n+1}{k}(n+1)^{-k}$ is equal to $(1-k/(n+1))(1-1/(n+1))^k$. In view of 8.5.1 (10), the ratio does not exceed 1. Therefore, each term of expansion of $(1+1/n)^n$ does not exceed the corresponding term in the expansion of $(1+1/(n+1))^{n+1}$.

Besides, in the latter expansion we also have the $(n+2)$-th term $(n+1)^{-(n+1)}>0$. Hence follows a strict inequality.

8.5.2. (1) We put $a_k=\left(\frac{2(n-k)}{k+1}\right)^k\Big/\binom{n}{k}$ and verify that $a_{k+1}/a_k<1$ for $k\geqslant 1$. Since $a_1=1$, it follows that the first inequality holds. The second inequality follows from Problem 8.5.1 (9). (2) $(n/k)^k\leqslant (n)_k/k!$ for $1\leqslant k\leqslant n$. The second inequality can be obtained by using the ratio a_{k+1}/a_k, where $a_k=\binom{n}{k}k^k(n-k)^{n-k}n^{-n}$. We have $a_{k+1}/a_k=(1+(n-k-1)^{-1})^{n-k-1}(1+1/k)^{-k}$. Since $(1+1/m)^m$ increases monotonically with m (see Problem 8.5.1 (11)), $a_{k+1}/a_k\geqslant 1$ for $k\leqslant (n-1)/2$. Considering that $a_1<1$, we find that the inequality $a_k<1$ is valid for $k\leqslant (n-1)/2$. For $k>n/2$, the inequality follows from symmetry considerations. (3) We prove the first inequality by induction on n. For $n=1$, the inequality holds. Assuming that it is valid for a certain $n\geqslant 1$, we have

$$\binom{2(n+1)}{n+1}=2\,\frac{2n+1}{n+1}\binom{2n}{n}>2\,\frac{2n+1}{n+1}\,\frac{4n}{2\sqrt{n}}$$

$$>\frac{2(2n+1)}{\sqrt{n+1}\,n}\,4^n>\frac{1}{2\sqrt{n+1}}\,4^{n+1}.$$

The second inequality can be proved by induction and using the result of Problem 8.5.1 (7).

8.5.3. (1) $(2n-1)!!=(2n)!\,(n!)^{-1}2^{-n}\backsim\sqrt{2\pi(2n)}\,(2n)^{2n}\times e^{-2n}(2\pi n)^{-1/2}n^{-n}e^n2^{-n}\backsim\sqrt{2}\,n^ne^{-n}2^n$. (2) $\binom{2n}{n}=(2n)!\,(n!)^{-2}\sim\sqrt{4\pi n}\,(2n)^{2n}e^{-2n}\,(2\pi n)^{-1}n^{-2n}e^{2n}\sim(\pi n)^{-1/2}4^n$. (3) $n!\,([n/3]!)^{-2}\times((n-2[n/3])!)^{-1}\sim(2\pi n)^{1/2}n^ne^{-n}(2\pi\,[n/3])^{-1}\,[n/3]^{-2[n/3]}\,e^{2[n/3]}\times(2\pi\,(n-2[n/3]))^{-1/2}\,(n-2[n/3])^{-n+2[n/3]}\,e^{n-2[n/3]}\sim(2\pi)^{-1}\times[n/3]^{-1}\,n\,(n-2[n/3])^{-1/2}\,n^n\,[n/3]^{-2[n/3]}\,(n-2[n/3])^{-n+2[n/3]}$. We note that $(n/3)-1<[n/3]\leqslant n/3$, $n/3<n-2[n/3]<n/3+2$. We have $[n/3]^{-1}(n-2[n/3])^{-1/2}\geqslant((n/3)-1)^{-1}(n/3)^{-1/2}\geqslant(n/3)^{-3/2}\times(1-3/n)^{-1}\sim(n/3)^{-3/2}$; $[n/3]^{-1}(n-2[n/3])^{-1/2}\leqslant(n/3)^{-1}((n/3)-2)^{-1/2}=(n/3)^{-3/2}(1-6/n)^{-1/2}\sim(n/3)^{-3/2}$. We put $\alpha=n/3-[n/3]$, $0\leqslant\alpha<1$. This gives $[n/3]^{2[n/3]}(n-2[n/3])^{n-2[n/3]}=((n/3)-2)^{2[n/3]}\,((n/3)+2\alpha)^{n-2[n/3]}=(n/3)^n\,(1-3\alpha/n)^{2[n/3]}\,(1+6\alpha/n)^{n-2[n/3]}$. Further $(1-3\alpha/n)^{2[n/3]}=(1-3\alpha/n)^{2((n/3)-\alpha)}=(1-3\alpha/n)^{2n/3}(1-3\alpha/n)^{-2\alpha}\sim(1-3\alpha/n)^{2n/3}\sim e^{-2\alpha}$. Similarly, $(1+6\alpha/n)^{n-2[n/3]}\sim e^{2\alpha}$. Hence we obtain the required asymptotic equality. (4) $\frac{(m+1)\ldots(m+n)}{(k+1)\ldots(k+n)}=\frac{k!}{m!}\,\frac{(n+m)!}{(n+k)!}\sim\frac{k!}{m!}\times\sqrt{\frac{n+m}{n+k}}\,(n+m)^{n+m}\,(n+k)^{-(n+k)}\,e^{k-m}\backsim\frac{k!}{m!}\,n^{m-k}e^{k-m}\,(1+m/n)^{n+m}\,(1+k/n)^{-n-k}\sim\frac{k!}{m!}\,n^{m-k}e^{k-m}\left(1+\frac{m}{n}\right)^n\left(1+\frac{k}{n}\right)^{-n}\times$

$\left(1+\frac{m}{n}\right)^m\left(1+\frac{k}{n}\right)^{-k}\sim\frac{k!}{m!}n^{m-k}$. In the last case, we have used asymptotic equalities of form $\left(1+\frac{m}{n}\right)^n\sim e^m$, $(1+(m/n))^m\sim e^{m^2/n}\sim 1$ for $m^2=o(n)$. (5) $(2n)!!\,((2n-1)!!)^{-1}=(2n)!\,((2n-1)!!)^{-2}$. Then we can use Stirling's formula and the result of 8.5.2 (1).

8.5.4. (1) $\int\limits_0^1(1+t)^n\,dt=\frac{1}{n+1}(1+t)^{n+1}\Big|_0^1=\frac{2^{n+1}}{n+1}-\frac{1}{n+1}$.

On the other hand, $\int\limits_0^1[(1+t)^n\,dt=\int\limits_0^1\sum\limits_{k=0}^n\binom{n}{k}t^k=\sum\limits_{k=0}^n\binom{n}{k}\times\int\limits_0^1 t^k\,dt=\sum\limits_{k=0}^n\frac{1}{k+1}\binom{n}{k}$. (2) Using the result of Problem 8.1.19 (3), we obtain

$$\sum_\nu\binom{n}{k\nu+r}=\frac{1}{k}\sum_{\nu=0}^{k-1}e^{-\frac{2\pi ir\nu}{k}}\left(1+e^{\frac{2\pi i\nu}{k}}\right)^n$$

$$=\frac{1}{k}\,2^n+\sum_{\nu=1}^{k-1}e^{\frac{-2\pi ir\nu}{k}}\left(1+e^{\frac{2\pi i\nu}{k}}\right)^n.$$

It should be noted that $|\exp\{-2\pi ir\nu/k\}|=1$ and $|1+\exp\{2\pi i\nu/k\}|=\left|1+\cos\frac{2\pi\nu}{k}+i\sin\frac{2\pi\nu}{k}\right|=\left|2\cos\frac{\pi\nu}{k}\left(\cos\frac{\pi\nu}{k}+i\sin\frac{\pi\nu}{k}\right)\right|=2\left|\cos\frac{\pi\nu}{k}\right|\leqslant 2\cos\frac{\pi}{k}$. Hence

$$\left|\sum_{\nu=1}^{k-1}e^{-\frac{2\pi ir\nu}{k}}\left(1+e^{\frac{2\pi i\nu}{k}}\right)^n\right|\leqslant(k-1)\left(2\cos\frac{\pi}{k}\right)^n=o\left(\frac{2^n}{k}\right).$$

(3) Use the result of Problem 8.1.20 (1), noting that $|1+\alpha e^{2\pi\nu/k}|\leqslant 1-\alpha+2\alpha\cos\frac{\pi\nu}{k}<1+\alpha$ for $0<\nu<k$. (5) Using Problems 8.1.18 (9) and 8.5.3 (2), we obtain

$$\sum_{k=1}^n k\binom{n}{k}^2=n\sum_{k=1}^n\binom{n-1}{k-1}\binom{n}{n-k}=n\binom{2n-1}{n-1}$$

$$=\frac{n}{2}\binom{2n}{n}\sim\frac{1}{2}\sqrt{\frac{n}{\pi}}\,4^n.$$

8.5.5. (1) Yes, they are. The validity of the inequalities follows from the fact that

$$\binom{n}{k} a^k \leqslant \binom{n}{k} b_k \leqslant \binom{n}{k} c^k.$$

(2) No, they are not. A counterexample is a sequence $\{b_k\}$ such that $b_k = 2 \times 3^{-k}$ for even k and $b_k = 3^{-k}$ for odd k. Putting $a = 1/3$ and $c = 2/3$, we obtain $0 < a^k \leqslant b_k < c^k < 1$. However, in view of 8.1.20 (1), we find that

$$\sum_{k=0}^{n} (-1)^k \binom{n}{k} b_k = \sum_{k=0}^{n} (-1)^k \binom{n}{k} \left(\frac{1}{3}\right)^k$$

$$+ \sum_{s} \binom{n}{2s} \left(\frac{1}{3}\right)^{2s} = (1-1/3)^n$$

$$+ \frac{1}{2}\left((1+1/3)^n + \left(1-\frac{1}{3}\right)^n\right) > (1-a)^n.$$

8.5.6. (1) We have $Da = \frac{1}{n} \sum_{i=1}^{n} (a_i - \bar{a}) \geqslant \frac{1}{n} \sum_{a_i : |a_i - \bar{a}| \geqslant t} (a_i - \bar{a})^2 \geqslant \delta_t t^2$. Hence $\delta_t \leqslant Da/t^2$. (2) Let us consider a set $A = \{a_0, a_1, \ldots, a_{2^n-1}\}$ in which a_ν is the number of unities in the binary vector $\tilde{\alpha} = (\alpha_1, \ldots, \alpha_n)$ having the number ν. It should be noted that

$$\bar{a} = 2^{-n} \sum_{0 \leqslant \nu < 2^n} a_\nu = 2^{-n} \sum_{0 \leqslant k < n} k \binom{n}{k} = n/2.$$

Further, we note that the number on the left-hand side of the inequality to be proved is equal to the number of a_ν for which $|a_\nu - \bar{a}| \geqslant t\sqrt{n}$. In view of Problem 8.5.6 (1), this number does not exceed $2^n Da (t\sqrt{n})^{-2}$. But $2^n Da = \sum_{0 \leqslant \nu < 2^n} (a_\nu - \bar{a})^2 = \sum_{0 \leqslant \nu < 2^n} (a_\nu^2 - \bar{a}^2) = -2^n \bar{a}^2 + \sum_{0 \leqslant k \leqslant n} k^2 \binom{n}{k} = -2^n (n/2)^2 + (n^2+n) 2^{n-2} = n 2^{n-2}$. Therefore, $2^n Da/(t\sqrt{n})^2 = 2^{n-2}/t^2$. (3) Minorization. We put $t = \ln n$. Taking into account 8.5.6 (2), we can write $\sum_{k=1}^{n} \frac{1}{k^2} \binom{n}{k} > \sum_{k:\, |k-n/2| < t\sqrt{n}} k^{-2} \binom{n}{k} \geqslant \frac{1}{(n/2 + t\sqrt{n})^2} \times \sum_{|k-n/2| < t\sqrt{n}} \binom{n}{k} \geqslant \frac{4}{(n-4t\sqrt{n})^2} (2^n - 2^n/t^2) \sim 2^{n+2}/n^2$. Majorization. We first note that $\binom{n}{k} < \left(\frac{en}{k}\right)^n$ in view of 8.5.2 (1) and $\binom{n}{k-1} \Big/ \binom{n}{k} = \frac{k}{n-k+1} < 1/3$ for $k \leqslant n/4$. Therefore, $\sum_{k=1}^{n} k^{-2}\binom{n}{k} \leqslant \sum_{k \leqslant n/4} \binom{n}{k} + \frac{16}{n^2} \sum_{k:\,|k-n/2|>t\sqrt{n}} \binom{n}{k} + \left(\frac{n}{2} - t\sqrt{n}\right)^{-2} \sum_{k:\,|k-n/2| \leqslant t\sqrt{n}} \binom{n}{k} \leqslant (4e)^{n/4} \sum_{i=0}^{\infty} 3^{-i} + 16n^{-2}2^n/t^2 + \frac{4}{n^2}(1+tn^{-3/2})2^n \leqslant n^{-2}2^{n+2}$.

8.5.7. (1) Using formula (13), we obtain

$$\binom{n}{k} = \frac{\sqrt{n}\, n^n (1+O(1/n))}{\sqrt{2\pi k(n-k)}\, k^k (n-k)^{n-k}}.$$

We put $x = n/2 - k$. Then the obtained expression can be written as

$$\binom{n}{k} = \frac{(1+O(1/n))2^{n+1}}{\sqrt{2\pi n\left(1-\frac{2x}{n}\right)\left(1+\frac{2x}{n}\right)}\left(1-\frac{2x}{n}\right)^{n/2-x}\left(1+\frac{2x}{n}\right)^{n/2+x}}.$$

Further, we have

$$\ln\left[\left(1-\frac{2x}{n}\right)^{n/2-x}\left(1+\frac{2x}{n}\right)^{n/2+x}\right] = \left(\frac{n}{2}-x\right)\ln\left(1-\frac{2x}{n}\right)$$
$$+\left(\frac{n}{2}+x\right)\ln\left(1+\frac{2x}{n}\right) = \left(\frac{n}{2}-x\right)\left(-\frac{2x}{n}-\frac{1}{2}\left(\frac{2x}{n}\right)^2-\ldots\right)$$
$$+\left(\frac{n}{2}+x\right)\left(\frac{2x}{n}-\frac{1}{2}\left(\frac{2x}{n}\right)^2+\ldots\right) = \frac{2x^2}{n}+\frac{16x^4}{3n^3}+O\left(\frac{x^5}{n^4}\right).$$

Hence $\binom{n}{k} \sim \frac{2^{n+1}}{\sqrt{2\pi n}} e^{-\frac{(2k-n)^2}{n}}$. (2) As in 8.5.7 (1), by using Stirling's formula we find that

$$\binom{n}{k} a^k = \frac{\sqrt{n}\, a^k n^n (1+O(1/n))}{\sqrt{2\pi k(n-k)}\, k^k (n-k)^{(n-k)}}.$$

Putting $\alpha = a/(a+1)$, $\beta = 1/(a+1)$, $x = \alpha n - k$, we get

$$\binom{n}{k} a^k = \frac{(a+1)\, a^{n\alpha - x}(1+O(1/n))}{\sqrt{2\pi n a(1-x/\alpha n)(1+x/\beta n)}} \times$$
$$(\alpha - x/n)^{-\alpha n + x}(\beta + x/n)^{-\beta n - x}$$
$$= \frac{(a+1)^n (1+O((x+1)/n))}{\sqrt{2\pi n a}(1-x/\alpha n)^{\alpha n - x}(1+x/\beta n)^{\beta n - x}}.$$

Further, we have

$$\begin{aligned}\ln ((1-x/\alpha n)^{\alpha n-x}(1+x/\beta n)^{\beta n+x})\\ &=(\alpha n-x)(-x/\alpha n-x^2/2\alpha^2n^2-x^3/3\alpha^3n^3-\ldots)\\ &+(\beta n+x)(x/\beta n-x^2/2\beta^2n^2+x^3/3\beta^3n^3-\ldots)\\ &=x^2/2n\,(1/\alpha+1/\beta)+O(x^3/n^2)=x^2(a+1)^2/2an+o(1/n).\end{aligned}$$

Finally, we obtain

$$\binom{n}{k}a^k=\frac{(a+1)^{n+1}(1+O((x+1)/n))}{\sqrt{2\pi na}}\,e^{-(x(a+1))^2/2an}$$

where $x=na(a+1)^{-1}-k$. Since $x=o(n^{2/3})$, $\dfrac{x+1}{n}=o(n^{-1/3})$.

Hence it follows that

$$1-\Phi(x)\geqslant(2\pi)^{-1/2}e^{-x^2/2}(x^{-1}-x^{-3})\backsim(2\pi)^{-1/2}x^{-1}e^{-x^2/2}.$$

We put $D=\Phi(x_m+(h/2))-\Phi(x_k-(h/2))$ and $\Psi(x)=\dfrac{1}{x}e^{-x^2/2}$.

It remains for us to prove that

$$D\backsim(\sqrt{2\pi})^{-1}(\Psi(xk)-\Psi(x_m)). \tag{***}$$

Using (*), we obtain

$$D>\Phi(x_m)-\Phi(x_k)\sim(2\pi)^{-1/2}(\Psi(x_k)-\Psi(x_m)).$$

On the other hand, according to (*), we have

$$\begin{aligned}\sqrt{2\pi}\,D\sim\Psi(x_k-(h/2))-\Psi(x_m+(h/2))\\ &=\Psi(x_k)(1+O(x_mh))-\Psi(x_m).\end{aligned}$$

For $x_m-x_k\geqslant 1$, $x_k\to\infty$ and $x_kh\to 0$, we obtain $\Psi(x_m)=o(\Psi(x_k))$ and $\sqrt{2\pi}\,D\sim\Psi(x_k)\sim\Psi(x_k)-\Psi(x_m)$. If, however, $x_m-x_k<1$, we put $\eta=\eta(n,m,k)=(x_m-x_k)h^{-1}$. It follows from $m-k\to\infty$ as $n\to\infty$ that $\eta=\eta(n,m,k)\to\infty$. We note that $(x_m^2-x_k^2)/2=\dfrac{1}{2}(x_m-x_k)\times(x_m+x_k)>h\eta k_k$. Further $\sqrt{2\pi}\,D\sim\Psi(x_k)(1+O(x_mh))-\Psi(x_m)=\Psi(x_k)-\Psi(x_m)+\Delta$, where $\Delta=O(\Psi(x_k)x_mh)=O(\Psi(x_k)x_kh)$. But $\Psi(x_k)-\Psi(x_m)\geqslant\Psi(x_k)(1-\exp\{-(x_m^2-x_k^2)/2\})\geqslant\Psi(x_k)h\eta x_k$. Hence it follows that $\Delta=o(\Psi(x_k)-\Psi(x_m))$. Consequently, $\sqrt{2\pi}\,D\lessapprox\Psi(x_k)-$

$\Psi(x_m)$, Q.E.D. Thus, it follows from (*) and (***) that

$$\sum_{k\leqslant\nu\leqslant m}\binom{n}{\nu}a^{\nu}\sim(2\pi)^{-1/2}(\Psi(x_k)-\Psi(x_m))$$

$$=\frac{1}{\sqrt{2\pi}}\left(\frac{e^{-x_k^2/2}}{x_k}-\frac{e^{-x_m^2/2}}{x_m}\right).$$

8.5.7. (4) We put $x_1=x+2\sqrt{\ln n}$, $k(x)=\dfrac{na+x\sqrt{na}}{a+1}$. Then

$$\sum_{\nu\geqslant k(x)}\binom{n}{\nu}a^{\nu}=\sum_{k(x)\leqslant\nu\leqslant k(x_1)}\binom{n}{\nu}a^{\nu}+\sum_{\nu>k(x_1)}\binom{n}{\nu}a^{\nu}.$$

But

$$\sum_{\nu>k(x_1)}\binom{n}{\nu}a^{\nu}\leqslant n\binom{n}{\nu_0}a^{\nu_0},$$

where $\nu_0=]k(x_1)[$. In view of 8.5.7 (2), we hence obtain

$$\sum_{\nu>k(x_1)}\binom{n}{\nu}a^{\nu}\leqslant n(2\pi na)^{-1/2}(1+a)^{n+1}\exp\{-x_1^2/2\}$$

$$=n^{1/2}(2\pi)^{-1/2}(1+a)^{n+1}\exp\left\{-\frac{x^2}{2}-x\sqrt{\ln n}-2\ln n\right\}$$

$$\leqslant n^{-1.5}(2\pi)^{-1/2}(1+a)^{n+1}e^{-x^2/2}. \qquad (*)$$

On the other hand, it follows from 8.5.7 (3) that

$$\sum_{k(x)\leqslant\nu\leqslant k(x_1)}\binom{n}{\nu}a^{\nu}\sim\frac{(a+1)^n}{\sqrt{2\pi}}\left(\frac{e^{-x^2/2}}{x}-\frac{e^{-x_1^2/2}}{x_1}\right). \qquad (**)$$

The required estimate follows from (*) and (**).

8.5.8. (1) Using inequality (13) and the inequality $n!>\sqrt{2\pi n}\,n^ne^{-n}$ following from it, we obtain

$$\binom{n}{\lambda n}<\frac{\sqrt{2\pi n}\,n^n\exp(-n+1/(12n))}{2\pi n\sqrt{\lambda\mu}\,(\lambda n)^{\lambda n}(\mu n)^{\mu n}\exp(-\lambda n-\mu n)}$$

$$=\frac{e^{1/(12n)}}{\sqrt{2\pi n\lambda\mu}\,\lambda^{\lambda n}\mu^{\mu n}}\sim G(n,\lambda).$$

On the other hand,

$$\binom{n}{\lambda n}>\frac{\sqrt{2\pi n}\,n^n\exp\{-n\}}{2\pi n\sqrt{\lambda\mu}\,(\lambda n)^{\lambda n}(\mu n)^{\mu n}\exp\left\{-\lambda n-\mu n+\frac{1}{12\mu n}+\frac{1}{12\lambda n}\right\}}$$

$$=\frac{\exp\{-(1/(12n))(1/\lambda+1/\mu)\}}{\sqrt{2\pi n\lambda\mu}\,\lambda^{\lambda n}\mu^{\mu n}}\sim G(n,\lambda).$$

(2) The minorant follows from the solution of 8.5.8 (1). In order to obtain the majorant, using inequalities (13) we find that

$$\binom{n}{\lambda} G(n, \lambda) \exp\left\{\frac{1}{12n}-\frac{1}{12\lambda n}-\frac{1}{12\mu n}+\frac{1}{360(\lambda n)^3}+\frac{1}{360(\mu n)^3}\right\}.$$

Without loss of generality, we can assume that $\lambda \geqslant \mu$. Then $1/(12n) < 1/(12\lambda n)$ and $\frac{1}{360(\lambda n)^3}+\frac{1}{360(\mu n)^3}-\frac{1}{12\mu n} \leqslant \frac{1}{180\mu^3 n^3}-\frac{1}{12\mu n} < 0$. Consequently, $\binom{n}{k} < G(n, \lambda)$. (3) Using the minorant from 8.5.8 (2), we obtain $\binom{n}{k} > G(n, \lambda)\, e^{-\frac{1}{12n\lambda\mu}}$. But $\exp\{-(12n\lambda\mu)^{-1}\} \geqslant \exp\{-(3n)^{-1}\} \geqslant \exp\{-1/9\}$ for $n \geqslant 3$, $\exp\{-1/9\} > \sqrt{\pi}/2$. For $n=2$ and $\lambda n=\mu n=1$, the equality is observed. (4) Majorant

$$\sum_{k=\lambda n}^{n} \binom{n}{k} < \binom{n}{\lambda n} \sum_{i=0}^{\infty} \left(\frac{1-\lambda}{\lambda}\right)^i = \frac{1}{2\lambda-1}\binom{n}{\lambda n}.$$

(5) The inequality can be easily verified when (a) $\lambda = 1/2$ $\left(\text{since } \sum_{k\geqslant n/2} \binom{n}{k} < 2^n = \left(\frac{1}{2}\right)^{-n/2}\left(\frac{1}{2}\right)^{-n/2}\right)$; (b) $\lambda n = n-1$ (the left-hand side is $n+1$, and the right-hand side is $\left(\frac{n}{n-1}\right)^{n-1} n$); (c) $3 \leqslant n \leqslant 5$, $n/2 < \lambda n < n-1$. (This can be checked directly.) Let now $n \geqslant 5$ and $n/2 < \lambda n \leqslant n-2$. It follows from (4) and (1) that $\sum_{k\geqslant\lambda n} \binom{n}{k} \leqslant \lambda(2\lambda-1)^{-1}\binom{n}{\lambda n} \leqslant \lambda^{1/2} \times (2\lambda-1)^{-1}(2\pi n(1-\lambda))^{-1/2}\lambda^{-\lambda n}\mu^{-\mu n}$. We put $f(\lambda)=\lambda^{1/2}(2\lambda-1)^{-1} \times (2\pi n(1-\lambda))^{-1/2}$. We must prove that $f(\lambda) < 1$ for $\lambda \in [1/2+\varepsilon, (n-2)/n]$, where $\varepsilon = 1/n$ for even n and $1/(2n)$ for odd n. Differentiation with respect to λ reveals that the function $f(\lambda)$ is convex down on the segment under investigation. Therefore, the maximum values must be sought at the ends of the segment. We have $f\left(\frac{1}{2}+\varepsilon\right) = \sqrt{\frac{1}{2}+\varepsilon}\,(1+2\varepsilon)^{-1}(2\pi n(1/2-\varepsilon))^{-1/2} = ((1-4\varepsilon^2)\,2\pi n)^{-1/2} \leqslant ((1-16/n^2)\,2\pi n)^{-1/2} < 1$ for all $n \geqslant 5$. For

$n \geqslant 6$ we have: $f((n-2)/n) = \sqrt{1-2/n}\ (1-4/n)^{-1}\ (4\pi)^{-1/2} \leqslant (1-4/n)^{-1}\ (4\pi)^{-1/2} < 1$. If $\lambda < 1/2$, than $\mu = 1-\lambda > 1/2$.
(6) Since $\binom{n}{k} = \binom{n}{n-k}$, using 8.5.8 (5) we obtain

$$\sum_{0\leqslant k\leqslant \lambda n} \binom{n}{k} = \sum_{k=\mu n}^{n} \binom{n}{k} < \mu^{-\mu n}\lambda^{-\lambda n}.$$

8.5.9. (1) $(n)_k = n^k \prod_{i=0}^{k-1} (1-i/n) = n^k \exp \sum_{i=1}^{k-1} \ln(1-i/n)$,

$$\sum_{i=1}^{k-1} \ln(1-i/n) = -\sum_{i=1}^{k-1}\sum_{\nu=1}^{\infty} \frac{1}{\nu}\left(\frac{i}{n}\right)^{\nu} = -\sum_{\nu=1}^{\infty} \frac{1}{\nu n^{\nu}} \sum_{i=1}^{k-1} i^{\nu}.$$

(2) We have, on the one hand, $(n)_k \leqslant n^k$, which is obvious. On the other hand, $(n)_k = \prod_{i=0}^{k-1} (n-i) > (n-k)^k = n^k(1-(k/n))^k \geqslant n^k(1-k^2/n) \sim$ (because $k = o(\sqrt{n})$) $\sim n^k(1-o(1))$. (3) We use the fact (see Problem 8.5.13 (2)) that $\sum_{i=1}^{k-1} i^{\nu} = (\nu+1)^{-1}k^{\nu+1} + O(k^{\nu})$. Hence, in view of 8.5.8 (1), for $k \to \infty$ and $k = o(n)$ we have

$$(n)_k = n^k \exp\left\{-\sum_{\nu=1}^{\infty} (\nu n^{\nu})^{-1}((\nu+1)^{-1}k^{\nu+1} + O(k^{\nu}))\right\}$$

$$= n^k \exp\left\{-\sum_{\nu=1}^{\infty}\left(\frac{k^{\nu+1}}{\nu(\nu+1)n^{\nu}} + O\left(\frac{1}{\nu}\left(\frac{k}{n}\right)^{\nu}\right)\right)\right\}.$$

But

$$\sum_{\nu=1}^{\infty} \frac{k^{\nu+1}}{\nu(\nu+1)n^{\nu}} = \sum_{\nu=1}^{m-1} \frac{k^{\nu+1}}{\nu(\nu+1)n^{\nu}} + \sum_{\nu\geqslant m} \frac{k^{\nu+1}}{\nu(\nu+1)n^{\nu}}.$$

$$\sum_{\nu\geqslant m} \frac{k^{\nu+1}}{n^{\nu}} = k\sum_{\nu\geqslant m}\left(\frac{k}{n}\right)^{\nu} = k\left(\frac{k}{n}\right)^{m}\frac{1}{1-k/n} = O\left(\frac{k^{m+1}}{n^m}\right).$$

Hence we obtain the required equality. (4) Use 8.5.9 (3) for $m = 3$.
8.5.10. (1) In view of 8.1.13 (6) and 8.5.9 (1), we have

$$\binom{n-s}{k-s}\Big/\binom{n}{k} = \frac{(k)_s}{(n)_s} \sim \left(\frac{k}{n}\right)^s.$$

(2) Make use of 8.5.9 (3).

8.5.11. (1) We have

$$\binom{n-s}{k}\Big/\binom{n}{k}=\frac{(n-s)_k}{(n)_k}=\prod_{i=0}^{k-1}\left(1-\frac{s}{n-i}\right)$$

$$=\exp\left\{\sum_{i=0}^{k-1}\ln\left(1-\frac{s}{n-i}\right)\right\}=\exp\left\{-\sum_{i=0}^{k-1}\sum_{\nu=1}^{\infty}\frac{1}{\nu}\left(\frac{s}{n-i}\right)^{\nu}\right\}$$

$$=\exp\left\{-s\sum_{i=0}^{k-1}\frac{1}{n-i}-\frac{s^2}{2}\sum_{i=0}^{k-1}\frac{1}{(n-i)^2}-\sum_{i=0}^{k-1}\sum_{\nu\geqslant 3}\frac{1}{\nu}\left(\frac{s}{n-i}\right)^{\nu}\right\}.$$

Using 8.5.12, we obtain

$$\sum_{i=0}^{k-1}\frac{1}{n-i}=\ln\frac{n}{n-k+1}+O\left(\frac{1}{n}\right);\quad \sum_{i=0}^{k-1}\frac{1}{(n-i)^2}$$

$$=\frac{k}{n(n^2-k)}\left(1+O\left(\frac{1}{n}\right)\right);\quad \ln\frac{n}{n-k+1}=\sum_{\sigma=1}^{\infty}\frac{1}{\sigma}\left(\frac{k-1}{n}\right)^{\sigma};$$

$$\sum_{i=0}^{k-1}(n-i)^{-\nu}=\frac{1}{\nu-1}\left((n-k+1)^{-\nu+1}-n^{-\nu+1}\right)+O((n-k)^{-\nu})$$

for $\nu>2$. Therefore

$$\binom{n-s}{k}\Big/\binom{n}{k}\sim\exp\left\{-s\sum_{\sigma=1}^{\infty}\frac{1}{\sigma}\left(\frac{k-1}{n}\right)^{\sigma}-\frac{s^2k}{2n(n-k)}\right.$$

$$\left.-\sum_{\nu=0}^{\infty}\frac{s^{\nu}\left((n-k)^{-\nu+1}-n^{-\nu+1}\right)}{\nu(\nu-1)}\right\}$$

$$=\exp\left\{-\frac{sk}{n}-\frac{s(k-1)^2}{2n^2}-\frac{s^2k}{2n(n-k)}\right.$$

$$\left.-s\sum_{\sigma=3}^{\infty}\frac{1}{\sigma}\left(\frac{k-1}{n}\right)^{\sigma}-\sum_{\nu=3}^{\infty}\frac{s^{\nu}k\left(1+O\left(\frac{k}{n}\right)\right)}{\nu n(n-k)^{\nu-1}}\right\}$$

$$=\exp\left\{-\frac{sk}{n}-\frac{sk^2+s^2k}{2n^2}\left(1+O\left(\frac{k}{n}\right)\right)\right.$$

$$\left.-s\sum_{\sigma\geqslant 3}\frac{1}{\sigma}\left(\frac{k-1}{n}\right)^{\sigma}-k\sum_{\nu\geqslant 3}\frac{s^{\nu}(1+o(1))}{\nu n(n-k)^{\nu-1}}\right\}.$$

(2) It follows from 8.5.11 (1). (3) The majorant follows from the fact (see 8.5.11 (1)) that $\prod_{i=0}^{k-1}\left(1-\frac{s}{n-i}\right)\leqslant\left(1-\frac{s}{n}\right)^k<e^{\frac{-sk}{n}}$. In order to obtain the minorant, we shall use the relation (see 8.5.11 (1))

$$\binom{n-s}{k}\Big/\binom{n}{k}=\prod_{i=0}^{k-1}\left(1-\frac{s}{n-i}\right)\geqslant\left(1-\frac{s}{n-k}\right)^k$$

$$=\exp\left\{k\ln\left(1-\frac{s}{n-k}\right)\right\}=\exp\left\{-k\sum_{\nu=1}^{\infty}\frac{1}{\nu}\left(\frac{s}{n-k}\right)^{\nu}\right\}$$

$$\geqslant\exp\left\{-k\left(\frac{s}{n-k}+\frac{s^2}{2(n-k)^2}\sum_{\nu=0}^{\infty}\left(\frac{s}{n-k}\right)^{\nu}\right)\right\}$$

$$=\exp\left\{-\frac{sk}{n}\left(1+\frac{k}{n-k}+\frac{sn}{2(n-k)(n-k-s)}\right)\right\}.$$

8.5.12. Hint. $\sum_{k=n+1}^{m} f(k)$ is the upper and $\sum_{k=n}^{m-1} f(k)$ the lower integral sum for $\int_n^m f(x)\,dx$.

8.5.13. (1) We have (see 8.5.12) $\sum_{k=1}^{m}\ln k\leqslant\int_1^m \ln x\,dx+\ln m$, $\int\ln x\,dx=x\ln x-\int dx=x\ln x-x$. Hence $\sum_{k=1}^{n}\ln k\leqslant m\ln m-m+1+\ln m$. On the other hand (see 8.5.12), $\sum_{k=1}^{m}\ln k\geqslant\int_1^m\ln x\,dx=m\ln m-m+1$. (2)-(6) are similar to 8.5.13 (1).

8.5.14. (1) Induction on n. For $n=1$, we have $p_1=p_0-\alpha p_0^{\beta}=1-\alpha$. Hence $0<p_1<1$. Let $0<p_n<1$ for a certain $n\geqslant 1$. Then $p_{n+1}=p_n-\alpha p_n^{\beta}=p_n(1-\alpha p_n^{\beta-1})$. Since $0<\alpha<1$, $\beta>1$ and $0<p_n<1$, $0<p_{n+1}<1$. (2) It follows from the relation $p_{n+1}-p_n=-\alpha p_n^{\beta}<0$. (3) Majorant. We have $\frac{p_{k-1}-p_k}{\alpha p_{k-1}^{\beta}}=$

1. Hence $n = \sum_{k=1}^{n} \frac{p_{k-1} - p_k}{\alpha p_{k-1}^{\beta}} \leqslant \int_{p_n}^{1} \frac{dx}{\alpha x^{\beta}} = \frac{1}{\alpha(\beta-1)}(p_n^{1-\beta} - 1)$.

Therefore, $p_n^{1-\beta} \geqslant 1 + \alpha(\beta-1)n$ or, which is the same,

$$p_n \leqslant (1 + \alpha(\beta-1)n)^{1/1-\beta}. \qquad (*)$$

Minorant. It follows from (*) and the recurrence relation $p_n = p_{n-1}(1 - \alpha p_{n-1}^{\beta-1})$ that

$$p_n/p_{n-1} \geqslant 1 - \alpha/(1 + \alpha(\beta-1)(n-1)). \qquad (**)$$

Further,

$$\frac{1}{\alpha(\beta-1)}(p_n^{1-\beta} - 1) = \int_{p_n}^{1} \frac{dx}{\alpha x^{\beta}} \leqslant \sum_{k=1}^{n} \frac{p_{k-1} - p_k}{\alpha p_k^{\beta}}. \qquad (***)$$

It should be noted that, according to (**), $p_k \geqslant p_{k-1}(1-\alpha)$ for $1 \leqslant k \leqslant]\sqrt{n}[$ and $p_k \geqslant p_{k-1}(1 - (c\sqrt{n}))$ for $k >]\sqrt{n}[$, where c is a constant. Taking into account the relation $(p_{k-1} - p_k)/\alpha p_{k-1}^{\beta} = 1$ we obtain

$$\sum_{k=1}^{n} \frac{p_{k-1} - p_k}{\alpha p_k^{\beta}} \leqslant \sum_{k \leqslant]\sqrt{n}[} \frac{p_{k-1} - p_k}{\alpha p_{k-1}^{\beta}(1-\alpha)} + \sum_{]\sqrt{n}[< k \leqslant n} \frac{p_{k-1} - p_k}{\alpha p_{k-1}^{\beta}(1 - (c/\sqrt{n}))}$$

$$\leqslant (\alpha(1-\alpha))^{-1}]\sqrt{n}[+ n/(1 - (c/\sqrt{n})) = n + O(\sqrt{n}).$$

Taking into account (***), we find that for $\beta > 1$, $\frac{1}{\alpha(\beta-1)} \times (p_n^{1-\beta} - 1) \leqslant n + O(\sqrt{n})$. Hence $p_n \geqslant (\alpha(\beta-1)(n + O(\sqrt{n})))^{1/(1-\beta)}$. Therefore, for $n \to \infty$ $p_n \sim (\alpha(\beta-1)n)^{1/(1-\beta)}$.

8.5.15. (1) We write the equation in the form $x = \ln t - \ln x$ (*). Since $t \to \infty$, we can assume that $t > e$, and hence $x > 1$. Then it follows from (*) that $x < \ln t$, i.e. $1 < x < \ln t$. Therefore, $\ln x = O(\ln \ln t)$. Thus $x = \ln t + O(\ln \ln t)$ for $t \to \infty$. Taking logarithms, we obtain $\ln x = \ln \ln t + \ln(1 + O(\ln \ln t/\ln t)) = \ln \ln t + O(\ln \ln t/\ln t)$. Substituting this into (*), we obtain a new approximation: $x = \ln t - \ln \ln t + O(\ln \ln t/\ln t)$. Taking logarithms once again and substituting the result into the right-hand side of (*), we obtain the next approximation providing the required accuracy (see N.G. de Bruijn, "Asymptotic Methods in Analysis", North Holland, Amsterdam, 1958). (2) Since $t \to \infty$, $x \to \infty$ as well. Therefore, $e^x < t$ or, which is the same, $x < \ln t$. Hence $e^x = t - \ln x > t - \ln \ln t$ and

$$x > \ln t + \ln (1 - \ln \ln t/t) = \ln t - \sum_{\nu=1}^{\infty} \nu^{-1} (\ln \ln t/t)^{\nu} = \ln t -$$

$\ln \ln t/t + O((\ln \ln t/t)^2)$. Using this relation, we obtain $e^x = t - \ln x < t - \ln (\ln t - \ln \ln t/t + O((\ln \ln t/t)^2)) = t - \ln \ln t - \ln (1 - (\ln \ln t/(t \ln t) + O((\ln t)^{-1} (\ln \ln t/t)^2)) = t - \ln \ln t + O(\ln \ln t/(t \ln t))$. Taking logarithms, we obtain $x < \ln (t - \ln \ln t + O(\ln \ln t/(t \ln t))) = \ln t + \ln (1 - \ln \ln t/t + O(\ln \ln t/(t^2 \ln t))) = \ln t - \ln \ln t/t + O((\ln \ln t/t)^2)$. The majorant and minorant coincide with the required accuracy.

8.5.16. Hint. First prove that $f(t) = o(t)$ for $t \to \infty$. Then the initial equality can be written in the form $e^{tf(t)} = t + o(t) + O(1)$. Using this equality, prove that $f(t) = o(1)$ and transform the initial equality to $e^{tf(t)} = t + O(1)$. Finally, prove that $f(t) = \ln t/t + O\left(\frac{1}{t^2}\right)$.

8.5.17. (1) We expand $A(t)$ into simple fractions: $A(t) = \frac{C_1}{\lambda_1 - t} + \frac{C_2}{\lambda_2 - t} + \cdots + \frac{C_m}{\lambda_m - t} + B(t)$, where $B(t)$ is a polynomial. In order to determine the coefficient C_1, we multiply $A(t)$ by $\lambda_1 - t$. Then $(\lambda_1 - t) A(t) = \frac{-Q(t)}{(t-\lambda_2) \ldots (t-\lambda_m)}$. For $t = \lambda_1$, the left-hand side is equal to C_1, and the right-hand side to $\frac{-Q(\lambda_1)}{P'(\lambda_1)}$. Thus, $C_1 = \frac{-Q(\lambda_1)}{P'(\lambda_1)}$. Similarly, we can calculate the coefficients C_i $(i = \overline{2, m})$. The fraction $\frac{1}{1 - t/\lambda_k}$ can be expanded into the geometrical series $\left(1 - \frac{t}{\lambda_k}\right)^{-1} = \sum_{n=0}^{\infty} \times \left(\frac{t}{\lambda_k}\right)^n$. We obtain $A(t) = \sum_{i=1}^{m} \frac{C_i}{\lambda_i} \sum_{n=0}^{\infty} \left(\frac{t}{\lambda_i}\right)^n + B(t)$. Hence for large n $a_n \sim \frac{C_1}{\lambda_1^{n+1}} + \frac{C_2}{\lambda_2^{n+1}} + \cdots + \frac{C_m}{\lambda_m^{n+1}} \sim C_1 \lambda_1^{-n-1}$.

(2) Let us first consider the case when $P(t)$ has no roots other than λ_1 and $Q(t)$ has a power lower than that of $P(t)$ and λ_1 is not a root of $Q(t)$. Then $P(t) = (t - \lambda_1)^r$, $Q(t) = \sum_{i=0}^{r-1} q_i t^i$. The expan-

sion in powers of t $P^{-1}(t)$ has the form

$$P^{-1}(t)=(-1)^r\lambda_1^{-r}(1-t/\lambda_1)^{-r}=\left(-\frac{1}{\lambda_1}\right)^r\sum_{n=0}^{\infty}\binom{-r}{n}t^n/\lambda_1^n.$$

Hence

$$A(t)=Q(t)/P(t)=\left(-\frac{1}{\lambda_1}\right)^r\sum_{n=0}^{\infty}\left(\frac{t}{\lambda_1}\right)^n\sum_{i=0}^{r-1}g_i\binom{-r}{n-i}\lambda_1^i.$$

Therefore

$$a_n=(-1)^r\lambda_1^{-(r+n)}\sum_{i=0}^{r-1}q_i\binom{-r}{n-i}\lambda_1^i.$$

In the general case, $A(t)=A_1(t)+\sum_{s=1}^{k}\frac{Q_s(t)}{(t-\lambda_s)^{r_s}}$, where $A_1(t)$ is a polynomial, and $Q_s(t)$ is a polynomial of degree smaller than r_s. In this case, the asymptotic value of a_n is determined by the coefficient of t^n in the expansion of the fraction $Q_1(t)/(t-\lambda_1)^{r_1}$. The problem is reduced to the one considered earlier.

8.5.18. (1) 2×3^n. Use the results of Problem 8.5.17 (1). The polynomial $P_1(t)=3t^2-4t+1$ has the roots $\lambda_1=1/3$ and $\lambda_2=1$. We put $Q(t)=\frac{1}{3}(1+t)$ and $P(t)=t^2-\frac{4}{3}t+\frac{1}{3}$. Then

$$a_n\sim-\frac{Q(1/3)}{P'(1/3)}\left(\frac{1}{3}\right)^{-n-1}=-\frac{4/9}{2/3-4/3}3^{n+1}=2\times 3^n.$$

(2) $a_n\sim\left(-\frac{1}{13}\left(\frac{3}{2}\right)^{n+1}\right)$. The smaller (in magnitude) root of the polynomial $6t^2+5t-6$ is 2/3. Using 8.5.17 (1), we obtain the required result. (3) $a_n\sim\left(-\frac{8}{13}\left(\frac{3}{2}\right)^{n+1}\right)$. (4) $a_{2n}=0$, $a_{2n+1}\sim(-1)^n 2^{n+3}$. (5) **Hint.** $6t^4-17t^3+35t^2-22t+4=6\times\left(t-\frac{1}{2}\right)\left(t-\frac{1}{3}\right)(t-1+i\sqrt{3})(t-1-i\sqrt{3})$, $a_n\sim\frac{2}{31}3^n$. (6) $\frac{1}{15}2^{n+1}$. (7) $\binom{-2}{n}(\sqrt{3}-1)^{-n-2}$. The roots of the equation $t^2+2t-2=0$ are $\lambda_1=\sqrt{3}-1$ and $\lambda_2=-\sqrt{3}-1$. Representing $A(t)$ in the form

$$A(t)=\frac{at+b}{(t-\lambda_1)^2}+\frac{ct+d}{(t-\lambda_2)^2},$$

we find by the method of indeterminate coefficients that $a=0$ and $b=1$. Using 8.5.17 (2), we obtain

$$a_n=(-1)^2\lambda_1^{-n-2}\left(b\binom{-2}{n}+\lambda_1 a\binom{-2}{n-1}\right)=(\sqrt{3}-1)^2\binom{-2}{n}.$$

(8) $a_n\sim(10/7)^{n+2}\binom{-2}{n}$. The smallest in magnitude root of the denominator is 0.7 and has a multiplicity 2. We note that $A(t)=\dfrac{t+1}{(t-0.7)^2}-\dfrac{2}{2t^2+1}$. Using 8.5.17 (2), we obtain $a_n\sim\left(\dfrac{10}{7}\right)^{n+2}\binom{-2}{n}$.

8.5.19. (1) $a_n\sim(-3(-2)^n)$. Let $A(t)=\sum\limits_{n=0}^{\infty}a_nt^n$. Multiplying both sides of the relation by t^{n+2} and taking the sum, we obtain $A(t)-a_1t-a_0+3t(A(t)-a_0)+2t^2A(t)=0$. Since $a_0=1$ and $a_1=2$, we get $A(t)=(1-t)/(2t^2+3t+1)$. The roots of the denominator are $\lambda_1=-1/2$ and $\lambda_2=-1$. Using 8.5.17 (1) we find that $a_n\sim(-3(-2)^n)$. (2) 1/2. As in 8.5.19 (1), we can write $A(t)=(2(1-t))^{-1}+(2(1-(q-p)t))^{-1}$. Since $q+p=1$ and p, $q>0$, $|q-p|<1$; the roots of the denominators are $\lambda_1=1$ and $\lambda_2=q-p$, $|\lambda_1|<|\lambda_2|$. Using 8.5.17 (1) (or directly), we obtain $a_n\sim\dfrac{1}{2}$. (3) $a_n=\dfrac{2^{n+1}}{\sqrt{3}}\left(\sin\dfrac{2\pi n}{3}-\sin\dfrac{4\pi n}{3}\right)$; (4) $a_n\sim 3^n$. (5) $a_n\sim n^2 2^{n-5}$.

8.5.20. (1) **Hint.** If the limit of a_n exists and is equal to a, we have $a=(a+b/a)/2$ or $a=\sqrt{b}$ from the recurrence relation, since $a>0$. If $a_0=\sqrt{b}$, then $a_1=(a_0+b/a_0)/2=(\sqrt{b}+\sqrt{b})/2=\sqrt{b}$. It can be easily obtained by induction that $a_n=\sqrt{b}$. Let us consider the case when $a_0>b$ (the case $a_0<b$ is similar). We prove that a_n decreases and $a_n>\sqrt{b}$ for all $n\geqslant 0$. If $a_n>\sqrt{b}$ for a certain $n\geqslant 0$, then $a_{n+1}-a_n=((a_n+b/a_n)/2)-a_n=(b-a_n^2)/(2a_n)<0$. Thus, a_n decreases with increasing n. $a_{n+1}-\sqrt{b}=$

$$\frac{1}{2}(a_n+b/a_n)-\sqrt{b}=(a_n-\sqrt{b})^2/(2a_n)>0.$$

Consequently, a_n decreases and has a lower bound. This means that there exists a limit $a=\lim\limits_{n\to\infty}a_n$. As was shown above, this limit is equal to $\sqrt{b}$. (2) As in 8.5.20 (1), if the limit a_n exists, it is equal to $\sqrt[3]{b}$. Let us consider the case when $a_0<\sqrt[3]{b}$. As in 8.5.20 (1), we prove that a_n monotonically increases, and $a_n<\sqrt[3]{b}$ for all n. Hence it follows that there exists a limit of a_n. If $\lim\limits_{n\to\infty}a_n=a$, then proceeding to the limit in the recurrence relation, we obtain $a = \sqrt[3]{b}$. (3) $\sqrt{1+b} - 1$. Note that $a_1 - a_0 = (b - a_0^2)/2 - a_0 = (-b - b^2)/2 < 0$ and $a_1 = (b - b^2)/2 > 0$, i.e. $0 < a_1 < a_0$. Further, $a_2 - a_1 = (b - a_1^2)/2 + (b + b^2)/2 = (a_1^2 - b^2)/2 > 0$, $a_2 - a_0 = (b - a_1^2)/2 - b/2 = -a_1^2/2 < 0$. Hence $a_1 < a_2 < a_0$. In general, $a_{n+2} - a_n = (a_n^2 - a_{n+1}^2)/2$. By induction, we find that $\{a_{2n}\}$ increases, $a_{2n} < a_0$ for all $n \geqslant 1$, and $\{a_{2n+1}\}$ decreases, $a_{2n+1} > a_1$. Consequently, there exist the limits $c = \lim\limits_{n\to\infty} a_{2n}$ and $d = \lim\limits_{n\to\infty} a_{2n+1}$. Proceeding to the limit in the recurrence relation, we obtain $c = (b - d^2)$ and $d = (b - c^2)$. Hence $(c - d) \times (2 - c - d) = 0$; since $c < b/2 < 1/2$ and $d < b/2 < 1/2$, we have $2 - c - d > 0$; $c = d$. Therefore $c = \sqrt{1+b} - 1$.

8.5.21. (1) Using relations (14) and (15), we find that $a_3 = 9/128$ and $a_4 \leqslant 11/128$. For $n \geqslant 3$, inequality (15) can be written in the form $a_{n+2} \leqslant 17/1024 + a_n\,(323/1024 + a_n)$. Hence it follows that if $a_n \leqslant 1/8$, then $a_{n+2} \leqslant 1/8$ for $n \geqslant 3$. (2) Using the fact that $a_n \leqslant 1/8$, we derive from (15) a new inequality

$$a_{n+2} \leqslant \frac{1}{2^{n+3}} + \frac{n+1}{4^{n+1}} + \frac{1}{8}\left(\left(\frac{3}{4}\right)^{n+2} + \frac{n+2}{2^{n+2}} + 4a_n\right)$$
$$\leqslant \left(\frac{3}{4}\right)^{n+2} + \frac{1}{2}a_n.$$

Hence, by induction, $a_n \leqslant 9\left(\frac{3}{4}\right)^n$. (3) Using 8.5.21 (2), we obtain from (15) $a_{n+2} \leqslant 1/2^{n+3} + 66a_n\,(3/4)^{n+2}$. Using this inequality, we get $a_n = 2^{n-1}\,(1 + O\,((3/4)^n))$.

8.5.22. Induction on n. For $n=1$, we have $a_1 = a_1 \times 1$. If $a_n \leqslant a_1 n$, then $a_{n+1} \leqslant a_n + a_1 \leqslant a_1\,(n+1)$.

8.5.23. (1) Consider the relation $a_k = \dfrac{f(n, k+1)}{f(n, k)} = \dfrac{n-k}{k+1} 2^{-2^k}$. If $k < [\log_2 \log_2 n]$, then $a_k > 1$, and if $k > [\log_2 \log_2 n]$, then $a_k < 1$. Consequently, the maximum value of $f(n, k)$ is attained either for $k = [\log_2 \log_2 n]$, or for $k = [\log_2 \log_2 n] + 1$. (2) The same result as in 8.5.23 (1).

8.5.24. Note that $\lambda\,(n, r, k) = f\,(n, r+1, k)/f\,(n, r, k) = (k - r)\,2^{2^r - 1}/(r+1)$ and that $\lambda\,(n, r, k) > 1$ for $k \geqslant r > 0$. Consequently, $f\,(n, r, k)$ increases in r. Hence $\max f\,(n, r, k) = f(n, k, k) = \binom{n}{k} 2^{-k+2^k}$, $\min\limits_{k=0} f\,(n, r, k) = 2\binom{n}{k}^r$.

8.5.25. (1)

$$g(n) \sim \begin{cases} 2^{n+1}/n & \text{if } [\log_2 n] > \log_2 (n - \log_2 n), \\ (2^n/n)(1 + 2^{2^{[\log_2 n] - \log_2 (n - \log_2 n)}}) & \text{if } \log_2 (n - \log_2 n) \leqslant \\ [\log_2 n] \leqslant \log_2 (n - \log_2 n - \log_2 \log_2 n), & \\ 2^{n - [\log_2 n]} & \text{if } [\log_2 n] < \log_2 (n - \log_2 n - \log_2 \log_2 n). \end{cases}$$

The function $f(n, k)$, as a function of a real argument k, is convex down, and the minimum is attained at $k = k^* = \log_2\left(n - \log_2 n + O\left(\frac{\log_2 n}{n}\right)\right)$. We put $k_0 = [\log_2 n]$. Obviously, either $g(n) = f(n, k_0)$ or $g(n) = f(n, k_0 - 1)$. In order to find $g(n)$, we must choose the minimum value between $f(n, k_0)$ and $f(n, k_0 - 1)$. Let (a) $\log_2 n \geqslant k_0 > \log_2(n - \log_2 n)$. Then

$$f(n\, k_0 - 1) = 2^{n-k_0+1} + 2^{2^{k_0-1}} \sim 2^{n+1}/n,$$

$$f(n, k_0) = 2^{n-k_0} + 2^{2^{k_0}} \gtrsim 2^{n - \log_2 n} + 2^{2^{\log_2(n - \log_2 n)}} \sim 2^{n+1}/n.$$

Let (b) $\log_2(n - \log_2 n) \geqslant k_0 > \log_2(n - \log_2 n - \log_2 \log_2 n)$. Then $f(n, k_0) \sim \frac{2^n}{n} + 2^{2^{k_0}} = \frac{2^n}{n}(1 + 2^{2^{k_0} - \log_2(n - \log_2 n)})$, $f(n, k_0 - 1) \sim \frac{2^{n+1}}{n}$. Consequently, $g(n) \sim f(n, k_0)$. Let (c) $k_0 \leqslant \log_2(n - \log_2 n - \log_2 \log_2 n)$. Then $f(n, k_0) = 2^{n-k_0} + 2^{2^{k_0}} \sim 2^{n-k_0} \sim \frac{2^n}{n} 2^{\log_2 n - k_0}$, $f(n, k_0 - 1) \geqslant 2^{n-k_0+1} > f(n, k_0)$. Hence $g(n) = f(n, k_0) = 2^{n-k_0}$. (2) $g(n) \sim 2^n(2^{\alpha(n)} + 2^{-\alpha(n)})$, where $\alpha(n) = \log_2 n - [\log_2 n]$ if $\alpha(n) < 1/2$ and $g(n) \sim 2^n(2^{1-\alpha(n)} + 2^{\alpha(n)-1})$ if $\alpha(n) > 1/2$.

Bibliography

1. C. E. Shannon and J. McCarthy, *Automata Studies*, Princeton Univ. Press, Princeton, N.Y., 1956. (VI)
2. Z. V. Alferova, *Theory of Algorithms* (in Russian), Statistika, Moscow, 1973. (VII)
3. M. Arbib, *Brains, Machines and Mathematics*, McGraw-Hill, New York, 1965. (VI, VII)
4. C. Berge, *Theorie des graphes et ses applications*, Paris, 1958. (IV)
5. N. Ya. Vilenkin, *Combinatorial Analysis* (in Russian), Nauka, Moscow, 1969. (VIII)
6. *Discrete Mathematics and Mathematical Problems in Cybernetics*, Vol. 1 (in Russian), Nauka, Moscow, 1974. (I-V)
7. J. Kemeney, J. L. Snell, and J. G. Thompson, *Introduction to Finite Mathematics*, Prentice Hall, Englewood-Cliffs, 1957. (IV, VIII)
8. N. E. Kobrinskii and B. A. Trakhtenbrot, *Introduction to Theory of Finite Automata* (in Russian), Fizmatgiz, Moscow, 1962. (VI)
9. K. A. Rybnikov (ed.), *Combinatorial Analysis: Problems and Exercises* (in Russian), Nauka, Moscow, 1982. (IV, VIII)
10. N. Christofides, *Graph Theory: An Algorithmic Approach*, London, 1975. (IV)
11. A. I. Mal'tsev, *Algorithms and Recursive Functions* (in Russian), Nauka, Moscow, 1986. (VII)
12. M. Minsky, *Computation: Finite and Infinite Machines*, Prentice Hall, Englewood-Cliffs, 1967. (VI, VII)
13. O. Ore, *Theory of Graphs*, American Math. Soc., Providence, 1962. (IV)
14. Y. Peterson and E. Weldon, *Error Correcting Codes*, M.I.T. Press, Cambridge, Mass., 1972. (V)
15. J. Riordan, *An Introduction to Combinatorial Analysis*, J. Wiley, New York, 1958. (IV, VIII)
16. K. A. Rybnikov, *Introduction to Combinatorial Analysis* (in Russian), Moscow University Press, Moscow, 1985. (IV, VIII)
17. V. N. Sachkov, *Introduction to Combinatorial Methods in Discrete Mathematics* (in Russian), Nauka, Moscow, 1982. (IV, VIII)

---

[1] The Roman figures at the end of a reference indicate the chapters of this book to which this entry applies.

18. M. Swamy and K. Thulasiraman, *Graphs, Networks and Algorithms*, J. Wiley, New York, 1981. (IV)
19. B. A. Trakhtenbrot, *Algorithms and Computational Automata* (in Russian), Sov. Radio, Moscow, 1974. (VI, VII)
20. B. A. Trakhtenbrot and Ya. M. Barzdin', *Finite Automata* (in Russian), Nauka, Moscow, 1970. (VI)
21. R. J. Wilson, *Introduction to Graph Theory*, Oliver and Boyd, Edinburgh, 1972. (IV)
22. W. Feller, *Introduction to Probability Theory and its Applications*, Vol. 1-2, J. Wiley, New York, 1968-1971. (VIII)
23. F. Harary, *Graph Theory*, Addison-Wesley, London, 1969. (IV)
24. M. Hall, *Combinatorial Theory*, Blaisdell Pub. Co., Waltham, Mass., 1967. (IV, VIII)
25. S. V. Yablonsky, *Functional Constructions in k-Valued Logic*, in: Proceedings of V. A. Steklov Inst. of Maths., Vol. 51 (in Russian), Izd. Akad. Nauk SSSR, Moscow, 1958. (I-IV)
26. S. V. Yablonsky, *Introduction to Discrete Mathematics*, Mir, Moscow, 1989. (I-VIII)
27. S. V. Yablonsky, G. P. Gavrilov, and V. B. Kudryavtsev, *Boolean Functions and Post's Classes* (in Russian), Nauka, Moscow, 1966. (I, II)

Notations

Subject Index

TO THE READER

Mir Publishers would be grateful for your comments on the content, translation and design of this book. We would also be pleased to receive any other suggestions you may wish to make.

Our address is:
Mir Publishers
2 Pervy Rizhsky Pereulok
I-110, GSP, Moscow, 129820
USSR